Remote Sensing of Drought

Innovative Monitoring Approaches

Drought and Water Crises

Series Editor: Donald A. Wilhite

Published Titles:

Remote Sensing of Drought:
Innovative Monitoring Approaches
Editors: Brian D. Wardlow, Martha C. Anderson,
and James P. Verdin

Forthcoming Titles:

Drought, Risk Management, and Policy:
Decision-Making Under Uncertainty
Editors: Linda Courtenay Botterill and Geoff Cockfield

Remote Sensing of Drought

Innovative Monitoring Approaches

Edited by

Brian D. Wardlow

Martha C. Anderson

James P. Verdin

CRC Press
Taylor & Francis Group
Boca Raton London New York

CRC Press is an imprint of the
Taylor & Francis Group, an **informa** business

CRC Press
Taylor & Francis Group
6000 Broken Sound Parkway NW, Suite 300
Boca Raton, FL 33487-2742

First issued in paperback 2017

Version Date: 20120316

ISBN 13: 978-1-4398-3557-9 (hbk)
ISBN 13: 978-1-138-07520-7 (pbk)

Library of Congress Cataloging-in-Publication Data

Remote sensing of drought : innovative monitoring approaches / editors, Brian D. Wardlow, Martha C. Anderson, James P. Verdin.
p. cm. -- (Drought and water crises)
Includes bibliographical references and index.
ISBN 978-1-4398-3557-9 (hardback)
1. Droughts--Remote sensing. I. Wardlow, Brian D. II. Anderson, Martha C. III. Verdin, James P.

QC929.24.R46 2012
551.57'73--dc23 2012005883

Visit the Taylor & Francis Web site at
http://www.taylorandfrancis.com

and the CRC Press Web site at
http://www.crcpress.com

Contents

PART I Vegetation

PART II Evapotranspiration

PART III Soil Moisture/Groundwater

PART IV Precipitation

Series Preface

Drought is a normal part of the climate for virtually all climatic regimes. It is a complex, slow-onset phenomenon that affects more people than any other natural hazard and results in serious economic, social, and environmental impacts. Drought affects both developing and developed countries, but in substantially different ways. Society's ability to manage droughts more effectively in the future is contingent upon a paradigm shift—moving from a crisis management to a risk-based management approach directed at increasing the coping capacity or resilience of nations to deal effectively with extended periods of water shortage.

In 2005, I edited a book for CRC Press, *Drought and Water Crises: Science, Technology, and Management Issues*. The goal of this book series is to expand on the theme of the 2005 book by providing new information and innovative approaches to drought monitoring and early warning systems, mitigation, planning, and policy and the linkages between these challenges and important natural resources and environmental issues such as climate change, including increased climate variability, water scarcity, food security, desertification, transboundary water-related conflicts, and water management, to name just a few. There is an increasing demand for more information from scientists, natural resource managers, and policy makers on issues related to these challenges that are at the intersection of drought and water management issues as pressure on the world's finite water resources intensifies.

I trust this book series will not only heighten awareness of these issues but will also offer practical and adaptable solutions to address these challenges as they unfold in the years ahead.

Donald A. Wilhite

Foreword

This book is the first in the series to be published by CRC Press under the book series title Drought and Water Crises: Science, Technology, and Management Issues. In 2005, I edited a book by this same title that addressed the intersection of drought- and water-related issues. Since the publication of that book, concerns have continued to mount around the linkages between drought and water issues such as climate change, water scarcity, food security, and the sustainable use and management of natural resources. These are topics of frequent concern and debate in scientific literature, the policy arena, and the media. The opportunity to expand discussion on these and other topics through a series of books aimed at drought and water management is timely and will assist scientists, natural resource managers, and policy makers in gaining a better understanding of these important issues.

Drought is a slow-onset natural hazard that is often referred to as a creeping phenomenon. The challenge of monitoring drought's onset and evolution, and identifying its termination or end, is one that scientists, natural resource managers, and decision makers have been struggling with for decades. Although all drought events originate from a deficiency of precipitation, it is insufficient to rely solely on this element of climate to assess drought's severity and resultant impacts. Assessing the impacts of a drought episode is equally challenging. An effective drought early warning system must incorporate multiple indicators, and these indicators must be routinely monitored in order to analyze both the severity of drought and its potential economic, social, and environmental consequences. Some indicators to include in a comprehensive drought early warning system are climatic elements, such as temperature; water supply indicators, such as streamflow, snowpack, groundwater levels, reservoir and lake levels, and soil moisture; and other water use indicators, such as evapotranspiration.

Brian Wardlow, Martha Anderson, and James Verdin have prepared a seminal book on the interrelationship between drought management and the application of remote sensing technologies to the complex challenges associated with drought monitoring and early warning. Remote sensing can be an effective technology in assessing a wide range of critical drought indicators. This book explores the application of these new technologies and their potential contribution to the challenges of drought monitoring and early warning. Remote sensing provides the capability for high-resolution assessments of vegetation conditions and stress in concert with estimations of evapotranspiration, precipitation, soil moisture, and groundwater, particularly in data-sparse areas. These data can greatly enhance the effectiveness of drought early warning systems. I applaud the editors of this book for their efforts in assembling a timely and informative collection of contributions focused on new technologies that can address the challenges of drought monitoring referred to previously. Not only will this information aid many nations in the assessment of drought and its severity, but it will also contribute greatly to the efforts of many international organizations that are committed to the improved management and conservation of land and water resources and sustainable development.

The United States, through the leadership of the National Drought Mitigation Center at the University of Nebraska–Lincoln, the U.S. Department of Agriculture, and the National Oceanic and Atmospheric Administration, developed a state-of-the-art approach to drought monitoring and early warning through the creation of a web-based drought early warning system—the U.S. Drought Monitor (USDM)—in 1999. This system has been continuously evolving since its introduction and is conceptually based on a composite approach to drought monitoring by incorporating multiple drought indices and indicators into a weekly assessment of conditions. The USDM also calibrates the objective assessment of drought conditions based on multiple indices and indicators with field reports of impacts from several stakeholders and experts located throughout the country. Remote sensing tools have been integrated into this assessment process to an increasing degree as new capabilities have emerged and remotely sensed products have been developed. This drought early warning system is widely regarded as a model for other countries and is being promoted by the World Meteorological Organization and other international initiatives such as the Global Earth Observation System of Systems (GEOSS).

A comprehensive drought early warning system is an essential component of an effective drought mitigation plan and drought policy that exemplify a proactive, risk-based management approach to drought management. Indeed, it is the foundation of a drought mitigation plan because the goal of that system is to deliver reliable information to decision makers in a timely fashion. The information delivered by an early warning system when integrated into a well-designed decision support system can significantly enhance our capability to accurately assess conditions in a timely manner and, subsequently, to reduce the consequences of drought on multiple economic sectors and on the environment through the application of proactive risk-based mitigation measures. The intended outcome of this effort is to improve the drought-coping capacity of nations, leading to a more drought-resilient society. Remote sensing tools can be a significant contributor to improved drought management and this capacity-building effort.

Donald A. Wilhite
Series Editor
Drought and Water Crises: Science, Technology, and Management Issues
Professor and Director
School of Natural Resources
University of Nebraska–Lincoln
Lincoln, Nebraska

Preface

The motivation for this book on satellite remote sensing of drought stems from the increasing demand for drought-related information to address a wide range of societal issues (e.g., water scarcity, food security, and economic sustainability), the availability of unique Earth observations from many new satellite-based remote sensing instruments, and the advancement of analysis and modeling techniques. Collectively, the convergence of these factors has resulted in unprecedented new satellite-based estimates of evapotranspiration, rainfall, snow cover, subsurface moisture, and vegetation condition over large geographic areas that can support drought-monitoring activities. To date, a book solely dedicated to the topic of satellite remote sensing applications for drought monitoring has been lacking. The goal of this book is to provide a survey of many new, innovative remote sensing approaches that are either being applied or have the potential to be applied for operational drought monitoring and early warning. The book is by no means a comprehensive summary of all remote sensing–based methods that currently exist, but rather describes a representative set of the leading techniques that characterize major components of the hydrologic cycle related to drought.

The initial idea for this book was developed over the past five years as the National Drought Mitigation Center (NDMC) became increasingly involved in the development and application of information from several new remote sensing approaches in the U.S. Drought Monitor (USDM). These activities included the development of the Vegetation Drought Response Index (VegDRI) and the Vegetation Outlook (VegOut), as well as assisting other researchers to evaluate and integrate remote sensing–based evapotranspiration (ET), soil moisture, and terrestrial water storage (TWS) information. During this period, it became apparent that there was a wealth of new, innovative remote sensing–based tools emerging that offered new perspectives into the hydrologic cycle and could be applied to enhance drought monitoring and early warning. This was further reinforced at both the National Integrated Drought Information System (NIDIS) Contributions of Satellite Remote Sensing to Drought Monitoring Workshop in Boulder, Colorado (February 6–7, 2008), and the National Aeronautics and Space Administration (NASA) Drought Monitoring Tools Workshop in Silver Spring, Maryland (April 11–12, 2011), where a large number of new satellite-based remote sensing tools with a focus on drought applications were presented. At the same time, there is increasing interest from the drought community, both in the United States and internationally, to utilize satellite-based remote sensing information to improve our ability to characterize drought patterns at local scales, fill in data gaps for regions with sparse ground-based observation networks, and gain a more synoptic view of key environmental variables (e.g., evapotranspiration, groundwater, rainfall, and soil moisture) across large geographic areas than are captured in traditional, in situ measurement data sets. Several efforts are underway nationally and internationally to better unify and integrate remote sensing products in drought monitoring, including NIDIS, the Group on Earth Observations (GEO), and the World Meteorological Organization (WMO).

Based on this collection of activities, it is apparent that satellite remote sensing will be increasingly looked upon to enhance global drought monitoring and to help address key related issues associated with food security and water resources. However, understanding the diversity and breadth of the types of remote sensing applications poised to fill this role has been challenging because these tools are often initially developed for applications other than drought, and the scientists behind these innovative approaches are from a broad range of disciplines such as agriculture, climatology, engineering, geography, hydrology, meteorology, and numerical modeling. As a result, much of the work and literature related to these tools has been presented in various professional journals across many disciplines, which can limit their visibility within the broader remote sensing and drought communities.

In 2009, I set out to develop a book dedicated to the topics of remote sensing and drought monitoring with the goal of providing a single publication that summarized many of these new leading satellite-based tools providing new observations of many critical hydrologic cycle components related to drought. Dr. Martha Anderson of the U.S. Department of Agriculture (USDA) Agricultural Research Service (ARS) and Dr. James Verdin of the U.S. Geological Survey (USGS) agreed to serve as coeditors, bringing their extensive experience and expertise in hydrologic applications of satellite remote sensing to the development of this book. The specific remote sensing techniques described in this book were selected for two primary reasons. First, they are either currently operational or hold the potential to be applied operationally in the near future, which is a critical requirement for routine drought monitoring. Second, they have the potential to be globally applicable and of interest to the international community. This book contains 14 chapters on remote sensing techniques that are organized into 4 parts associated with different components of the hydrologic cycle: vegetation (reflecting plant uptake and root-zone moisture deficiencies), evapotranspiration, soil moisture/groundwater, and precipitation in the form of rainfall and snow. An introductory chapter by Dr. Michael Hayes (lead author), the director of the NDMC, and colleagues is included to frame the broader drought picture and the past, present, and current role of satellite remote sensing to support drought monitoring and decision support. The final chapter provides the perspective of several leading remote sensing scientists on current challenges that need to be addressed by the remote sensing community to support drought applications, as well as a look to the future regarding new opportunities for the further advancement of satellite-based Earth observations in this application area.

Brian D. Wardlow
Assistant Professor and GIScience Program Area Leader
National Drought Mitigation Center
School of Natural Resources
University of Nebraska–Lincoln
Lincoln, Nebraska

Acknowledgments

The development of this book was the culmination of efforts from many people. First and foremost, my coeditors, Dr. Martha C. Anderson (U.S. Department of Agriculture [USDA] Agricultural Research Service [ARS]) and Dr. James P. Verdin (U.S. Geological Survey [USGS]), provided the vision to develop a publication that highlights many of the leading new remote sensing tools and techniques that offer considerable potential to advance the application of satellite-based information for operational drought monitoring and early warning. Their countless hours of reviewing the content of this book and the support from their respective organizations to participate in this endeavor are greatly appreciated. I would like to acknowledge the book series editor, Dr. Donald A. Wilhite, for his encouragement to develop the first book dedicated to the topic of satellite remote sensing and drought within his book series Drought and Water Crises: Science, Technology, and Management Issues. A special thanks to Dr. Michael Hayes and my colleagues at the National Drought Mitigation Center (NDMC) for their support and for providing me the time to complete this book. The editorial guidance by Irma Shagla-Britton and Kari Budyk of CRC Press and Deborah Wood of the NDMC was invaluable throughout the book's development.

The contributions of the many chapter authors deserve a special acknowledgment because this book would not have been possible without their efforts. Each author was asked to provide a detailed description of satellite-based, drought monitoring tools they have worked to develop, along with examples of how these tools can be applied in practice and what new information remote sensing conveys in comparison with standard ground-based measurement networks. Each chapter was the product of an iterative development and review process involving the authors and editors to ensure consistency and completeness of the book's material. The editors wish to thank the authors for their patience and commitment during this process; it was greatly appreciated. Last, on behalf of my coeditors, I extend my gratitude to the numerous peer reviewers for ensuring that the remote sensing techniques were presented in a complete and understandable manner for readers. A number of individuals participated in the peer review, including Jude Kastens (Kansas Applied Remote Sensing [KARS] Program, University of Kansas), book chapter authors Jesslyn Brown (USGS Earth Resources and Observation Science [EROS] Center), Chris Hain (National Oceanic and Atmospheric Administration [NOAA] National Environmental Satellite, Data, and Information Service [NESDIS]), William Kustas (USDA-ARS), and John Mecikalski (University of Alabama-Huntsville).

Brian D. Wardlow

Editors

Brian D. Wardlow is an assistant professor and GIScience program area leader for the National Drought Mitigation Center (NDMC) in the School of Natural Resources at the University of Nebraska–Lincoln. His work focuses on the development of new remote sensing techniques for drought monitoring, transitioning new remote sensing tools from research to operational use in drought monitoring and early warning systems, and outreach and training in the United States and internationally on the application of these tools. Before joining the NDMC, he worked as a remote sensing specialist at the U.S. Geological Survey (USGS) Earth Resources and Observation Science (EROS) Center, working on the National Land Cover Dataset (NLCD) project. During his PhD program, he was also a graduate research fellow as part of the National Aeronautics and Space Administration's Earth System Science Graduate Fellowship program, investigating the application of time-series moderate resolution imaging spectroradiometer vegetation index data for regional-scale, crop-related land use/land cover mapping and monitoring in the U.S. Central Great Plains. His primary research interests are focused on the application of remote sensing and geographic information systems for drought monitoring and impact assessment, land use/land cover characterization, vegetation phenology, and agricultural and natural resources management and monitoring. He is currently a member of the NASA SMAP Application and National Phenology Network working groups. Dr. Wardlow received his BSc in geography from Northwest Missouri State University, an MA in geography from Kansas State University, and a PhD in geography from the University of Kansas.

Martha C. Anderson is a physical scientist in the Hydrology and Remote Sensing Laboratory of the U.S. Department of Agriculture's (USDA) Agricultural Research Service in Beltsville, Maryland. Prior to her position with the USDA, Dr. Anderson was a research scientist at the University of Wisconsin, working on integrating satellite and synoptic weather information into agricultural decision-making tools. Her current research interests focus on mapping water, energy, and carbon land-surface fluxes at field to continental scales using multisensor and multiwavelength remote sensing data, with applications in drought monitoring and soil moisture estimation. She is currently a member of the Landsat Science Team and the HyspIRI Science Study Group. Dr. Anderson received her BA in physics from Carleton College, Northfield, Minnesota, and her PhD in astrophysics from the University of Minnesota.

James P. Verdin is a physical scientist with the U.S. Geological Survey (USGS), Earth Resources Observation and Science (EROS) Center. He is currently assigned to work in the National Integrated Drought Information System (NIDIS) Program Office at National Oceanic and Atmospheric Administration (NOAA) in Boulder, Colorado. He coordinates activities for the NIDIS pilot drought early warning system for the Upper Colorado River Basin. Since 1995, he has led USGS activities in support of the USAID Famine Early Warning Systems Network. His research interests focus on the use of remote sensing and modeling to address questions of hydrology, agriculture, and hydroclimatic hazards. He has extensive experience in geographic characterization of drought hazards for food security assessment in Africa, Asia, and Latin America. He recently took on leadership of efforts to estimate crop water use with remote sensing for the USGS National Water Census. Before joining the USGS, he worked for 11 years with the U.S. Bureau of Reclamation, including a 3-year assignment in Brazil. Dr. Verdin received his BSc and MSc in civil and environmental engineering from the University of Wisconsin, Madison, and Colorado State University, respectively, and his PhD in geography from the University of California, Santa Barbara.

Contributors

Amir AghaKouchak
Department of Civil and Environmental Engineering
University of California, Irvine
Irvine, California

David Allured
Physical Science Division
Earth System Research Laboratory
National Oceanic and Atmospheric Administration
Boulder, Colorado

Martha C. Anderson
Hydrology and Remote Sensing Laboratory
Agricultural Research Service
U.S. Department of Agriculture
Beltsville, Maryland

Assaf Anyamba
Universities Space Research Association & Biospheric Sciences Laboratory
National Aeronautics and Space Administration
Goddard Space Flight Center
Greenbelt, Maryland

Stefanie Bohms
Earth Resources Observation and Science Center
U.S. Geological Survey
Sioux Falls, South Dakota

Jesslyn F. Brown
Earth Resources Observation and Science Center
U.S. Geological Survey
Sioux Falls, South Dakota

Karin Callahan
National Drought Mitigation Center
Center for Advanced Land Management Information Technology
School of Natural Resources
University of Nebraska–Lincoln
Lincoln, Nebraska

Steven K. Chan
Jet Propulsion Laboratory
California Institute of Technology
Pasadena, California

Bradley D. Doorn
Earth Science Division
National Aeronautics and Space Administration
Washington, District of Columbia

Michael Ek
Environmental Modeling Center
National Centers for Environmental Prediction
National Oceanic and Atmospheric Administration
Camp Springs, Maryland

Ralph Ferraro
National Environmental Satellite, Data, and Information Service
National Oceanic and Atmospheric Administration
Camp Springs, Maryland

Christopher Funk
Earth Resources Observation and Science Center
U.S. Geological Survey
Sioux Falls, South Dakota

and

Department of Geography
Climate Hazard Group
University of California, Santa Barbara
Santa Barbara, California

Xiaogang Gao
Department of Civil and Environmental Engineering
University of California, Irvine
Irvine, California

Christopher R. Hain
National Environmental Satellite, Data, and Information Service
National Oceanic and Atmospheric Administration
Camp Springs, Maryland

Michael J. Hayes
National Drought Mitigation Center
School of Natural Resources
University of Nebraska–Lincoln
Lincoln, Nebraska

Kuolin Hsu
Department of Civil and Environmental Engineering
University of California, Irvine
Irvine, California

Eric Hunt
National Drought Mitigation Center
School of Natural Resources
University of Nebraska–Lincoln
Lincoln, Nebraska

Bisher Imam
Department of Civil and Environmental Engineering
University of California, Irvine
Irvine, California

Felix Kogan
National Environmental Satellite, Data, and Information Service
National Oceanic and Atmospheric Administration
Camp Springs, Maryland

Cezar Kongoli
Earth System Science Interdisciplinary Center
University of Maryland
College Park, Maryland

and

National Environmental Satellite, Data, and Information Service
National Oceanic and Atmospheric Administration
Camp Springs, Maryland

William P. Kustas
Hydrology and Remote Sensing Laboratory
Agricultural Research Service
U.S. Department of Agriculture
Beltsville, Maryland

Doug LeComte
Climate Prediction Center
National Oceanic and Atmospheric Administration
Camp Springs, Maryland

Lifeng Luo
Department of Geography
Michigan State University
East Lansing, Michigan

Michael T. Marshall
Department of Geography
University of California, Santa Barbara
Santa Barbara, California

John R. Mecikalski
Department of Atmospheric Science
University of Alabama in Huntsville
Huntsville, Alabama

Joel Michaelsen
Department of Geography
University of California, Santa Barbara
Santa Barbara, California

Kenneth E. Mitchell
Environmental Modeling Center
National Centers for Environmental Prediction
National Oceanic and Atmospheric Administration
Camp Springs, Maryland

Gregory Neumann
Jet Propulsion Laboratory
California Institute of Technology
Pasadena, California

Son V. Nghiem
Jet Propulsion Laboratory
California Institute of Technology
Pasadena, California

Stefan Niemeyer
Institute for Environment and Sustainability
Joint Research Centre
European Commission
Ispra, Italy

Agustin Pimstein
Faculty of Agronomy and Forestry Engineering
Department of Fruit Production and Enology
The Pontifical Catholic University of Chile
Santiago, Chile

Matthew Rodell
Hydrological Sciences Laboratory
National Aeronautics and Space Administration
Goddard Space Flight Center
Greenbelt, Maryland

Peter Romanov
Earth System Science Interdisciplinary Center
University of Maryland
College Park, Maryland

and

National Environmental Satellite, Data, and Information Service
National Oceanic and Atmospheric Administration
Camp Springs, Maryland

Matthew Rosencrans
Climate Prediction Center
National Oceanic and Atmospheric Administration
Camp Springs, Maryland

Simone Rossi
Institute for Environment and Sustainability
Joint Research Centre
European Commission
Ispra, Italy

and

Food and Agriculture Organization of the United Nations
Rome, Italy

Gabriel B. Senay
Earth Resources Observation and Science Center
U.S. Geological Survey
Sioux Falls, South Dakota

Justin Sheffield
Department of Civil and Environmental Engineering
Princeton University
Princeton, New Jersey

Soroosh Sorooshian
Department of Civil and Environmental Engineering
University of California, Irvine
Irvine, California

Gregory J. Story
West Gulf River Forecast Center
National Weather Service
Fort Worth, Texas

Mark D. Svoboda
National Drought Mitigation Center
School of Natural Resources
University of Nebraska–Lincoln
Lincoln, Nebraska

Sharmistha Swain
National Drought Mitigation Center
School of Natural Resources
University of Nebraska–Lincoln
Lincoln, Nebraska

Tsegaye Tadesse
National Drought Mitigation Center
School of Natural Resources
University of Nebraska–Lincoln
Lincoln, Nebraska

Compton J. Tucker
Hydrological and Biospheric Sciences Laboratory
National Aeronautics and Space Administration
Goddard Space Flight Center
Greenbelt, Maryland

James P. Verdin
Earth Resources Observation and Science Center
U.S. Geological Survey
Sioux Falls, South Dakota

Brian D. Wardlow
National Drought Mitigation Center
School of Natural Resources
University of Nebraska–Lincoln
Lincoln, Nebraska

Eric F. Wood
Department of Civil and Environmental Engineering
Princeton University
Princeton, New Jersey

Youlong Xia
Environmental Modeling Center
National Centers for Environmental Prediction
National Oceanic and Atmospheric Administration
Camp Springs, Maryland

Xiwu Zhan
Center for Satellite Applications and Research
National Environmental Satellite, Data, and Information Service
National Oceanic and Atmospheric Administration
Camp Springs, Maryland

1 Drought Monitoring
Historical and Current Perspectives

Michael J. Hayes, Mark D. Svoboda, Brian D. Wardlow, Martha C. Anderson, and Felix Kogan

CONTENTS

1.1 INTRODUCTION

Drought is a normal, recurring feature of climate throughout the world, with characteristics and impacts that can vary from region to region. Figure 1.1 illustrates the regular occurrence of drought within the United States between 1895 and 2010 with approximately 14% of the country, on average (plotted by black dotted line), experiencing severe to extreme drought conditions during any given year. Drought conditions can persist in a region for several years, as occurred in the United States in the 1930s, 1950s, and early 2000s, and tree ring and other proxy records confirm that multiple-year droughts are part of the long-term climate history for the United States and most other regions around the world (Woodhouse and Overpeck, 1998; Dai et al., 2004; Jansen et al., 2007). Drought has wide-ranging impacts on many sectors of society (e.g., agriculture, economics, ecosystems services, energy, human health, recreation, and water resources) and ranks among the most costly of

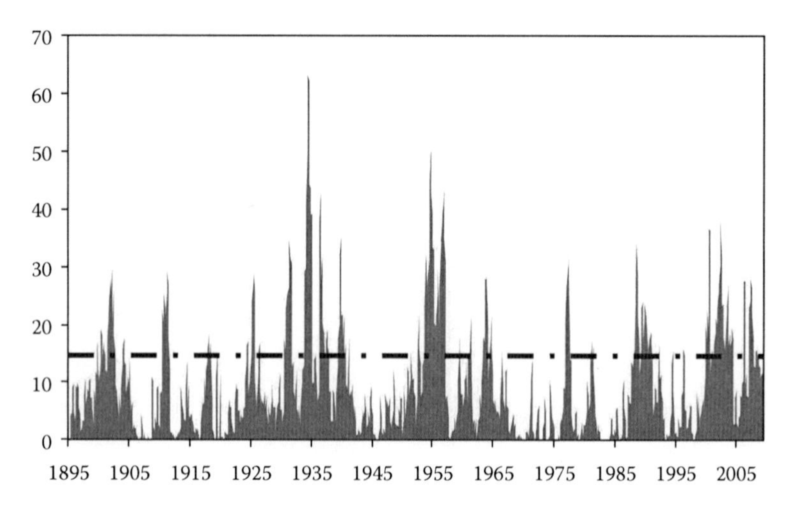

FIGURE 1.1 Percent area of the continental United States that experienced severe to extreme drought between 1895 and 2009 (histogram plots monthly values over this historical period). (Based on data from the National Climatic Data Center/NOAA, Asheville, NC.)

all natural disasters. For example, in the United States, drought affects more people than any other hazard (NSTC, 2005) and has resulted in 14 "billion-dollar" events since 1980 totaling more than $180 billion (U.S.) in damages and losses (NCDC, 2011). This amount represents 25% of all losses from billion-dollar weather disasters, including hurricanes and floods. Globally, drought along with other natural disasters affects more than 255 million people each year (Guha-Sapir et al., 2004), with an estimated $932 billion (U.S.) in losses since 2001 in the 42 countries ranked highest by the United Nations in terms of the combination of life expectancy, education, and income (Guha-Sapir, 2011). In developing nations, drought impacts can transcend economic losses, triggering severe famine and potentially human mortality.

Traditionally, drought response has been conducted in a very reactive, post-event fashion referred to as a "crisis" management approach. This book, *Remote Sensing of Drought: Innovative Monitoring Approaches*, is part of a book series that focuses on proactive, risk-based approaches to dealing with drought known as "risk" management. Risk management in drought response includes preparedness planning, mitigation, monitoring and early warning, and prediction to reduce the impacts of drought—both now and into the future. Drought monitoring, defined as tracking the severity and location of drought, is a critical piece in this risk-based paradigm. A substantial evolution in drought monitoring strategies has taken place during the past 50 years around the world, greatly improving the ability to provide relevant and timely drought information in terms of early warning to decision makers. Many of these efforts have emphasized climate- and hydrologic-based indicators and indices that track changes in components of the hydrologic cycle (e.g., precipitation, temperature, streamflow, and soil moisture), derived primarily from point-based, in situ observations.

Satellite remote sensing offers a unique perspective for operational drought monitoring that complements the in situ–based climate and hydrologic data traditionally used for this application. Spaceborne sensors provide synoptic, repeat coverage of spatially continuous spectral measurements collected in a consistent, systematic, and objective manner. In addition, satellite data are increasingly being looked upon to fill in or supplement data from existing observation networks, even in regions with abundant point-based data.

In 1960, the first meteorological satellite—the Television Infrared Observation Satellite (TIROS-1)—was launched, ushering in the era of satellite-based environmental monitoring and providing the basis for development of land observation satellites such as the Advanced Very High Resolution Radiometer (AVHRR), Landsat, and the Geostationary Operational Environmental Satellites (GOES). In the 1980s, satellite-based indices such as the Normalized Difference Vegetation Index (NDVI) became increasingly used for various environmental monitoring applications including drought monitoring (Goward et al., 1985; Tucker et al., 1986, 1991; Kogan, 1995a,b, 1997; Peters et al., 2002). During the past decade, a number of new satellite-based instruments have been launched that, accompanied by major advances in computing and analysis/modeling techniques, have resulted in the rapid emergence of many new remote sensing tools and products applicable for drought monitoring. These tools have moved beyond traditional vegetation index (VI) data, which provide information primarily related to vegetation health, to a suite of new environmental variables (precipitation, evapotranspiration [ET], soil moisture, and snow cover) that enable a more comprehensive view of drought conditions.

This chapter will highlight the complex nature of drought and the challenges it presents for effective monitoring and early warning efforts. The historical evolution of traditional drought monitoring techniques will be discussed, as well as the emergence of drought early warning systems (DEWS). Finally, the role—both past and present—of satellite remote sensing in drought monitoring is introduced, along with opportunities for improved monitoring capabilities provided by the new tools presented in this book.

1.2 DROUGHT MONITORING

1.2.1 Complexity of Drought: Definitions and Characteristics

Drought is a slow-onset natural hazard with effects that accumulate over a considerable period of time (e.g., weeks to months). Drought does not have a single universally accepted definition, which makes the identification and monitoring of key characteristics such as duration, severity, and spatial extent difficult. Drought originates from a deficiency of precipitation that results in a water shortage for some activity (e.g., crop production) or group (e.g., users of water resources) (Wilhite and Glantz, 1985). Drought can be characterized into three physically based perspectives—meteorological, agricultural, and hydrological (Figure 1.2). Timing, impact, and recovery rate differ between these three perspectives of drought, with shorter-term dryness being reflective of meteorological drought, an intermediate period of precipitation deficit representative of agricultural drought (i.e., the relationship between plant water demands and the amount of available water, particularly within the soil environment), and longer-term dryness

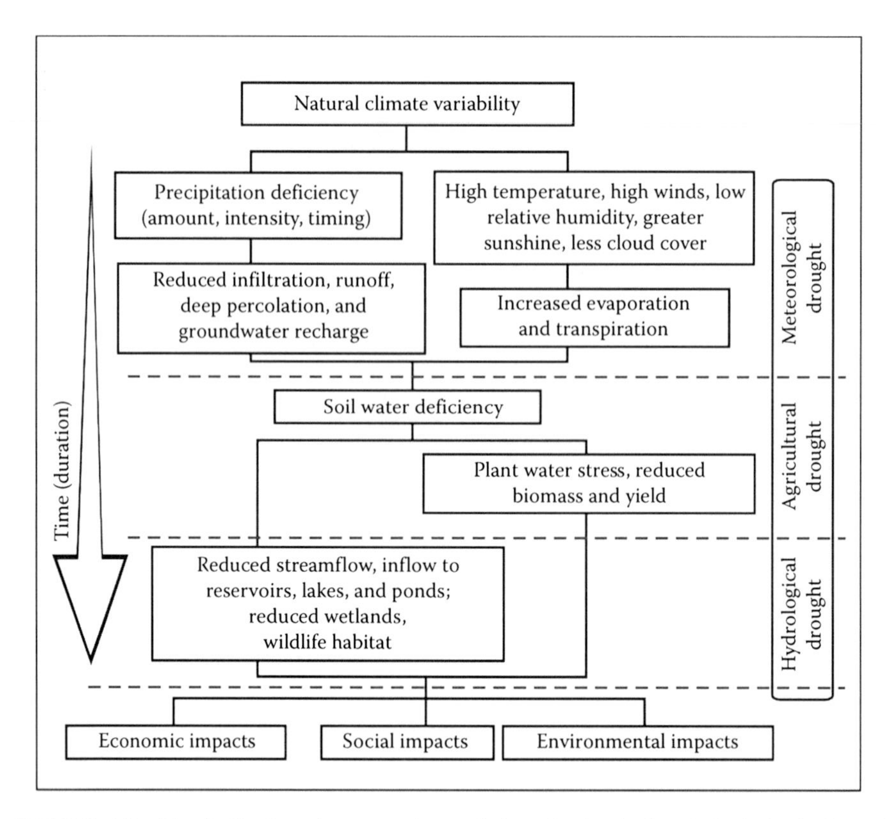

FIGURE 1.2 Physically based perspectives of drought (agricultural, hydrological, and meteorological) and associated impacts that define them. Many of these impacts have the capacity to be monitored by remote sensing techniques. (From Wilhite, D.A. In *Drought Volume I: A Global Assessment*, ed. D.A. Wilhite, New York, Routledge, 2000.)

indicative of hydrological drought (e.g., streamflow and/or groundwater reduction). As a result, drought designations among sectors may or may not coincide in space and time. For example, several weeks of dryness may result in vegetation stress triggering an agricultural drought classification but may have little effect on streamflow and groundwater, which would not result in a hydrological drought classification for the same event.

1.2.2 Monitoring and Early Warning for Drought Risk Management

Drought monitoring involves the continuous assessment of the natural indicators of drought severity, spatial extent, and impacts (Wilhite and Buchanan-Smith, 2005). Using that information to elicit an appropriate response is called *early warning*. Wilhite and Buchanan-Smith (2005) argue that a DEWS, which combines both assessment and decision-maker response, is integral to effective drought risk management. Decision makers require accurate early warning information to implement effective drought policies and response and recovery programs. An example of an early warning system in which drought is a primary consideration is the well-established

Famine Early Warning Systems Network (FEWS NET) that is in place to address food security issues for specific locations around the world. Components of a DEWS can vary and be adapted for any region to account for the needs and resources available within that region, but generally these components include a drought monitoring network, access to timely data, and analysis, synthesis, and dissemination of data that can then be used in decision support tools, communication strategies, educational efforts, and often a forecast element if one is available.

An important feedback loop occurs as drought monitoring and risk management strategies evolve, where better drought management drives the need for improved drought monitoring and, in turn, improved drought monitoring encourages more effective drought management (Wilhite, 2009). As drought risk management plans become more specific in space and time, the need for information at higher spatial and temporal resolutions increases. One example of this type of coevolution in drought monitoring and risk management has occurred over the past decade in the United States, where improvements in the U.S. Drought Monitor (USDM) (discussed in Section 1.2.4) have led to shifts in national agricultural policies, inspiring additional advancements in the spatiotemporal resolution of drought monitoring to support implementation of these policies at a local scale.

Satellite remote sensing can play an important role within a DEWS by providing synoptic, rapid repeat, and spatially continuous information about regional drought conditions. Remote sensing products will be looked upon to

1. Provide information at spatial scales required for local-scale drought monitoring and decision making that cannot be adequately supported from information derived traditional, point-based data sources (e.g., single area-based value over administrative geographic unit or spatially interpolated climate index grids)
2. Fill in informational gaps on drought conditions for locations between in situ observations and in areas that lack (or have very sparse) ground-based observational networks
3. Enable earlier drought detection in comparison to traditional climatic indices
4. Collectively provide a suite of tools and data sets geared to meet the observational needs (e.g., spatial scale, update frequency, and data type) for a broad range of decision support activities related to drought

1.2.3 Early Monitoring Strategies

Drought indicators and indices are variables that are used to describe the physical characteristics of drought severity, spatial extent, and duration (Steinemann et al., 2005). Within the drought monitoring community, the terms *indicator* and *index* are often used interchangeably. *Indicator* is a broader term that includes parameters such as precipitation, temperature, streamflow, groundwater levels, reservoir levels, soil moisture levels, snowpack, and drought indices. Indices are typically a computed numerical representation of a drought's severity or magnitude, using combinations of the climatic or hydrometeorological indicators listed earlier.

The earliest drought monitoring efforts tended to focus on either absolute precipitation amounts or on precipitation deficiencies, represented in terms of the departure from normal or the percent of normal (Heim, 2002). Other typical drought indicators were based on in situ measurements of related hydrologic variables such as streamflow, reservoir, soil moisture, and snowpack levels. As the need grew to provide historical context for assessing the relative severity of specific drought events, attempts were made to develop new indices and indicators that would robustly encapsulate the complexity of comparing droughts in different regions and climatic conditions (Steinemann et al., 2005).

The Palmer Drought Severity Index (PDSI; Palmer, 1965) was one of the first drought indices developed to enable assessment of relative drought severity at the national scale. The PDSI was subsequently adopted for policy implementation by a variety of federal drought programs, beginning in 1976 (GAO, 1979; Wilhite and Rosenberg, 1986; Heim, 2002). Alley (1984) identified three positive characteristics of the PDSI that contributed to its popularity: (1) it provided decision makers with a current assessment of the severity of drought for a region, (2) it provided an opportunity to place current conditions in historical perspective, and (3) it provided spatial and temporal representations of historical droughts. As attempts to use the PDSI in drought monitoring applications expanded, multiple limitations of the PDSI were recognized (Alley, 1984; Hayes et al., 1999; Heim, 2002). These limitations, among others, include the fact that the PDSI values are inconsistent across diverse climatological regions for spatial comparisons and that the empirical constants used by Palmer to represent climate and duration characteristics were determined from measurements made at a relatively small number of locations (Wells et al., 2004). Some of these limitations were addressed by Wells et al. (2004) in the formulation of the Self-Calibrated PDSI (SC-PDSI) that is now widely used (e.g., Dai, 2011), but other limitations still unique to the PDSI remain. One of these limitations is the built-in lag within the PDSI, identified by Hayes et al. (1999), which makes it difficult to use in rapidly evolving drought conditions.

Because of these limitations, other indices such as the Surface Water Supply Index (SWSI; Shafer and Dezman, 1982) and the Standardized Precipitation Index (SPI; McKee et al., 1995) were subsequently developed. The SWSI was originally calculated for the state of Colorado, taking into account historical hydrological factors within a basin (i.e., precipitation, streamflow, reservoir levels, and snowpack), and is now being calculated for many of the river basins across the western United States. The SPI, also developed in Colorado, was designed to quantify the precipitation deficit for multiple timescales with the understanding that a deficit has different impacts on groundwater, reservoir storage, soil moisture, snowpack, and streamflow. McKee et al. (1993) originally calculated the SPI for 3, 6, 12, 24, and 48 month timescales to address these different impacts, but the index can be calculated for any weekly or monthly timescale. Because the SPI is normalized based on the statistical representation of the historical record at every location, wetter and drier climates can both be represented in the same way. An SPI value also places the severity of a current event (either dry or wet) into an historical perspective because the frequency of each value is known. This is one feature that differentiates the SPI from the PDSI. This same technique can also be used to represent other standardized indicators and has led to the development of the Standardized Precipitation Evapotranspiration Index (SPEI, Vicente-Serrano et al., 2010), which incorporates both temperature and precipitation. The SPI is now the accepted standard worldwide and was recommended at a World Meteorological Organization (WMO) meeting held in December 2009 in Lincoln, Nebraska, as the primary meteorological drought index to be used by national meteorological and hydrological agencies to track meteorological drought.

The use of satellite-based remote sensing for drought monitoring began in the 1980s with the application of NDVI data from the operational NOAA AVHRR instrument (Tucker et al., 1986, 1991; Hutchinson, 1991; Eidenshink and Hass, 1992). The NDVI (Rouse et al., 1974; Tucker, 1979) was a simple mathematical transformation of two commonly available spectral bands (visible red and near infrared) on AVHRR and other satellite instruments and had been shown to have a strong relationship with several biophysical parameters of vegetation (e.g., leaf area index and green biomass) (Asrar et al., 1989; Baret and Guyot, 1991). Early work by Hutchinson (1991) and Tucker et al. (1991) among others, discussed in more detail in Chapter 2 (Anyamba and Tucker, 2012), demonstrated the value of NDVI data for drought monitoring, serving as a simple metric to assess vegetation conditions (Tucker, 1979). Time series of AVHRR NDVI data have readily been used to support operational drought monitoring systems worldwide, including FEWS NET and national efforts in countries such as Australia. Kogan (1995a) developed a set of drought indices from time series of AVHRR NDVI and thermal infrared data, available in near-real time at http://www.star.nesdis.noaa.gov/smcd/emb/vci/VH/index.php. The NDVI-based Vegetation Condition Index (VCI) and the thermal-based Temperature Condition Index (TCI) are indicators of land-surface vegetation and moisture conditions, respectively, and are expressed as an anomaly or departure of the current index value relative to its longer-term climatology (historical minimum and maximum boundary values). The Vegetation Health Index (VHI) is a weighted combination of the VCI and TCI reflecting the integrated effects of moisture and temperature on vegetation. These indices have been widely used to assess drought conditions (Kogan, 1995a,b, 1997, 2002; Unganai and Kogan, 1998; Rojas et al., 2011) and have been integrated into operational monitoring systems (Heim, 2002). The use of these satellite-based VIs over the past 20+ years has demonstrated the valuable complementary information that remotely sensed data can provide for drought monitoring, but their application has been limited primarily to the characterization of agricultural drought.

This process of evolution in the development and application of drought indicators and indices has led to the understanding that, in the majority of cases, no single indicator or index can represent the diversity and complexity of drought conditions across the temporal and spatial dimensions affected by drought (Hayes et al., 2005; Mizzell, 2008). In most applications, it is best to use a combination of indicators when monitoring drought. However, this option can be very confusing for the decision makers, who often do not know about the characteristics of each indicator (Mizzell, 2008).

1.2.4 U.S. and North American Drought Monitors: Hybrid Monitoring Approaches

With the recognition that no one indicator or index can fully capture the multi-scale, multi-impact nature of drought in all its complexity, there has been a movement to develop a "composite" index approach that is "hybrid" in nature and combines many parameters, indicators, and/or indices into a single product. The presentation of drought information in a single map with a simple classification system is preferred by decision makers and the general public in contrast to multiple maps depicting various indicators with differing classification schemes. In order for tools and indices to be accepted and readily used by decision and policy makers, it is important to understand and follow this simple premise. Naturally, the ability also exists to extract and analyze the inputs to the composite indicator individually to determine how a specific indicator is contributing to the hybrid product.

The most prominent composite indicator approach used within the United States is the weekly USDM (Svoboda et al., 2002) (Figure 1.3), which was initiated in 1999 and is globally considered the current state-of-the-art drought monitoring tool. The USDM is not a forecast, but rather an assessment or snapshot of current drought conditions. The USDM product is not an index in and of itself, but rather a combination of indicators and indices that are synthesized using a simple D0–D4 drought severity classification scheme that utilizes a percentile ranking methodology (Table 1.1) and addresses both short- and long-term drought across the United States. The key indicators/indices used in the USDM focus on monitoring major components of the hydrologic cycle, including precipitation, temperature, streamflow, soil moisture, snowpack, and snow water equivalent. Various indices, such as the SPI and PDSI, are incorporated with other in situ data (e.g., streamflow) and remotely sensed VIs and are collectively analyzed by experts using a "convergence of evidence" approach to determine a drought severity classification. In fact, the VHI (Kogan, 1995a) discussed earlier was one of the first indicators included in the production of the USDM. Recently, the USDM has also experimented with or integrated several remotely sensed indicators or indices including the Vegetation Drought Response Index (Brown et al., 2009; presented in

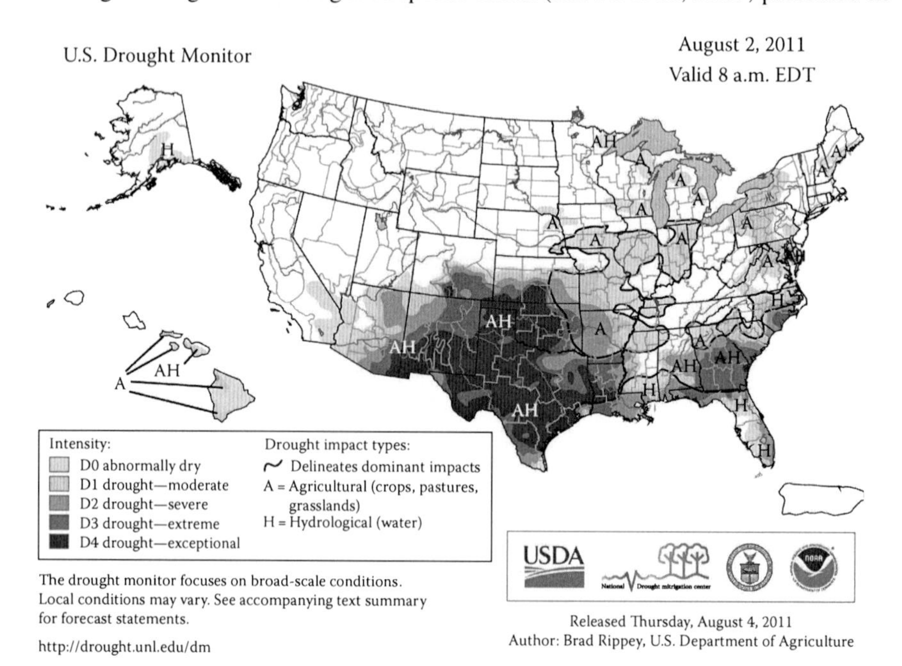

FIGURE 1.3 USDM map for August 2, 2011.

TABLE 1.1
USDM Classification and Ranking Percentile Scheme

USDM Category	Description	Ranking Percentile
D0	Abnormally dry	30
D1	Moderate	20
D2	Severe	10
D3	Extreme	5
D4	Exceptional	2

Source: Svoboda et al., 2002. The Drought Monitor. *Bulletin of the American Meteorological Society*, 83(8): 1181–1190.

Chapter 3 by Wardlow et al., 2012) and the multisensor hybrid rainfall estimate tool that combines radar with rain gauge data made available by the National Oceanic and Atmospheric Administration (NOAA)–National Weather Service (NWS) Advanced Hydrological Prediction Service (AHPS) program (presented in Chapter 12 by Story 2012). Despite these efforts, the integration of remote sensing information into the USDM has been limited, but tremendous opportunity exists to incorporate data from new satellite-based tools retrieving various parameters of the hydrologic cycle.

The USDM has adopted a ranking percentile approach to drought classification (further discussed in Svoboda et al., 2002), which allows the user to directly compare and contrast indicators originally having different units and periods of record into one comprehensive indicator that addresses the customized needs of any given user. The approach also allows for flexibility and adaptation to the latest indices, indicators, and data that become available. The overall USDM "process" can be described as a blending of objective science, based on the ranking percentile approach, and subjective expert experience, as well as guidance provided through the integration of impacts, reports, and other data from approximately 300 local experts throughout the United States. In addition, a set of short- and long-term Objective Drought Indicator Blend maps that combine different sets of indicators and indices with variable weightings (depending on region and type of drought) are also used to guide the USDM map development process. The impacts labeled on the USDM map are (A) for agricultural and (H) for hydrological drought. As another sign of its continual evolution and response to needed change by the user community, the USDM is currently going through a process to replace the current (A) and (H) impact labels to a broader (S) for short-term and (L) for long-term drought impacts. This change was a response to the need for an accounting of stress on the unmanaged (nonagricultural) environment (i.e., ecosystems), as well as moisture deficits during the winter in agricultural regions (agricultural drought designations are not assigned during dry winter months under the current USDM classification scheme). The new labels allow for more flexibility as to which sectors are impacted by both short- and long-term droughts. More details and information on the USDM, its classification scheme, and the objective blends can be found at http://droughtmonitor.unl.edu.

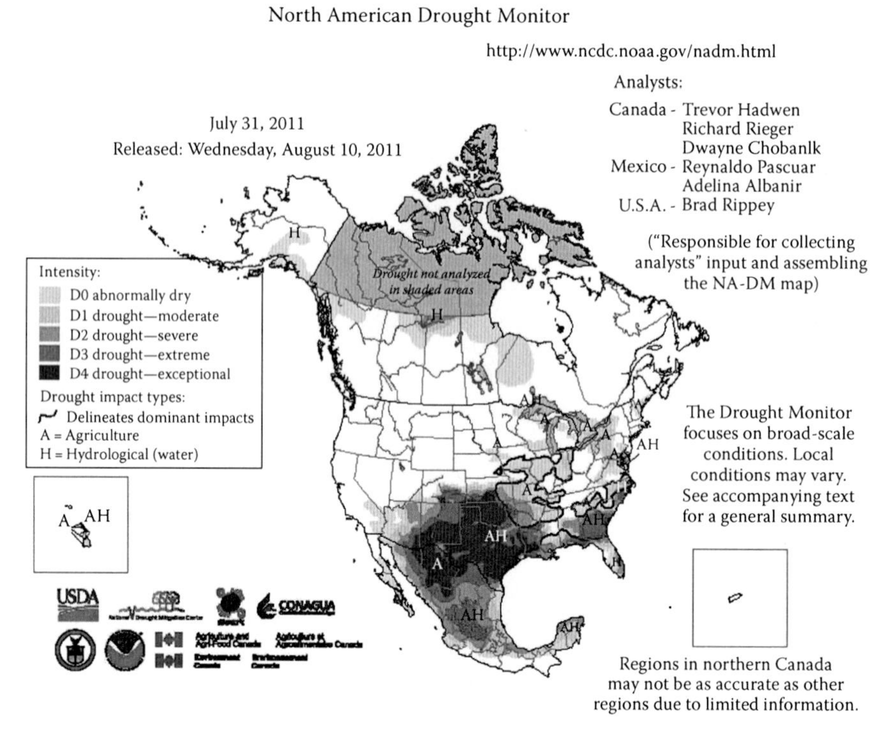

FIGURE 1.4 NADM map for February 2011.

The North American Drought Monitor (NADM) (Lawrimore et al., 2002) (Figure 1.4) was developed 3 years after the USDM in 2002 and is a monthly product forged from a partnership between several entities in Canada, Mexico, and the United States. As with the USDM, the NADM blends science and art. The ranking percentile principal used in the USDM is also applied to the NADM, but the inputs vary slightly depending on which indicators are readily available to the respective agencies involved in each country. As the process stands now, each country follows the same basic methodology, utilizing their own indicators to create a depiction of drought within their borders. A variety of data inputs are used among the countries to develop their respective national drought monitor maps, which are then merged into the continental NADM map. However, the specific types of variables, as well as the general data quality and density of in situ observations, vary widely from country to country and can lead to transboundary issues in terms of the agreement in drought patterns depicted at the international border. To date, the application of remote sensing products with continental coverage within the NADM has been extremely limited. Satellite remote sensing has the capability to acquire seamless, consistent, and objective data that traverse these three countries to provide common data inputs into the NADM map development process and help address transboundary drought depiction discrepancies that may arise. Other regional, multi-country monitoring efforts, such as FEWS NET and the European Drought Observatory being developed by the European Union's Joint Research Centre, are heavily reliant on satellite remote sensing to address discontinuities in ground-based data resources from country to country. Currently, the monthly NADM author (which rotates between the three countries) is responsible for working out the merging of the geographic information system (GIS) shape files and reconciling any disputes along the borders. Impact and data information are exchanged in working out any differences in an interactive fashion until all are resolved. More information and details on the NADM can be found at http://www.ncdc.noaa.gov/temp-and-precip/drought/nadm/index.html.

1.2.5 Emerging Drought Early Warning System Initiatives

1.2.5.1 National Integrated Drought Information System

The National Integrated Drought Information System (NIDIS), signed into U.S. Public Law in December 2006, is the first coordinated effort to develop a U.S. DEWS. It is a NOAA-led partnership among federal, state, tribal, and local organizations with the goal of improving the nation's capacity to proactively manage drought-related risks by providing decision makers with the best available information and tools to assess the impact of drought and to better prepare for and mitigate its effects. NIDIS has three general tasks:

1. Provide an effective DEWS that
   a. Collects and integrates information on the key indicators of drought and drought severity
   b. Provides timely information that reflects state and regional differences in drought conditions
2. Coordinate federal research in support of a DEWS
3. Build on existing forecasting and assessment efforts

The development of regional DEWS is a major effort of NIDIS, which has the goal of developing expert-driven, issue-based decision support tools and informational products that characterize local-scale drought conditions within the region and address key drought-related decision-making activities. Experts within each region were collectively tasked with identifying key decisions and issues associated with drought, assessing their current monitoring capacities and data gaps, and developing and implementing the customized DEWS. Currently, four regional DEWS efforts are underway: the Upper Colorado River Basin (UCRB), the Four Corners Tribal Lands area, the state of California in the western United States, and the Apalachicola-Chattahoochee-Flint (ACF) River Basin in the southeastern United States (Figure 1.5). By design, the NIDIS regional DEWS are tailored for local-scale monitoring (i.e., county, state, watershed, and/or region) in support of specific applications as compared to the USDM, which is national in scope and intended to provide a holistic view of drought conditions. As a result, the regional DEWS will have different data requirements than the USDM, and the informational needs among the DEWS can vary in terms of the spatial scale and types of variables. For example, the UCRB

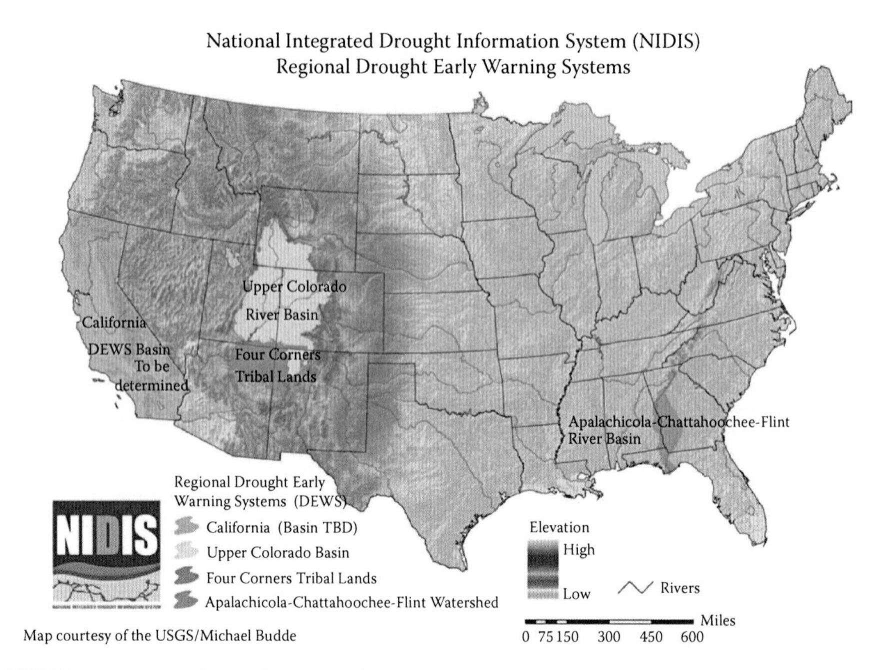

FIGURE 1.5 Locations of the NIDIS regional drought early warning systems.

DEWS has a primary focus on water resource management (e.g., hydroelectric power, agricultural/commercial/municipal water use, and interbasin transfers), particularly related to water supply from mountain snow melt. Despite having considerable in situ measurement capabilities (e.g., rain, snowpack, and stream gauges) in the UCRB, large data gaps still exist with poor spatial coverage over many areas of this large region and insufficient information for key variables (e.g., precipitation, ET, and snow water equivalent) at certain locations (e.g., low elevations) and spatial scales. Similarly, the other regional DEWS have their own unique data gaps for certain variables and data sets at relevant spatial scales.

1.2.5.2 Efforts to Build a Global Drought Early Warning System

Over the past 3–5 years, several efforts have begun to focus on the establishment of a virtual-based Global DEWS (GDEWS). The National Drought Mitigation Center (NDMC) has been an advocate of such a system for more than a decade now in what was originally proposed to the WMO as a Global Drought Preparedness Network (GDPN). The concept was based on the idea that individually, many countries lack or are unable to improve their capacity to cope with drought but collectively, through global, regional, and national partnerships, resources, information, and experiences can be shared and leveraged to reduce the impacts of drought. The idea would involve setting up a series of continental-based virtual regional networks, starting with North America, which already has the NADM in operation, and expanding to other efforts in Australia, Europe, South America, Africa, and Asia. The initial NADM activity in North America would serve as a blueprint for these future efforts throughout the world.

More recently, NIDIS, the Group on Earth Observations (GEO), and WMO have promoted similar yet different approaches that could ultimately be combined in part to produce a GDEWS. Through their work in the United States, NIDIS has already established a potential framework for a Global Integrated Drought Information System (GIDIS) on their web portal (drought.gov) and has begun to integrate various regional and continental drought monitors, remote sensing data sets, and other data resources from around the world. By capitalizing on agreed-on interoperability standards, services, and deliverables, the potential to collaborate virtually in the global arena to establish a GDEWS is growing rapidly.

The WMO has taken a somewhat more holistic approach toward the development of an Integrated Drought Management Programme (http://www.wmo.int/agm) that would target governmental, intergovernmental, and nongovernmental organizations in drought monitoring, prediction, risk reduction, and management. The goal would be to develop a global coordination of efforts to strengthen drought monitoring, prediction, early warning, and risk identification and develop a drought management knowledge base that captures mitigation and best practices in dealing with this hazard.

GEO (http://www.earthobservations.org) has also begun to look at integrating drought and DEWS into their 10 year action plan and via the Global Earth Observation System of Systems (GEOSS) information within their societal benefit areas through data and information sharing, communication, and capacity building in order to address the growing worldwide threat of drought. At the GEO Ministerial Summit in 2007, it was agreed to build on existing GEO programs to work toward establishing a GDEWS within the coming decade to provide regular drought warning assessments as frequently as possible during a drought crisis. The GEO vision is to build a global drought community of practice with an end goal of producing a global drought monitor.

Although considerable progress has been made, there are still several challenges to overcome in establishing DEWS at any scale or in any region of the world. Some of the primary challenges include the following (partial list taken from WMO, 2006):

1. Meteorological/hydrological data networks are inadequate in terms of spatial distribution and/or density, and quality long-term data records are lacking for many networks and/or stations.
2. Data sharing within government agencies/ministries and between countries and regions is inadequate, although some regions (e.g., European Commission's Joint Research Centre, European Drought Observatory [EDO], and the Drought Management Centre for Southeast Europe [DMCSEE]) are beginning to leverage resources and work together on this topic.
3. Information delivered through DEWS is often too technical or in complex formats, limiting its use by decision/policy makers and the general public.
4. Existing drought indices are sometimes inadequate for detecting the early onset and end of drought.

Although this is not a complete list of all challenges faced in developing a GDEWS, satellite remote sensing holds considerable potential to begin addressing these specific challenges, as will be discussed further in Section 1.3. Lastly, the development of an effective GDEWS and necessary data inputs requires a commitment of resources, which can be a challenge for all regions of the world. Typically, support for such a system many times comes down to political will and capitalizing on a severe drought as a focusing event to spur action, albeit in a reactive crisis management fashion.

1.3 SATELLITE REMOTE SENSING OF DROUGHT: A NEW ERA

As drought monitoring and DEWS activities continue to evolve around the world, the demand will continue to increase for consistent, high-quality data sets and observations in support of applications across a range of spatial scales (i.e., local, national, regional, and global). Given the general limited density of stream and rain gauges and soil moisture measurements in existing observational networks and the lack of resources globally to enhance, maintain, and expand these networks, as well as institutional barriers that limit data sharing, in situ–based data sets by themselves will be unable to meet these growing demands. The unique characteristics of satellite remote sensing data will be looked upon to assist in closing this informational gap, improving our capacity to track drought, particularly at the global scale.

The suite of new instruments on both global imagers and geostationary platforms that have become available since the late 1990s has ushered in a new era in the remote sensing of drought. Satellite-based sensors and missions such as the Moderate Resolution Imaging Spectroradiometer (MODIS), Advanced Microwave Scanning Radiometer for Earth Observing System (AMSR-E), Tropical Rainfall Measuring Mission (TRMM), and Gravity Recovery and Climate Experiment (GRACE) are collecting near-daily global data at varying spatial resolutions ranging from 250 m to hundreds of kilometers across the visible, infrared (near, middle, and thermal), and microwave portions of the electromagnetic spectrum, as well as for the earth's gravitational field. This has resulted in a rapid development of many new innovative remote sensing techniques characterizing various components of the hydrologic cycle relevant to drought. The chapters in this book will describe many of these new tools that monitor key hydrologic components (Figure 1.6), including ET (Chapters 6 through 8 in this book), groundwater (Chapter 11), rainfall (Chapters 12 through 14), snow cover (Chapter 15), soil moisture (Chapters 9 through 11), and vegetation health (Chapters 3 through 5). In addition, other remote sensing efforts are underway to estimate or retrieve other components such as surface water levels (Durand et al., 2008; Lee et al., 2011) and streamflow (Durand et al., 2010) that will not be covered in detail in this book. Unlike the past, where AVHRR VI data were the primary remote sensing products used to assess drought (primarily the effects of agricultural drought), the capability now exists to analyze various components of the hydrologic cycle either individually or collectively to define drought conditions. These new types of remote sensing data sets, in combination with traditional in situ–based, hydroclimate index and indicator data, should provide a more complete depiction of current drought conditions for decision makers.

A number of efforts, highlighted in several chapters of this book, are currently underway to assess the accuracy of these new satellite remote sensing tools and their utility within drought monitoring systems, such as the USDM. It is clear that satellite remote

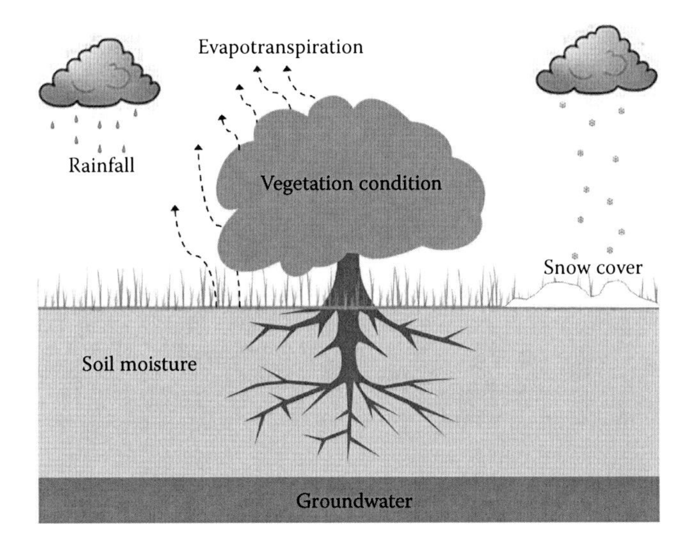

FIGURE 1.6 Components of the hydrologic cycle relevant to drought that can be estimated or retrieved from satellite remote sensing tools presented in this book.

sensing represents both a time- and cost-efficient means for collecting key information for drought monitoring over large geographic areas and in data-poor regions lacking gauge-based observational networks. However, there is a clear need to better bridge the research-to-operations hurdle in integrating remote sensing products from both a historical and a real-time operational basis. Drought monitoring and early warning systems have key operational requirements in terms of the latency, update interval, and format of the data inputs, as well as the specific metric(s) (e.g., percent of historical average) used to summarize the retrieved variable from the satellite data (e.g., soil moisture). Given that limited resources are often allocated to maintain and synthesize the input data into useful information for DEWS, clear communication and coordination between the remote sensing and drought monitoring communities is needed to maximize the value of these new remote sensing products. Many examples of these efforts to customize data products from various remote sensing tools for drought monitoring systems are presented throughout this book, including the adoption of commonly accepted cartographic color schemes to depict varying drought conditions, the development of techniques to downscale coarse resolution data to higher spatial resolutions to accommodate more local-scale monitoring (in Chapter 7, Anderson et al., 2012; and Chapter 11, Rodell, 2012), and the temporal extension of remotely sensed time-series variables calculated from limited satellite observational records using land data assimilation techniques to provide a longer historical context for anomaly detection (in Chapter 10, Sheffield et al., 2012; and Chapter 11, Rodell, 2012).

1.4 CONCLUSIONS AND OUTLOOK

Given the complex dimensions of drought and the challenges they pose for routine drought monitoring, it is essential that we continue to find innovative and robust ways to quantify and more effectively communicate the impacts of this hazard as part of an operational DEWS approach. For those who are only concerned with one aspect of drought (e.g., impacts of hydrological drought on reservoir levels), then monitoring, analysis, and assessment may be much more focused on one or a few indicators. In most cases, however, a much more comprehensive and multifaceted approach will be necessary when monitoring drought events. Composite indicators such as the USDM allow a user to remain flexible in utilizing new tools, indices, and indicators as they become available and/or useful for a particular region or a particular season.

Integration of remotely sensed indicators and indices continues to gain acceptance and traction as their history and application expands, and as drought monitoring experts and other decision makers become more familiar with their strengths and limitations. The current capability of remote sensing to estimate and retrieve an increasing number of variables related to the hydrologic cycle (as shown in Figure 1.6) illustrates the tremendous potential these new remote sensing tools have to support a wide range of drought applications that vary in terms of the specific variables and the spatial scale of the data that are required. Supplementing the current set of traditional drought monitoring tools with remote sensing data and products is logical given the limitations associated with ground-based observational networks, and the use of remotely sensed data will continue to grow as the satellite period of record lengthens and more people within the drought community become comfortable with fully utilizing such information in an operational setting. A DEWS that nests a global-to-local scale drill-down approach, which utilizes composite indicators that contain remotely sensed variables, will assist in moving the global drought community forward to better monitor drought at multiple spatial scales and for various sectors well into the twenty-first century.

REFERENCES

Alley, W.M. 1984. The Palmer Drought Severity Index: Limitations and assumptions. *Journal of Climate and Applied Meteorology* 23:1100–1109.

Anderson, M.C., C. Hain, B.D. Wardlow, A. Pimstein, J.R. Mecikalski, and W.P. Kustas. 2012. A drought index based on thermal remote sensing of evapotranspiration. In *Remote Sensing of Drought: Innovative Monitoring Approaches*, eds. B.D. Wardlow, M.C. Anderson, and J.P. Verdin. Boca Raton, FL: CRC Press.

Anyamba, A. and C.J. Tucker. 2012. Historical perspective of AVHRR NDVI and vegetation drought monitoring. In *Remote Sensing of Drought: Innovative Monitoring Approaches*, eds. B.D. Wardlow, M.C. Anderson, and J.P. Verdin. Boca Raton, FL: CRC Press.

Asrar, G., R.B. Myneni, and E.T. Kanemasu. 1989. Estimation of plant canopy attributes from spectral reflectance measurements. In *Theory and Applications of Optical Remote Sensing*, ed. G. Asrar, pp. 252–296. New York: Wiley.

Baret, F. and G. Guyot. 1991. Potentials and limits of vegetation indices for LAI and APAR assessment. *Remote Sensing of Environment* 35:161–173.

Brown, J.F., S. Maxwell, and S. Pervez. 2009. Mapping irrigated lands across the United States using MODIS satellite imagery. In *Remote Sensing of Global Croplands for Food Security*, eds. P.S. Thenkabail, J.G. Lyon, H. Turral, and C.M. Biradar, pp. 177–198. Boca Raton, FL: CRC Press.

Dai, A. 2011. Characteristics and trends in various forms of the Palmer Drought Severity Index (PDSI) during 1900–2008. *Journal of Geophysical Research—Atmosphere*, doi:10.1029/2010JD015541.

Dai, A., K.E. Trenberth, and T. Qian. 2004. A global dataset of Palmer Drought Severity Index for 1870–2002: Relationship with soil moisture and effects of surface warming. *Journal of Hydrometeorology* 5:1117–1130.

Durand, D., K.M. Andreadis, D.E. Alsdorf, D.P. Lettenmaier, D. Moller, and M. Wilson. 2008. Estimation of bathymetric depth and slope from data assimilation of swath altimetry into a hydrodynamic model. *Geophysical Research Letters* 35:L20401, doi:10.1029/2008GL034150.

Durand, M., E. Rodriguez, D.E. Alsdorf, and M. Trigg. 2010. Estimating river depth from remote sensing swath interferometry measurements of river height, slope, and width. *IEEE Journal of Selected Topics in Applied Earth Observations and Remote Sensing* 3(1):20–31.

Eidenshink, J.C. and R.H. Hass. 1992. Analyzing vegetation dynamics of land systems with satellite data. *GeoCarto International* 1:53–61.

General Accounting Office (GAO). 1979. Federal response to the 1976–77 drought: What should be done next? Report to the Comptroller General, Washington, DC, p. 29.

Goward, S.N., C.J. Tucker, and D.G. Dye. 1985. North American vegetation patterns observed with the NOAA-7 advanced very high resolution radiometer. *Plant Ecology* 64:3–14.

Guha-Sapir, D. 2011. Natural disasters in countries with very high human development. *CRED Crunch* 24:1–2.

Guha-Sapir, D., D. Hargitt, and P. Hoyois. 2004. Thirty years of natural disasters 1974–2003: The numbers. Centre for Research on the Epidemiology of Disasters, Belgium. http://www.emdat.be/old/Documents/Publications/publication_2004_emdat.pdf (accessed August 26, 2011).

Hayes, M.J., M.D. Svoboda, D.A. Wilhite, and O.V. Vanyarkho. 1999. Monitoring the 1996 drought using the Standardized Precipitation Index. *Bulletin of the American Meteorological Society* 80(3):429–438.

Hayes, M., M. Svoboda, D. LeComte, K. Redmond, and P. Pasteris. 2005. Drought monitoring: new tools for the 21st century. In: *Drought and Water Crises: Science, Technology, and Management Issues*, Ed. D. Wilhite, Boca Raton, FL: CRC Press, pp. 53–69.

Heim, R.R. 2002. A review of twentieth-century drought indices used in the United States. *Bulletin of the American Meteorological Society* 83:1149–1165.

Hutchinson, C.F. 1991. Use of satellite data for famine early warning in sub-Saharan Africa. *International Journal of Remote Sensing* 12:1405–1421.

Jansen, E., J. Overpeck, K.R. Briffa, J.-C. Duplessy, F. Joos, V. Masson-Delmotte, D. Olago, B. Otto-Bliesner, W.R. Peltier, S. Rahmstorf, R. Ramesh, D. Raynaud, D. Rind, O. Solomina, R. Villalba, and D. Zhang. 2007. Paleoclimate. In *Climate Change 2007: The Physical Science Basis. Contribution of Working Group I to the Fourth Assessment Report of the Intergovernmental Panel on Climate Change*, eds. S. Solomon, D. Qin, M. Manning, Z. Chen, M. Marquis, K.B. Averyt, M. Tignor, and H.L. Miller. Cambridge, MA: Cambridge University Press.

Kogan, E.N. 1995a. Application of vegetation index and brightness temperature for drought detection. *Advances in Space Research* 15:91–100.

Kogan, F.N. 1995b. Droughts of the late 1980s in the United States as derived from NOAA polar orbiting satellite data. *Bulletin of the American Meteorological Society* 76:655–668.

Kogan, F.N. 1997. Global drought watch from space. *Bulletin of the American Meteorological Society* 78:621–636.

Kogan, F. 2002. World droughts in the new millennium from AVHRR-based vegetation health indices. *EOS, Transactions of the American Geophysical Union* 83(48):557–564.

Lawrimore, J., R.R. Heim, Jr., M. Svoboda, V. Swail, and P.J. Englehart. 2002. Beginning a new era of drought monitoring across North America. *Bulletin of the American Meteorological Society* 83(8):1191–1192.

Lee, H., C.K. Shum, K.-H. Tseng, J.-Y. Guo, and C.-Y. Kuo. 2011. Present-day lake level variations from Envisat altimetry over the northeastern Qinghai-Tibetan Plateau: Links with precipitation and temperature. *Terrestrial, Atmospheric, and Oceanic Sciences* 22(2):169–175.

McKee, T.B., N.J. Doesken, and J. Kleist. 1993. The relationship of drought frequency and duration to time scales. Preprints, *8th Conference on Applied Climatology*, Anaheim, CA, pp. 179–184.

McKee, T.B., N.J. Doesken, and J. Kleist. 1995. Drought monitoring with multiple time scales. Preprints, *9th Conference on Applied Climatology*, Dallas, TX, pp. 233–236.

Mizzell, H. 2008. Improving drought detection in the Carolinas: Evaluation of local, state, and federal drought indicators. Dissertation, Department of Geography, University of South Carolina, Columbia, SC, p. 149.

NCDC (National Climatic Data Center). 2011. Billion dollar U.S. weather disasters. http://www.ncdc.noaa.gov/oa/reports/billionz.html (accessed December 10, 2011).

NSTC (National Science and Technology Council). 2005. *Grand Challenges for Disaster Reduction: A Report of the Subcommittee on Disaster Reduction.* http://www.sdr.gov/SDRGrandChallengesforDisasterReduction.pdf (accessed on December 10, 2011).

Palmer, W.C. 1965. Meteorological drought. Research Paper No. 45, U.S. Department of Commerce Weather Bureau, Washington, DC.

Peters, A.J., E.A. Walter-Shea, L. Ji, A. Vina, M. Hayes, and M. Svoboda. 2002. Drought monitoring with NDVI-based Standardized Vegetation Index. *Photogrammetric Engineering and Remote Sensing* 68(1):71–75.

Rodell, M. 2012. Satellite gravimetry applied to drought monitoring. In *Remote Sensing of Drought: Innovative Monitoring Approaches*, eds. B.D. Wardlow, M.C. Anderson, and J.P. Verdin. Boca Raton, FL: CRC Press.

Rojas, O., A. Vrieling, and F. Remold. 2011. Assessing drought probability for agricultural areas in Africa with coarse resolution remote sensing imagery. *Remote Sensing of Environment* 115(2):343–352.

Rouse, J.W. Jr., R.H. Haas, J.A. Schell, D.W. Deering, and J.C. Harlan. 1974. Monitoring the vernal advancement and retrogradation (green wave effect) of natural vegetation. NASA/GSFC Type III Final Report, Greenbelt, MD.

Shafer, B.A. and L.E. Dezman. 1982. Development of a Surface Water Supply Index (SWSI) to assess the severity of drought conditions in snowpack runoff areas. In *Proceedings of the Western Snow Conference*, Fort Collins, CO, pp. 164–175.

Sheffield, J., Y. Xia, L. Luo, E.F. Wood, M. Ek, K.E. Mitchell, and NLDAS team. 2012. The North American land data assimilation system (NLDAS): A framework for merging model and satellite data for improved drought monitoring. In *Remote Sensing of Drought: Innovative Monitoring Approaches*, eds. B.D. Wardlow, M.C. Anderson, and J.P. Verdin. Boca Raton, FL: CRC Press.

Steinemann, A., M. Hayes, and L. Cavalcanti. 2005. Drought indicators and triggers. In *Drought and Water Crises: Science, Technology, and Management Issues*, ed. D. Wilhite. Boca Raton, FL: CRC Press.

Story, G.J. 2012. Estimating precipitation from WSR-88D observations and rain gauge data—Potential for drought monitoring. In *Remote Sensing of Drought: Innovative Monitoring Approaches*, eds. B.D. Wardlow, M.C. Anderson, and J.P. Verdin. Boca Raton, FL: CRC Press.

Svoboda, M., D. LeComte, M. Hayes, R. Heim, K. Gleason, J. Angel, B. Rippey, R. Tinker, M. Palecki, D. Stooksbury, D. Miskus, and S. Stephens. 2002. The Drought Monitor. *Bulletin of the American Meteorological Society* 83(8):1181–1190.

Tucker, C.J. 1979. Red and photographic infrared linear combinations for monitoring vegetation. *Remote Sensing of Environment* 8:127–150.

Tucker, C.J., C.O. Justice, and S.D. Prince. 1986. Monitoring the grasslands of the Sahel 1984–1985. *International Journal of Remote Sensing* 7:1571–1581.

Tucker, C.J., W.W. Newcomb, S.O. Los, and S.D. Prince. 1991. Mean and inter-year variation of growing-season normalized difference vegetation index for Sahel 1981–1989. *International Journal of Remote Sensing* 12:1133–1135.

Unganai, L.S. and F.N. Kogan. 1998. Drought monitoring and corn yield estimation in southern Africa from AVHRR data. *Remote Sensing of Environment* 63:219–232.

Vicente-Serrano, S.M., S. Begueria, and J.I. Lopez-Moreno. 2010. A multi-scalar drought index sensitive to global warming: The Standardized Precipitation Evapotranspiration Index—SPEI. *Journal of Climate* 23:1696–1718.

Wardlow, B.D., T. Tadesse, J.F. Brown, K. Callahan, S. Swain, and E. Hunt. 2012. The vegetation drought response index (VegDRI): An integration of satellite, climate, and biophysical data. In *Remote Sensing of Drought: Innovative Monitoring Approaches*, eds. B.D. Wardlow, M.C. Anderson, and J.P. Verdin. Boca Raton, FL: CRC Press.

Wells, N., S. Goddard, and M.J. Hayes. 2004. A self-calibrating Palmer Drought Severity Index. *Journal of Climate* 17(12):2335–2351.

Wilhite, D.A. 2000. Drought as a natural hazard: concepts and definitions. In *Drought Volume I: A Global Assessment*, ed. D.A. Wilhite, New York: Routledge.

Wilhite, D.A., 2009. Personal communication. *The National Integrated Drought Information System and Climate Services Workshop for the Midwest United States* sponsored by the Western Governors' Association and Western States Water Council Western States Federal Agency Support Team, Lincoln, NE, October 13–14, 2009.

Wilhite, D.A. and M.H. Glantz. 1985. Understanding the drought phenomenon: The role of definitions. *Water International* 10:111–120.

Wilhite, D.A. and M. Buchanan-Smith. 2005. Drought as hazard: Understanding the natural and social context. In *Drought and Water Crises: Science, Technology, and Management Issues*, ed. D.A. Wilhite, pp. 3–29, Boca Raton, FL: Taylor and Francis.

Wilhite, D.A. and N.J. Rosenberg. 1986. Improving federal response to drought. *Journal of Climate and Applied Meteorology* 25:332–342.

WMO (World Meteorological Organization). 2006. *Drought Monitoring and Early Warning: Concepts, Progress and Future Challenges*, WMO-No. 1006, ISBN 92-63-11006-9, Geneva, Switzerland.

Woodhouse, C.A. and J.T. Overpeck. 1998. 2000 years of drought variability in the central United States. *Bulletin of the American Meteorological Society* 79(12):2693–2714.

Part I

Vegetation

2 Historical Perspectives on AVHRR NDVI and Vegetation Drought Monitoring

Assaf Anyamba and Compton J. Tucker

CONTENTS

2.1 INTRODUCTION

Satellite measurements of the biosphere have now become common place in various aspects of large-scale environmental monitoring, including drought and crop monitoring. This was not the case until the launch of the Advanced Very High Resolution Radiometer (AVHRR) instrument on June 27, 1979, on board the first National Oceanic and Atmospheric Administration (NOAA) Advanced Television

Infrared Observation Satellite (TIROS-N/NOAA-6) polar-orbiting satellite. Initially, the NOAA AVHRR satellites were designed to observe the earth's weather patterns: primarily cloud dynamics, vertical soundings of the atmosphere, and sea surface temperatures. Early studies on remote sensing of vegetation were focused on understanding seasonality. The vernal advancement and retrogradation of vegetation (e.g., spring green-up, summer abundance, and fall dry-down) was first studied over the north–south expanse of the U.S. Great Plains using data from the Earth Resources Technology Satellite (ERTS) Multi-Spectral Scanner (MSS) instrument (Rouse et al., 1974a,b). Rouse et al. (1974a) and others demonstrated that biophysical characteristics of vegetation over this rangeland and cropland region could be inferred from satellite spectral measurements despite solar zenith angle differences across a long latitudinal gradient (Deering et al., 1975). Rouse et al. (1974b) developed a difference ratio metric between the red and near-infrared (NIR) radiances over their sum to normalize the effects of the solar zenith angle. This derivation is based on the unique spectral response function of vegetated surfaces compared to other surface matter in the visible and NIR portion of the electromagnetic spectrum, as shown in Figure 2.1.

Spectral reflectances and radiances of green vegetation canopies in the red region of the electromagnetic spectrum are inversely related to in situ chlorophyll density because of photosynthetic chlorophyll absorption by vegetation in this band. In contrast, energy in the NIR region is scattered and reflected by the canopy structure

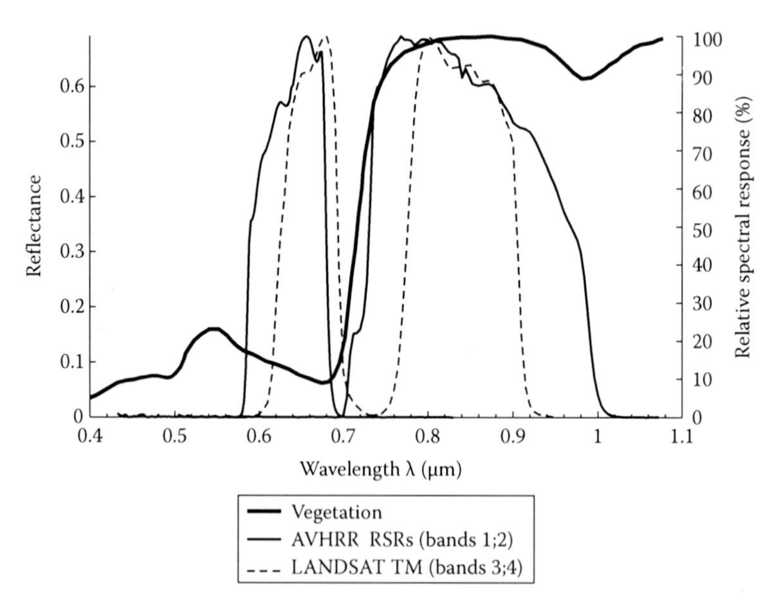

FIGURE 2.1 Spectral response curve of vegetation and the relative spectral response (RSR) of Landsat Thematic Mapper (TM) and AVHRR bands in the visible and infrared portions of the electromagnetic spectrum. This characteristic response pattern of vegetation has defined the design of remote sensing instruments and the derivation of various metrics for vegetation monitoring. The AVHRR's wide spectral bandwidths are subject to atmospheric interference (specifically aerosols and water vapor).

of the vegetation, and hence NIR reflectance is directly related to the green leaf density (Gates et al., 1965; Knipling, 1970; Woolley, 1971). These characteristics drive the spectral response of plants in these two spectral regions, and the reflectances vary with seasonal changes in vegetation condition (phenology) and/or stress (e.g., drought). When captured remotely through time, such variations can be exploited for vegetation drought-monitoring purposes. The ratio between red and NIR reflectances was named the Vegetation Index (VI). There were other variants of the VI such as the Transformed Vegetation Index (square-root transformation of difference–sum ratio), the Simple Ratio (red/infrared), and the Perpendicular Vegetation Index (Rouse et al., 1974a,b; Deering et al., 1975; Richardson and Weigand, 1977). Eventually, researchers agreed upon the Normalized Difference Vegetation Index (NDVI) as the most efficient and simple metric to identify vegetated areas and their condition (Tucker, 1979). Normalization had many advantages, including minimizing directional reflectance and off-nadir viewing effects; reducing sun-angle, shadow, and topographic variation effects; and minimizing aerosol and water-vapor effects (Holben, 1986). This normalization enabled large-scale vegetation monitoring, allowing comparison of different regions through time.

The NDVI is computed as follows:

$$\mathrm{NDVI} = \frac{\mathrm{NIR} - \mathrm{RED}}{\mathrm{NIR} + \mathrm{RED}} \tag{2.1}$$

where RED and NIR are the spectral reflectance measurements in the red and NIR regions of the electromagnetic spectrum, respectively. These spectral reflectances are ratios of reflected radiation to incoming radiation in each spectral band, with values ranging between 0.0 and 1.0. Theoretically, NDVI values can range between −1.0 and +1.0. However, the typical range of NDVI measured from vegetation and other earth surface materials is between about −0.1 (NIR less than RED) for nonvegetated surfaces and as high as 0.9 for dense green vegetation canopies (Tucker, 1979). The NDVI increases with increasing green biomass, changes seasonally, and responds to favorable (e.g., abundant precipitation) or unfavorable climatic conditions (e.g., drought). This early research was based on the ERTS program of the National Aeronautics and Space Administration (NASA), now known as the Landsat program. However, because of the low repeat cycle of the satellite (18 days) and persistent cloud cover, ERTS could not by itself provide the temporal frequency of measurements for systematic monitoring of the land surface for drought and other environmental applications that need cloud-free measurements.

2.2 BRIEF HISTORY OF AVHRR

From 1980 to 1982, independent teams of researchers with an interest in land-surface monitoring demonstrated that the visible channels on the NOAA AVHRR could be exploited for vegetation monitoring (Tucker et al., 1983). For the first time, vegetation could be monitored at a global scale from a satellite platform with a high temporal frequency of repeat observations (near-daily global coverage). This meant that aspects of vegetation seasonality and health could be studied and monitored over time. However, measurements for land-surface monitoring were not originally planned in the instrumental design of the NOAA polar-orbiting satellite program. The first-ever AVHRR was flown on the TIROS-N meteorological satellite in 1978. This AVHRR was configured with four spectral channels (0.55–0.90, 0.73–1.1, 3.5–3.9, and 10.5–11.5 μm) customized for meteorological observations and applications. Table 2.1 shows the spectral bandwidths of various AVHRR platforms and their possible uses and applications. Note that some channel bandwidths have changed over time with observational and technological requirements.

After this pioneer mission, it became apparent that future AVHRR sensors required modifications to increase effectiveness for snow mapping and vegetation monitoring, primarily by narrowing the first channel in the red spectral region to 0.55–0.68 μm (Schneider et al., 1981). Since chlorophyll absorption of solar radiation is confined to these wavelengths, narrowing the first channel made detecting and mapping vegetation

TABLE 2.1
AVHRR Spectral Channels, Resolution, and Primary Uses

Channel Number	Wavelength (μm) TIROS-N	Wavelength (μm) NOAA-6, 8,10,12,14,16,18	Wavelength (μm) NOAA-7, 9,11,15,17	IFOV (mrad)	Primary Uses
1	0.550–0.90	0.580–0.68	0.58–0.68	1.39	Daytime cloud/surface and vegetation mapping
2	0.725–1.10	0.725–1.10	0.725–1.0	1.41	Surface water, ice, snow melt, and vegetation mapping
3A		1.58–1.64 (NOAA-16,18)	1.58–1.64 (NOAA-15,17)	1.3	Snow and ice detection
3B	3.550–3.93	3.550–3.93	3.55–3.93	1.51	Sea surface temperature, nighttime cloud mapping
4	10.500–11.50	10.500–11.50	10.3–11.3	1.41	Sea surface temperature, day and night cloud mapping
5	Ch4 repeated	Ch4 repeated	11.5–12.5	1.30	Sea surface temperature, day and night cloud mapping

Source: Compiled from NOAA Satellite and Information Services, *NOAA Polar Orbiter Data User's Guide*, Silver Spring, MD, http://psbcw1.nesdis.noaa.gov/terascan/home_basic/polarsats_sensors_tables.html (accessed December 12, 2011).

Note: The instantaneous field of view (IFOV) of each channel is approximately 1.4 mrad, leading to a resolution at the satellite subpoint of 1.1 km for a nominal altitude of 850 km (528 miles).

more effective, as opposed to the wider channel on the first TIROS, which constrained vegetation monitoring because of atmospheric attenuation. After the launch of the NOAA-6 platform in 1979, a new unintended measurement became possible from meteorological satellites: the NDVI. A major advantage of AVHRR is its daily global coverage. Because of the instrument's wide field of view (±55°) and polar-orbiting, sun-synchronous orbit, the AVHRR images the earth's land surface twice daily (day and night). Data collected over consecutive days minimize the effects of cloud cover and other unfavorable atmospheric conditions. The high temporal resolution of AVHRR, coupled with continual operational data acquisitions over many years, provides a historical context for long-term monitoring and comparison of land-surface conditions. Metrics obtained through image differencing or other decomposition techniques can be used to identify different types of ecosystem anomalies that can be used for drought detection (Tucker et al., 1986, 1991; Eastman and Fulk, 1993; Tucker, 1996). At present, a 30 year history of global AVHRR NDVI measurements (Table 2.2) sets a climate-scale benchmark for land-surface studies and applications (Figure 2.2). This long-term data set can be used to study aspects of drought frequency and extent, as well as relationships between drought and climate variability.

2.3 NORMALIZED DIFFERENCE VEGETATION INDEX

2.3.1 NDVI Derived from AVHRR Measurements

The NDVI from broadband AVHRR data is calculated from channels 1 (0.55–0.70 μm) and 2 (0.73–1.1 μm) using the following equation:

$$\text{NDVI} = \frac{\text{Channel 2} - \text{Channel 1}}{\text{Channel 2} + \text{Channel 1}} \tag{2.2}$$

Figure 2.1 shows a comparison between AVHRR and Landsat TM spectral bandwidths for vegetation mapping and their respective spectral responses. Often, AVHRR's channel 4 or 5 (10.3–11.3, 11.5–12.5 μm) is used as a thermal cloud mask. The thermal cloud mask eliminates NDVI data below a set brightness temperature threshold, which is usually around 285 K. In most tropical latitudes, it is assumed that the surface brightness temperatures will be above the threshold value (even at high elevations during the afternoon AVHRR overpasses). Since cloud pixels typically have brightness temperatures less than the land surface, the corresponding NDVI pixels are masked (Holben, 1986). Kimes (1983) and Holben and Fraser (1984) found that several factors unrelated to vegetation condition can affect NDVI over green vegetation, including atmospheric effects, cloud detection, and bidirectional reflectance effects. Different compositing techniques had to be investigated to form solutions to these issues (Kimes, 1983; Gatlin et al., 1984; Holben and Fraser, 1984; Kimes et al., 1984; Holben, 1986; Holben et al., 1986). One such technique was to form time-composite images of maximum NDVI values over periods of several days, such as 7-day, 10-day, 15-day, or monthly intervals (Holben and Fraser, 1984; Holben, 1986). The use of a thermal cloud mask combined with maximum value compositing (MVC) reduces the effects of cloud contamination, atmospheric

TABLE 2.2
NOAA POES Launch Dates and Periods of Operation

Name after Launch	Name before Launch	Date of Launch	AVHRR Version	Spacecraft Mission Operational Status
TIROS-N	TIROS-N	October 1978		?
NOAA-6	NOAA-A	June 27, 1979		End of mission: November 11, 1986
NOAA-7	NOAA-C	June 23, 1981	2	End of mission: June 7, 1986
NOAA-8	NOAA-E	March 28, 1983	1	End of mission: October 31, 1985
NOAA-9	NOAA-F	December 12, 1984	2	End of mission: November 5, 1994
NOAA-10	NOAA-G	September 17, 1986	1	Decommissioned: September 17, 1991
NOAA-11	NOAA-H	September 24, 1988	2	Decommissioned: June 16, 2004
NOAA-12	NOAA-D	May 15, 1991	1	Decommissioned: August 10, 2007, after setting an extended lifetime record of over 16 years
NOAA-13	NOAA-I	August 1993		Failure after launch: 2 weeks after launch, the spacecraft suffered a catastrophic power system anomaly
NOAA-14	NOAA-J	December 30, 1994	2	Decommissioned: May 23, 2007
NOAA-15	NOAA-K	May 13, 1998	3	AM secondary
NOAA-16	NOAA-L	September 21, 2000	3	PM secondary
NOAA-17	NOAA-M	June 2002	3	AM backup
NOAA-18	NOAA-N	May 20, 2005	3	PM secondary
NOAA-19	NOAA-O	February 6, 2009	3	PM primary
METOP-A		October 19, 2006	3	AM primary

Source: Compiled from NOAA Satellite and Information Service, Silver Spring, MD, http://www.oso.noaa.gov/history/first-launched.htm (accessed on December 12, 2011).

Note: The AVHRR instrument is a cross-track scanning system similar to the former VHRR on TIROS-N but features four (AVHRR/1), five (AVHRR/2), or six (AVHRR/3) spectral channels. Note that METOP mission is owned and operated by EUMETSAT to provide data for the AM orbit.

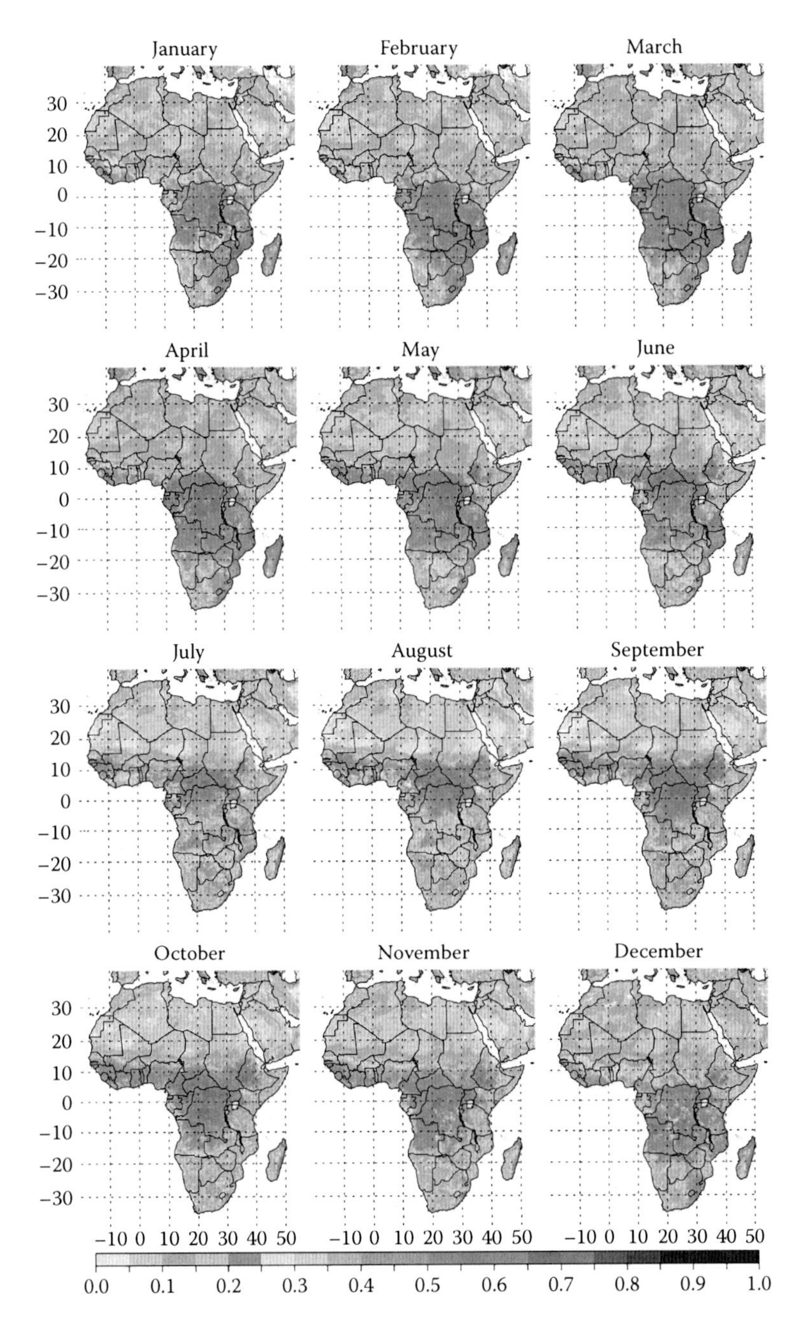

FIGURE 2.2 **(See color insert.)** Example of typical monthly NDVI time-series data for Africa and the Middle East. In general, areas of high NDVI or high vegetation density are represented in shades of green while areas of low NDVI/low vegetation density such as semi-arid lands and Sahara and Arabian deserts are shown in shades of yellow to brown. The patterns change seasonally from January through December. (Data produced by GIMMS Group at NASA/GSFC, Greenbelt, MD.)

attenuation, view and illumination geometry, and surface directional reflectance. This is because maximum NDVI values are found to be associated with a clear atmosphere, while compositing tends to minimize other angular effects.

MVC was enacted as standard operating procedure in 1983 for the production of global AVHRR products generated by the Global Inventory Monitoring and Modeling Studies (GIMMS) group at the NASA Goddard Space Flight Center (GSFC). Reasons for enacting this as standard protocol were (a) the AVHRR channels 1 and 2 are spectrally very wide, (b) channel 2 contains a water absorption band, (c) directional effects were not well understood for broadband sensors, and (d) detailed atmospheric data were not available for explicit atmospheric correction. Therefore, since atmospheric conditions tend to suppress NDVI values in a nonclear atmosphere, MVC was the preferred solution.

2.3.2 Interpretation of NDVI

Radiation measurements of the earth's surface from satellites are complex functions of not only the state and properties of the surface itself but also the conditions and dynamics of the atmosphere through which the reflected radiation is sensed. NDVI and other derived VIs are an attempt to provide the best estimates of the state and condition of vegetation while minimizing (or eliminating) the influence of other factors, as mentioned in the previous section. Derived metrics have to be universal and reliable in time and space, irrespective of these extraneous factors. A detailed analysis of the interpretation of VIs is given by Myneni et al. (1995), who concluded that NDVI represents the energy that drives photosynthesis. Over the past two decades, the NDVI has been widely used in many terrestrial applications. Some examples include drought early warning, locust monitoring, vector disease risk assessment, estimation of forage, agricultural monitoring, land cover classification, and as an input to land-surface and biophysical models. Early proof-of-concept studies showed NDVI to be a nondestructive measure of intercepted photosynthetically active radiation (PAR) (Asrar et al., 1984, 1985; Hatfield et al., 1984; Wiegand and Richardson, 1984), photosynthetic capacity, and primary production (Kumar and Monteith, 1982; Sellers, 1985, 1987; Asrar et al., 1986; Tucker and Sellers, 1986). Other studies used NDVI to study total biomass production for a wide range of vegetation types, including grasslands, agricultural crops, and salt marshes (Steven et al., 1983; Tucker et al., 1983, 1985b; Hardisky et al., 1984). Intensive field studies using a combination of time-series AVHRR NDVI data and ground-based spectral measurements were used to estimate total biomass production in savanna ecosystems (Tucker et al., 1983, 1985b; Hiernaux and Justice, 1986; Prince and Tucker, 1986). Results from these studies showed that the cumulative NDVI was linearly related with the total aboveground dry biomass sampled at the end of the growing season over the Sahelian zone of Africa, as shown in Figure 2.3.

2.3.3 Early Applications of NDVI

Researchers began to build on these early findings with large-scale studies using coarse-resolution AVHRR NDVI data to map regional-to-continental scale vegetation types (Norwine and Greegor, 1983; Justice et al., 1985; Townshend et al., 1985,

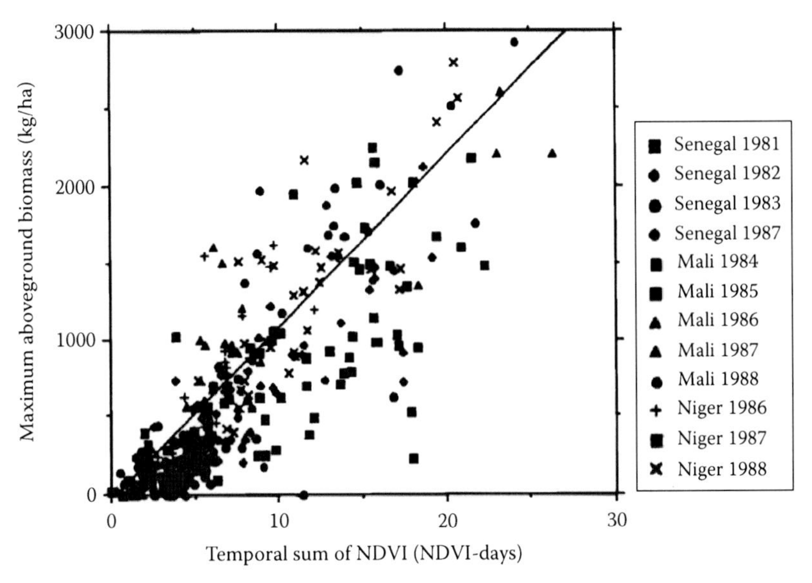

FIGURE 2.3 Summary figure for the NOAA AVHRR 1 km NDVI-Sahelian biomass relationship from 1981 to 1988. The figure presents a comparison between ground measurements of aboveground total dry herbaceous biomass sampled at the end of the growing season and time-integrated NDVI data for each country/year combination. (From Prince, 1991: [biomass (kg/ha) = −86 + 114 * NDVI-days; Confidence Intervals: @3 NDVI-days, ±61 kg/ha; @10 NDVI-days, ±51 kg/ha.]) (Adapted from Tucker, C.J., History of the use of AVHRR data for land applications, in *Advances in the Use of NOAA AVHRR Data for Land Applications*, Kluwer Academic Publishers, Dordrecht, the Netherlands, 1996, pp. 1–19.)

1987; Tucker et al., 1985a; Dyer and Crossley 1986; Loveland et al., 1991; Tateishi and Kajiwara, 1992; Eastman and Fulk, 1993; Stone et al., 1994). These studies assumed that mapping the photosynthetic capacity of vegetation would lead to disaggregation of the land surface into land cover groupings based on vegetation function. Pertinent examples of such successful studies were (1) land cover mapping, (2) investigating the relationship between photosynthetic capacity and rainfall in semiarid lands (Figure 2.4) (Nicholson et al., 1990; Tucker et al., 1991), and (3) monitoring ecological conditions favorable for insects and birds in arid and semiarid ecosystems (Tucker et al., 1985c; Hielkema et al., 1986). These studies led to the first use of AVHRR NDVI data in drought and desert locust monitoring through cooperation among NASA GSFC, the Food and Agricultural Organization (FAO), and the U.S. Agency for International Development (USAID) (Hutchinson, 1991; Tucker, 1996).

2.4 AVHRR NDVI DROUGHT APPLICATIONS

The absence of reliable, continuous, and high-density time series of terrestrial weather and climate observations for most parts of the world made it difficult to monitor the spatial patterns of drought and other climate-related anomalies in the past (Janowiak, 1988; Nicholson, 1989). The 1983–1985 large-scale drought that affected

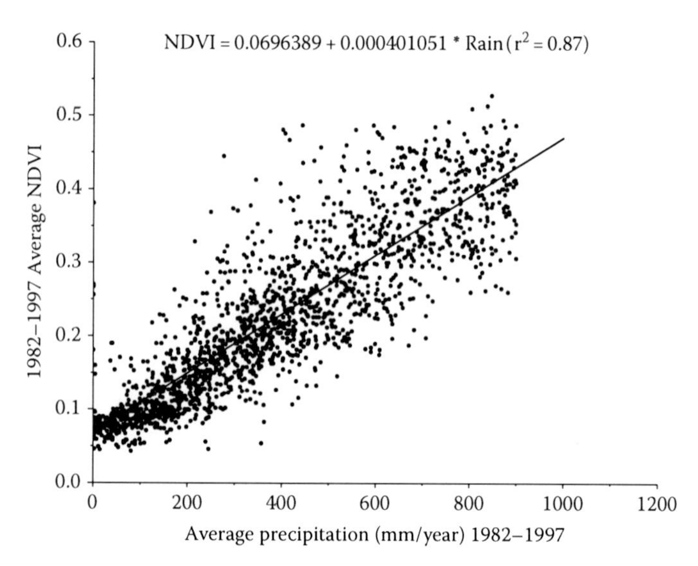

FIGURE 2.4 Comparison between average precipitation from 1982 to 1987 for 1794 stations from West Africa and coincident NDVI. (Data updated and adapted from Tucker, C.J., History of the use of AVHRR data for land applications, in *Advances in the Use of NOAA AVHRR Data for Land Applications*, Kluwer Academic Publishers, Dordrecht, the Netherlands, 1996, pp. 1–19.)

the Sahel region of Africa was the signature event that eventually led to the use of coarse spatial resolution AVHRR NDVI data for drought monitoring (Figure 2.5).

High temporal resolution data are a key factor for drought monitoring in order to capture the frequency of rainfall events (Tucker et al., 1986). During the early 1980s drought, most rainfall data records that were available through

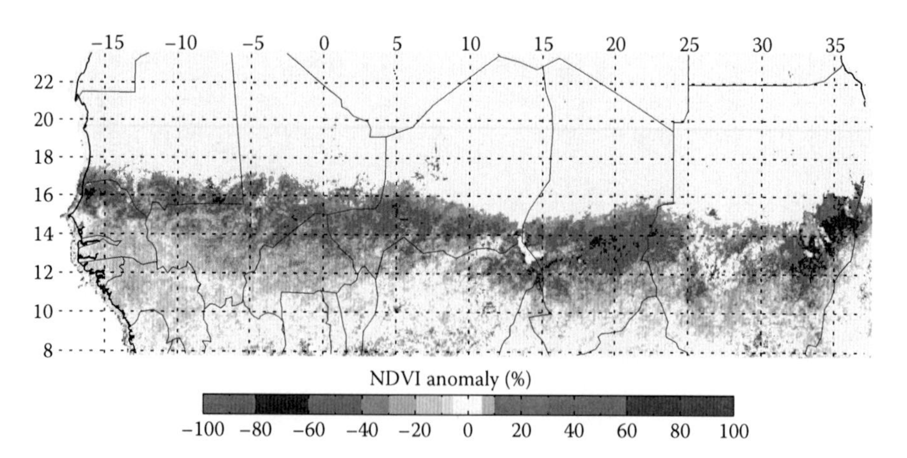

FIGURE 2.5 **(See color insert.)** Growing season (July–October) NDVI anomaly for the Sahel region showing the large areal extent of the Sahelian drought in 1984. Before AVHRR NDVI data became available, such regional-to-continental mapping of drought extent and patterns was not possible. (Adapted from Anyamba, A. and Tucker, C.J., *J. Arid Environ.*, 63, 596, 2005.)

the World Meteorological Organization (WMO) and national meteorological services were compiled as weekly or 10 day cumulative records. Comparisons with near-real-time rainfall data could be made by temporally sampling the AVHRR NDVI data to form weekly and decadal composites for drought monitoring, particularly with regard to agricultural conditions. Tucker et al. (1991) demonstrated that intercomparisons of extended time series of NDVI data can provide useful information for drought monitoring in the Sahel region. Baseline vegetation conditions for the growing season (i.e., July–October) were defined as the mean NDVI calculated over several years, and the coefficient of variation in NDVI was used to represent variation between growing seasons. Drought and areas of high interannual vegetation variability were reflected in the multitemporal AVHRR NDVI data sets, especially in the Sahel between the 1984–1985 drought years and the wet year in 1988 (Tucker et al., 1991). With a baseline established for vegetation (by month or growing season), current conditions could be assessed as above, below, or near normal, which is particularly important for monitoring agricultural conditions and determining agricultural production estimates. This was a pioneering first step toward using interannual and/or anomaly analysis of NDVI for drought monitoring of vegetation. Other related efforts have applied NDVI for drought monitoring (Gallo, 1990; Kogan, 1990, 1995a; Eidenshink and Hass, 1992; Burgan and Hartford, 1993; Burgan et al., 1996; Unganai and Kogan, 1998, among many others). Several NDVI-based indices developed for monitoring drought are discussed in the following section, including vegetation condition conveyed by remotely sensed land-surface temperature (LST).

2.4.1 Drought Monitoring Using NDVI

2.4.1.1 NDVI Anomalies

The simplest and most common NDVI-based methods of detecting and mapping drought use NDVI anomalies. Anomalies are calculated as the difference between the NDVI composite value for a specified time period (e.g., week, biweek, or month) and the long-term mean value for that period. This isolates the variability in the vegetation signal and establishes a meaningful historical context for the current NDVI to determine relative drought severity. Anyamba and Tucker (2005) found that negative NDVI anomalies could identify and map the spatial extent of drought response in vegetation with a baseline period of 20 years for the Sahel region (Figure 2.5). Other related studies have demonstrated strong relationships on an interannual time scale between NDVI anomalies and El Niño/La Niña-Southern Oscillation (ENSO) phenomena for east Africa (Anyamba and Eastman, 1996; Anyamba et al., 2001), southern Africa (Verdin et al., 1999; Martiny et al., 2006), the southeastern United States (Mennis, 2001), Brazil (Liu and Negron Juarez, 2001; Barbosa et al., 2006), the northern hemisphere (Lotsch et al., 2005), and globally (Myneni et al., 1996; Los et al., 2001). By understanding the relationship between ENSO events and drought occurrence, the dynamics of NDVI anomalies can be used to predict an oncoming drought (Liu and Negron Juarez, 2001).

2.4.1.2 Standardized Vegetation Index

Building on the NDVI anomaly concept, the Standardized Vegetation Index (SVI) developed by Peters et al. (2002) describes the probability of variation from normal NDVI over multiple years of data (e.g., 12 years), on a weekly time step. The SVI is calculated as a *Z*-score deviation from the mean in units of the standard deviation, calculated from the NDVI values for each pixel location of a composite period for each year during a given reference period. This is expressed in equation form as follows:

$$Z_{ijk} = \frac{\text{NDVI}_{ijk} - \overline{\text{NDVI}_{ij}}}{\sigma_{ij}} \tag{2.3}$$

where
- Z_{ijk} is the *Z*-score for year k
- NDVI_{ijk} is the weekly NDVI value
- NDVI_{ij} is the mean NDVI value
- σ_{ij} is the standard deviation in NDVI for pixel i during week j

The SVI was found to provide useful drought-related vegetation condition information over the U.S. Great Plains in near real time (Peters et al., 2002). Due to the weekly time step of the SVI maps, this method can capture the rapidly changing patterns of drought and its severity during the growing season over a large area.

2.4.2 Drought Monitoring Using NDVI and Land-Surface Temperature

2.4.2.1 Vegetation Condition Index

Kogan and Sullivan (1993) introduced a vegetation index-based drought metric called the Vegetation Condition Index (VCI) and developed a global drought-watch system using this index derived from AVHRR smoothed weekly NDVI data. The VCI is defined as follows:

$$\text{VCI} = \frac{(\text{NDVI} - \text{NDVI}_{\text{min}}) \times 100}{\text{NDVI}_{\text{max}} - \text{NDVI}_{\text{min}}} \tag{2.4}$$

where NDVI, NDVI_{max}, and NDVI_{min} are values of the smoothed weekly NDVI and the multiple-year NDVI maximum and minimum, respectively. The smoothed weekly data are scaled relative to the amplitude of their range at each given pixel location and then linearly scaled with a minimum of 0 and maximum of 100. Low values of VCI indicate poor/stressed vegetation due to unfavorable weather conditions, and high values of VCI indicate healthy vegetation conditions. The pixel-based normalization is performed to minimize the effect of spurious or short-term signals in the data and to amplify the long-term ecological signal. In a VCI study conducted by Liu and Kogan (1996), both NDVI anomalies and the VCI were shown to be correlated with rainfall anomalies. However, VCI was found to be more useful for seasonal and interannual comparisons of drought conditions over the South American continent. A study in India found that the utility of the VCI for drought monitoring was improved when used in conjunction with the Temperature Condition Index (TCI) (Singh et al., 2003), which is calculated from AVHRR's thermal channels (10.3–11.3 μm). The TCI is defined as follows:

$$TCI = \frac{100(BT_{max} - BT)}{BT_{max} - BT_{min}} \quad (2.5)$$

where BT, BT_{max}, and BT_{min} are the smoothed weekly and multiple-year maximum and minimum thermal brightness temperatures, respectively. Liu and Kogan (1996) found that the TCI performed better than NDVI and VCI, especially in cases where soil moisture is excessive because of heavy rainfall or persistent cloudiness. Under such conditions, NDVI is depressed, and VCI values are low, which can be interpreted erroneously as drought. To address the issue of false positives for drought, a third VCI (VHI, Vegetation Health Index) was developed by Kogan (1995a), combining the VCI and TCI. VHI is expressed mathematically as follows:

$$VHI = \alpha \times VCI + (1 - \alpha) \times TCI \quad (2.6)$$

where α is a coefficient determining the relative contribution of the TCI and VCI. Thus, VHI is a proxy characterizing vegetation health by combining estimation of moisture and thermal conditions. Global maps of VCI, TCI, and VHI are routinely produced and distributed by NOAA-NESDIS at http://www.star.nesdis.noaa.gov/smcd/emb/vci/VH/vh_ftp.php.

2.4.2.2 Temperature-NDVI Ratio Index

Another drought index called the temperature-NDVI ratio was made by integrating LST and NDVI data. In a study to assess drought impacts over Papua New Guinea, McVicar and Bierwirth (2001) developed this drought index as a ratio of LST and NDVI, which they defined as T_s/NDVI. During a large-scale drought in 2007, the integral of this ratio over the period from January to December showed a strong positive correlation ($r^2 = 0.82$) with severe drought conditions in most of the provinces experiencing food shortages. Additionally, the index had an inverse relationship ($r^2 = 0.81$) when plotted against cumulative rainfall from various meteorological stations in areas experiencing drought. The results from the study demonstrated that the composite AVHRR T_s/NDVI ratio provides an effective and rapid way to assess drought conditions. Under conditions of vegetation stress, T_s increases because of lower transpiration rates and decreases at high NDVI values because of higher transpiration rates associated with increased photosynthetic activity.

2.4.3 Limitations of NDVI as a Drought-Monitoring Tool

The studies previously described illustrate the wide range of applications of NDVI and NDVI-thermal-based indices for drought monitoring. They also bring to light some possible limitations and shortcomings of using NDVI, NDVI-derived indices, and combined NDVI-thermal indices for drought applications. Of particular relevance are the limitations of NDVI over dense vegetation canopy areas. For example, in areas such as tropical forests and the boreal regions of the northern hemisphere, NDVI saturates, and the relationship between NDVI and canopy dynamics will break down. In very wet ecosystems, where soil moisture does not limit vegetation growth, the relationship does not hold at the peak of the growing season when NDVI reaches its maxima, although rainfall may still be increasing (Nicholson et al., 1990; Baret and Guyot, 1991; Wang et al., 2005). In such areas, the seasonal variation of NDVI is too small to discern significant drought events. Furthermore, many of these areas have persistent cloud cover throughout the year (Holben, 1986; Fensholt et al., 2006b,c), which can lead to biases in anomaly analysis; this is because of the limited cloud-free observations available to calculate the NDVI and LST long-term means.

In semiarid areas with sparse vegetation canopies, soil background conditions exert considerable influence on partial canopy spectra and the calculated VIs. Such soil background conditions include primary variations associated with the brightness of bare soil and secondary variations associated with "color" differences among bare soils, as well as soil-vegetation spectral mixing. For example, a brighter soil background results in higher NDVI values than a dark soil background for the same quantity of partial vegetation cover (Huete et al., 1985; Huete, 1988; Huete and Tucker, 1991). In a study over the Sahel, secondary soil variance was responsible for the Saharan desert "artifact" areas of increased VI response in AVHRR NDVI imagery. Some bare soil areas on the margins of deserts result in high NIR reflectance relative to the red reflectance, which artificially enhances the system NDVI. On the other hand, in the Negev Desert, high NDVI values are shown to be associated with photosynthetic activity of microphytes (lower plants, consisting of mosses, lichens, algae, and cyanobacteria), which cover most of the rock and soil surfaces in this semiarid region (Karnieli et al., 1996). Therefore, both soil characteristics and reflectance of lower plant communities may lead to misinterpretation of the vegetation dynamics and overestimation of ecosystem productivity and drought conditions in some semiarid environments.

Indices that incorporate NDVI and LST, such as VHI and the Vegetation Temperature Condition Index (VTCI), rely on a strong inverse relationship between NDVI and LST. Increasing LST is assumed to negatively impact vegetation vigor and consequently cause plant stress. However, this hypothesis does not hold across all global ecosystems. For example, in northern hemisphere and high-altitude ecosystems (like Mongolia), where temperature is a limiting factor on vegetation growth, a positive correlation is found between these two variables. Therefore, neither of the two indices can truly indicate drought in such places (Karnieli et al., 2006, 2010). In these ecosystems, warmer temperatures usually mean more favorable rather than adverse growing conditions for vegetation. Furthermore, even in areas where this NDVI-LST relationship is assumed to be predominantly positive, it has been shown that the relationship varies with location, season, and vegetation type (Lambin and Ehrlich, 1996; Tateishi and Ebata, 2004). Therefore, the application of empirical NDVI-LST-based indices such as VHI and VTCI must be restricted to areas and periods where negative correlations are observed and not on a global scale (Karnieli et al., 2010).

2.5 OPERATIONAL NDVI-BASED DROUGHT-MONITORING SYSTEMS

Ecosystems such as the grasslands of East Africa, the Sahel, and North America are excellent examples of where NDVI can be effectively used to monitor vegetation and drought conditions. This is because the phenology of vegetation closely reflects the seasonal cycle of rainfall (Justice et al., 1985; Nicholson et al., 1990; Ji and Peters, 2004). Using this knowledge, a drought-monitoring product was prototyped for the central United States using AVHRR NDVI data as a primary input (Brown et al., 2002, 2008) with the purpose of providing vegetation-specific drought information. This product, known as the Vegetation Drought Response Index (VegDRI), has since been expanded to cover the rest of the continental United States (Wardlow et al., 2009). VegDRI integrates satellite-based observations of vegetation (AVHRR NDVI) with climate-based drought indices, as well as other biophysical information (such as land use/land cover type, soil characteristics, elevation, and ecological conditions). With these data, drought severity maps are produced to indicate any drought-related vegetation stress.

Another drought-monitoring risk-based system was developed for East and Southeast Asia. The system analyzes current vegetation conditions (inferred from AVHRR NDVI) and precipitation data by comparing 10-day intervals to long-term means, in an effort to detect areas of drought and its effects on agriculture (Song et al., 2004). This and many other national early warning and drought-monitoring systems (e.g., VegDRI) are based largely upon the pioneering efforts of the USAID to establish the Famine Early Warning System Network (FEWS NET) in 1985. FEWS NET brought together the U.S. government agencies NASA, U.S. Geological Survey (USGS), and NOAA to provide technical expertise, data (including NDVI), and data systems integration for drought monitoring over sub-Saharan Africa. This system has now grown to include Haiti, Central America, and Afghanistan. At the global level, FAO created the Global Information and Early Warning System on Food and Agriculture (GIEWS). GIEWS relies on FAO's Africa Real-Time Environmental Monitoring Information System (ARTEMIS), which has been in existence since 1988. ARTEMIS provides analysis of near-real-time AVHRR NDVI data and European Meteosat satellite cold cloud duration (CCD) images (as a proxy for rainfall) over Africa every 10 days. With a historical satellite record since 1988, GIEWS analysts can pinpoint areas experiencing anomalously low rainfall. Around the world, several national and regional-level drought-monitoring centers use NDVI data from AVHRR or other satellite systems as major input to their drought-monitoring activities. Two such examples are the AVHRR NDVI-based greenness products produced operationally for the continental United States by the USGS Earth Resources Observation and Science (EROS) Center (http://ivm.cr.usgs.gov/) and Australia's AVHRR NDVI-based drought-monitoring system (http://www.bom.gov.au/sat/NDVI/NDVI2.shtml).

2.6 RECENT DEVELOPMENTS IN NDVI ANALYSES

In earlier studies, a major impediment to using AVHRR NDVI in environmental and drought monitoring was the lack of a long-term time series of measurements to establish historically meaningful baselines. By the early 1990s, a sufficiently long time series (~20 years) of NDVI observations had been collected for research to start utilizing the data set for large-scale drought monitoring and studying the relationship between vegetation and large-scale climate variability. Such long-term analysis studies have employed time-series decomposition techniques such as principal component analysis (PCA) (Eastman and Fulk, 1993), which decompose the NDVI image time series into various spatial and temporal components (Figure 2.6). In general, the first four principal components explain most of the variance in the data set and characterize long-term conditions and different seasonality regimes. Lower-order components were primarily related to interannual climate events (such as ENSO), localized patterns, and nonvegetation-related noise from changes in different satellite platforms over time (Anyamba and Eastman, 1996). Fourier analysis is another decomposition technique that can detect temporal variability patterns by breaking NDVI into phase and amplitude components (Azzali and Menenti, 2000). However, Fourier analyses require stationary data, whereby statistical parameters describing the data series (such as the mean and variance) do not change over space or time. AVHRR NDVI data series, in particular, do not have these characteristics because NDVI is subject to external factors like weather, climate, and human influences (land cover transformation), which change the nature of its statistics over time. Therefore, applying a Fourier transform to nonstationary NDVI time series results in spurious signals. Such signals are not related to ecosystem dynamics because the technique assumes harmonic behavior; the results are time-dependent periodic signals, each defined by a unique phase and amplitude value. Although regular events such as the seasonality of vegetation can be extracted, interannual patterns (and hence long-term trends) like drought are not easily resolved. As a result, the use of Fourier analyses on nonstationary NDVI data should be performed with caution, with a focus on extracting patterns of seasonality.

Another technique is anomaly analysis, where departures from a base mean period are used to detect periodic temporal patterns in NDVI. These types of analyses have been used to evaluate the relationships between NDVI anomaly patterns and ENSO in different regions of the world (Anyamba and Eastman 1996; Myneni et al., 1996; Anyamba et al., 2002), linking vegetation response to variations in large-scale climate mechanisms (see Figure 2.6 as an example). The understanding of such linkages or teleconnection patterns, especially the driving mechanisms, can be used in predicting areas that are likely to be impacted by drought (Verdin et al., 1999; Funk et al., 2008).

Unlike floods, drought is a creeping phenomenon, and some studies are now attempting to use time-integrated NDVI data to represent the cumulative aspect of drought. As shown in the example for Australia in Figure 2.7, the impacts of drought (A and B) can be detected as increasing cumulative severity (negative NDVI anomalies) through the 2006–2007 growing season. This drought pattern is a sharp contrast to the period of excess moisture or inferred above-normal rainfall (C and D) illustrated by the positive cumulative NDVI anomaly over the 2008–2009 growing season. Such comparisons are useful in examining drought severity from year to year across different regions and are especially useful in drought and agricultural real-time monitoring applications.

These and other studies have demonstrated the utility of AVHRR NDVI data in drought monitoring and mapping and also illustrated the contribution that

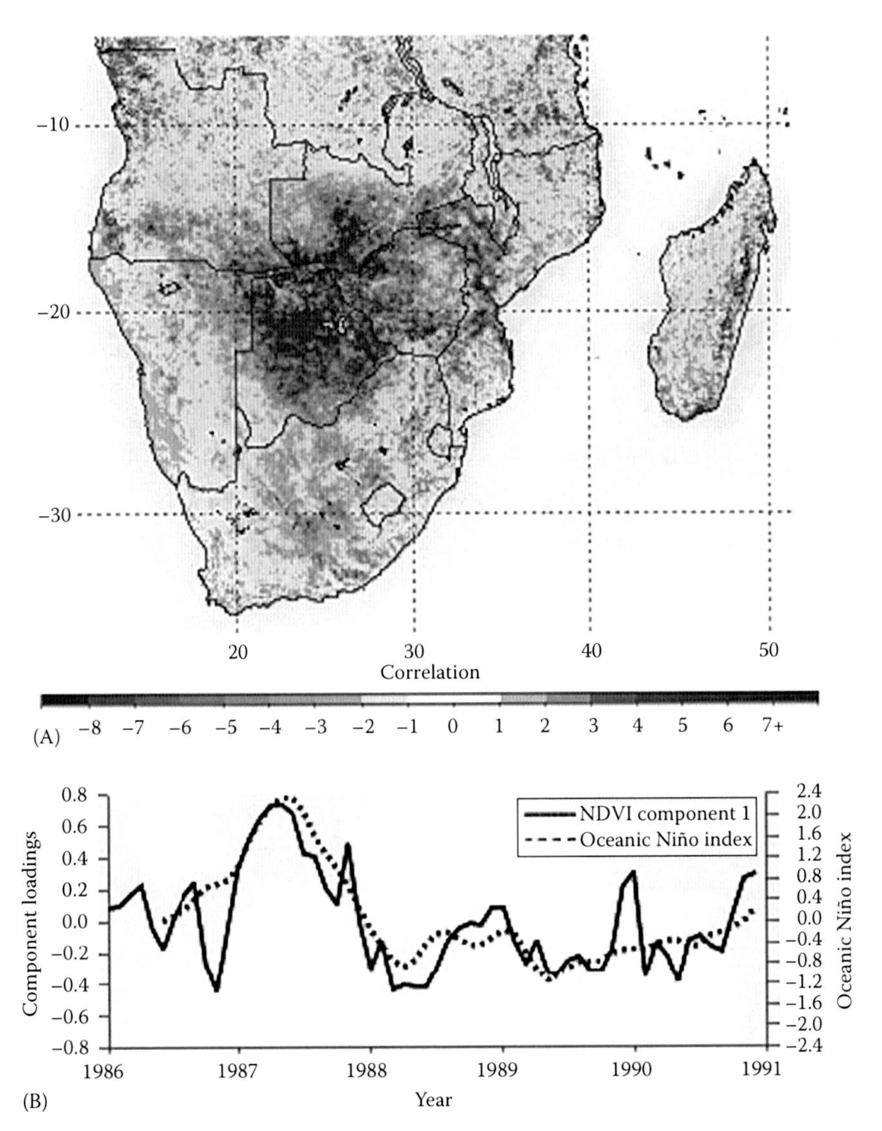

FIGURE 2.6 **(See color insert.)** Principal component analysis results of monthly NDVI anomaly time series for southern Africa for the period 1986–1990, showing the drought spatial pattern in (A) and the associated temporal loadings for the first principal component in (B). This component accounts for 9.73% of the total variance of the anomaly time series. The temporal loadings (B) represent the correlation between each image in the time series with the component spatial pattern in the map (A). The component loadings show a positive correlation with the drought (negative) spatial component pattern in the map (A) between late 1986 and late 1987 and negative correlation (wetter or greener than normal conditions) between 1988 and 1990 with the spatial component pattern (A). This component pattern is related to interannual variability rainfall associated with the ENSO phenomenon. The temporal loadings are highly correlated ($r = 0.80$) with ENSO that is represented by the Oceanic Niño Index (ONI). (Reconstructed after Anyamba, A. and Eastman, J.R., *Int. J. Rem. Sens.*, 17, 2533, 1996.)

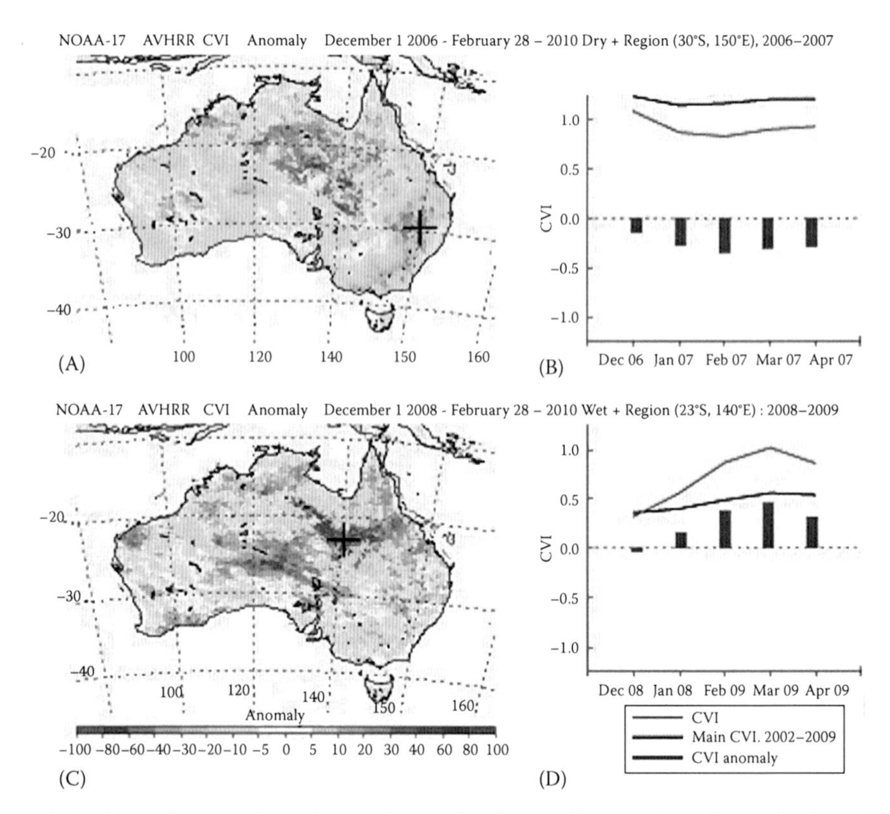

FIGURE 2.7 (See color insert.) Cumulative NDVI anomalies (CVI) for Australia, showing the cumulative nature of drought from December 2006 to 2007 (A) and the wetter/greener-than-normal conditions from December 2008 to February 2009 (C). Cumulative time-series profiles are shown for a drought location (B) and a wet location (D).

these time-series data have made toward a better understanding of the processes that lead to drought and the land-surface response to short-term and long-term climate variability.

2.7 INSTRUMENTAL CHALLENGES AND NEXT-GENERATION NDVI SENSORS

Although the AVHRR NDVI data have provided unprecedented information for drought and environmental monitoring, challenges remain for data usage. First, the AVHRR has a large pixel footprint ranging from 1 to 8 km spatial resolution (8 km for most available global data sets). Second, the instrument uses wide spectral bandwidths that are subject to atmospheric interference (specifically aerosols and water vapor) and therefore not ideally suited for vegetation monitoring. Finally, given that AVHRR is an optical sensor, major cloud contamination problems remain, especially in the tropical regions (Holben, 1986; Fensholt et al., 2009). For most areas of the world, the growing season is persistently cloudy, with cloud cover present 30%

or more of the time. As a result, it is not possible to monitor the dynamics of land-surface conditions at a high temporal frequency in such areas.

Presently, NDVI data from the Moderate Resolution Imaging Spectroradiometer (MODIS) on board NASA's Terra (AM) and Aqua (PM) are now available and widely used in drought monitoring and agricultural applications (Gu et al., 2008; Becker-Reshef et al., 2010; Pittman et al., 2010). MODIS offers a generational improvement over AVHRR. The narrower spectral bandwidths for the "red" band (band 1: 620–670 nm), which has increased chlorophyll sensitivity, and the NIR band (band 2: 841–876 nm), which is less influenced by water-vapor absorption, are marked advancements over AVHRR. Coupled with state-of-the-art atmospheric correction techniques, the MODIS spectral bands offer improved sensitivity to vegetation conditions/changes and provide data at several temporal (8, 16, and 32 days) and spatial (250, 500, and 1000 m) resolutions (Huete et al., 2002; Justice and Townshend, 2002). However, even with these improvements, cloud cover is still an impediment in some areas. Currently, there are various attempts to employ data from the geostationary Meteosat Second Generation (MSG) satellite to fill coverage gaps (Fensholt et al., 2006a) and to exploit radar/microwave systems (uninhibited by clouds) for alternative remote sensing observations of land-surface conditions. The MSG can provide cloud-free imagery over cloud-contaminated areas in the tropics with a composite period of less than 5 days because of high-frequency imaging of the instrument (every 15 min), and therefore cloud-free NDVI can be generated over areas with persistent cloud cover (e.g., West Africa) during the growing season (Fensholt et al., 2006b,c). Radar also provides cloud-free imaging capabilities that can be particularly useful in studying vegetation dynamics of northern dense canopy forests (Ranson and Sun, 1994) and the Amazon (Hess et al., 1995). Some studies have begun to investigate the use of these data for drought and vegetation monitoring in a research mode, but have yet to transition the data and techniques to operational production and application. Another method of cloud screening uses Fourier analysis and empirical mode decomposition (EMD) to derive cloud-free NDVI data by approximating NDVI values of cloudy pixels (Roerink et al., 2000; Pinzon et al., 2005). However, such techniques are better suited to long time-series research data sets than to real-time operational applications because they require intensive analysis of baseline data for processing a corrected data set.

Another challenge is the intersatellite instrument calibration among the series of AVHRR instruments that have been used over the past 29+ years to develop a historical NDVI time series. On average, the AVHRR sensors have a lifespan of 5 years (see Figure 2.8, Table 2.2). Therefore, the existing long-term NDVI data set is made of a compilation of observations from several different AVHRR instruments with different calibration characteristics. The orbital decay and intersensor differences between these instruments introduce bias in the derived NDVI time series, which must be compensated for in order to develop a long-term data set appropriate for environmental monitoring. There have been several attempts to produce a coherent long-term NDVI time series, including the GIMMS NDVI version g data set (Pinzon et al., 2005; Tucker et al., 2005) (shown in Figure 2.8), and the Long-Term Data Records (LTDR) project (Pedelty et al., 2007). Biases that emerge in using AVHRR

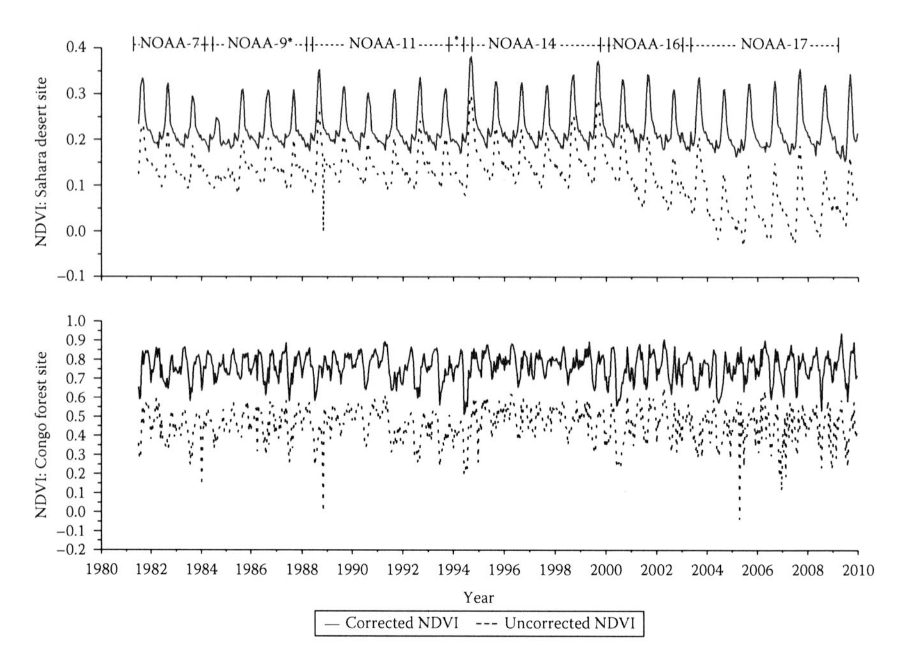

FIGURE 2.8 Uncorrected (dotted line) and corrected (thick line) NDVI time-series data plots for a Sahel site and Congo Forest site. The EMD method was applied to the original NDVI data (uncorrected), which eliminates satellite discontinuities (e.g., in 1984, 1988, and 1994) as well as spurious trends. (From Pinzon, J.E. et al., Satellite time series correction of orbital drift artifacts using empirical mode decomposition, in *EMD and Its Applications*, eds. N.E. Huang and S.S.P. Shen, Part II, Chap. 10, World Scientific Publishers, Singapore.)

instruments from the different sensing platforms are apparent in Figure 2.8 for two different land surfaces: the desert of the semiarid Sahel and Congo forest. Following the uncorrected data series, the discontinuities or dramatic step-change (decrease or increase) in the NDVI time series at different points in time (e.g., 1984, 1988, and 1995) indicate the change from one AVHRR instrument to another. Each instrument has slightly different calibration characteristics; once in orbit, it usually takes a couple of months for a given instrument to be tested and cleared for operational use. Over time, the AVHRR is subject to orbital drift (e.g., 1992–1994 Congo forest site)—hence the degradation of sensor performance.

The GIMMS and LTDR data sets are attempts to remove and correct the effects of the aforementioned factors by creating a standardized and coherent time series, as shown in Figure 2.8. Using the lessons learned from MODIS (i.e., processing techniques, calibration, atmospheric correction, and directional effect determination), the LTDR project applies these methods to the overlapping AVHRR and MODIS periods. The results gleaned from the overlap period are then applied to the AVHRR data preceding the MODIS instrument to produce higher-quality AVHRR data products. However, these data sets are more appropriate for historical, retrospective studies rather than near-real-time operational activities (such as drought monitoring) because of the 1- to 3-year time lag between data updates.

Although MODIS data are available, delays in data processing render them inadequate for real-time drought monitoring. MODIS is classified as a science mission, and, therefore, its data processing chain was not designed for operational use. To help overcome this problem, the Land Atmosphere Near real-time Capability for EOS (LANCE) is providing near-real-time (<less than 3 h from observation) access to processed products such as eMODIS from USGS EROS. These real-time products are used for drought monitoring by FEWS and for VegDRI. In addition, MODIS data from LANCE are utilized for forest threat early warning (USFS Eastern Forest Environmental Threat Assessment Center: http://www.forestthreats.org/) and Global Land Agricultural Monitoring (GLAM: http://www.pecad.fas.usda.gov/glam.cfm), which is a joint project between GSFC and USDA Foreign Agricultural Service (FAS) to monitor global agricultural conditions.

2.8 CONCLUSIONS AND THE FUTURE

The AVHRR instrument was originally designed solely for meteorological applications. A significant and unintended use for these data has been land-surface monitoring through use of the NDVI. The AVHRR NDVI data set from July 1981 to present has created a record of land-surface conditions that has never been available before to the scientific and environmental monitoring communities. As illustrated by the many examples in this chapter, and this book as a whole, these data will continue to be important to a wide range of users, including the drought monitoring community. In the short term, there will be continued availability of NDVI data from the NOAA series of satellites (NOAA-18 and NOAA-19) and the Meteorological Operational polar-orbiting satellite (MetOp) AVHRR series from the European community that also fly AVHRR instruments. In addition, the availability of data since 2001 from MODIS has ensured that there is a redundancy in NDVI data availability. MODIS also provides a more spatially detailed global record of NDVI observations (250 and 500 m). These observations are better suited for a full suite of applications that require more landscape-level observations than the AVHRR 1 km/8 km data could provide. The follow-on satellites in the Joint Polar Satellite System (JPSS) with the Visible/Infrared Imager/Radiometer Suite (VIIRS) instruments (first launched in October 2011) will guarantee that the history of coarse- to medium-scale global remote sensing data will continue to be available to support operational activities such as drought monitoring. However, commitments by governments and space agencies to continue these missions into the future and lessen the turn-around time between science missions are of paramount importance to the drought-monitoring community and other societal applications.

ACKNOWLEDGMENTS

We acknowledge the contributions of the Earth Sciences Directorate, NASA/HQ, and USAID through the FEWS NET and FAO of the United Nations for support during the early phase of the development of the NDVI data set by the GIMMS group.

Current support is provided by the USDA FAS and Agricultural Research Service and Department of Defense Global Emerging Infections Surveillance and Response System (GEIS). We thank Kathrine Collins, Jennifer Small, Edwin Park, Robert Mahoney, Ronald J. Eastman, and Jorge E. Pinzon for contributing illustrations and figures, Kathrine Collins for dedicating her time to assist in many ways on research for this chapter, and Judith Strohmaier for other logistical assistance. We appreciate the assistance of reviewers in making this chapter possible.

REFERENCES

Anyamba, A. and J.R. Eastman. 1996. Interannual variability of NDVI over Africa and its relation to El Niño/Southern Oscillation. *International Journal of Remote Sensing* 17:2533–2548.

Anyamba, A. and C.J. Tucker. 2005. Analysis of Sahelian vegetation dynamics using NOAA-AVHRR NDVI data from 1981–2003. *Journal of Arid Environments* 63:596–614.

Anyamba, A., C.J. Tucker, and J.R. Eastman. 2001. NDVI anomaly patterns over Africa during the 1997/98 ENSO warm event. *International Journal of Remote Sensing* 22:1847–1859.

Anyamba, A., C.J. Tucker, and R. Mahoney. 2002. From El Niño to La Niña: Vegetation response patterns over East and Southern Africa during the 1977–2000 period. *Journal of Climate* 15:3096–3103.

Asrar, G., M. Fuchs, E.T. Kanemasu, and J.L. Hatfield. 1984. Estimating absorbed photosynthetically active radiation and leaf area index from spectral reflectance in wheat. *Agronomy Journal* 76:300–306.

Asrar, G., E.T. Kanemasu, R.D. Jackson, and P.J. Pinter. 1985. Estimation of total above-ground phytomass production using remotely sensed data. *Remote Sensing of Environment* 17:211–220.

Asrar, G., E.T. Kanemasu, G.P. Miller, and R.L. Weiser. 1986. Light interception and leaf area estimates from measurements of grass canopy reflectance. *IEEE Transactions on Geoscience and Remote Sensing* GE24:76–82.

Azzali, S. and M. Menenti. 2000. Mapping vegetation-soil-climate complexes in southern Africa using Fourier analysis of NOAA-AVHRR NDVI data. *International Journal of Remote Sensing* 21:973–996.

Barbosa, H.A., A.R. Huete, and W.E. Baethgen. 2006. A 20-year study of NDVI variability over the Northeast Region of Brazil. *Journal of Arid Environments* 67:288–307.

Baret, F. and G. Guyot. 1991. Potentials and limits of vegetation indices for LAI and APAR assessment. *Remote Sensing of Environment* 35:161–173.

Becker-Reshef, I., C. Justice, M. Sullivan, E. Vermote, C. Tucker, A. Anyamba, J. Small, E. Pak, E. Masuoka, J. Schmaltz, M. Hansen, K. Pittman, C. Birkett, D. Williams, C. Reynolds, and B. Doorn. 2010. Monitoring global croplands with coarse resolution earth observations: The Global Agriculture Monitoring (GLAM) Project. *Remote Sensing* 2:1589–1609.

Brown, J.F., B.C. Reed, M.J. Hayes, D.A. Wilhite, and K. Hubbard. 2002. A prototype drought monitoring system integrating climate and satellite data. *Pecora 15/Land Satellite Information IV/ISPRS Commission I/FIEOS 2002 Conference Proceedings*, Denver, CO, November 10–14.

Brown, J.F., B.D. Wardlow, T. Tadesse, M.J. Hayes, and B.C. Reed. 2008. The Vegetation Drought Response Index (VegDRI): A new integrated approach for monitoring drought stress in vegetation. *GIScience and Remote Sensing* 45:16–46.

Burgan, R.E. and R.A. Hartford. 1993. Monitoring vegetation greenness with satellite data. Gen. Tech. Rep. DNT-297, U.S. Department of Agriculture, Forest Service, Intermountain Research Station, Ogden, UT.

Burgan, R.E., R.A. Hartford, and J.C. Eidenshink. 1996. Using NDVI to assess departure from average greenness and its relation to fire business. General Technical Report INT-GTR-333, U.S. Department of Agriculture, Forest Service, Intermountain Research Station, Ogden, UT.

Deering, D.W., J.W. Rouse Jr., R.H. Haas, and J.A. Schell. 1975. Measuring "forage production" of grazing units from Landsat MSS data. *10th International Symposium on Remote Sensing of Environment Proceedings*, Ann Arbor, MI, October 6–10.

Dyer, M.I. and D.A. Crossley. 1986. Coupling of ecological studies with remote sensing: Potentials at four biosphere reserves in the United States. Publication 9504, U.S. State Department's U.S. Man and the Biosphere Program, Washington, DC.

Eastman, R.R. and M. Fulk. 1993. Long sequence time series evaluation using standardized principal components. *Photogrammetric Engineering and Remote Sensing* 59:991–996.

Eidenshink, J.C. and R.H. Hass. 1992. Analyzing vegetation dynamics of land systems with satellite data. *GeoCarto International* 1:53–61.

Fensholt, R., A. Anyamba, S. Stisen, I. Sandholt, E. Pak, and J. Small. 2006a. Comparisons of compositing period length for vegetation index data from polar-orbiting and geostationary satellites for the cloud-prone region of West Africa Special Issue: Cloud-prone and Rainy areas Remote Sensing (CARRS). *Photogrammetric Engineering and Remote Sensing* 73:297–310.

Fensholt, R., T.T. Nielsen, and S. Stisen. 2006b. Evaluation of AVHRR PAL and GIMMS 10-day composite NDVI time series products using SPOT-4 vegetation data for the African continent. *International Journal of Remote Sensing* 27:2719–2733.

Fensholt, R., K. Rasmussen, T.T. Nielsen, and C. Mbow. 2009. Evaluation of earth observation based long-term vegetation trends—Intercomparing NDVI time series trend analysis consistency of Sahel from AVHRR GIMMS, Terra MODIS and SPOT VGT data. *Remote Sensing of Environment* 113:1886–1898.

Fensholt, R., I. Sandholt, S. Stisen, and C.J. Tucker. 2006c. Analyzing NDVI for the African continent using the geostationary Meteosat second generation SEVIRI sensor. *Remote Sensing of Environment* 101:212–229.

Funk, C.M., M.D. Dettinger, J.C. Michaelsen, J.P. Verdin, M.E. Brown, M. Barlow, and A. Hoell. 2008. Warming of the Indian Ocean threatens eastern and southern African food security but could be mitigated by agricultural development. *Proceedings of the National Academy of Science USA* 105:11081–11086.

Gallo, K.P. 1990. Satellite derived vegetation indices: A new climatic variable? *Proceedings of the Symposium on Global Change Systems, Special Sessions on Climate Variations and Hydrology*, American Meteorological Society, Anaheim, CA, February 5–9, pp. 133–137.

Gates, D.M., H.J. Keegan, J.C. Schleter, and V.P. Weldner. 1965. Spectral properties of plants. *Applied Optics* 4:11–20.

Gatlin, J.A., R.J. Sullivan, and C.J. Tucker. 1984. Considerations of and improvements to large-scale vegetation monitoring. *IEEE Transactions on Geoscience and Remote Sensing* GE22:496–502.

Gu, Y., E. Hunt, B. Wardlow, J.B. Basara, J.F. Brown, and J.P. Verdin. 2008. Evaluation of MODIS NDVI and NDWI for vegetation drought monitoring using Oklahoma Mesonet soil moisture data. *Geophysical Research Letters* 35:L22401.

Hardisky, M.A., F.C. Daiber, C.T. Roman, and V. Klemas. 1984. Remote sensing of biomass productivity of a salt marsh. *Remote Sensing of Environment* 16:91–106.

Hatfield, J.L., G. Asrar, and E.T. Kanemasu. 1984. Intercepted photosynthetically active radiation in wheat canopies estimated by spectral reflectance. *Remote Sensing of Environment* 14:65–76.

Hess, L.L., J.M. Melack, S. Filoso, and Y. Wang. 1995. Delineation of inundated area and vegetation along the Amazon floodplain with the SIR-C synthetic-aperture radar. *IEEE Transactions on Geoscience and Remote Sensing* 33:896–904.

Hielkema, J.U., J. Roffey, and C.J. Tucker. 1986. Assessment of ecological conditions associated with the 1980/1981 desert locust plague upsurge in West Africa using environmental satellite data. *International Journal of Remote Sensing* 7:1609–1622.

Hiernaux, P.H.Y. and C.O. Justice. 1986. Suivi du developpement vegetal au cours de l'ete 1984 dans le Sahel Malien. *International Journal of Remote Sensing* 7:1515–1532.

Holben, B.N. 1986. Characteristics of maximum-value composite images from temporal AVHRR data. *International Journal of Remote Sensing* 7:1417–1434.

Holben, B.N. and R.S. Fraser. 1984. Red and near-infrared sensor response to off-nadir viewing. *International Journal of Remote Sensing* 5:145–160.

Holben, B.N., D. Kimes, and R.S. Fraser. 1986. Directional reflectance response in AVHRR red and near-IR bands for three cover types and varying atmospheric conditions. *Remote Sensing of Environment* 19:213–236.

Huete, A.R. 1988. A soil-adjusted vegetation index (SAVI). *Remote Sensing of Environment* 25:295–309.

Huete, A., K. Didan, T. Miura, and E. Rodriguez. 2002. Overview of the radiometric and biophysical performance of the MODIS vegetation indices. *Remote Sensing of Environment* 83:195–213.

Huete, A.R., R.D. Jackson, and D.F. Post. 1985. Spectral response of a plant canopy with different soil backgrounds. *Remote Sensing of Environment* 17:37–53.

Huete, A.R. and C.J. Tucker. 1991. Investigation of soil influences in AVHRR red and near-infrared vegetation index imagery. *International Journal of Remote Sensing* 12:1223–1242.

Hutchinson, C.F. 1991. Use of satellite data for famine early warning in sub-Saharan Africa. *International Journal of Remote Sensing* 12:1405–1421.

Janowiak, J.E. 1988. An investigation of interannual rainfall variability in Africa. *Journal of Climate* 1:240–255.

Ji, L. and A.J. Peters. 2004. A spatial regression procedure for evaluating the relationship between AVHRR-NDVI and climate in the northern Great Plains. *International Journal of Remote Sensing* 25:297–311.

Justice, C.O. and J.R.G. Townshend. 2002. Special issue on the moderate resolution imaging spectroradiometer (MODIS): A new generation of land surface monitoring. *Remote Sensing of Environment* 83:1–2.

Justice, C.O., J.R.G. Townshend, B.N. Holben, and C.J. Tucker. 1985. Analysis of the phenology of global vegetation using meteorological satellite data. *International Journal of Remote Sensing* 6:1271–1318.

Karnieli, A., N. Agam, R.T. Pinker, M. Anderson, M.L. Imhoff, G.G. Gutman, N. Panov, and A. Goldberg. 2010. Use of NDVI and land surface temperature for drought assessment: Merits and limitations. *Journal of Climate* 23:618–633.

Karnieli, A., M. Bayasgalan, Y. Bayarjargal, N. Agam, S. Khudulmur, and C.J. Tucker. 2006. Comments on the use of the Vegetation Health Index over Mongolia. *International Journal of Remote Sensing* 27:2017–2024.

Karnieli, A., M. Shachak, H. Tsoar, E. Zaady, Y. Kaufman, A. Danin, and W. Porter. 1996. The effect of microphytes on the spectral reflectance of vegetation in semiarid regions. *Remote Sensing of Environment* 2:88–96.

Kimes, D.S. 1983. Dynamics of directional reflectance factor distributions for vegetation canopies. *Applied Optics* 22:1364–1372.

Kimes, D.S., B.N. Holben, C.J. Tucker, and W.W. Newcomb. 1984. Optimal directional view angles for remote sensing missions. *International Journal of Remote Sensing* 5:877–891.

Knipling, E.B. 1970. Physical and physiological basis for the reflectance of visible and near infrared radiation from vegetation. *Remote Sensing of Environment* 1:155–159.

Kogan, E.N. 1990. Remote sensing of weather impacts on vegetation. *International Journal of Remote Sensing* 11:1405–1419.

Kogan, E.N. 1995a. Application of vegetation index and brightness temperature for drought detection. *Advances in Space Research* 15:91–100.

Kogan, E.N. 1995b. Droughts of the late 1980s in the United States as derived from NOAA polar-orbiting satellite data. *Bulletin of the American Meteorological Society* 76:655–668.

Kogan, F. and J. Sullivan. 1993. Development of global drought-watch system using NOAA/AVHRR data. *Advances in Space Research* 13:219–222.

Kumar, M. and J.L. Monteith. 1982. Remote sensing of plant growth. In *Plants and the Daylight Spectrum*, ed. H. Smith, pp. 133–144. London, U.K.: Academic Press Inc.

Lambin, E.F. and D. Ehrlich. 1996. The surface temperature–vegetation index space for land cover and land-cover change analysis. *International Journal of Remote Sensing* 17:463–478.

Liu, W.T. and F.N. Kogan. 1996. Monitoring regional drought using the Vegetation Condition Index. *International Journal of Remote Sensing* 17:2761–2782.

Liu, W.T. and R.I. Negron Juarez. 2001. ENSO drought onset prediction in northeast Brazil using NDVI. *International Journal of Remote Sensing* 22:3483–3501.

Los, S.O., G.J. Collatz, L. Bounoua, P.J. Sellers, and C.J. Tucker. 2001. Global inter-annual variations in sea surface temperature and land surface vegetation, air temperature, and precipitation. *Journal of Climate* 14:1535–1549.

Lotsch, A., M.A. Friedl, B.T. Anderson, and C.J. Tucker. 2005. Response of terrestrial ecosystems to recent Northern Hemisphere drought. *Geophysical Research Letters* 32:L06705.

Loveland, T.R., J.W. Merchant, J.F. Brown, and D.O. Ohlen. 1991. Development of a land-cover characteristics database for the conterminous U.S. *Photogrammetric Engineering and Remote Sensing* 57:1453–1463.

Martiny, N., P. Camberlin, Y. Richard, and N. Philippon. 2006. Compared regimes of NDVI and rainfall in semi-arid regions of Africa. *International Journal of Remote Sensing* 27:5201–5223.

McVicar, T.R. and P.N. Bierwirth. 2001. Rapidly assessing the 1997 drought in Papua New Guinea using composite AVHRR imagery. *International Journal of Remote Sensing* 22:2109–2128.

Mennis, J. 2001. Exploring relationships between ENSO and vegetation vigor in the south-east USA using AVHRR data. *International Journal of Remote Sensing* 22:3077–3092.

Myneni, R.B., F.B. Hall, P.J. Sellers, and A.L. Marshak. 1995. The interpretation of spectral vegetation indices. *IEEE Transactions on Geoscience and Remote Sensing* 33:481–486.

Myneni, R.B., S.O. Los, and C.J. Tucker. 1996. Satellite-based identification of linked vegetation index and sea surface temperature anomaly areas from 1982–1990 for Africa, Australia and South America. *Geophysical Research Letters* 23:729–732.

Nicholson, S.E. 1989. African drought: Characteristics, casual theories, and global teleconnections. In *Understanding Climate Change*, eds. A. Berger, R.E. Dickinson, and J.W. Kidson, pp. 79–100. Washington, DC: American Geophysical Union.

Nicholson, S.E., M.L. Davenport, and A.R. Malo. 1990. A comparison of the vegetation response to rainfall in the Sahel and East Africa, using normalized difference vegetation index from NOAA AVHRR. *Climatic Change* 17:209–242.

NOAA Satellite and Information Service. *NOAA Polar Orbiter Data User's Guide*, Silver Spring, MD. http://www.ncdc.noaa.gov/oa/pod-guide/ncdc/docs/podug/figures.htm (accessed on December 12, 2011).

NOAA Satellite and Information Service, Silver Spring, MD, http://www.oso.noaa.gov/history/first-launched.htm (accessed on December 12, 2011).

Norwine, J. and D.H. Greegor. 1983. Vegetation classification based on Advanced Very High Resolution Radiometer (AVHRR) satellite imagery. *Remote Sensing of Environment* 13:69–87.

Pedelty, J., S. Devadiga, E. Masuoka, M. Brown, J. Pinzon, C. Tucker, D. Roy, J. Ju, E. Vermote, S. Prince, J. Nagol, C. Justice, C. Schaaf, J. Liu, J. Privette, and A. Pinheiro. 2007. Generating a long-term land data record from the AVHRR and MODIS instruments. *Proceedings of Geoscience and Remote Sensing Symposium IGARRS*, New York.

Peters, A.J., E.A. Walter-Shea, L. Ji, A. Vina, M. Hayes, and M.D. Svoboda. 2002. Drought monitoring with NDVI-Based Standardized Vegetation Index. *Photogrammetric Engineering and Remote Sensing S* 68:71–75.

Pinzon, J.E, M.E. Brown, and C.J. Tucker. 2005. Satellite time series correction of orbital drift artifacts using empirical mode decomposition. In *EMD and Its Applications*, eds. N.E. Huang and S.S.P. Shen, Part II, Chap. 10. Singapore: World Scientific Publishers.

Pittman, K., M.C. Hansen, I. Becker-Reshef, P.V. Potapov, and C.O. Justice. 2010. Estimating global cropland extent with multi-year MODIS data. *Remote Sensing* 2:1844–1863.

Prince, S.D. and C.J. Tucker. 1986. Satellite remote sensing of rangelands in Botswana II: NOAA AVHRR and herbaceous vegetation. *International Journal of Remote Sensing* 7:1555–1570.

Ranson, K.J. and G. Sun. 1994. Mapping biomass of a northern forest using multifrequency SAR data. *IEEE Transactions on Geoscience and Remote Sensing* 32:388–396.

Richardson, A.J. and C.L. Wiegand. 1977. Distinguishing vegetation from soil background information. *Photogrammetric Engineering and Remote Sensing* 43:1541–1552.

Roerink, G.J., M. Menenti, and W. Verhoef. 2000. Reconstructing cloud free NDVI composites using Fourier analysis of time series. *International Journal of Remote Sensing* 21:1911–1917.

Rouse Jr., J.W., R.H. Haas, J.A. Schell, and D.W. Deering. 1974a. Monitoring vegetation systems in the Great Plains with ERTS. Paper presented at the *3rd ERTS-1 Symposium*, Greenbelt, MD.

Rouse Jr., J.W., R.H. Haas, J.A. Schell, D.W. Deering, and J.C. Harlan. 1974b. Monitoring the vernal advancement and retrogradation (green wave effect) of natural vegetation. NASA/GSFC Type III Final Report, Greenbelt, MD.

Schneider, S.R., D.F. McGinnis, and J.A. Gatlin. 1981. Use of NOAA AVHRR visible and near-infrared data for land remote sensing. NOAA Technical Report NESS 84 for the NOAA National Earth Satellite Service, Washington, DC.

Sellers, P.J. 1985. Canopy reflectance, photosynthesis, and transpiration. *International Journal of Remote Sensing* 6:1335–1372.

Sellers, P.J. 1987. Canopy reflectance, photosynthesis, and transpiration II: The role of biophysics in the linearity of their interdependence. *Remote Sensing of Environment* 21:143–183.

Singh, R.P., S. Roy, and F. Kogan. 2003. Vegetation and temperature condition indices form NOAA AVHRR data for drought monitoring over India. *International Journal of Remote Sensing* 24:4393–4402.

Song, X., G. Saito, M. Kodama, and H. Sawada. 2004. Early detection system of drought in East Asia using NDVI from NOAA/AVHRR data. *International Journal of Remote Sensing* 25:3105–3111.

Steven, M.D., P.V. Biscoe, and K.W. Jaggard. 1983. Estimation of sugar beet productivity from reflection in the red and near-infrared spectral bands. *International Journal of Remote Sensing* 4:325–334.

Stone, T.A., P. Schlesinger, R.A. Houghton, and G.M. Woodwell. 1994. A map of the vegetation of South America based on satellite imagery. *Photogrammetric Engineering and Remote Sensing* 60:541–551.

Tateishi, R. and M. Ebata. 2004. Analysis of phenological change during 1982–2000 Advanced Very High Resolution Radiometer (AVHRR) data. *International Journal of Remote Sensing* 25:2287–2300.

Tateishi, R. and K. Kajiwara. 1992. Global land cover monitoring by AVHRR NDVI data. *Earth Environment* 7:4–14.

Townshend, J.R.G., T.E. Goff, and C.J. Tucker. 1985. Multi-temporal dimensionality of images of normalized difference vegetation index at continental scales. *IEEE Transactions on Geoscience Remote Sensing* GE-23:888–895.

Townshend, J.R.G., C.O. Justice, and V. Kalb. 1987. Characterization and classification of South American land cover types using satellite data. *International Journal of Remote Sensing* 8:1189–1207.

Tucker, C.J. 1979. Red and photographic infrared linear combinations for monitoring vegetation. *Remote Sensing of Environment* 8:127–150.

Tucker, C.J. 1996. History of the use of AVHRR data for land applications. In *Advances in the Use of NOAA AVHRR Data for Land Applications*, eds. G. D'Souza, A.L. Belward, and J. Malingreau, pp. 1–19. Dordrecht, the Netherlands: Kluwer Academic Publishers.

Tucker, C.J., J.U. Hielkema, and J. Roffey. 1985c. The potential of satellite remote sensing of ecological conditions for survey and forecasting desert-locust activity. *International Journal of Remote Sensing* 6:127–138.

Tucker, C.J., C.O. Justice, and S.D. Prince. 1986. Monitoring the grasslands of the Sahel 1984–1985. *International Journal of Remote Sensing* 7:1571–1581.

Tucker, C.J., W.W. Newcomb, S.O. Los, and S.D. Prince. 1991. Mean and inter-year variation of growing-season normalized difference vegetation index for Sahel 1981–1989. *International Journal of Remote Sensing* 12:1133–1135.

Tucker, C.J., J.E. Pinzon, M.E. Brown, D.A. Slayback, E.W. Pak, R. Mahoney, E. Vermote, and N. El Saleous. 2005. An extended AVHRR 8-km NDVI dataset compatible with MODIS and SPOT vegetation NDVI data. *International Journal of Remote Sensing* 26:4485–4498.

Tucker, C.J. and P.J. Sellers. 1986. Satellite remote sensing of primary production. *International Journal of Remote Sensing* 7:1395–1416.

Tucker, C.J., J.R.G. Townshend, and T.E. Goff. 1985a. African land-cover classification using satellite data. *Science* 227:369–375.

Tucker, C.J., C.L. Vanpraet, E. Boerwinkel, and A. Gaston. 1983. Satellite remote sensing of total dry matter production in the Senegalese Sahel: 1980–1984. *Remote Sensing of Environment* 13:461–474.

Tucker, C.J., C.L. Vanpraet, M.J. Sharman, and G. Van Ittersum. 1985b. Satellite remote sensing of total herbaceous biomass production in the Senegalese Sahel: 1980–1984. *Remote Sensing of Environment* 17:233–249.

Unganai, L.S. and F.N. Kogan. 1998. Drought monitoring and corn yield estimation in southern Africa from AVHRR data. *Remote Sensing of Environment* 63:219–232.

Verdin, J., C. Funk, R. Klaver, and D. Roberts. 1999. Exploring the correlation between Southern Africa NDVI and Pacific sea surface temperatures: Results for the 1998 maize growing season. *International Journal of Remote Sensing* 20:2117–2124.

Wang, Q., S. Adiku, J. Tenhunen, and A. Granier. 2005. On the relationship of NDVI with leaf area index in a deciduous forest site. *Remote Sensing of Environment* 94:244–255.

Wardlow, B.D., M.J. Hayes, M.D. Svoboda, T. Tadesse, and K.H. Smith. 2009. Sharpening the focus on drought—New monitoring and assessment tools at the National Drought Mitigation Center. *Earthzine*. http://www.earthzine.org/2009/03/30/sharpening-the-focus-on-drought-%E2%80%93-new-monitoring-and-assessment-tools-at-the-national-drought-mitigation-center/ (accessed June 20, 2011).

Wiegand, C.L. and A.J. Richardson. 1984. Leaf area, light interception, and yield estimates from spectral components analysis. *Agronomy Journal* 76:543–548.

Woolley, J.T. 1971. Reflectance and transmittance of light by leaves. *Plant Physiology* 47:656–662.

3 Vegetation Drought Response Index

An Integration of Satellite, Climate, and Biophysical Data

Brian D. Wardlow, Tsegaye Tadesse, Jesslyn F. Brown, Karin Callahan, Sharmistha Swain, and Eric Hunt

CONTENTS

3.1 INTRODUCTION

Drought is a normal, recurring feature of climate in most parts of the world (Wilhite, 2000) that adversely affects vegetation conditions and can have significant impacts on agriculture, ecosystems, food security, human health, water resources, and the economy. For example, in the United States, 14 billion-dollar drought events occurred between 1980 and 2009 (NCDC, 2010), with a large proportion of the losses coming from the agricultural sector in the form of crop yield reductions and degraded hay/pasture conditions. During the 2002 drought, Hayes et al. (2004) found that many individual states across the United States experienced more than $1 billion in agriculture losses associated with both crops and livestock. The impact of drought on vegetation can have serious water resource implications as the use of finite surface and groundwater supplies to support agricultural crop production competes against other sectoral water interests (e.g., environmental, commercial, municipal, and recreation). Drought-related vegetation stress can also have various ecological impacts. Prime examples include widespread piñon pine tree die-off in the southwest United States due to protracted severe drought stress and associated bark beetle infestations (Breshears et al., 2005) and the geographic shift of a forest-woodland ecotone in this region in response to severe drought in the mid-1950s (Allen and Breshears, 1998). Tree mortality in response to extended drought periods has also been observed in other parts of the western United States (Guarin and Taylor, 2005), as well as in boreal (Kasischke and Turetsky, 2006), temperate (Fensham and Holman, 1999), and tropical (Williamson et al., 2000) forests. Droughts have also served as a catalyst for changes in wildfire activity (Swetnam and Betancourt, 1998; Westerling et al., 2006) and invasive plant species establishment (Everard et al., 2010).

Monitoring drought stress of vegetation is a critical component of proactive drought planning designed to mitigate the impact of this natural hazard. Approaches that characterize the spatial extent, intensity, and duration of drought-related vegetation stress provide essential information for a wide range of management and planning decisions. For example, such information could be used by agricultural producers and water resource managers to adjust crop irrigation schedules and by ranchers to determine stocking rates and grazing rotations for cattle. In addition, this knowledge allows natural resource managers to implement best management practices under drought conditions and other decision makers to better target assistance and response activities (e.g., release of Conservation Reserve Program grasslands for emergency grazing or early detection of hot spots for wildfires) in a timely manner.

For more than 20 years, satellite-based remote sensing has been widely used for many large-area vegetation characterization applications (e.g., land cover classification, biophysical estimates, and phenology) including drought monitoring. Satellite-based observations from global imagers such as the Advanced Very High Resolution Radiometer (AVHRR) and the more recent Medium Resolution Imaging Spectrometer (MERIS), Moderate Resolution Imaging Spectroradiometer (MODIS), and SPOT (Satellite Pour l'Observation de la Terre) Vegetation instruments have provided a near-daily, global coverage of spatially continuous spectral measurements to complement point-based weather station observations that have been used to generate traditional, climate-based drought indices such as the Palmer Drought Severity Index (PDSI) (Palmer, 1965) and the Standardized Precipitation Index (SPI) (McKee et al., 1995).

Over this period, a number of remote sensing-based vegetation indices (VIs) have been developed from various spectral band combinations to monitor vegetation health.

The Normalized Difference Vegetation Index (NDVI) (Rouse et al., 1974) has been the most widely used VI for large-area vegetation monitoring (e.g., Tucker et al., 1985; Townshend et al., 1987; Reed et al., 1996; Jakubauskas et al., 2002). NDVI is a simple, two-band mathematical transformation that capitalizes on the differential response of chlorophyll absorption and internal spongy mesophyll layer reflectance from plant leaves in the visible red and near infrared (NIR) spectral regions, respectively. A large body of research has found that NDVI fluctuations over time are strongly correlated with climate variations (Peters et al., 1991; Yang et al., 1998; McVicar and Bierwirth, 2001; Ji and Peters, 2003), indicating that this index is an effective measure of climate-related vegetation changes. Over the past two decades, several operational AVHRR-derived NDVI products have been developed for large-area vegetation monitoring, including the Global Inventory Modeling and Mapping Studies (GIMMS) global NDVI data set (Tucker et al., 2005), the Famine and Early Warning System Network (FEWS NET) regional NDVI data sets (e.g., Africa, Afghanistan, and Latin America), and national NDVI products over Australia and United States (Eidenshink, 2006) produced by the Australian Bureau of Meteorology and U.S. Geological Survey (USGS), respectively.

The Vegetation Health Index (VHI) (Kogan, 1995), which incorporates both NDVI and brightness temperature (BT) data collected by AVHRR, is another index that has been applied to assess national- to continental-scale drought conditions (Liu and Kogan, 1996; Kogan, 1997; Seiler et al., 1998; Unganai and Kogan, 1998; Kogan, 2002). The VHI concept assumes an inverse relationship between NDVI and BT because higher land surface temperatures (LSTs) tend to negatively impact vegetation vigor (and decrease NDVI), which can be indicative of a drought stress signal because of reduced evapotranspiration (ET). However, Karnieli et al. (2006, 2010) found VHI had limited utility in "energy limited" environments (e.g., high latitude or elevation locations) where LST and NDVI exhibit a positive relationship and was most useful for locations where water was the primary limiting factor of vegetation growth. Several other methods of integrating NDVI and LST data from AVHRR and MODIS have also been tested for drought monitoring that include simple division (McVicar and Bierwirth, 2001), two-dimensional geometric expressions (Karnieli and Dall'Olmo, 2003), and ratios (Wan et al., 2004) between these two variables.

The launch of MODIS, with an increased number of land-related spectral bands and expanded spectral coverage into the shortwave-infrared region (SWIR), led to the development of several new VIs incorporating SWIR observations. MODIS has two SWIR bands that are sensitive to changes in plant (Band 6: 1628–1652 nm) and soil (Band 7: 2105–2155 nm) water content, respectively. Gao (1996) developed the Normalized Difference Water Index (NDWI), which capitalizes on the differential response of the NIR (i.e., high reflectance by intercellular spaces) and the SWIR (i.e., high absorption by plant water content) reflectances in healthy vegetation. In a study over grasslands, Gu et al. (2007) found NDWI to be slightly more sensitive than NDVI to the onset of drought stress. Gu et al. (2008) extended both the NDVI and NDWI concepts by integrating both into an index called the Normalized Difference Drought Index (NDDI). Wang et al. (2007) built upon the original NDWI concept by developing a three-band index called the Normalized Multi-band Drought Index (NMDI), which incorporates data from both of MODIS' SWIR bands, as well as the NIR band. The NMDI utilizes the difference between the two SWIR bands, which are sensitive to soil and plant water content, respectively. The relative difference between these two SWIR bands changes according to fluctuation in both the soil and plant water content.

Collectively, this body of work illustrates the value of satellite-based VI observations for assessing vegetation conditions and the considerable emphasis that has been placed on developing new VIs in support of drought monitoring. However, two major challenges exist among all these satellite-based VIs in terms of applying them for drought monitoring. The first challenge is establishing the appropriate threshold(s) that discriminates between drought and nondrought conditions, as well as varying levels of drought stress (e.g., moderate, severe, and extreme). Typically, a relative VI value or a departure of a VI value from a baseline (e.g., low percentage of the average historical VI value) is used as an indicator of drought stress instead of classifying specific levels of drought severity. Selection of thresholds to classify drought conditions using VI information is difficult because they can vary by land cover type, geographic location, and season. The second challenge is the ability to discriminate drought-impacted areas from other locations experiencing vegetation stress due to other causes solely from remotely sensed VI information. A number of environmental factors (e.g., fire, flooding, hail, pests, plant disease, and human-induced land cover/use changes) can produce negative VI anomalies (Peters et al., 2000; Domenikiotis et al., 2003; Wang et al., 2003; Goetz et al., 2006; Franke and Menz, 2007) that mimic a drought stress signal. Ancillary information such as climate data or ground observations (e.g., field reports of crop conditions) is needed to better define these negative VI anomalies within a drought context.

This chapter presents a new hybrid index called the Vegetation Drought Response Index (VegDRI) that integrates traditional remote sensing–based VI observations and climate-based drought index data with several general biophysical characteristics of the environment to characterize "drought-related" vegetation stress (Brown et al., 2008). VegDRI was designed to capitalize on the valuable spatiotemporal vegetation condition information contained in multitemporal NDVI data while focusing on the drought component of these conditions through the addition of climate and biophysical data. VegDRI overcomes the interpretation difficulties encountered using traditional remote sensing–based VIs and classifies vegetation drought severity using an objective, quantitative classification scheme. A review of VegDRI's specific data inputs, classification scheme, and modeling approach is presented in this chapter along with case examples of VegDRI results from 2009 across the United States to illustrate the performance and utility of this new drought VI.

3.2 VegDRI DATA INPUTS AND METHODOLOGY

3.2.1 Overview of the VegDRI Concept

VegDRI targets the effects of drought on vegetation by collectively analyzing general vegetation conditions as observed in satellite-derived VI data and the level of dryness expressed in climate-based drought indices for a specific location.

Additional biophysical/environmental characteristics such as ecoregion, elevation, land use/land cover (LULC) type, and soil type are also considered because they can influence climate-vegetation interactions. This integrated approach was developed to capitalize on the strengths of both satellite- and climate-based indices that have been traditionally used for drought monitoring. The set of data inputs used to calculate VegDRI can be categorized into three components: satellite, climate, and biophysical. The satellite component provides spatially detailed information about the distribution and general health of vegetation from 1 km AVHRR NDVI data. The climate component consists of two commonly used drought indices, the PSDI and SPI, which provide a measure of dryness. Specifically, the PDSI is used to train the empirically-based VegDRI models, providing an eight-category drought severity classification system widely recognized by the drought community that ranges from extremely moist to extreme drought conditions. The biophysical component comprises several biophysical variables that reflect different terrestrial characteristics that can influence the response of vegetation to drought. Table 3.1 lists the specific VegDRI input variables, which will be further described in this section along with a detailed description of the VegDRI methodology. This methodology consists of four primary steps: (1) creation of a historical database of input variables for model development, (2) generation of biweekly, empirically based VegDRI models, (3) generation of near-real-time gridded data inputs, and (4) application of model to gridded inputs to produce 1 km VegDRI maps (Figure 3.1).

TABLE 3.1
Input Variables for the Biophysical, Climate, and Satellite Components of VegDRI

Data Set	Source	Format	Temporal Resolution
Climate component variables			
SPI	ACIS/NADSS	ASCII (at sites)	Biweekly
PDSI—self-calibrated	ACIS/NADSS	ASCII (at sites)	Biweekly
Satellite component variables			
PASG	AVHRR NDVI	1 km raster	Biweekly
SOSA	AVHRR NDVI	1 km raster	Annual
Biophysical component variables			
NLCD	National Land Cover Database	1 km raster	Static
Soil AWC	STATSGO	1 km raster	Static
IrrAg	USGS MIrAD	1 km raster	Static
Ecological regions (ECO)	EPA ecoregions	1 km raster	Static
Elevation (DEM)		1 km raster	Static

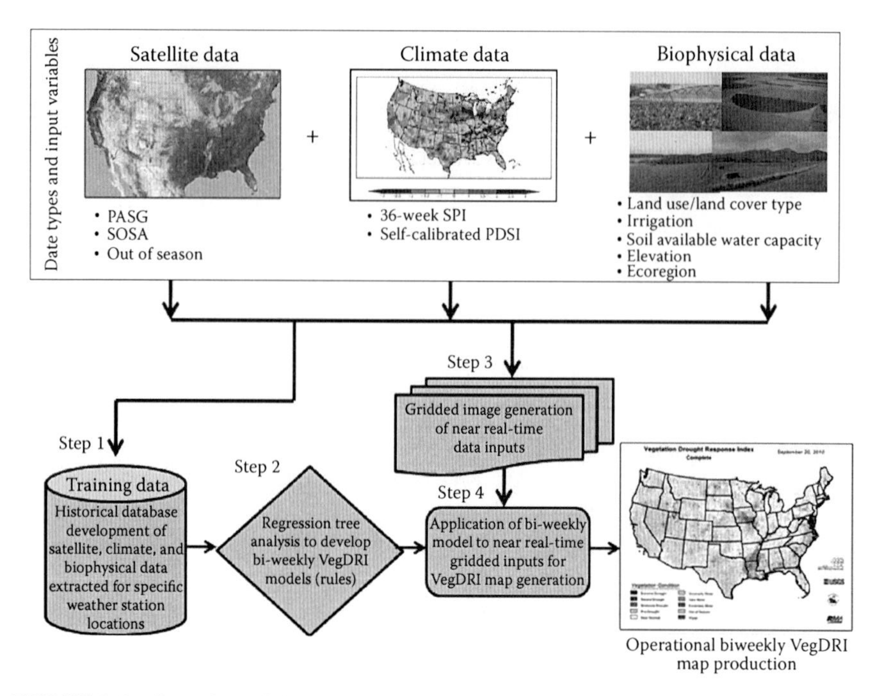

FIGURE 3.1 Overview of the data inputs and methodology of VegDRI.

3.2.2 VegDRI Classification Scheme

VegDRI has eight vegetation condition classes (Table 3.2) based on a modified version of the PDSI classification system (Palmer, 1965). There are three classes of drought severity (moderate, severe, and extreme), as well as a predrought class that represents the dry side of near-normal class value range. The predrought class was included to highlight areas that may be nearing initial drought conditions. VegDRI also has four nondrought classes (normal, unusually, very, and extremely moist) that characterize

TABLE 3.2
VegDRI Classification Scheme and Class Value Ranges

VegDRI Class Names	Value Range
Extreme drought	<−4
Severe drought	−4 to −3
Moderate drought	−3 to −2
Predrought	−2 to −1
Near normal	−1 to +2
Unusually moist	+2 to +3
Very moist	+3 to +4
Extremely moist	>+4

locations with normal to better than normal vegetation conditions, as well as areas of excessively wet conditions that could result in poor vegetation conditions due to flooding or water logging. An "out of season" (OS) class is also included to identify time periods for a given location when the vegetation is dormant (e.g., winter months) and VegDRI values are not calculated. OS is objectively defined through the historical analysis of time-series AVHRR NDVI data, which is discussed later in this section.

3.2.3 Data Inputs

3.2.3.1 Satellite Variables

A 20 year time series of biweekly, composited 1 km AVHRR NDVI data (Eidenshink, 2006) is used to calculate three vegetation-related metrics used in the VegDRI model, which include Percent Annual Seasonal Greenness (PASG), Start of Season Anomaly (SOSA), and OS. Prior to the calculation of these metrics, the NDVI time series is smoothed using a weighted least squares regression technique (Swets et al., 1999) to minimize noise and other artifacts (e.g., residual clouds) commonly found in the AVHRR data (Los et al., 1994) while maintaining the major multitemporal features of the original NDVI data.

3.2.3.1.1 Percent Annual Seasonal Greenness

The PASG provides a measure of how vegetation conditions for a specific biweekly period in a given year compare to the historical average conditions for the same period over the 20 year record of AVHRR NDVI observations. In order to calculate the PASG for each period, a historical median growing season window for each 1 km pixel in the AVHRR imagery is determined by identifying the Start and End Of Season Time (SOST and EOST) day of year (DOY) from annual AVHRR NDVI time series data using a moving-window averaging technique (Reed et al., 1994). A seasonal greenness (SG) metric, which represents the accumulated NDVI above a background NDVI baseline (i.e., nongrowing season or "latent" NDVI contributed from the soil background and/or atmospheric effects that has little to no biophysical meaning related to vegetation) across each 14 day period, is then calculated starting from the historical SOST DOY for each year in the historical record. SG is calculated sequentially for each period across the year until the EOST DOY, and the SG value for a specific period (SG_{PnYn}, the SG for biweekly period $n(P_n)$ in year $n(Y_n)$) represents the sum of the SG for the current and all preceding biweekly periods in the growing season. For each biweekly period, a historical mean SG (μSG_{Pn}) is calculated from the 20 yearly SG values. The 20 year record of PASG values for each period-year combination is then produced using the following equation:

$$PASG_{PnYn} = \left(\frac{SG_{PnYn}}{\mu SG_{Pn0}} \right) \times 100. \quad (3.1)$$

Brown et al. (2008) provide additional details regarding the PASG calculations. A low PASG value (e.g., <50%) for a specific biweekly period indicates below-normal (stressed) vegetation conditions compared to the historical conditions for that period, while high PASG values greater than 100% reflect above-average (or nonstressed) vegetation conditions.

3.2.3.1.2 Start of Season Anomaly

The SOSA represents the departure in the SOST for a specific year ($SOST_n$) from the median historical SOST ($SOST_{med}$) for a given pixel. For each year in the 20 year time series, the pixel-level SOSA ($SOSA_n$) expressed in number of days is calculated using the following equation:

$$SOSA_n = SOST_n - SOST_{med}. \tag{3.2}$$

The SOSA is included in the VegDRI model to distinguish areas that have a normal start of season and are experiencing low PASG because of interannual climatic variations (e.g., drought or cold early-season temperatures) from areas that experience an unusually late SOST because of nonclimate-related factors (e.g., LULC change or changes in management practices) that might result in a comparably low PASG.

3.2.3.1.3 Out of Season

The OS metric represents the nongrowing season period when vegetation is dormant. A historical median OS period is determined for each pixel using the SOST and EOST DOYs calculated for the PASG. The OS is defined as the period from EOST DOY (e.g., DOY 305 or November 1) to the SOST DOY of the next year (e.g., DOY 90 or March 31). During the OS for a given pixel, historical data are excluded from VegDRI model development, and no VegDRI values are calculated in the maps for biweekly periods within this temporal window. Excluding VegDRI calculations during the OS was implemented to avoid "false positive" drought signals from being depicted in the maps during periods of the year when the vegetation is not photosynthetically active, resulting from fluctuations in the NDVI (and resultant PASG) associated with nonvegetation-related factors (e.g., soil background and angular effects).

3.2.3.2 Climate Variables

The self-calibrated PDSI and the SPI were incorporated into VegDRI as indicators of climatic dryness. Historical data for both indices from 2417 weather station locations across the United States (3.2) were acquired from the Applied Climate Information System (ACIS) (http://www.rcc-acis.org/) (Hubbard et al., 2004). To ensure that high-quality historical time-series data are incorporated into VegDRI model development, only data from stations with a minimum 30 year data record and less than 10% missing observations are used. For each station, a 20 year time series of self-calibrated PDSI and SPI was calculated on a biweekly time step consistent with PASG calculations.

3.2.3.2.1 Standardized Precipitation Index

The SPI was designed to quantify precipitation anomalies over multiple time intervals (e.g., 1–12 month periods) based on fitting a long-term precipitation record at a given location over a specified interval to a probability distribution, which is then transformed into a gamma distribution so that the mean SPI value for that location and time period is 0 (McKee et al., 1995). SPI values are positive if the precipitation over a specific time period is higher than the historical average precipitation over that same period and negative if precipitation is less than the historical mean. The strength of the SPI is its temporal flexibility to assess conditions over short, intermediate, and long time intervals. A 36 week SPI was selected for VegDRI after exhaustive statistical testing of all SPI time intervals spanning from 1 to 52 weeks for the 2417 stations. Selection was based on the SPI that had a consistently high correlation coefficient value across all growing season periods.

3.2.3.2.2 Palmer Drought Severity Index

The PDSI is a prominent drought index that has been widely used to assess agricultural drought in the United States (Keyantash and Dracup, 2002). The PDSI is calculated from a simple supply-and-demand model of water balance that integrates precipitation and temperature information, as well as the available water holding capacity of the soil at a given location (Palmer, 1965). A new self-calibrated PDSI (Wells et al., 2004) is used in VegDRI, which calibrates the constants and duration factors in the PDSI computations to the local environmental characteristics of a specific location while still retaining the objectives of the original PDSI. These local adjustments improve the spatial comparability of PDSI values and calibrate the index so that extreme dry and wet events have a comparable rate of occurrence at any location (Guttman et al., 1992), providing a more consistent national PDSI data input for the VegDRI models.

3.2.3.3 Biophysical Variables

3.2.3.3.1 Land Use/Land Cover

The LULC variable was incorporated into VegDRI to reflect the variety of seasonal cycles and climate-vegetation responses exhibited by different LULC types. A 1 km LULC map was developed from the USGS 30 m National Land Cover Dataset (NLCD) circa 2001 (Homer et al., 2004). For each 1 km pixel in the AVHRR grid over the conterminous United States (CONUS), the majority LULC class among the 30 m NLCD data was determined and assigned to that 1 km pixel. Some thematic classes in the original NLCD classification scheme were merged (e.g., emergent herbaceous and woody wetland classes assigned to a single wetland class) to create more general LULC classes for model development.

3.2.3.3.2 Irrigated Agriculture

An irrigated agriculture (IrrAg) variable was integrated into VegDRI to differentiate irrigated locations, which are less susceptible to drought stress because of targeted water applications, from rainfed agricultural areas. A 1 km map depicting the spatial distribution of IrrAg was generated from a 250 m MODIS Irrigated Agriculture Dataset (MIrAD) developed from a combination of MODIS NDVI data, USDA county irrigation statistics, and LULC information (Brown et al., 2009). The 1 km IrrAg map represents the percentage of irrigated 250 m MIrAD pixels contained within each 1 km pixel footprint.

3.2.3.3.3 *Soil Available Water Capacity*

The available water capacity (AWC) variable is used to reflect the potential of the soil to hold moisture that is available to plants, which influences the susceptibility of vegetation to drought stress. A 1 km AWC map was developed by extracting the AWC values for the total soil column from the State Soil Geographic (STATSGO) database for each soil map unit (USDA, 1994) and converting the map unit polygons to a 1 km raster grid.

3.2.3.3.4 *Ecoregion*

The ecoregion variable provides a geographic framework to account for the considerable variability in environmental conditions encountered across the CONUS that can influence the level of drought stress experienced at a given location. For example, two locations (e.g., High Plains versus Flint Hills) may be assigned to same general grassland class by the LULC variable but may have differing responses to drought because they represent different general grassland types (e.g., shortgrass versus tallgrass prairie) with different dominant species compositions (e.g., cool- versus warm-season grasses) that have acclimated to the collective environmental conditions of the area (e.g., climate, soils, and topography). A 1 km ecoregion grid was created from Omernik Level III ecoregion vector data (Omernik, 1987), which divides the CONUS into a series of geographic regions with similar ecosystems and environmental resources defined using both abiotic (e.g., physiography) and biotic (e.g., plant species) criteria.

3.2.3.3.5 *Elevation*

A digital elevation model (DEM) consisting of a 1 km raster grid of evenly spaced elevation values derived from the USGS 30 m DEM is included to account for influences of elevation on vegetation types and their sensitivity to drought.

3.2.4 VegDRI Training Database Development

A training database of all climate, satellite, and biophysical data discussed earlier was extracted and assembled for the 2417 weather station locations in Figure 3.2. Historical, point-based PDSI and SPI data were calculated for each station and sequentially ordered by biweekly period for each year in the database. Data were also extracted from the gridded satellite and biophysical data sets at the location of each weather station. A 3 × 3 pixel window centered on each station location was used to calculate the average value across all pixels in the window for continuous variables (e.g., PASG) and the majority value for categorical variables (e.g., LULC). Pixels within the window classified as urban or water in the LULC grid or flagged to be OS for a specific biweekly period were excluded from the mean or majority zonal calculations for that period. Data values associated with both urban and water locations were excluded because both LULC types are representative of primarily nonvegetated areas that are not the monitoring target for VegDRI. Pixels flagged to be OS for a specific biweekly period were also removed or excluded from the window average for that period because of the nonvegetated spectral signal detected from

FIGURE 3.2 Geographic location of the 2417 weather station locations used to develop the empirical VegDRI models.

the pixel at that time. A 20 year historical time series of biweekly PASG and OS values was calculated for each station using this approach and sequentially ordered in the same manner as the climate data. For each year in the time series, a SOSA value was calculated for each station location and held constant for all "in-season" biweekly periods for that year. The biophysical variable values calculated for each station were held static across the 20 year period. Historical records of all stations in the database were then temporally subset into 26 biweekly periods (e.g., biweekly period 1: January 1–14) across the calendar year to develop a series of separate, biweekly VegDRI models.

3.2.5 VegDRI Model Development

For each biweekly period, a commercial Classification and Regression Tree (CART) algorithm called Cubist (Quinlan, 1993) was used to analyze the historical data in the training database for that specific period and generate a rule-based, piecewise linear regression VegDRI model. Each model incorporates historical data for the "dynamic" climate and satellite-based variables while holding the biophysical variables constant over a four-biweek window that includes the current biweekly period (e.g., biweek 10) plus the three prior biweekly periods (e.g., biweeks 7, 8, and 9) in the calendar year. As discussed earlier, the self-calibrated PDSI serves as the dependent variable in these empirical-based models, providing a well-established classification system for VegDRI to categorize varying levels of drought severity on vegetation based on the analysis of the other biophysical, climate, and satellite variables. Twenty-six period-specific VegDRI models were developed. The Cubist-derived models consist of an unordered set of rules, with each rule having the syntax "if *x* conditions are met then use the associated linear regression model" to calculate the VegDRI value. The following is an example of one of many rules generated for a specific biweekly period:

Rule 1:
If land cover in {Grassland, Pasture/Hay, Row Crops}
Ecoregion in {western High Plains, central Great Plains}
36 week SPI ≤ -1.4
AWC ≤ 4.5
PASG ≤ 50
then VegDRI = −3.5 + 0.6 PASG + 1.48 SPI − 0.14 AWC + 0.25 percent irrigated.

In other words, if the data associated with a case (i.e., pixel) meet the threshold criteria for the three continuous variables and are represented by one of the three land cover types and either ecoregion, then the following linear regression equation is used to calculate a VegDRI value. Most period-specific VegDRI models comprise 30–40 rules. If two or more rules apply to a case, then all linear regression equations are used to calculate a series of values that are averaged to determine the final VegDRI value. It should be noted that some rules and/or associated linear regression equations may not use all the independent variables. For example, in the rule shown earlier, elevation and SOSA are not used. However, each independent variable is incorporated into a subset of the multiple rules and regression equations that are collectively utilized to calculate the final VegDRI value.

3.2.6 VegDRI Model Implementation and Mapping

The rules from a biweekly VegDRI model are then applied to the set of gridded image data inputs (listed in Table 3.1) for the corresponding period using MapCubist software developed at the USGS Earth Resources Observation and Science (EROS) Center to produce a 1 km VegDRI map. For the SPI variable, which is acquired as a point-based index value from weather station data, a 1 km raster image is generated using an inverse distance weighting (IDW) interpolation method. During model implementation to the gridded image data, the values of all the input variables associated with each pixel are considered to determine the specific rule(s) and corresponding linear regression equation(s) to be applied at the pixel level. This process is repeated until all 1 km pixels in the image domain have been assigned a VegDRI value.

The VegDRI map is the result of inverting the empirically based regression models, which describe the historical relationship between PDSI and the other climate, environmental, and satellite input variables for known locations (i.e., weather station locations). Although the models are applied at the pixel level during the mapping phase, collectively the integration of these variables results in landscape-level drought depictions across the image domain. In VegDRI's conceptual design, the remote sensing inputs provide high spatial resolution inputs of vegetation patterns and conditions across the landscape, which add structure to 1 km VegDRI maps when combined with the coarser precipitation patterns represented in the interpolated SPI grid. The spatial patterns represented by the other static environmental variables in the image domain also add spatial structure to maps by providing geographic stratification of the relationship between the remote sensing inputs (i.e., PASG and SOSA) and SPI to estimate PDSI values across the CONUS in the 1 km VegDRI map generated in this final mapping step.

3.3 RESULTS AND DISCUSSION

3.3.1 Statistical Analysis of Historical VegDRI Model Performance across the United States

Assessment of the statistical accuracy of the national-level, biweekly VegDRI models over the CONUS for a 20 year study period (1989–2008) is presented in Figure 3.3. This analysis was conducted to determine how well the VegDRI model was able to reproduce self-calibrated PDSI classifications at weather station locations across the CONUS. An *x*-fold cross-validation technique (Kohavi, 1995) using "hold out" years was used to assess VegDRI's historical performance across the growing season for the 2417 weather station locations in Figure 3.2. For each biweekly period, 20 validation iterations (or folds) were performed by using 19 years of historical data to train a model (e.g., 1990–2008) and one independent "hold out" year (e.g., 1989) to determine VegDRI's predictive accuracy across all stations. A different hold-out year was selected for each iteration, allowing every year in the 20 year record to be withheld for testing. Correlation coefficient results for the primary growing season periods over the CONUS are presented for each biweekly period in Figure 3.3. The Pearson correlation coefficient (r) values for each date represent the mean correlation between the predicted and observed VegDRI values across the 20 year period for all station locations.

The correlation results show that the VegDRI models had a relatively high predictive accuracy across the growing season with r values greater than 0.75 for all

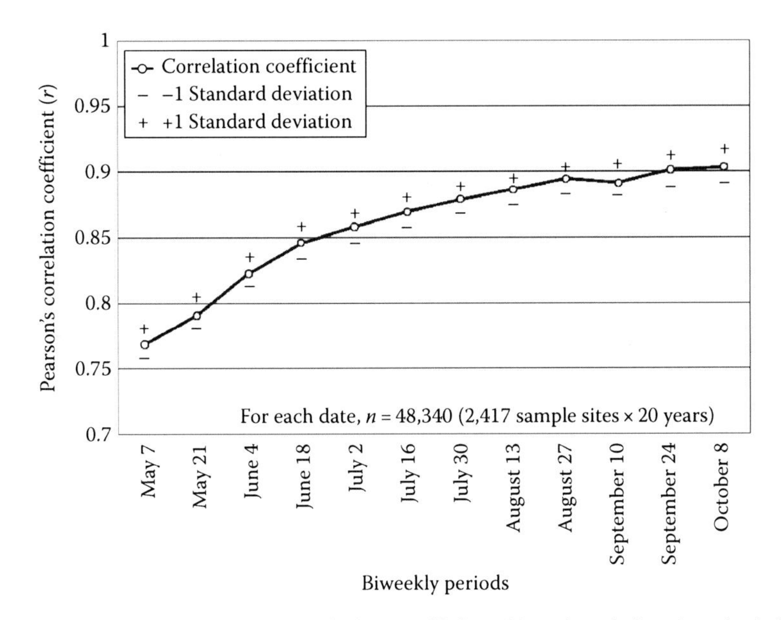

FIGURE 3.3 Average Pearson correlation coefficient (r) and variation (standard deviation) between the observed and predicted VegDRI values over a 20 year period (1989–2008) across 2,417 weather station locations of the CONUS for each biweekly period of the growing season.

biweekly periods. The predictive accuracy gradually increased from May 7 ($r = 0.77$) to October 8 ($r = 0.90$). An increase in predictive accuracy early in the growing season might be expected given the inherent interannual variations in the emergence and initial growth rates of vegetation because of varying climatic conditions (e.g., air and soil temperature). Subtle variations in early growing season conditions when the vegetation has relatively low green biomass can result in dramatic changes in PASG values (due to small dynamic range of NDVI values at that time) compared to later in the growing season when the vegetation has much higher biomass and the PASG are less influenced by the same level of variation (due to larger dynamic range of NDVI values). As a result, small changes in vegetation conditions earlier in the year can result in larger PASG changes and thus increased VegDRI error compared to later dates as the growing season progresses when the PASG and resulting VegDRI values are less sensitive to such variations. The period-specific VegDRI models were also found to have a stable predictive accuracy across the 20 year period with relatively low interannual variability among the annual r values for each biweekly period. This is reflected by the small range of the average ±1 standard deviation (σ) values ($1\sigma = \sim 0.01$) that bounded the mean r values for all biweekly periods in Figure 3.3. These results indicate that the performance of VegDRI was reasonably robust over the CONUS across the growing season and relatively uninfluenced by interannual climate variability over two decades.

Because this testing used in situ meteorological observations at each station location to calculate the PDSI validation data sets, these results should be viewed as a "best case" accuracy of VegDRI because calculations in the VegDRI maps for locations between stations are based on spatially interpolated PDSI values (rather than from observed station data). As a result, correlation values in the map lacking in situ observations will likely be lower than those reported here, with the accuracy being highly dependent on the accuracy of the spatial interpolation technique and density of weather station locations in close proximity to that specific location. Further testing is needed to fully assess the overall accuracy of VegDRI for locations lacking in situ–based PDSI data. This could be accomplished by assessing spatially interpolated PDSI grids generated from the station data used in this study or using Parameter-elevation Regressions on Independent Slopes Model (PRISM) or radar-based observed precipitation data from the National Weather Service (NWS) to calculate PDSI values for nonstation locations to compare with the VegDRI results.

3.3.2 National-Level VegDRI: An Example from 2009

Figure 3.4a and b compares the national VegDRI map for July 13, 2009, with the U.S. Drought Monitor (USDM) map for July 14, 2009, to illustrate the national-level drought patterns and improved spatial resolution of information provided by this index. The USDM (http://drought.unl.edu/dm/monitor.html) represents an appropriate benchmark to compare the performance of VegDRI because it is the current state-of-the-art drought monitoring tool for the United States. The USDM map represents a broad-scale depiction of national agricultural and hydrological drought conditions based on the collective analysis of an array of climate-, hydrologic-, and satellite-based indicators, as well as input from climate and water experts across the

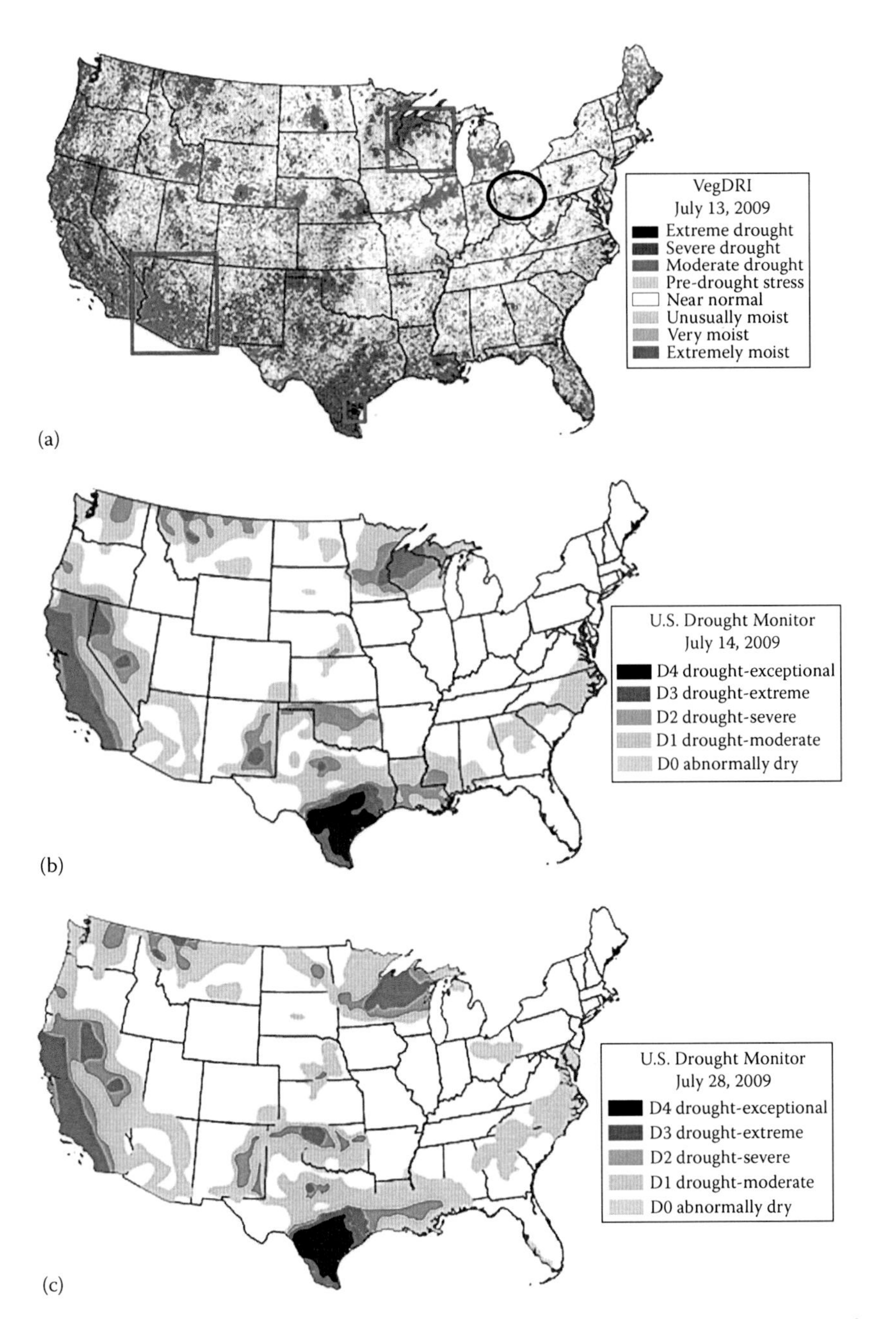

FIGURE 3.4 **(See color insert.)** VegDRI map (a) for July 13, 2009, and USDM maps for July 14 (b) and July 28 (c), 2009, over the continental United States. The black circle highlights an area of central Ohio that was classified as predrought stress in the VegDRI map but lagged by 2 weeks in the USDM maps, which did not show abnormally dry conditions until late July. The red boxes on the VegDRI map delineate the geographic extent of the local case study areas presented later in Section 3.4.

country (Svoboda et al., 2002). Lower resolution PDSI is one variable commonly used in the construction of the USDM.

VegDRI and the USDM depicted similar drought patterns across the United States for this mid-July 2009 date. Major drought areas such as the severe to extreme conditions in south Texas and moderate to severe conditions in the Oklahoma Panhandle, eastern Minnesota and northern Wisconsin, northwest Montana, and California and western Nevada are seen in both maps. A predrought signal appeared in VegDRI over north Georgia and the Carolinas that was consistent with the abnormally dry areas depicted in the USDM. Small areas of predrought and moderate drought appeared in VegDRI in central Ohio that were absent from the USDM map. However, by late July, similar drought conditions were expressed in the USDM (Figure 3.4c), suggesting that VegDRI may have provided an early indicator of dryness that was not represented in the USDM. Clearly the use of remote sensing information in VegDRI provides higher spatial resolution drought information than that currently conveyed in the USDM. The hope is that indices that incorporate satellite-based observations such as VegDRI will allow the USDM to improve the spatial precision of the drought patterns represented in their maps in the future. The ability of VegDRI to characterize substate to county-level drought patterns over a range of climate regimes and land cover types is further illustrated by the case examples presented in the next section.

3.3.3 Local-Scale VegDRI Information: Examples from across the United States in 2009

3.3.3.1 South Texas

In 2009, south Texas suffered from severe to extreme drought conditions, with many locations, particularly along the Gulf Coast, experiencing their driest year in the modern climatic data record. Figure 3.5a shows the extreme drought conditions detected by VegDRI over a three-county area centered on Corpus Christi in Nueces County. As the national VegDRI map in Figure 3.4a shows, most of south Texas experienced very severe to extreme drought conditions, but the focal point of the most intense drought signal in VegDRI emerged in the local area of Kleberg, Nueces, and San Patricio counties along the Gulf of Mexico. Precipitation records for several weather stations in these counties (Corpus Christi, Kingsville, Mathis, and Robstown) revealed the magnitude of the 2009 drought, with each station recording its driest year in more than 50 years of precipitation observations. Precipitation deficits were significant, with each station receiving less than 15% of mean annual rainfall. On average, these humid tropical locations receive more than 54 in. of rain annually, but in 2009, they received between 5 and 10 in. of precipitation. Agricultural production was devastated with a near-complete failure of two primary crops: cotton and sorghum. For example, USDA (2010a) reported that ~29% of 168,000 planted acres of sorghum were harvested in Nueces County with an average yield of 40 bushels per acre, which ranked as the third lowest production total since 1962. In addition, local media reports collected by the Drought Impact Reporter (http://droughtreporter.unl.edu/) stated that more than 90% of the cotton and sorghum crops in these counties were destroyed by drought, and the harvestable crop was of poor quality.

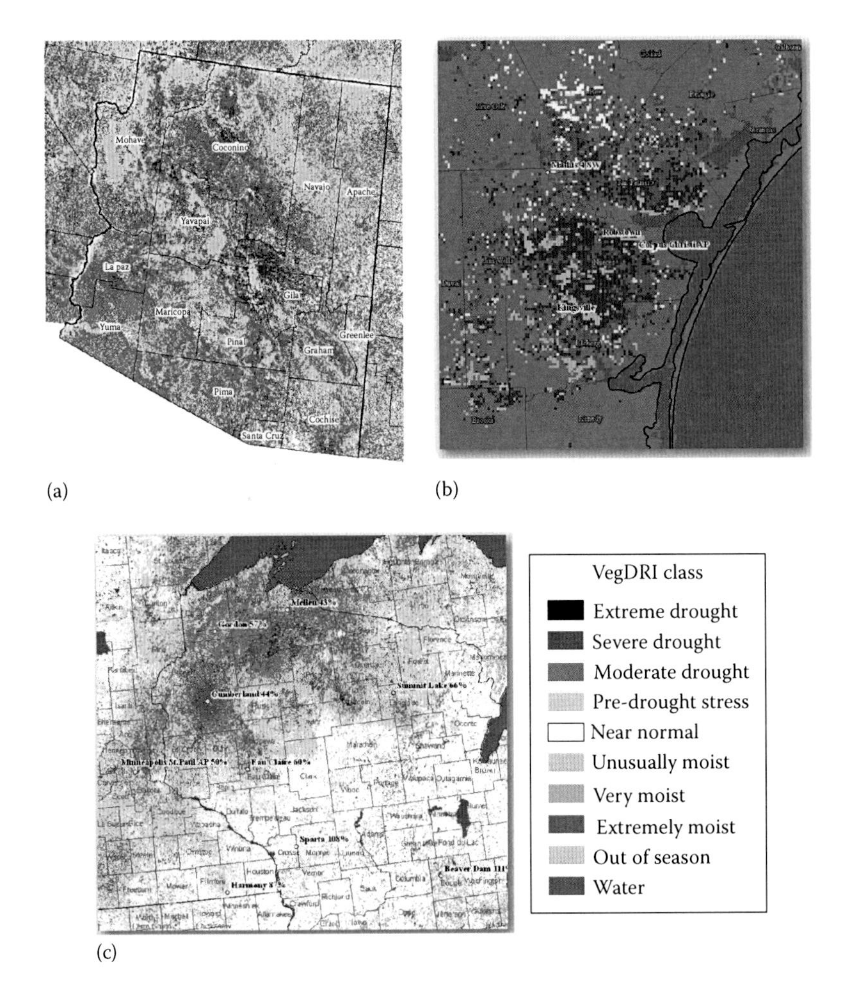

FIGURE 3.5 **(See color insert.)** Local-scale VegDRI results on June 29, 2009, over south Texas (a), on November 2, 2009, over the state of Arizona (b), and on August 10, 2009, over eastern Minnesota and northern Wisconsin (percentages for highlighted locations represent the percent of historical average precipitation received at those locations in 2009) (c).

3.3.3.2 Arizona

The severe to extreme drought conditions that were prevalent across Arizona throughout the second half of 2009 are shown in the VegDRI map for November 2 (Figure 3.5b). Drought conditions in Arizona rapidly intensified during the summer and fall because of a lack of rainfall during the monsoon season (July–September), which traditionally accounts for most of the state's annual precipitation in an otherwise arid climate. In 2009, the state of Arizona experienced its third driest June–August period in more than a century (USDA, 2010b). By late October, most of Arizona had received less than 50% of average precipitation for the year, with the exception of far eastern Arizona, where near-average precipitation was received. VegDRI characterized this rapid progression in drought intensity from predominately near-normal to predrought conditions on July 13 (Figure 3.4a) to the severe to extreme drought conditions on November 2 (Figure 3.5b) across most of the state. At the substate level, some of the driest conditions occurred in central Arizona in Coconino, Navajo, and Gila counties, where many locations received <25% of average annual precipitation. In Figure 3.5b, severe to extreme drought conditions over these counties were reflected in VegDRI. In addition, the more favorable conditions in eastern Arizona in Apache and Greenlee counties are classified in the predrought to moderate drought categories.

The pronounced drought conditions across the state were reflected by the numerous reports of crop losses, degraded rangeland conditions, and negative impacts on forest health. USDA assigned a natural disaster declaration to 13 of 15 counties in Arizona because of substantial agricultural production losses. La Paz and Yuma counties in southwest Arizona were not assigned a disaster declaration because their production losses were not as substantial. The reduced drought severity in this area was depicted by VegDRI in Figure 3.5b, with the majority of Yuma County and much of La Paz County experiencing moderate drought. A time series of VegDRI maps from September through December 2009 (complete VegDRI time series available at http://drought.unl.edu/vegdri/VegDRI_archive.htm) revealed that any severe drought conditions in either county were short lived, and a weaker, moderate drought signal persisted over this period compared to the other western and central counties in Arizona.

3.3.3.3 Minnesota and Wisconsin

A snapshot of the moderate to severe drought conditions that persisted over east-central Minnesota and northwest Wisconsin throughout the 2009 growing season is presented in the VegDRI map for August 10 (Figure 3.5c). A band of dry conditions spanning an area from Minneapolis, Minnesota, northeastward to Lake Superior (near Mellen, Wisconsin) began to emerge by early June and continued to intensify to moderate to severe drought conditions by midsummer (mid-July to early August). This example illustrates the local-scale variations in drought patterns depicted by VegDRI, which were consistent with ground observations and impacts reported for this area. The percent average growing season precipitation received by selected weather stations in Figure 3.5c shows that the spatial variations in drought conditions depicted for VegDRI agreed with the rainfall deficit patterns recorded at weather stations across this area. For example, the transition from severe drought conditions in Wisconsin surrounding Cumberland to moderate drought near Eau Claire to near-normal conditions at Sparta classified by VegDRI reflects the localized precipitation gradient recorded during the 2009 growing season between these locations. During the 3 months before August 10, the percent of average precipitation received during that period (typically between 11 and 12 in.) increased along this drought severity gradient from 44% to 60% to 108% for these three locations, respectively. In addition, the core area of moderate to severe drought delineated by VegDRI from Minneapolis to Mellen was consistent with the weather station observations over this area, which typically recorded less than 50% of average rainfall.

The majority of the drought-stricken area classified as moderate to severe drought in Figure 3.5c is densely forested, and the impact of these dry conditions on vegetation was reflected by an increased number of burn bans and wildfires reported in 2009. Foresters in northern Wisconsin reported an increased rate of mortality among several tree species (e.g., oak and maple) primarily attributed to the increased susceptibility of drought-weakened trees to many native insects and pathogens (Schwingle, 2009). Only a small area of extensive cropland between Cumberland, Eau Claire, and Minneapolis was located within the core drought area defined by VegDRI. However, USDA county officials within this area reported dry soil moisture conditions and stressed crops and grasslands by early July that eventually lead to a USDA drought declaration for most counties in east-central Minnesota and northern Wisconsin. Locations classified by VegDRI to have near-normal vegetation conditions south of the core drought area (near stations such as Beaver Dam, Harmony, and Sparta) were not assigned a drought declaration by USDA. This was consistent with USDA National Agricultural Statistics Service (NASS) Crop Progress reports for Wisconsin, which reported adequate rainfall to support agricultural production for this area.

3.4 ENHANCING VegDRI WITH MODIS SATELLITE DATA

Work is ongoing to transition the satellite inputs for VegDRI from AVHRR-based NDVI data to a MODIS-based expedited NDVI data stream produced by the USGS eMODIS system (Jenkerson et al., 2010), which has the flexibility to accommodate the production schedule of a specific application (e.g., daily, weekly, or biweekly). The current biweekly AVHRR NDVI composite production schedule is rigid; composites are updated at a 2 week interval on Tuesdays, which restricts the operational production of new VegDRI maps to the middle of the week (i.e., Tuesday or Wednesday) once every 2 weeks. In contrast, the USGS eMODIS system provides a near-real-time, rolling 7 day NDVI composite for the CONUS that allows VegDRI to be updated weekly on Mondays to accommodate the schedule of users such as the USDM authors. In addition, the satellite observations from MODIS used to generate the NDVI data are expected to provide higher-quality information for VegDRI because of improved instrument calibration and higher geolocational accuracy, as well as the rigorous atmospheric and radiometric corrections applied to the spectral data. The eMODIS-based VegDRI will use empirical models incorporating the SOSA and PASG calculated from historical AVHRR NDVI observations that are translated to a "MODIS-like" NDVI time series in order to be consistent with eMODIS NDVI images to which the models are applied for map generation. Development of an AVHRR-to-MODIS NDVI translation algorithm and application within a phenological-based geographic framework (Gu et al., 2010) is nearing completion. eMODIS VegDRI is currently produced at USGS EROS and available via a web map interface (http://vegdri.cr.usgs.gov/viewer/viewer.htm). The transition to operational eMODIS VegDRI production for the CONUS is scheduled for 2011.

3.5 CONCLUSIONS AND FUTURE WORK

VegDRI represents a new "hybrid" index for operational vegetation drought monitoring in the United States, incorporating traditional satellite-based VI observations and climate-based drought index data with general biophysical information about the environment to produce 1 km resolution national maps that depict "drought-related" vegetation stress. VegDRI is designed to characterize county to subcounty level drought patterns, which is an appropriate spatial scale to support a wide range of local-scale decision-making activities. Historical testing of the VegDRI models for a 20 year period across the CONUS showed that this index maintained a high predictive accuracy when compared with station-based, self-calibrated PDSI across both the growing season and diverse environmental conditions. Case examples from 2009 over Arizona, south Texas, and northern Minnesota and Wisconsin further illustrated the ability of VegDRI to characterize local-scale variations in drought conditions across a wide range of climatic regimes (i.e., arid to humid) and different land cover types (shrubs, grass, crops, and forest). In addition, model performance was relatively unaffected by interannual climate variations over the two-decade study period. From a national perspective, the major drought patterns classified by VegDRI were consistent with those mapped by the nation's state-of-the-art drought monitoring tool, the USDM, as shown in Figure 3.4a and b. The improved spatial resolution of the 1 km VegDRI map compared to the USDM map is evident, suggesting that higher resolution inputs such as VegDRI could be used to enhance the spatial precision of the drought patterns depicted in the USDM.

Currently, VegDRI is only operationally produced across the CONUS, but the potential exists to expand this hybrid-based index method to other parts of the world. Satellite-based NDVI observations comparable to those used for VegDRI in the United States are globally available from AVHRR, MERIS, MODIS, and SPOT Vegetation. However, the specific variables used in the biophysical and climate components of VegDRI would be unique for each country or region and depend on the specific data sets that are available. A strength of the VegDRI approach is its flexibility to be customized to the data resources of a given location and its ability to integrate new data inputs as they become available. For example, a temperature component is currently lacking from the VegDRI approach presented in this chapter. However, the potential exists to develop a historical time series of AVHRR thermal observations (or derived ET estimates) that can be integrated into VegDRI to better represent the influence of LST on vegetation conditions. In addition to geographic expansion of VegDRI beyond the United States, the development of a higher spatial resolution VegDRI using MODIS 250 m NDVI observations is an area of future work to accommodate the needs of local-scale decision makers, who require more detailed landscape-level information that is not contained in the current 1 km VegDRI products.

Continued validation of VegDRI using multiple information sources (e.g., soil moisture observations, biophysical vegetation measurements, crop/grass production data, and impact reports) is also needed to better characterize index performance over an extended period of time for locations with different environment conditions. Efforts are currently underway to evaluate VegDRI's spatiotemporal performance across the CONUS over two decades (1989–2009) using statistical cross-validation. This work will assess the historical accuracy and variability of VegDRI and investigate the index's performance for major land cover types and different ecological regions of the United States. Comparisons between VegDRI and other drought-related indices and indicators such as the Evaporative Stress Index (ESI) (Anderson et al., 2007, 2010) and the USDM are also being conducted to better understand the complementary drought information that VegDRI can provide. Quantitative validation of VegDRI trends with in situ–based biophysical measures of vegetation (e.g., biomass) is also planned, but such long-term data sets are sparse and typically limited to a few long-term ecological reserve sites and research plots maintained by organizations such as USDA's Agricultural Research Service. As a result, VegDRI validation work will utilize a "convergence of evidence" approach that incorporates a range of qualitative and quantitative assessments applied to the broad range of information sources that have been discussed in this chapter to establish the relative strengths and weaknesses of this hybrid drought index.

ACKNOWLEDGMENTS

Support was provided by the USDA Risk Management Agency (RMA) under USDA partnership 02-IE-0831-0208 with the National Drought Mitigation Center (NDMC), University of Nebraska–Lincoln (UNL), and the U.S. Geological Survey (USGS) (USGS contract 03CRCN0001). The authors thank Deborah Wood of the NDMC for her editorial suggestions.

REFERENCES

Allen, C.D. and D.D. Breshears. 1998. Drought-induced shift of a forest-woodland ecotone: Rapid landscape response to climate variation. *Proceedings of the National Academy of Science* 95:14389–14842.

Anderson, M.C., W.P. Kustas, J.M. Norman, C.R. Hain, J.R. Mecikalski, L. Schultz, M.P. Gonzalez-Dugo, C. Cammalleri, G. d'Urso, A. Pimstein, and F. Gao. 2010. Mapping daily evapotranspiration at field to continental scales using geostationary and polar orbiting satellite imagery. *Hydrology and Earth System Sciences Discussions*, 7:5957–5990.

Anderson, M.C., J.M. Norman, J.R. Mecikalski, J.A. Otkin, and W.P. Kustas. 2007. A climatological study of evapotranspiration and moisture stress across the continental U.S. based on thermal remote sensing: 2. Surface moisture climatology. *Journal of Geophysical Research* 112: D11112, doi: 11110.11029/12006JD007507.

Breshears, D.D., N.S. Cobb, P.M. Rich, K.P. Price, C.D. Allen, R.G. Balice, W.H. Romme, J.H. Kastens, M.L. Floyd, J. Belnap, J.J. Anderson, O.B. Myers, and C.W. Meyer. 2005. Regional vegetation die-off in response to global-change-type drought. *Proceedings of the National Academy of Science* 102:15144–15148.

Brown, J.F., S. Maxwell, and S. Pervez. 2009. Mapping irrigated lands across the United States using MODIS satellite imagery. In *Remote Sensing of Global Croplands for Food Security*, eds. P.S. Thenkabail, J.G. Lyon, H. Turral et al., pp. 177–198. Boca Raton, FL: CRC Press.

Brown, J.F., B.D. Wardlow, T. Tadesse, M.J. Hayes, and B.C. Reed. 2008. The Vegetation Drought Response Index (VegDRI): A new integrated approach for monitoring drought stress in vegetation. *GIScience and Remote Sensing* 45(1):16–46.

Domenikiotis, C., A. Loukas, and N.R. Dalezios. 2003. The use of NOAA/AVHRR satellite data for monitoring and assessment of forest fires and floods. *Natural Hazards and Earth System Science* 3(1–2):115–128.

Eidenshink, J.C. 2006. A 16-year time series of 1 km AVHRR satellite data of the conterminous United States and Alaska. *Photogrammetric Engineering and Remote Sensing* 72:1027–1035.

Everard, K., E.W. Seabloom, W.S. Harpole, and C. de Mazancourt. 2010. Plant water use affects competition for nitrogen: Why drought favors invasive species in California. *The American Naturalist* 175(1):85–97.

Fensham, R.J. and J.E. Holman. 1999. Temporal and spatial patterns in drought-related tree dieback in Australian savanna. *Journal of Applied Ecology* 36:1035–1050.

Franke, J. and G. Menz. 2007. Multi-temporal wheat disease detection by multi-spectral remote sensing. *Precision Agriculture* 8(3):161–172.

Gao, G. 1996. NDWI—A normalized difference water index for remote sensing of vegetation liquid water from space. *Remote Sensing of Environment* 58:257–266.

Goetz, S.J., G.J. Fiske, and A.G. Bunn. 2006. Using satellite time-series data sets to analyze fire disturbance and forest recovery across Canada. *Remote Sensing of Environment* 101:352–365.

Gu, Y., J.F. Brown, T. Muira, W.J.D. van Leewen, and B.C. Reed. 2010. Phenological classification of the United States: A geographic framework for extending multi-sensor time-series data. *Remote Sensing* 2(2):526–544.

Gu, Y., J.F. Brown, J.P. Verdin, and B. Wardlow. 2007. A five-year analysis of MODIS NDVI and NDWI for grassland drought assessment over the central Great Plains of the United States. *Geophysical Research Letters* 34, doi: 10.1029/2006GL029127.

Gu, Y., E. Hunt, B. Wardlow, J.B. Basara, J.F. Brown, and J.P. Verdin. 2008. Evaluation of MODIS NDVI and NDWI for vegetation drought monitoring using Oklahoma Mesonet soil moisture data. *Geophysical Research Letters* 35, doi: 10.1029/2008GL035772.

Guarin, A. and A.H. Taylor. 2005. Drought triggered tree mortality in mixed conifer forests in Yosemite National Park, California, USA. *Forest Ecology and Management* 218:229–244.

Guttman, N.B., J.R. Wallis, and J.R.M. Hosking. 1992. Spatial comparability of the Palmer Drought Severity Index. *Water Resources Bulletin* 28:1111–1119.

Hayes, M.J., M.D. Svoboda, C.L. Knutson, and D.A. Wilhite. 2004. Estimating the economic impact of drought. *Proceedings of the 14th Conference on Applied Climatology*, Seattle, WA. ams.confex.com/ams/pdfpapers/73004.pdf

Homer, C., C. Huang, L. Yang, B. Wylie, and M. Coan. 2004. Development of a 2001 national land cover database for the United States. *Photogrammetric Engineering and Remote Sensing* 70:829–840.

Hubbard, K.G., A.T. DeGaetano, and K.D. Robbins. 2004. SERVICES: A modern applied climate information system. *Bulletin of the American Meteorological Society* 85(6):811–812.

Jakubauskas, M.E., D.L. Peterson, J.H. Kastens, and D.R. Legates. 2002. Time series remote sensing of landscape-vegetation interactions in the Southern Great Plains. *Photogrammetric Engineering and Remote Sensing* 68:1021–1030.

Jenkerson, C., T. Maiersperger, and G. Schmidt. 2010. eMODIS—A User-Friendly Data Source. U.S. Geological Survey Open-File Report 2010-1055. http://pubs.usgs.gov/of/2010/1055/ (accessed June 10, 2011).

Ji, L. and A.J. Peters. 2003. Assessing vegetation response to drought in the northern Great Plains using vegetation and drought indices. *Remote Sensing of Environment* 87:85–98.

Karnieli, A., N. Agam, R.T. Pinker, M. Anderson, M.L. Imhoff, G.G. Gutman, N. Panov, and A. Goldberg. 2010. Use of NDVI and land surface temperature for drought assessment: Merits and limitations. *Journal of Climate* 23:618–633.

Karnieli, A., M. Bayasgalan, Y. Bayarjargal, N. Agam, S. Khudulmur, and C.J. Tucker. 2006. Comments on the use of the Vegetation Health Index over Mongolia. *International Journal of Remote Sensing* 27(20):2017–2024.

Karnieli, A. and G. Dall'Olmo. 2003. Remote-sensing monitoring of desertification, phenology, and droughts. *Management of Environmental Quality: An International Journal* 41(1):22–38.

Kasischke, E.S. and M.R. Turetsky. 2006. Recent changes in the fire regime across the North American boreal region—Spatial and temporal patterns of burning across Canada and Alaska. *Geophysical Research Letters* 33:L09070, doi: 10.1029/2006GL0256767.

Keyantash, J. and J.S. Dracup. 2002. The quantification of drought: An evaluation of drought indices. *Bulletin of the American Meteorological Society* 83:1167–1180.

Kogan, F.N. 1995. Application of vegetation index and brightness temperature for drought detection. *Advances in Space Research* 11:91–100.

Kogan, F.N. 1997. Global drought watch from space. *Bulletin of the American Meteorological Society* 78(4):621–636.

Kogan, F.N. 2002. World droughts in the new millennium from AVHRR-based vegetation health indices. *EOS, Transactions of the American Geophysical Union* 83:557–564.

Kohavi, R. 1995. A study of cross-validation and bootstrap for accuracy estimation and model selection. In *Proceedings of the 14th International Joint Conference on Artificial Intelligence*, ed. C.S. Mellish, pp. 1137–1143. San Francisco, CA: Morgan Kaufmann Publishers Inc.

Liu, W.T. and F.N. Kogan. 1996. Monitoring regional drought using the Vegetation Condition Index. *International Journal of Remote Sensing* 17(14):2761–2782.

Los, S.O., C.O. Justice, and C.J. Tucker. 1994. A global 1 by 1 km NDVI data set for climate studies derived from the GIMMS continental NDVI data. *International Journal of Remote Sensing* 15(17):3493–3518.

McKee, T.B., N.J. Doesken, and J. Kleist. 1995. Drought monitoring with multiple time scales. In *Preprints, 9th Conference on Applied Climatology*, Dallas, TX, pp. 233–236. Boston, FL: American Meteorological Society.

McVicar, T.R. and P.B. Bierwirth. 2001. Rapidly assessing the 1997 drought in Papua New Guinea using composite AVHRR imagery. *International Journal of Remote Sensing* 22:2109–2128.

NCDC (National Climatic Data Center). 2010. Billion dollar U.S. weather disasters. Disasters chart, http://lwf.ncdc.noaa.gov/img/reports/billion/disasters2009.pdf (accessed June 11, 2011).

Omernik, J.M. 1987. Ecoregions of the conterminous United States. *Annals of the Association of American Geographers* 77(1):118–125.

Palmer, W.C. 1965. Meteorological drought. Research Paper No. 45, U.S. Department of Commerce Weather Bureau, Washington, DC.

Peters, A.J., S.C. Griffin, A. Vina, and L. Ji. 2000. Use of remotely sensed data for assessing crop hail damage. *Photogrammetric Engineering and Remote Sensing* 66(11):1349–1355.

Peters, A.J., D.C. Rundquist, and D.A. Wilhite. 1991. Satellite detection of the geographic core of the 1988 Nebraska drought. *Agricultural and Forest Meteorology* 57:35–47.

Quinlan, J.R. 1993. *C4.5 Programs for Machine Learning*. San Mateo, CA: Morgan Kaufmann Publishers.

Reed, B.C., J.F. Brown, D. VanderZee, T.R. Loveland, J.W. Merchant, and D.O. Ohlen. 1994. Measuring phenological variability from satellite imagery. *Journal of Vegetation Science* 5:703–714.

Reed, B.C., T.R. Loveland, and L.L. Tieszen. 1996. An approach for using AVHRR data to monitor U.S. Great Plains grasslands. *Geocarto International* 11(3):13–22.

Rouse, J.W., R.H. Haas, J.A. Schell, and D.W. Deering. 1974. Monitoring vegetation systems in the Great Plains with ERTS. In *Third Earth Resources Technology Satellite-1 Symposium Proceedings*, NASA SP351, NASA, Greenbelt, MD, pp. 370–371.

Schwingle, B. 2009. Northern Region Forest Insect and Disease Report. Wisconsin Department of Natural Resources, p. 2, http://dnr.wi.gov/forestry/FH/intheNews/2009/NOR_Dec2009.pdf (accessed August 21, 2010).

Seiler, R.A., F. Kogan, and J. Sullivan. 1998. AVHRR-based vegetation and temperature condition indices for drought detection in Argentina. *Advances in Space Research* 21(3):481–484.

Svoboda, M., D. LeComte, M. Hayes, R. Heim, K. Gleason, J. Angel, B. Rippey, R. Tinker, M. Palecki, D. Stooksbury, D. Miskus, and S. Stephens. 2002. The Drought Monitor. *Bulletin of the American Meteorological Society* 83(8):1181–1190.

Swetnam, T.W. and J.L. Betancourt. 1998. Mesoscale disturbance and ecological response to decadal climatic variability in the American Southwest. *Journal of Climate* 11:3128–3147.

Swets, D.L., B.C. Reed, J.R. Rowland, and S.E. Marko. 1999. A weighted least-squares approach to temporal smoothing of NDVI. In *Proceedings of the 1999 ASPRS Annual Conference*, Portland, OR, May 17–21. Bethesda, MD: American Society for Photogrammetry and Remote Sensing.

Townshend, J.R.G., C.O. Justice, and V. Kalb. 1987. Characterization and classification of South American land cover types using satellite data. *International Journal of Remote Sensing* 8:1189–1207.

Tucker, C.J., J.E. Pinzon, M.E. Brown, D. Slayback, E.W. Pak, R. Mahoney, E. Vermote, and N. El Saleous. 2005. An extended AVHRR 8-km NDVI data set compatible with MODIS and SPOT Vegetation NDVI data. *International Journal of Remote Sensing* 26(20):4485–4598.

Tucker, C.J., C.L. Vanpraet, M.J. Sharman, and G. Van Ittersum. 1985. Satellite remote sensing of total herbaceous biomass production in the Senegalese Sahel: 1980–1984. *Remote Sensing of Environment* 17:233–249.

Unganai, L.S. and F.N. Kogan. 1998. Drought monitoring and corn yield estimation in southern Africa from AVHRR data. *Remote Sensing of Environment* 63(3):219–232.

USDA (United States Department of Agriculture). 1994. State Soil Geographic (STATSGO) Data Base: Data Use Information. USDA Miscellaneous Publication 1492:1–113.

USDA (United State Department of Agriculture). 2010a. USDA National Agricultural Statistics Service (NASS)—Quick Stats, http://www.nass.usda.gov/QuickStats/Create_County_All.jsp (accessed June 5, 2010).

USDA (United State Department of Agriculture). 2010b. *Weekly Weather and Crop Bulletin*. NOAA/USDA Joint Agricultural Weather Facility, Washington, DC., http://www.usda.gov/oce/weather/pubs/Weekly/Wwcb/index.htm (accessed June 11, 2011).

Wan, Z., P. Wang, and X. Li. 2004. Using MODIS Land Surface Temperature and Normalized Difference Vegetation Index products for monitoring drought in the southern Great Plains, USA. *International Journal of Remote Sensing* 25:61–72.

Wang, L., J.J. Qu, and X. Hao. 2007. Forest fire detection using the normalized multi-band drought index (NMDI) with satellite measurements. *Agricultural and Forest Meteorology* 148:1767–1776.

Wang, Q., M. Watanabe, S. Hayashi, and S. Murakami. 2003. Using NOAA AVHRR data to assess flood damage in China. *Environmental Monitoring and Assessment* 82(2):118–148.

Wells, N., S. Goddard, and M.J. Hayes. 2004. A self-calibrating Palmer Drought Severity Index. *Journal of Climate* 17(12):2335–2351.

Westerling, A.L., H.G. Hidalgo, D.R. Cayan, and T.W. Swetnam. 2006. Warming and earlier spring increase western U.S. forest wildfire activity. *Science* 313:940–943.

Wilhite, D.A. 2000. Preparing for drought: A methodology. In *Drought: A Global Assessment*, ed. D.A. Wilhite, pp. 89–104. London, U.K.: Routledge.

Williamson, G.B., W.F. Laurance, A.A. Oliveira, P. Delamonica, C. Gascon, T.E. Lovejoy, and L. Pohl. 2000. Amazonian tree mortality during the 1997 El Niño drought. *Conservation Biology* 14:1538–1542.

Yang, L., B. Wylie, L.L. Tieszen, and B.C. Reed. 1998. An analysis of relationships among climatic forcing and time-integrated NDVI of grasslands over the U.S. northern and central Great Plains. *Remote Sensing of Environment* 65:25–37.

4 Vegetation Outlook (VegOut)

Predicting Remote Sensing–Based Seasonal Greenness

Tsegaye Tadesse, Brian D. Wardlow, Mark D. Svoboda, and Michael J. Hayes

CONTENTS

4.1 INTRODUCTION

Accurate and timely prediction of vegetation conditions enhances knowledge-based decision making for drought planning, mitigation, and response. This is very important in countries that are highly dependent on rainfed agriculture. For example, studies show that remote sensing–based observations and vegetation condition prediction have great potential for estimating crop yields (Verdin and Klaver, 2002; Ji and Peters, 2003; Seaquist et al., 2005; Tadesse et al., 2005a, 2008; Funk and Brown, 2006), which in turn may help to address agricultural development and food security issues, as well as improve early warning systems.

Many studies have demonstrated the value of Vegetation Indices (VIs), such as the Normalized Difference Vegetation Index (NDVI), calculated from satellite observations for assessing vegetation cover and conditions (Tucker et al., 1985; Roerink et al., 2003; Anyamba and Tucker, 2005; Seaquist et al., 2005), and such data have become a common source of information for vegetation monitoring. The term *vegetation condition* in this chapter refers to vegetation greenness or vegetation health, as inferred from canopy reflectance values measured by satellite observations (Mennis, 2001; Anyamba and Tucker, 2005). The vegetation greenness metric is commonly calculated from time-series NDVI (Reed et al., 1994) and represents the seasonal, time-integrated NDVI at a specific date, which has been shown to be representative of indicators of general vegetation health including net primary production (NPP) and green biomass (Tucker et al., 1985; Reed et al., 1996; Yang et al., 1998; Eklundh and Olsson, 2003; Hill and Donald, 2003). As a result, VIs and VI derivatives such as time-integrated VI can be used to characterize the temporal and spatial relationships between climate and vegetation and improve our understanding of the lagged relationship between climate (e.g., precipitation and temperature) and vegetation response (Roerink et al., 2003; Anyamba and Tucker, 2005; Seaquist et al., 2005; Camberlin et al., 2007; Groeneveld and Baugh, 2007). Quantitative descriptions of climate-vegetation response lags can then be used to identify and predict vegetation stress during drought.

Predicting vegetation conditions over large geographic areas is imperative for a wide range of applications such as crop and rangeland condition assessments, drought monitoring, fire risk potential, and ecological studies. However, predicting vegetation conditions and understanding the impact of drought on vegetation are challenging because vegetation health is dependent not only on climatic patterns but also on complex relationships involving soil characteristics, land use/land cover (LULC), topography, and other ecological characteristics. Improvements in our predictive capabilities in this area are becoming possible with the increasing availability of many high-quality environmental data sets (e.g., climate, ocean, and remote sensing observations), longer historical records of observations, improved computing capabilities, and the emergence of advanced data analysis techniques useful for data mining.

Several studies have shown significant associations between indices of large-scale oceanic/atmospheric variables and climate over North America (e.g., Panu and Sharma, 2002; Tadesse et al., 2005b; Baigorria et al., 2008; Martinez et al., 2009). For example, Tadesse et al. (2005b) indicated a connection between the occurrences of drought over Nebraska and the Southern Oscillation Index (SOI), Multivariate El Niño Southern Oscillation Index (MEI), Pacific North American Oscillation (PNA), Pacific Decadal Oscillation (PDO), and North Atlantic Oscillation (NAO). The importance of the Atlantic and tropical Pacific sea surface temperature (SST) on past and present drought occurrences across North America is emphasized by Feng et al. (2008) in understanding North American drought variability and predictability. Martinez et al., (2009) showed a strong correlation between the climate indices and oceanic indices that are derived from the Pacific–North American pattern and tropical North Atlantic and eastern tropical Pacific SSTs to predict corn yields in the southeastern United States. Since ocean–atmosphere interactions can drive precipitation patterns affecting vegetation health, a suite of variables should be incorporated into predictive models of vegetation conditions to capture the teleconnected climate–vegetation response linkages.

In addition, the integration of satellite data with climate and oceanic data holds considerable potential for improving our capabilities to predict future vegetation conditions, as demonstrated in the work of Ji and Peters (2003) and Funk and Brown (2006). Using climate (monthly precipitation and relative humidity) and satellite (Advanced Very High Resolution Radiometer (AVHRR) NDVI) data, these studies indicated that NDVI is an effective indicator of vegetation-moisture conditions, but seasonal timing should be taken into consideration when monitoring drought with NDVI (Ji and Peters, 2003). At present, various high-quality climate, ocean, and remote sensing data sets with increasing length of records of more than 20 years are available to provide the historical basis to develop predictive techniques. In addition, the availability of advanced statistical data mining techniques such as regression tree analysis allows these diverse data sets to be effectively integrated into new vegetation-related models such as the Vegetation Drought Response Index (VegDRI) (Brown et al., 2008), upon which similar predictive models could be developed.

The National Drought Mitigation Center (NDMC), in partnership with the USDA Risk Management Agency (RMA), has developed a new drought monitoring tool called the Vegetation Outlook (VegOut) (Tadesse et al., 2005a, 2010). VegOut provides outlooks of general vegetation conditions based on prior climate and ocean index measurements, satellite-based observations of current vegetation conditions, and other environmental information including ecological setting, elevation, soil characteristics, and LULC type. Regression tree modeling was used to analyze historical time-lag relationships between satellite-observed vegetation conditions and oceanic and climatic observations and to develop empirically-based models, which are applied to a suite of "current" observations to predict future vegetation conditions at multiple time steps such as 2, 4, and 6 week outlooks.

In this chapter, the VegOut methodology will be presented in terms of the specific data inputs and predictive modeling approach. Results from two contrasting growing seasons (the 2008 drought and the 2009 nondrought years) over the central United States will be presented to demonstrate the utility and potential of VegOut for vegetation and drought monitoring. Future work to improve the current VegOut method and other possible alternative approaches that have potential use for operational drought monitoring will also be discussed.

4.2 DATA AND METHODS

4.2.1 Study Area

The 15-state region of the central United States (Figure 4.1) provides a geographically diverse study area in terms of land cover types, land use practices, and climate across which to test the capability of VegOut. Land cover in this study area varies from alpine forests along the Rocky Mountains in the west and the forested regions of northern Minnesota to the west–east transition of shortgrass to tallgrass prairie across the Great Plains states (e.g., Kansas, Nebraska, North Dakota, and South Dakota) and the sparsely vegetated shrubland of southern Texas and New Mexico. In addition, many parts of

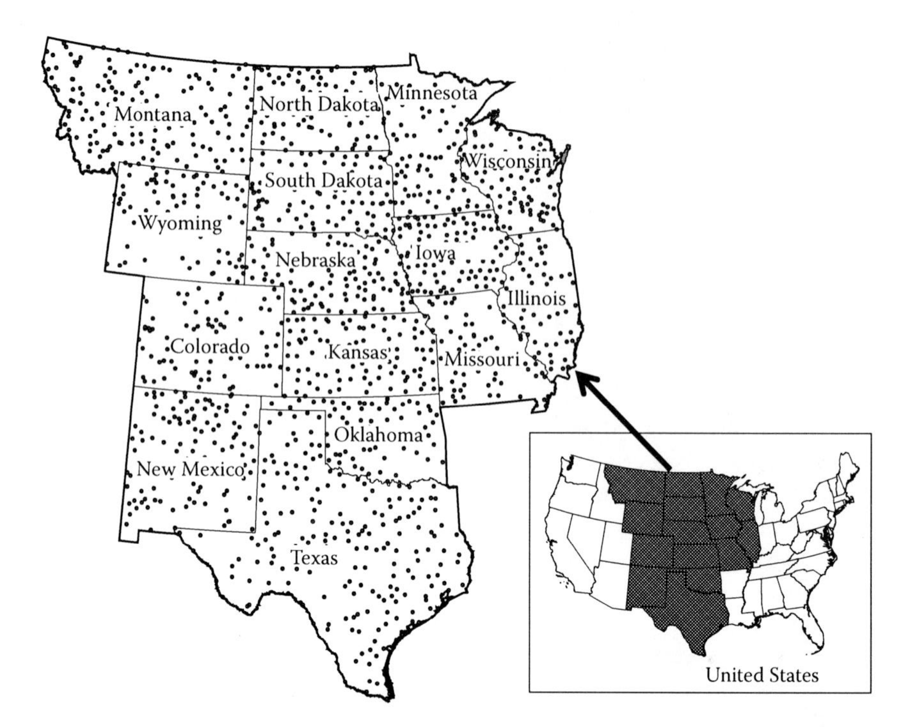

FIGURE 4.1 Central U.S. study area, showing 1420 weather stations providing training and testing data used to develop the VegOut models.

the study area are intensively cultivated, including the corn–soybean–dominated Corn Belt (central Nebraska eastward through Illinois and northward into Minnesota), the Winter Wheat Belt (northern Texas, central Oklahoma, and south-central Kansas), and extensive tracts of irrigated and rainfed cropland stretching the length of the Great Plains from North Dakota to Texas. The study area also has a marked precipitation gradient ranging from 255 to 510mm in the semiarid western locations to more than 1020mm in the east. Growing season length is also highly variable, ranging from ~125 days (mid-May to late September) in the extreme northern part of the study area to more than 250 days (late February to late November) in southern Texas.

4.2.2 General Overview of VegOut

The fundamental basis for developing a predictive vegetation condition tool such as VegOut is building a comprehensive and integrated database of long-term historical records of key observed variables (e.g., climate-based indices, ocean teleconnections, remote sensing–based VI, and other environmental characteristics) that contribute information regarding the complex nature of vegetation growth. These data sets must be readily available over large areas to accurately represent the range of conditions that might be encountered over the spatial modeling domain. In addition, access to these data sets in near real time is essential to the application of VegOut as an operational tool that is capable of generating informational products in a timely manner to support a variety of decision-making activities. Figure 4.2 shows a

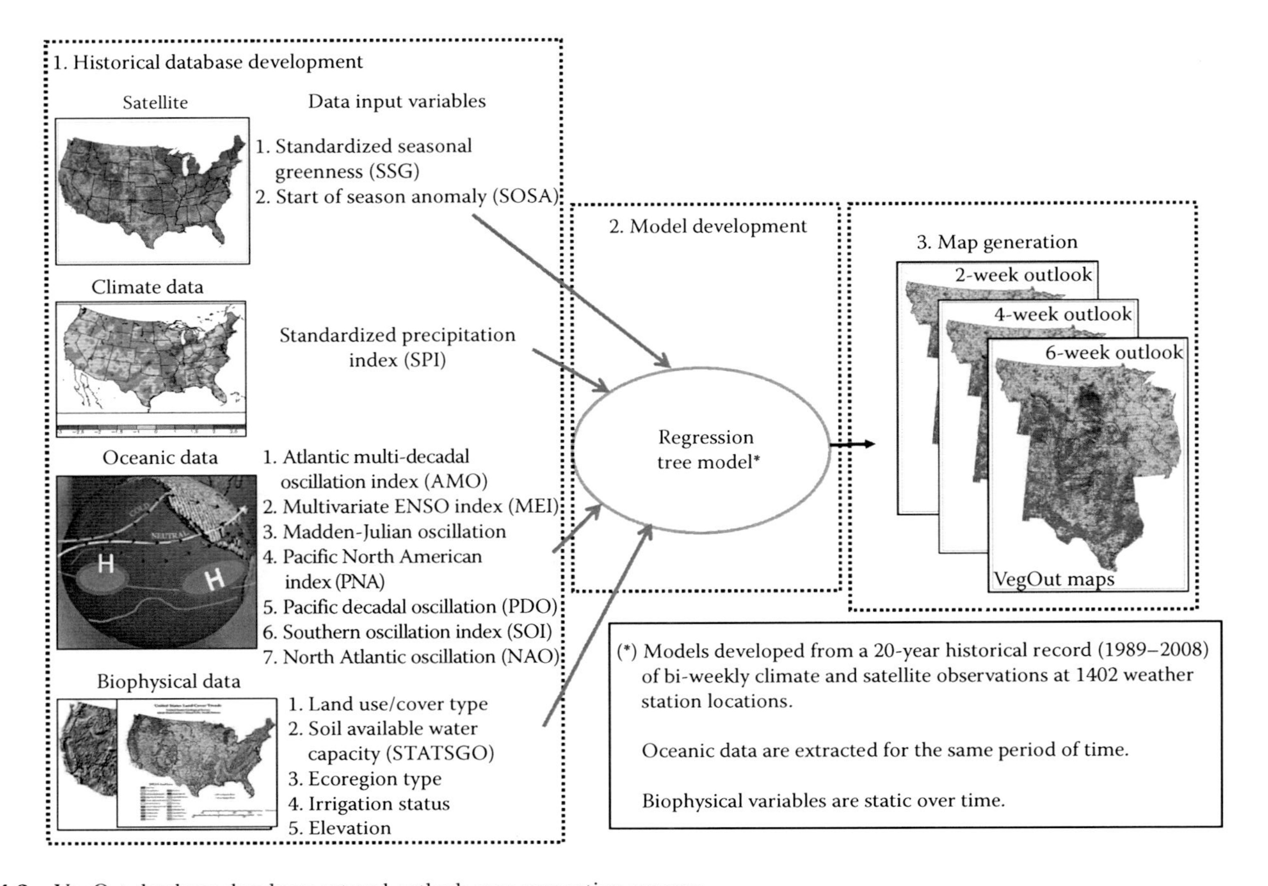

FIGURE 4.2 VegOut database development and outlook map generation process.

graphical overview of the specific variables incorporated into VegOut as well as the general methodology, which includes historical database development, rule generation for the model, and model application in the gridded image domain to produce the predicted seasonal greenness (*SG*) maps. VegOut uses rule-based regression tree models that make predictions of future vegetation conditions based on historical temporal and spatial relationships and patterns identified among satellite-derived VIs, climatic drought indices, oceanic indices, and other environmental variables.

The capability exists to predict *SG* several months into the future using VegOut. However, the predictive accuracy of VegOut decreases linearly as the prediction time interval increases (see Figure 4.3). The predictive accuracy of VegOut has been found to be greater than $R^2 = 0.8$ for model outlook periods of 6 weeks or less. Although testing across a variety of potential predictive outlook intervals is important for understanding model capabilities and limitations, only the results from VegOut models for 2, 4, and 6 week vegetation outlook periods are presented in this chapter to illustrate the potential of this tool for predicting the vegetation *SG*.

SG is defined as the NDVI accumulated through time from the start of the growing season (*SOS*) to a specific date in the growing season, with the accumulation continuing until the end of the growing season (*EOS*). Both the *SOS* and *EOS* are determined at the pixel level over the image domain from satellite-based time-series NDVI data. Reed et al. (1994) and Brown et al. (2008) provide more detailed descriptions of the *SG*, *SOS*, and *EOS* calculations from time-series NDVI data. For each 1 km grid cell and for a given biweekly period, the predicted *SG* patterns produced by the VegOut model are based on the analysis of patterns observed in the historical records of satellite, climate, and oceanic observations over a 20 year period (1989–2008). Also considered in the model are a set of general environmental characteristics that remain static over time but provide a baseline geographic framework to facilitate spatial differentiation of the dynamic patterns. These historical records of patterns provide the basis to forecast future *SG* in the VegOut model. For example, if *SG* patterns are being predicted on July 11 for July 25 (i.e., a 2 week vegetation outlook), then records in the historical database that exhibited similar relationships between the climatic, oceanic, satellite, and environmental variables would be used to predict the *SG* values. In short, a forward/backward time step approach (i.e., forward or backward time-lag relationship) involving the historical record among these variables is used to build the model to predict the *SG*. The specific input variables in the VegOut model (shown in Figure 4.2) are described in the following section.

4.2.3 Historic Database Development

4.2.3.1 Satellite Data

A time series of standardized *SG* (SSG) observations produced from biweekly AVHRR 1 km NDVI data was calculated for each year in a 20 year historical record (1989–2008) to provide vegetation condition information for the VegOut models. First, *SG* is calculated, representing the accumulated NDVI through time from the *SOS* to the last day of each biweek using the following formula:

$$SG = \sum_{p=SOS(=P_1)}^{EOS(=P_n)} \left(NDVI_p - NDVI_b\right) \tag{4.1}$$

where

SG is the seasonal greenness
$P_1, P_2, \ldots, P_n$ refer to individual biweekly periods
$NDVI_p$ is the observed value in the AVHRR composite data
$NDVI_b$ is the latent (or baseline) NDVI value (representative of the nonvegetated background signal) defined at the SOS for each pixel (Reed et al., 1994)

The SSG metric is then calculated for each biweekly time step across each year in the historical record using the following standardization formula:

$$SSG_i = \frac{SG_i - \overline{SG_i}}{\sigma_i} \tag{4.2}$$

where

SG_i is the current SG
$\overline{SG_i}$ is the average SG observed in the historical record up to time period i
σ is the standard deviation of these historical SG_i values (Tadesse et al., 2010)

The result is a 20 year historical time series of SSG images, which have zero-centered values in deviation units (ranging from −4.0 to +4.0) reflecting general vegetation conditions that are spatiotemporally comparable over both space and time because of the standardization process.

The other satellite-derived variable used in VegOut is the SOS anomaly (SOSA), which is the difference between the current year SOS date and the median 20 year historical SOS date. For each year in the historical record, a single SOSA value is calculated at the pixel level across the image domain using a delayed moving average approach developed by Reed et al. (1994). SOSA is used to distinguish areas with low SSG attributable to a shift to a substantially later SOS date, which results in a shorter interval of accumulated NDVI and thus lower SSG. Such shifts often result from human-induced LULC change, and they also frequently occur in areas of low SSG that have an SOS date similar to the historical average but much lower NDVI values over the same period due to some type of environmental stress that can include drought and late-spring freeze.

4.2.3.2 Climate Data

The Standardized Precipitation Index (SPI) (McKee et al., 1995) is used to identify climatic patterns of meteorological dryness at 2 week intervals corresponding to the biweekly periods of the satellite data across the 20 year record. The SPI is based on precipitation data and has the flexibility to detect both short- and long-term precipitation deficits. Because the SPI has the inherent flexibility to be calculated over various time spans, an optimal SPI interval had to be selected for the central United States.

Exhaustive testing of all SPI intervals ranging from 1 to 51 weeks was conducted for each biweekly period across the growing season to determine the specific interval that provided the best predictive accuracy within VegOut. For each SPI interval, a 20 year analysis of the 2, 4, and 6 week VegOut model results incorporating that specific SPI was conducted for each growing season biweekly period across 1420 weather station locations within the 15-state study region (Figure 4.1). The test results showed the 36 week SPI consistently provided the highest VegOut model accuracy across most of the growing season, and it was selected as the SPI input for the VegOut models.

Point-based, tabular SPI data for each weather station location shown in Figure 4.1 were used to develop the empirically-based VegOut models. For model implementation and VegOut map production, the point-based SPI data were spatially interpolated using an inverse distance weighting technique to create 1 km gridded SPI images.

4.2.3.3 Oceanic/Atmospheric Index Data

As stated earlier in this chapter, several studies have shown a teleconnective link between the oceanic and climate indices (Asner et al., 2000; Los et al., 2001; Barnston et al., 2005; Tadesse et al., 2005b; Baigorria et al., 2008; Martinez et al., 2009). Understanding these relationships and using oceanic/atmospheric data help improve vegetation monitoring and prediction by incorporating various complex parameters that influence vegetation health. Schubert et al. (2007) stated that modeling work for drought prediction has largely attributed the major North American droughts of the last 150 years to global circulation anomalies that were forced by tropical SST. Based on the correlation coefficient values, however, it was observed that not all oceanic indices had a strong relationship with climate and vegetation response over the central United States (Tadesse et al., 2009).

Seven of the most commonly used oceanic/atmospheric indices were selected for integration into the VegOut predictions of SSG to account for the temporal and spatial relationships between ocean–atmosphere dynamics and climate–vegetation interactions (i.e., teleconnection patterns) that have been observed over the Central United States. These indices include the Atlantic Multidecadal Oscillation (AMO), MEI, Madden–Julian Oscillation (MJO), PNA, PDO, SOI, and NAO. Data for each of these oceanic indices are freely available online from different sources (Tadesse et al., 2009). For each oceanic index, a single value is reported in a tabular format for each time interval across the historical record, which can vary from bimonthly to monthly updates. The historical, tabular oceanic index data adapted to the biweekly time step were used to develop VegOut models. For the mapping portion of VegOut, the single oceanic index value for a specific biweek was gridded as a constant value over the 15-state study area to produce the series of 1 km oceanic index raster images.

4.2.3.4 Environmental Data

A set of five general environmental variables that describe aspects of the environment that influence climate–vegetation interactions were incorporated into the VegOut model. These variables include LULC type, soil available water holding capacity (AWC), ecosystem type (Eco), percent of irrigated land (Percent_Irrig), and elevation (Elev). The LULC input was derived from the 2001 National Land Cover

Dataset (NLCD, Vogelmann et al., 2000) and is essential because the climate–vegetation *SG* of different land cover types such as crops, forest, and grassland (DeBeurs and Henebry, 2004) may also have different response in vegetation condition. Soil AWC, which was derived from the USDA STATSGO data set (USDA, 1994), is also a critical parameter because it defines available moisture for plant growth, and variations in soil AWC can result in different responses from the same land cover type under similar climatic conditions. The percent irrigated agriculture variable derived from the Moderate Resolution Imaging Spectroradiometer (MODIS) Irrigated Agriculture Dataset (MIrAD) (Brown et al., 2009) was incorporated to stratify the landscape into areas of rainfed vegetation that are more sensitive to climate variations such as drought versus those irrigated areas that are less affected because of targeted water applications. Elevation derived from the U.S. Geological Survey (USGS) 30 m digital elevation model (DEM) was also included to account for the different altitudinal climate regions across which a land cover type may be found (e.g., alpine versus coastal evergreen forest). An ecoregion input (Omernik, 1987) was included to account for regional differences in the collective environmental setting (e.g., climate, topography, soils, and vegetation types) that a specific land cover type (e.g., grassland) might be located (e.g., mountains versus plains). Each of these environmental variables was held constant over the 20 year study period because consistent, seamless data sets reflecting changes in these variables over time are not available for the study area.

Zonal calculations within a 3-by-3 km square window (snapped to the 1 km AVHRR pixel grid) surrounding each weather station were used to extract data from the gridded climate, environmental, and satellite variables for inclusion in the historical training database used to develop the VegOut models. For continuous variables such as percent irrigated agriculture, the mean value within each station window was calculated, and for categorical variables such as land cover type, the majority class within the window was used. Once the data were extracted for all station locations, they were merged with the tabular SPI and oceanic index data in a database used to train the VegOut models.

4.2.4 VegOut Model Development and Implementation

For VegOut model development, the dynamic climate, oceanic, and satellite variables in the training database were organized into a continuous time series of biweekly observations across the 20 year historical record, while the environmental variables were assumed static over this period. Biweekly VegOut models were developed from historical data extracted for 1420 weather station locations across the study area (Figure 4.1) using commercial classification and regression tree (CART) modeling software called Cubist (Quinlan, 1993; Rulequest, 2010). These models serve to identify historical relationships over the 20 year record at each training site between observed vegetation *SG*, climate, and oceanic conditions for a specific biweekly period and the corresponding vegetation *SG* that occurred after that date at some future biweekly time period (e.g., 2 weeks). Individual models for the 2, 4, and 6 week vegetation outlooks were generated for each biweekly period across the May–October growing season.

4.2.5 Model Development

The algorithm underlying the VegOut model to predict SSG is based on a series of multiple linear regression equations defined by the CART-based Cubist software through the analysis of the historical data discussed in the previous section. The model calculates the SSG value for future biweekly period $t = i$ (e.g., $t = 2$ weeks into the future) by applying a set of linear regression equations associated with historical periods in the database that exhibited similar patterns (or behavior) among the set of independent variables. To calculate the predicted value, the regression equation(s) is applied using the conditions of the current week ($t = 0$). The following is the general form of the linear regression equation defined by Cubist that is applied to calculate the SSG for a future biweekly time period $t = i$:

$$\begin{aligned}\text{VegOut}(i) = {} & f_{1,i}(\text{SSG}, \text{SOSA})_{t=0} + f_{2,i}(\text{SPI})_{t=0} \\ & + f_{3,i}(\text{LULC}, \text{Eco}, \text{Percent\_Irrig}, \text{AWC})_{t=0} \\ & + f_{4,i}(\text{MEI}, \text{MJO}, \text{NAO}, \text{PDO}, \text{SOI}, \text{AMO}, \text{PNA})_{t=0} \end{aligned} \quad (4.3)$$

where VegOut(i) is the predicted SSG at future biweekly time period i as a function of the current ($t = 0$) values of the input variables. The equation shows that the VegOut is defined as four functions (f_1, f_2, f_3, and f_4) of the current (i.e., the date on which the *SG* prediction is made) climate, environmental, and satellite variables and the values of the oceanic indices, respectively.

4.2.6 VegOut Map Generation

For VegOut map production, the regression tree rules in the VegOut model for a specific biweekly period in the growing season are applied to the gridded image input data (as shown in Figure 4.2) for the corresponding biweekly period in a given year (e.g., June 10, 2010) using MapCubist software developed at the USGS Center for Earth Resources Observation and Science (EROS). The capability exists to apply the model in near real time to current observational inputs to produce an up-to-date VegOut map or to apply it retrospectively to generate a map for that biweekly period for any year in the historical record. During model implementation, the values of all input variables for that specific period at each pixel are considered to identify which rule(s) in the VegOut rule set should be used, which in turn determines the linear regression equation(s) that will be applied to input data values to calculate a VegOut SSG value for each pixel across the study area. In many instances, multiple rules may apply to each pixel, resulting in multiple linear regression equations being applied, and the average value across all regression calculations is used as the predicted SSG. Operationally, the period-specific VegOut models can be sequentially applied for each biweekly period across the year to generate a complete time series of 2, 4, and 6 week VegOut maps for the growing season.

4.3 RESULTS AND DISCUSSION

4.3.1 VegOut Predictive Accuracy across the Growing Season

Figure 4.3 shows the average correlation between predicted and observed SSG across all periods of the growing season for 20 years (1989–2008). Examination of the predictive accuracy of the VegOut model across the growing season shows that the model's accuracy decreases linearly as the forecast interval increases. Based on these analyses, VegOut predictions presented in this chapter are limited to forecast intervals associated with historical R^2 values of 0.8 or higher (which were observed for 2, 4, and 6 week forecasts) to illustrate the potential of this new predictive approach.

Individual 2, 4, and 6 week VegOut forecasts for each biweekly period across the growing season (Figure 4.4) were constructed to assess their accuracy across the year as vegetation progresses through its various phenological stages. The results of this evaluation showed that the lowest predictive accuracy (R^2 = 0.7–0.8) occurred in the early spring (April and early May) for all three outlooks. By late May, the accuracy of the outlooks exceeded an R^2 value of 0.8 and was relatively stable for the remainder of the growing season. The lower R^2 values during the spring phase may be due to low green biomass associated with early stages of vegetation green-up, resulting in greater fluctuation of the SSG values during this part of the year that is magnified by early season interannual temperature variations (e.g., late spring freeze) and land management decisions (e.g., crop planting times). The relatively high and stable predictive accuracy of VegOut throughout the late spring and summer is encouraging for

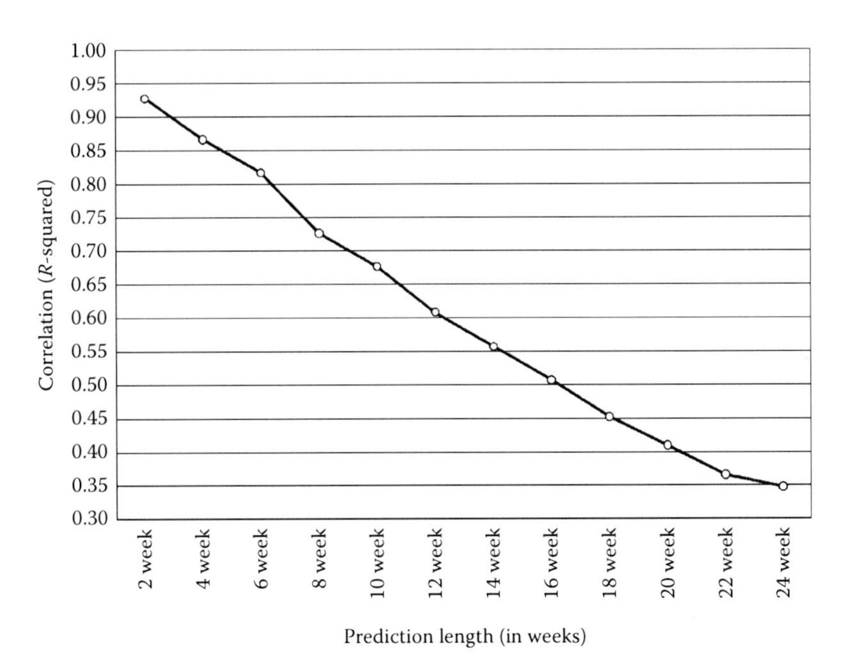

FIGURE 4.3 Twenty-year (1989–2008) average R^2 between the observed and predicted SSG values across the May–October growing season for outlook periods ranging from 2 to 24 weeks.

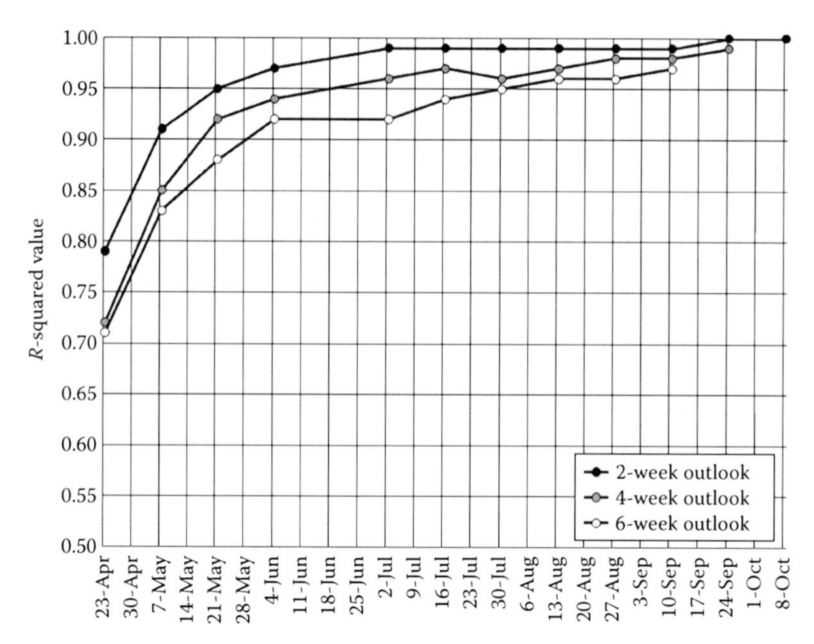

FIGURE 4.4 Twenty-year (1989–2008) average R^2 between the observed and predicted SSG values for the 2-, 4-, and 6-week outlooks for each biweekly period during the growing season.

drought monitoring because this is an important period that determines crop yields and grassland production. The ability of VegOut to provide outlooks of vegetation SSG with reasonable accuracy over these critical months could provide new insights into the early-stage identification of emerging agricultural drought conditions.

The predictive accuracy across the growing season was consistently highest for the 2-week outlook, with the R^2 values slightly declining as the outlook interval increased. This is expected because uncertainty in future SSG values will generally increase with longer prediction intervals because of the increasing uncertainty of future states of the complex land–atmosphere system being modeled (Cushman-Roisin and Beckers, 2008).

4.3.2 Spatial Pattern Assessment for Drought and Nondrought Years

VegOut maps showing 2, 4, and 6 week outlooks generated for a midsummer biweekly period during a drought year (2008) and nondrought year (2009) over the study area are examined to demonstrate the capabilities of VegOut to predict SSG patterns under contrasting climatic conditions. During these 2 years, with the exception of southern Texas, a large portion of the 15-state study area experienced drought conditions in 2008 and nondrought conditions in 2009, as shown by the U.S. Drought Monitor (USDM) maps in Figure 4.5a and b. In 2008, for example, large areas of extreme drought (D4 classification in the USDM) over western North Dakota and moderate

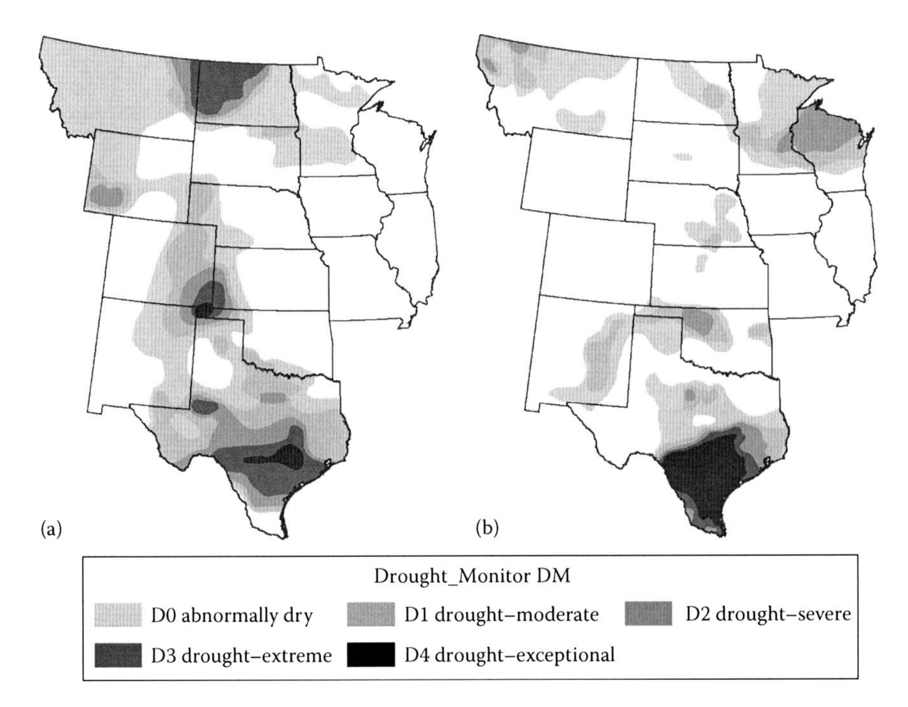

FIGURE 4.5 **(See color insert.)** USDM maps over the study area for (a) July 29, 2008 and (b) July 28, 2009.

to extreme drought (D2–D4 designation in the USDM) from western Nebraska southward through eastern Colorado, western Kansas, northwest Oklahoma, and parts of northern Texas were observed in the USDM (Figure 4.5a). However, in 2009, the conditions had improved to a nondrought or abnormally dry (D0) classification over most these areas (Figure 4.5b). In contrast, eastern Minnesota, northern Wisconsin, northwest Montana, and parts of eastern New Mexico and central Oklahoma were drier in 2009 than 2008.

Maps of the predicted SSG values for the 2, 4, and 6 week outlooks as forecast on June 30, 2008, and June 29, 2009, are shown in Figures 4.6b through d and 4.7b through d, respectively. The initial SSG conditions observed from AVHRR NDVI image data on forecast submission dates in 2008 and 2009 are presented in Figures 4.6a and 4.7a, respectively. Figures 4.6e through g and 4.7e through g show the observed SSG patterns from AVHRR NDVI on the targeted dates of the three vegetation outlooks in 2008 and 2009. The broadscale spatial patterns of SSG depicted in the 2, 4, and 6 week VegOut forecasts produced across both summer seasons were in general agreement with observed SSG patterns across the 15-state area in each corresponding period. Generally, the most substantial differences between the predicted and observed SSG were limited to small, localized areas in both years (Figures 4.6h through j and 4.7h through j). In an effort to highlight major differences between the predicted and observed SSG patterns in the difference maps, ±1 standard deviation thresholds were used to indicate pixels with excessive error. In 2008, there was good spatial agreement between the SSG patterns predicted in the three outlook maps

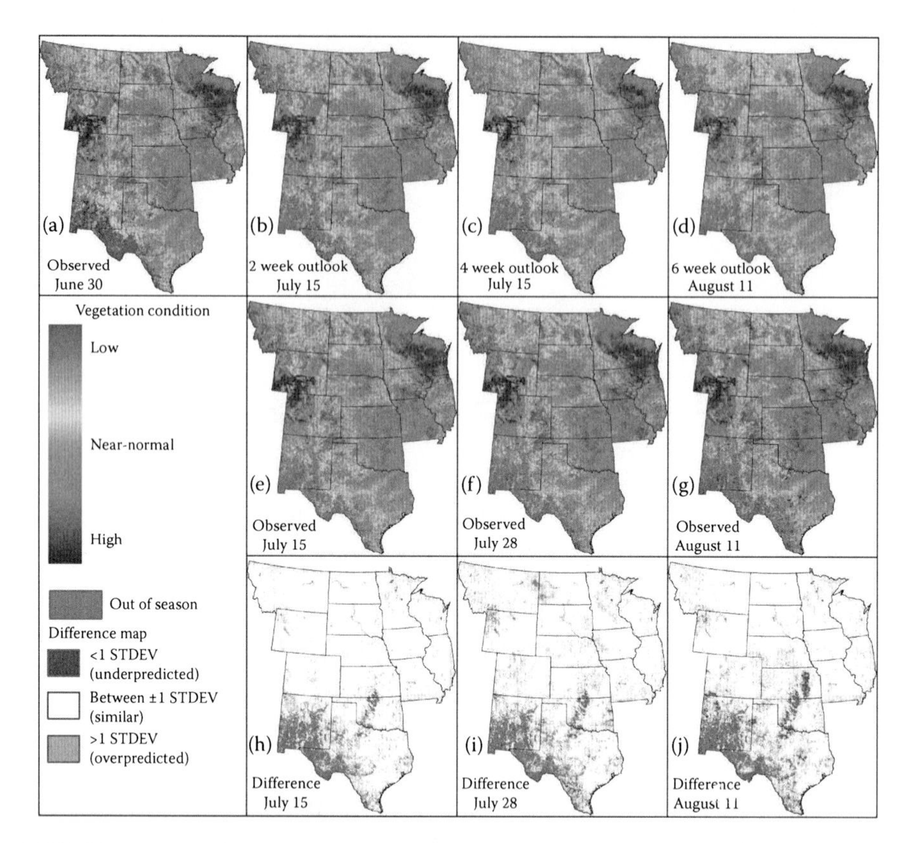

FIGURE 4.6 (See color insert.) (a) Observed SSG for June 30, 2008; (b), (c), and (d) are 2, 4, and 6 week outlooks; (e), (f), and (g) are observed SSG for July 15, July 28, and August 11 that correspond to the 2, 4, 6 week outlooks, respectively; and (h), (i), and (j) show the difference between the predicted and observed greenness for the corresponding 2, 4, and 6 week outlooks, respectively.

and those observed from satellite over the drought-impacted areas. For example, the spatial extent and evolution of the lower SSG values observed by satellite over North Dakota and the High Plains (i.e., eastern Colorado and western Kansas) (Figure 4.6e through g) were consistent with those predicted by VegOut for the July and August dates (Figure 4.6b through d). In 2009, the spatial extent and magnitude of the SSG values were comparable between the VegOut results for the three predictive periods and the satellite observations over the drought-impacted areas in northern Wisconsin and eastern New Mexico. These results suggest that the information presented in the series of vegetation outlooks could be used as an early indicator of the impact of drought conditions on vegetation in the near future.

In addition, the major high SSG landscape features observed for these 2008 dates in Wisconsin and southwest Wyoming were also depicted in the series of VegOut maps. The most notable difference during the 2008 drought year was the slight underestimation of SSG values over some locations with either extremely high (Wisconsin) or low (south-central North Dakota) SSG values, particularly in the

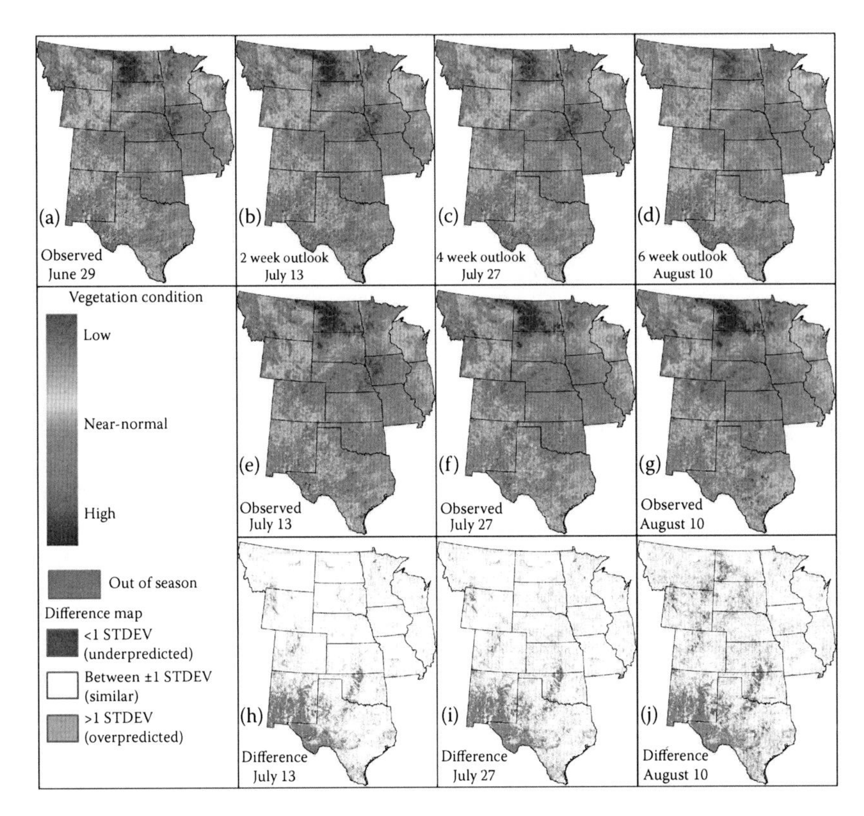

FIGURE 4.7 **(See color insert.)** (a) Observed SSG for June 29, 2009; (b), (c), and (d) are 2, 4, and 6 week outlooks; (e), (f), and (g) are observed SSG for July 13, July 27, and August 10 that correspond to the 2, 4, 6 week outlooks, respectively; and (h), (i), and (j) show the difference between the predicted and observed greenness for the corresponding 2, 4, and 6 week outlooks, respectively.

longer 6 week outlook maps (Figure 4.6d and g). In 2009, strong spatial agreement was found between the predicted and observed SSG values across the 15 states for all three outlook periods. The main exceptions were the underestimation of some high SSG values observed over western North Dakota and intermediate SSG values over Wyoming and south-central Montana in the 4 and 6 week VegOut maps.

In general, the performance of VegOut was fairly robust over most of the central United States under both drought and nondrought conditions. The exception was over sparsely vegetated areas of New Mexico and southwest Texas, where persistent differences were found in the form of both under- and over-predicted SSG values for the longer 4 and 6 week outlooks in 2008 and 2009. This discrepancy could be due to the dynamic range of SSG values (i.e., minimum to maximum SSG value range over the year), which is quite low in these sparsely vegetated landscapes; thus, a minimal difference between the predicted and observed SSG values will often exceed the one standard deviation threshold used to detect differences. However, in reality, those values may be very similar in comparison with the complete SSG value range for the entire study area.

A visual comparison of the series of difference maps showed that the majority of the study area did not have a notable difference between the observed and predicted SSG across all three outlook intervals for both study years. The majority of differences identified in the series of difference maps were distributed as small pockets for varying locations across the study area in both years. However, a closer inspection of these differences revealed several notable patterns and trends. First, the majority of marked differences flagged were associated with the underprediction of SSG values by VegOut. Second, the areal extent of these differences slightly increased as the outlook period lengthened. This would be expected given the decline in predictive accuracy of the VegOut models observed from the 2 to 6 week interval shown in Figure 4.3. Third, there was tendency by VegOut in some locations with extreme SSG values to underpredict the highest values and overpredict the lowest values. For example, the extremely low SSG values observed for western North Dakota and eastern Montana on July 28, 2008, (Figure 4.6f) were substantially underpredicted in the 4 week outlook (see the green areas in Figure 4.6i). Another example for high SSG values occurred over western North Dakota on August 10, 2009, where the high observed SSG values (dark blue area in Figure 4.7g) were substantially underestimated in the 6 week outlook (red area in Figure 4.7j). In general, the prediction of extreme values is challenging for any type of empirically based forecasting because there may not be representative events in the historical record used to develop the predictive models. However, these results show that the VegOut predictions do not contain a consistent bias to underpredict or overpredict extreme SSG values. Overall, the VegOut was found to predict comparable SSG values to those observed from satellite over the majority of the central United States, with the most substantial differences isolated to small geographic areas that had little impact on the overall SSG patterns depicted in the VegOut maps.

4.4 FUTURE DIRECTIONS

In an effort to enhance and extend VegOut as a predictive tool for mapping future vegetation conditions, several research activities are currently underway or planned in the near future. These include (1) expanded testing of VegOut over both the western and eastern United States and other regions of the world; (2) continued testing of longer outlook periods ranging from 3 to 6 months; (3) incorporating new variables such as remote sensing–based evapotranspiration, soil moisture, and land surface temperature, as well as refined sets of climate and oceanic indicators; and (4) testing and transitioning to VI data collected from new satellite sensors such as the MODIS and the Visible/Infrared Imager/Radiometer Suite (VIIRS).

A "scenario-based" VegOut modeling approach called Scenario-VegOut is also being developed to complement the "diagnostic-based" VegOut approach presented in this chapter. Scenario-VegOut is designed to predict *SG* for different climatic episodes (i.e., dry, normal, and wet conditions) using the same regression tree–based modeling and input variables as the diagnostic VegOut model. This approach provides users the flexibility to project future vegetation *SG* under these different precipitation scenarios during defined outlook periods (e.g., 2 week interval of the 2 week outlook). In this approach, Scenario-VegOut predictions are calculated for three possible scenarios over 2, 4, and 6 week outlook periods that represent below-normal, near-normal, and above-normal precipitation conditions. Scenario-VegOut will predict *SG* based on scenarios over each outlook period that represent dry conditions (e.g., 0%–50% of average precipitation), near normal conditions (e.g., 75%–125% of normal precipitation), and wet conditions (e.g., more than 150% of average precipitation). For each scenario, the SPI grids will be generated from historical, station-based precipitation values for each of these targeted precipitation percentages. This approach has the flexibility to base outlooks on different percentages if desired.

4.5 SUMMARY AND CONCLUSION

Because of the varied and potentially costly losses caused by drought events, better tools for monitoring and predicting general vegetation conditions are needed to more effectively deal with this natural hazard. VegOut attempts to fill this need by predicting vegetation *SG* patterns based on analysis of satellite, climate, and oceanic data sets and other general environmental variables using an advanced data mining technique. VegOut capitalizes on historical climate–vegetation interactions and teleconnections between the ocean and climate (such as El Niño and Southern Oscillation [ENSO]) to generate these outlooks, while considering several static environmental characteristics such as LULC type, irrigation status, soil characteristics, and ecological setting, which can influence vegetation's response to weather conditions. The goal of VegOut is to provide timely information about future vegetation conditions across large geographic areas, which can be used by drought experts to identify the early stages of vegetation drought stress and gain insight into the possible near-term trends in vegetation conditions. In addition to drought monitoring, VegOut information could also be used by agricultural producers, natural resource managers, and policy makers to make more informed decisions at local to regional scales.

The evaluation of the spatiotemporal performance of VegOut presented in this chapter across the 2008 and 2009 growing seasons found the models to have high predictive accuracy ($R^2 > 0.8$) for the central United States and predicted SSG patterns in 2, 4, and 6 week outlook maps to have strong spatial agreement with observed SSG patterns. The comparisons of the predicted and observed SSG patterns of the VegOut maps of the 2008 and 2009 summer seasons in this study showed that major differences between the predicted and observed SSG values (both underprediction and overprediction) occurred primarily at a local scale over sparsely vegetated areas. This discrepancy occurs because the narrower, possibly more tail-heavy dynamic ranges of *SG* values that frequently characterize sparsely vegetated areas reflect the more rapid changes in actual *SG* values for these regions (i.e., these regions are more sensitive to short-term climate fluctuations), which leads to increased predictive uncertainty in the models and thus more frequent, larger model "misses." Some disagreement was also found for some locations exhibiting both extreme high and low SSG values, but was restricted to relatively isolated locations within the longer outlook periods. Although the examples shown in this chapter illustrate the potential value of VegOut for predicting large-area vegetation conditions, additional validation work is needed to fully understand this new predictive tool's performance. This study was restricted to a limited number of midsummer dates, and similar evaluations should be performed across the entire growing season and across the entire 20+ year historical record of observations that are available to generate the VegOut maps. In addition, further assessment of VegOut results under varying levels of drought severity should be carried out to determine its ability to characterize both rapid and slow onset drought stress events.

Because VegOut maps and products integrate climate, satellite, and oceanic data as well as incorporate the environmental characteristics of the local areas to predict the vegetation condition with a reasonable accuracy, they can be used by agricultural producers, extension agents, early warning institutes, policy makers, and other stakeholders to make more informed decisions at the local levels.

REFERENCES

Anyamba, A. and C.J. Tucker. 2005. Analysis of Sahelian vegetation dynamics using NOAA-AVHRR NDVI data from 1981–2003. *Journal of Arid Environments* 3(3):596–614.

Asner, G.P., A.R. Townsend, and B.H. Braswell. 2000. Satellite observation of El Niño effects on Amazon forest phenology and productivity. *Geophysical Research Letters* 27(7):981–984.

Baigorria, G.A., J.W. Hansen, N. Ward, J.W. Jones, and J.J. O'Brien. 2008. Assessing predictability of cotton yields in the southeastern USA based on atmospheric circulation and surface temperatures. *Journal of Applied Meteorology and Climatology* 47:76–91.

Barnston, A.G., A. Kumar, L. Goddard, and M.P. Hoerling. 2005. Improving seasonal prediction practices through attribution of climate variability. *Bulletin of the American Meteorological Society* 86:59–72.

Brown, J.F., S. Maxwell, and M.S. Pervez. 2009. Mapping irrigated lands across the United States using MODIS satellite imagery. In *Remote Sensing of Global Croplands for Food Security*, eds. P.S. Thenkabail, J.G. Lyon, H. Turral, and C.M. Biradar, pp. 177–198. Boca Raton, FL: Taylor & Francis.

Brown, J.F., B.D. Wardlow, T. Tadesse, M.J. Hayes, and B.C. Reed. 2008. The vegetation drought response index (VegDRI): A new integrated approach for monitoring drought stress in vegetation. *GIScience and Remote Sensing* 45(1):16–46.

Camberlin, P., N. Martiny, N. Philippon, and Y. Richard. 2007. Determinants of the interannual relationships between remote sensed photosynthetic activity and rainfall in tropical Africa. *Remote Sensing of Environment* 106(2):199–216.

Cushman-Roisin, B. and J-M. Beckers. 2008. *Introduction to Geophysical Fluid Dynamics: Physical and Numerical Aspects* 2nd edn. Amsterdam, the Netherlands: Academic Press, Elsevier.

DeBeurs, K.M. and G.M. Henebry, 2004. Land surface phenology, climatic variation, and institutional change: Analyzing agricultural land cover change in Kazakhstan. *Remote Sensing of Environment* 89(4):497–509.

Eklundh, L. and L. Olsson, 2003. Vegetation index trends for the African Sahel 1982–1999. *Geophysical Research Letters* 30:131–134.

Feng, S., R.J. Oglesby, C.M. Rowe, D.B. Loope, and Q. Hu. 2008. Atlantic and Pacific SST influences on Medieval drought in North America simulated by the Community Atmospheric Model. *Journal of Geophysical Research* 113:D11101, doi:10.1029/2007JD009347.

Funk, C.C. and M.E. Brown. 2006. Intra-seasonal NDVI change projections in semi-arid Africa. *Remote Sensing of Environment* 101:249–256.

Groeneveld, D.P. and W.M. Baugh. 2007. Correcting satellite data to detect vegetation signal for eco-hydrologic analyses. *Journal of Hydrology* 344:135–145.

Hill, M.J. and G.E. Donald. 2003. Estimating spatio-temporal patterns of agricultural productivity in fragmented landscapes using AVHRR NDVI time series. *Remote Sensing of Environment* 84:367–384.

Ji, L. and A.J. Peters. 2003. Assessing vegetation response to drought in the northern Great Plains using vegetation and drought indices. *Remote Sensing of Environment* 87:85–98.

Los, S.O., G.J. Collatz, L. Bounoua, P.J. Sellers, and C.J. Tucker. 2001. Global interannual variations in sea surface temperature and land surface vegetation, air temperature, and precipitation. *Journal of Climate* 14(7):1535–1549.

Martinez, C.J, G.A. Baigorria, and J.W. Jones. 2009. Use of indices of climate variability to predict corn yields in the Southeast USA. *International Journal of Climatology* 29:1680–1691.

McKee, T.B., N.J. Doesken, and J. Kleist. 1995. Drought monitoring with multiple time scales. *Preprints, 9th Conference on Applied Climatology*, Dallas, TX, pp. 233–236. Boston, MA: American Meteorological Society.

Mennis, J. 2001. Exploring relationships between ENSO and vegetation vigour in the south-east USA using AVHRR data. *International Journal of Remote Sensing* 22(16):3077–3092.

Omernik, J.M. 1987. Ecoregions of the conterminous United States. *Annals of the Association of American Geographers* 77(1):118–125.

Panu, U.S. and T.C. Sharma. 2002. Challenges in drought research: Some perspectives and future directions. *Hydrological Sciences Journal* 47(S):19–29.

Quinlan, J.R. 1993. *C4.5 Programs for Machine Learning*. San Mateo, CA: Morgan Kaufmann Publishers.

Reed, B.C., J.F. Brown, D. VanderZee, T.R. Loveland, J.W. Merchant, and D.O. Ohlen. 1994. Measuring phenological variability from satellite imagery. *Journal of Vegetation Science* 5:703–714.

Reed, B.C., T.R. Loveland, and L.L. Tieszen. 1996. An approach for using AVHRR data to monitor U.S. Great Plains grasslands. *Geocarto International* 11(3):13–22.

Roerink, G.J., M. Menenti, W. Soepboer, and Z. Su. 2003. Assessment of climate impact on vegetation dynamics by using remote sensing. *Physics and Chemistry of the Earth* 28:103–109.

Rulequest. 2010. An overview of Cubist. http://www.rulequest.com/cubist-win.html (accessed on December 12, 2011).

Schubert, S., R. Koster, M. Hoerling, R. Seager, D. Lettenmaier, A. Kumar, and D. Gutzler. 2007. Predicting drought on seasonal-to-decadal time scales. *Bulletin of the American Meteorological Society* 88:1625–1630.

Seaquist, J.W., L.L. Olsson, and J. Ardö. 2005. A remote sensing-based primary production model for grassland biomes. *Ecological Modeling* 169(1):131–155.

Tadesse, T., J.F. Brown, and M.J. Hayes. 2005a. A new approach for predicting drought-related vegetation stress: Integrating satellite, climate, and biophysical data over the U.S. central plains. *ISPRS Journal of Photogrammetry and Remote Sensing* 59(4):244–253.

Tadesse, T., M. Haile, G. Senay, C. Knutson, and B.D. Wardlow. 2008. Building integrated drought monitoring and food security systems in sub-Saharan Africa. *Natural Resources Forum* 32(4):245–279.

Tadesse, T., B.D. Wardlow, M.J. Hayes, M.D. Svoboda, and J.F. Brown. 2010. The vegetation condition outlook (VegOut): A new method for predicting vegetation seasonal greenness. *GIScience and Remote Sensing* 47(1):25–52.

Tadesse, T., B.D. Wardlow, and J.H. Ryu. 2009. Discovering the spatial and temporal relationships between vegetation condition and climate in monitoring drought. *Proceedings of the 2009 Climate Prediction Applications Science Workshop*, Norman, OK, March 23–25.

Tadesse, T., D.A. Wilhite, M.J. Hayes, S.K. Harms, and S. Goddard. 2005b. Discovering associations between climatic and oceanic parameters to monitor drought in Nebraska using data-mining techniques. *Journal of Climate* 18(10):1541–1550.

Tucker, C.J., J.R.G. Townshend, and T.E. Goff. 1985. African land cover classification using satellite data. *Science* 9227(4685):369–375.

USDA (U.S. Department of Agriculture), Natural Resource Conservation Service. 1994. State Soil Geographic (STATSGO) Data Base: Data Use Information, Washington, DC: U.S. Soil Conservation Service. *Miscellaneous Publication Number* 1492.

Verdin, J. and R. Klaver. 2002. Grid cell based crop water accounting for the famine early warning system. *Hydrological Processes* 16:1617–1630.

Vogelmann, J.E., S.M. Howard, L. Yang, C.R. Larson, B.K. Wylie, and N. Van Driel. 2000. Completion of the 1990s National Land Cover Dataset for the coterminous United States from Landsat Thematic Mapper data and ancillary data sources. *Photogrammetric Engineering & Remote Sensing* 67(6):650–661.

Yang, L., B.K. Wylie, L.L. Tieszen, and B.C. Reed. 1998. An analysis of relationships among climate forcing and time-integrated NDVI of grasslands over the U.S. northern and central Great Plains. *Remote Sensing of the Environment* 65:25–37.

5 Drought Monitoring Using Fraction of Absorbed Photosynthetically Active Radiation Estimates Derived from MERIS

Simone Rossi and Stefan Niemeyer

CONTENTS

5.1 INTRODUCTION

Precipitation and net radiation are fundamental parameters controlling plant transpiration, a key process in vegetation development. Either of these two meteorological parameters can be a limiting factor for vegetation growth. In humid areas such as northwest Europe, precipitation is abundant or even excessive, and the primary limiting factor for vegetation development in these "energy-limited" environments is net radiation, which can be reduced by seasonality (variations in sun angle and day length) and cloud cover. In contrast, in semiarid areas such as the Mediterranean region where net radiation is usually abundant, precipitation is the limiting factor for vegetation growth. As a result, the sensitivity of vegetation to precipitation variations is higher in these "moisture-limited" environments. In general, precipitation is an important limiting factor for vegetation development over most of Europe, with the exclusion of very humid areas (Roerink et al., 2003). Negative precipitation anomalies and subsequent soil moisture deficits have a direct impact on the state of vegetation, causing plant water stress and wilting. These deficits have various effects on vegetation growth and vigor, depending on the vegetation growth phase in which they occur. Monitoring vegetation status is therefore important from a drought monitoring perspective, since vegetation stress provides information about the severity of a drought event and its impact on vegetation and food security.

Remote sensing is widely used for monitoring vegetation conditions because it facilitates retrieval of consistent and spatially distributed information over large areas at regular time intervals. Several approaches have been proposed to monitor vegetation water stress related to droughts (Jeyaseelan, 2004), many of which characterize the state of vegetation using empirical-based spectral vegetation indices (VIs). Time series of the most commonly used VI, the Normalized Difference Vegetation Index (NDVI) (Rouse et al., 1974), have been extensively used for drought monitoring (e.g., Ji and Peters, 2003; Song et al., 2004; Gonzalez Loyarte and Menenti, 2008). The NDVI is responsive to changes in chlorophyll content and the internal spongy mesophyll layers of the plant's leaves through the integration of the visible red and near infrared (NIR) spectral bands into the index. NDVI data from the National Oceanic and Atmospheric Administration's (NOAA) Advanced Very High Resolution Radiometer (AVHRR) are also included as one input parameter among multisource drought indices such as the Vegetation Health Index (VHI) (Kogan, 1995) and the Vegetation Drought Response Index (VegDRI) (Brown et al., 2008). Spectral VIs focusing on leaf water content, such as the Normalized Difference Water Index (NDWI) (Gao, 1996) and the Multi-Band Drought Index (MBDI) (Wang and Qu, 2007), have also been used for drought monitoring (Gu et al., 2008). Other spectral VIs have been proposed specifically to improve drought detection, such as the Perpendicular Drought Index (PDI) (Ghulam et al., 2006), based on the spatial distribution of moisture at different fractions of vegetation cover in the NIR-red two-dimensional spectral space. Remote sensing–based approaches have also been proposed to estimate evapotranspiration (e.g., Su et al., 2003) within energy balance models and to estimate soil moisture using radar data (e.g., Wagner et al., 1999).

In this chapter, we examine the utility of a drought indicator based on the fraction of Absorbed Photosynthetically Active Radiation (fAPAR), which is a physically based vegetation state indicator that can be estimated from satellite data. In particular, we outline the fAPAR estimation approach followed within the European Drought Observatory (EDO), which employs this indicator operationally for drought monitoring and detection. The fAPAR performance is also evaluated in a case study over the Iberian Peninsula to identify the relative strengths and limitations of this indicator in different contexts.

5.2 EUROPEAN DROUGHT OBSERVATORY

Recent drought events during the 2000s in Europe caused considerable economic losses (European Commission, 2007b). In the summer of 2003, warm and dry conditions persisted over Europe, resulting in a major drought that showed the dramatic impact of this type of climate variability on water availability and consequently on human society. Drought and heat wave conditions claimed more than 20,000 lives in Europe (UNEP, 2004) with US$15 billion in economic losses across the continent (ABUHRC, 2004). Vegetation growth was greatly reduced in an unprecedented way (about 30%; Ciais et al., 2005) across the continent. The European agriculture and forestry sectors were particularly affected by the drought. Overall, cereals production in the European Union (EU) in 2003 was reduced by 11% compared to the average annual production for the decade (1999–2009). In Italy, for example, the drought resulted in a 60% reduction in agricultural production for some crops (CIA, 2003). Fodder production deficits varied from 30% (Spain, Germany, and Austria) to 60% (France), with a financial impact on beef production of US$2 billion. The financial impact at the national level was considerable across the continent, ranging from US$269 million in Austria to US$5.5 and US$6.8 billion in France and Italy, respectively (COPA-COGECA, 2003).

Another example is the severe drought in Catalonia (Spain) that persisted from 2004–2008. This event affected the Barcelona area and was considered the most severe drought episode for this area since the beginning of the instrumental climate data record (dating back to 1916). With reference to the biennium 2007–2008, the Catalan Water Agency (Agencia Catalana del'Aigua), which oversees the 16,600 km^2 of the Catalan inland basins (52% of the Catalan territory), calculated the damages from direct (e.g., agricultural and hydroelectric production) and indirect (e.g., banking) drought impacts at almost US$1.2 billion. These monetary figures increased to $2.3 billion (almost 1% of the Catalan gross domestic product) when nonmarket welfare losses and social costs (e.g., restrictions of water supply to private households and environmental costs) were also included (Martin-Ortega and Markandya, 2009).

These recent drought events and their impacts across Europe illustrate the need for new tools to support the decision-making process of policy makers and authorities in charge of water management. The European Commission Communication on Water Scarcity and Drought (European Commission, 2007a) underlined the increased need for consistent and timely information on droughts at the European scale to support policy makers in defining adequate strategies for sustainable use of water resources. In this context, the Joint Research Centre (JRC) of the European Commission is developing a prototype of the EDO (http://edo.jrc.ec.europa.eu). The EDO is envisaged as a web-based platform for drought detection and monitoring, forecasting, and information exchange that integrates data from different geographic units ranging from the continental to EU member state level.

A multidisciplinary set of indicators has therefore been developed within the EDO to constantly monitor the various environmental components (e.g., soil moisture and vegetation health) potentially affected by this natural hazard in order to obtain a comprehensive picture of the drought situation. These indicators are produced from

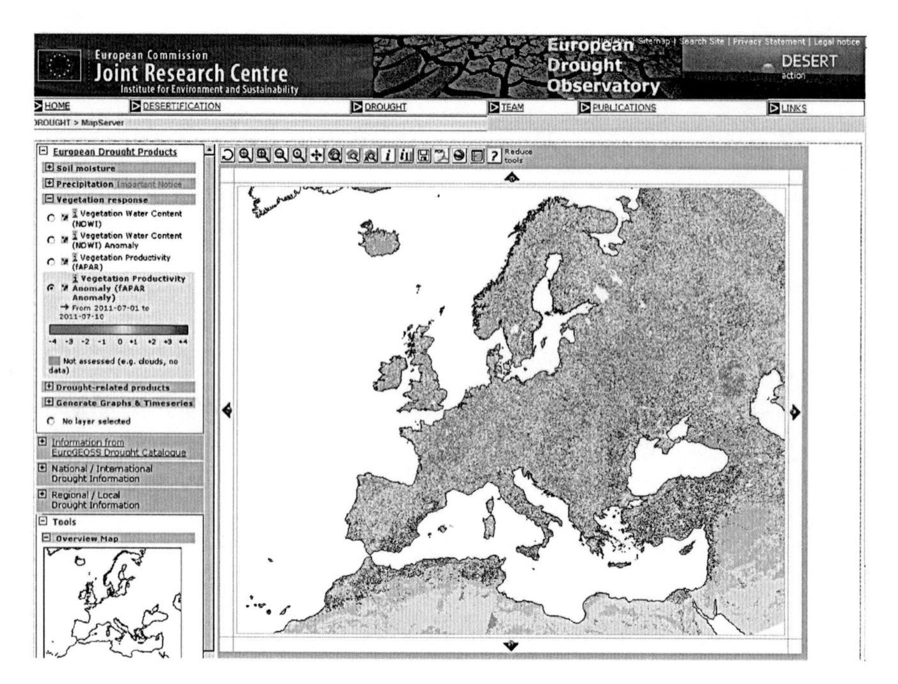

FIGURE 5.1 **(See color insert.)** The EDO prototype map server (http://edo.jrc.ec.europa.eu), showing the fAPAR anomaly map over Europe for the July 1–10, 2011, period. Red indicates areas of anomalously low fAPAR.

different data sources such as in situ meteorological observations, hydrological models, and satellite-based remote sensing, and are displayed in map form on the EDO map server (Figure 5.1).

5.3 MERIS fAPAR FOR DROUGHT DETECTION WITHIN THE EDO

5.3.1 fAPAR and Drought

This chapter illustrates the potential of one specific indicator produced by the EDO for drought monitoring: the anomalies in fAPAR. fAPAR is a biophysical variable that represents the fraction of the solar energy absorbed by vegetation within the visible portion of the optical spectral range (400–700 nm) that is used for photosynthesis. fAPAR values range from 0.0 (e.g., nonvegetated land surfaces) to 1.0 (densely vegetated areas), but most observed values generally range between 0.0 and 0.9. This biophysical variable is directly correlated with the primary productivity of vegetation since the intercepted radiation by the canopy is the energy available for important biochemical processes of plants such as photosynthesis. fAPAR is also related to other biophysical parameters such as Leaf Area Index (LAI). The relationship between LAI and fAPAR is near linear for LAI values less than 3, while fAPAR approaches an asymptotic value of 0.95 for larger LAI values depending on the canopy, soil, and atmospheric parameters (Myneni and Williams, 1994). Like NDVI, fAPAR can also be used to monitor vegetation phenology because seasonal changes in fAPAR follow the phenological cycle of the plants (Verstraete et al., 2008).

The fAPAR is one of the essential climate variables (ECVs) recognized by the United Nations (UN) Global Climate Observing System (GCOS) as holding great potential for characterizing climate variations since it is a physically based, quantitative, and clearly defined measure that is directly related to the maintenance of life systems on the planet (GCOS, 2006). GCOS issued specific recommendations for the systematic monitoring of this variable, considering it useful for a number of applications ranging from agriculture and forestry production to environmental stress and carbon assimilation assessments. The Food and Agriculture Organization (FAO) Global Terrestrial Observing System (GTOS) also recognizes fAPAR among its 13 ECVs (Gobron and Verstraete, 2009). In addition, fAPAR is an important component within the family of diagnostic terrestrial carbon models known as Production Efficiency Models (PEMs) used to calculate gross and net primary productivity (GPP/NPP), and it is often the only satellite-derived variable used in PEMs (McCallum et al., 2010).

Given the importance of fAPAR as a critical variable related to terrestrial vegetation, it has also been proposed as a drought indicator (Gobron et al., 2005, 2007). In particular, anomalies in fAPAR compared to the normal conditions are useful for detecting anomalous vegetation development caused by water stress or other factors (Rossi et al., 2008). From a conceptual point of view, one advantage of using a physically based indicator such as fAPAR for drought monitoring rather than empirical VIs is that the relevance of fAPAR for this application is grounded in biophysical meaning. In fact, water stress can cause changes in a plant's ability to intercept solar radiation and degrade light use efficiency (LUE), both of which can result in reduced vegetation growth rates during droughts. Different effects can occur depending on the growth stage(s) of the plant at the time of the drought. Droughts occurring in the early stage of development close to emergence decrease the amount of radiation intercepted by the vegetation and cause long-term changes in LUE. Droughts occurring during a later stage of development can accelerate plant senescence and decrease radiation interception, but do not significantly decrease LUE (Legg et al., 1979). Monteith (1977) observed that radiation interception is much more sensitive to water stress than is radiation use efficiency. In a study of barley, Jamieson et al. (1995) found that a reduction in intercepted radiation (and therefore in fAPAR) is a consequence of drought occurring in both the early and late growth stages and is the main mechanism for reducing biomass production in this crop.

Measuring fAPAR in the field is a complex task (GCOS, 2006), and no standards or protocols have been formally defined to date. Field methods are often closely related to those used to measure LAI and require simultaneous measurements of PAR both above and within the canopy at different heights in order to establish an fAPAR profile within the canopy (Gobron and Verstraete, 2009). For large areas, space-based remote sensing can be used to monitor fAPAR. Optical satellite data have to be interpreted based on the understanding of the physical processes underlying the transmission of light through the atmosphere and the canopy to estimate such biophysical parameters. Different algorithms have been proposed, but are generally based on the exploitation of the red and NIR spectral bands. Empirical fAPAR-VI relationships have been developed based on field measurements in combination with high-resolution satellite images (Chen and Cihlar, 1999). However, physically-based approaches are now the main methodology used to retrieve fAPAR, since they are able to take into account the various physical processes that influence the properties of light during its travel through the atmosphere and the canopy, minimizing the influence of atmospheric disturbances and other effects related to changes in soil brightness and the geometry of illumination and observation (Gobron and Verstraete, 2009). In these physically based approaches, fAPAR is generally inferred from radiative transfer models using remote sensing observations as model constraints.

Different organizations such as the National Aeronautics and Space Administration (NASA) and the European Space Agency (ESA) operationally produce global fAPAR products at varying temporal and spatial resolutions using various estimation methods. Table 5.1 summarizes these various activities, including the remote sensing inputs and fAPAR retrieval methods. The next section describes in more detail the ESA/JRC Medium Resolution Imaging Spectrometer (MERIS) fAPAR product, which was selected as a drought indicator for the EDO because of its performance and because it is produced using an algorithm developed at the JRC.

5.3.2 Derivation of fAPAR from MERIS Data: The MGVI Algorithm

The production of fAPAR anomalies as a drought indicator within EDO relies on products prepared by ESA from data collected by the MERIS instrument onboard the ESA Environmental Satellite (Envisat). ESA operationally produces daily fAPAR estimations from MERIS data as a MERIS Level 2 land product by means of the MERIS Global VI (MGVI) algorithm developed at the JRC (Gobron et al., 1999, 2004). MGVI is part of a suite of algorithms to consistently retrieve fAPAR using data from different satellite sensors such as MERIS, the Moderate Resolution Imaging Spectroradiometer (MODIS) (Gobron et al., 2006b), and the Sea-viewing Wide Field-of-view Sensor (SeaWiFS) (Gobron et al., 2002). The MERIS instrument has 15 spectral channels distributed within the visible and NIR spectral domain and collects a complete global coverage of data at 300 m spatial resolution every 3 days in the equatorial regions, while providing a more frequent revisit time over the polar regions because of the convergence of the satellite orbits. A detailed description of the MGVI algorithm is included in the MGVI Algorithm Theoretical Basis Document (Gobron et al., 2004). Here, a short summary of the algorithm is provided to highlight the added value offered by these data for vegetation monitoring compared to simple VIs.

The MGVI algorithm was developed following an approach proposed by Verstraete and Pinty (1996) and Govaerts et al. (1999). According to this approach, an optimal spectral index for environmental monitoring should be very sensitive to the desired information (e.g., vegetation status) while minimizing the sensitivity to other unrelated factors such as atmospheric effects, variations in soil background color and brightness, and variations in the geometry of the sun-target-sensor relationship. In the design of the MGVI, a coupled surface-atmosphere model is used to simulate MERIS data over different representative land surface conditions (e.g., LAI, canopy height, leaf angle distribution, and soil reflectance). A plane-parallel, structurally homogeneous canopy model (Gobron et al., 1997) representing the land component is used to simulate bidirectional reflectance factors (BRFs) and fAPAR, while the 6S model (Vermote et al., 1997) is used to minimize the atmospheric contribution to the detected spectral signal by converting top of atmosphere (TOA)

TABLE 5.1
Main fAPAR Products Currently Available

Projects/Institution	Sensor/Input Data	Retrieval Method	References	Spatial Resolution	Temporal Resolution	Years Available
ESA/JRC MERIS fAPAR	MERIS TOA BRFs in blue, red, and NIR bands	Optimization formula based on radiative transfer models	Gobron et al. (2000, 2006a, 2008)	• 300 m (full resolution) • 1.2 km (reduced resolution)	• 1–3 days or • 10 day composites	2003–now (extendable back to 1997 with SeaWiFS data)
NASA MODIS LAI/fPAR	MODIS Surface reflectance (seven spectral bands) and land cover map	Inversion of a 3D model versus land cover type with backup solution based on NDVI relationship	Knyazikhin et al. (1999)	1 km	8 day composites	2000–now
GlobCarbon	Vegetation, (A)ATSR, MERIS Surface reflectance (red, NIR, and shortwave infrared (SWIR))	Parametric relation with LAI as a function of land cover type	Plummer et al. (2006)	0.5°	Daily	1998–2007
Cyclopes	Vegetation Surface reflectance (blue, red, NIR, and SWIR)	Neural network	Baret et al. (2007)	1 km	10 day composites	1999–2003

(continued)

TABLE 5.1 (continued)
Main fAPAR Products Currently Available

Projects/Institution	Sensor/Input Data	Retrieval Method	References	Spatial Resolution	Temporal Resolution	Years Available
LANDSAF (land surface application facility)	SEVIRI Red and NIR bands	Relationship between VI and simulated fAPAR	Camacho-de Coca et al. (2007)	3 km	10 day composites	2005–now
NASA MISR LAI/fPAR	MISR Surface products (blue, green, red, and NIR bands)	Inversion of 3D model versus land cover type with backup solution based on NDVI relationship	Knyazikhin et al. (1998)	1.1 km	Monthly	2005–now
JRC-TIP	MODIS MISR Broadband surface albedo in visible and NIR bands	Inversion of two-stream model using the Adjoint and Hessian codes of a cost function	Pinty et al. (2007)	1 km	No operational production yet	No operational production yet

Source: Gobron, N. and Verstraete, M., *fAPAR—Fraction of Absorbed Photosynthetically Active Radiation*, Essential Climate Variables Report, Version 8, Global Terrestrial Observing System, Rome, Italy, 2009.

BRFs data into top of canopy (TOC) BRFs. A parametric BRF model (Rahman et al., 1993) was implemented to produce training data sets accounting for the main variations in the signal due to geometry changes (anisotropy of the radiance field) without making any a priori assumption about land cover.

The algorithm can be described in three main steps. In the first step, three MERIS spectral bands (the 442 nm [blue], 681 nm [red], and 865 nm [NIR] bands) are used to generate two "rectified bands" in the red and NIR wavelengths, reducing atmospheric and angular effects. The goal is to minimize the difference between the rectified channels and the TOC spectral reflectances under a standard illumination and viewing geometry. This rectification process is based on polynomial coefficients optimized such that values correspond to the BRFs that would be measured at the TOC, which are normalized by the spectrally appropriate anisotropy reflectance function. In the second step, the sensitivity of the MGVI to soil and atmospheric effects is minimized by the simulation of the radiation transfer through the coupled surface-atmosphere model described earlier. A lookup table (LUT) of simulated BRFs under the different conditions was generated and the fAPAR computed for each of them. In the third step, these simulations provide the basic information to optimize and constrain the MGVI by defining the optimal values in the coefficients of the rational polynomial of the MGVI, thus obtaining the mathematical combination of spectral bands that best accounts for the variations in the simulated fAPAR. The resulting equation is then used to obtain the fAPAR estimations from the MERIS reflectance values in the different bands.

The MGVI algorithm and other comparable algorithms implemented with the same approach at the JRC for other sensors have been tested and validated in different studies. The JRC SeaWiFS fAPAR product has been tested against ground-based observations, and the results show the estimates are well within a range of ±0.1 compared to the ground-based fAPAR estimations (Gobron et al., 2006a). The test also highlighted the ability of SeaWiFS estimations to detect vegetation stress and represent the seasonal vegetation development cycle. McCallum et al. (2010) compared the same product with three other global fAPAR data sets (NASA/MODIS fAPAR, Cyclopes, and Globcarbon) over northern Eurasia and found higher agreement among the data sets over deciduous broadleaf forests and croplands than over mixed and needleleaf forests. In the same study, the authors concluded that the SeaWiFS fAPAR was the best in capturing fAPAR across most land cover types, while the performance of other data sets varied depending on the land cover type. Seixas et al. (2009) compared MERIS and MODIS fAPAR (collection 4.1) over six biomes in Iberia to evaluate their impacts on ecosystem modeling applications. MODIS fAPAR values were found to be systematically higher than MERIS fAPAR, with higher MODIS-MERIS correlations obtained during the summer. MERIS fAPAR revealed spatial homogeneity over large areas characterized by homogeneous biomes, thus better capturing vegetation characteristics compared to MODIS fAPAR, which showed a higher spatial variability and therefore a higher spatial standard deviation. Gobron et al. (2008) found that the differences among the operational MERIS fAPAR and the fAPAR products obtained with the JRC algorithms for SeaWiFS and MODIS were in the range of 5%–10%. In the same study, a comparison between MERIS fAPAR and ground-based fAPAR measurements for six sites of different land cover types showed a good agreement between the estimations, with larger discrepancies occurring during the senescent phase of the growing season.

5.3.3 EDO fAPAR Processing Chain

The MERIS Level 2 land product data are further processed to remove residual perturbations and then aggregated at a lower spatial resolution (1.2 km), obtaining MERIS Level 3 products. Time series products derived from optical satellite data often still contain residual artifacts (e.g., clouds) even after the series of corrections that were applied by the MGVI algorithm. A temporal compositing procedure is used to obtain the most representative value over the compositing period (i.e., 10 consecutive days). For fAPAR products, ESA uses the algorithm proposed by Pinty et al. (2002), which defines the most representative value within the compositing period as the sample that is closest to the temporal average of the 10 day period, as identified through a two-step procedure aimed at rejecting outliers caused by various disturbances. In the first step, the temporal average μ and the corresponding average deviation Δ are calculated, and daily values that are outside the range $\mu \pm \Delta$ are considered probable outliers and rejected for consideration in the next step. In the second step, a new average is calculated from the remaining values, and the most representative value for the 10 day period is identified as the value closest to this average. The MERIS land products resulting from the time-composite processing chain are in the original MERIS L2 300 m spatial resolution and are spatially aggregated by ESA to a lower 1.2 km spatial resolution (Aussedat et al., 2007). In December 2008, the ESA began operational production of MERIS fAPAR Level 3 aggregated 10 day composite products generated for the entire European continent. Level 2 data from previous years were retrospectively processed to build a Level 3 archive dating back to the beginning of the MERIS data acquisitions in 2002. Data are delivered via the ESA Service Support Environment (SSE). Typically, new fAPAR data are delivered 6–10 days after the end of each 10 day composite period. At the moment, a similar product for the African continent is in the testing phase on the SSE platform.

The fAPAR anomalies ($Anomaly_{dy}$) are then produced at JRC for inclusion into EDO by comparing the current fAPAR value for a specific 10 day period d and year y (X_{dy}) to the historical mean fAPAR ($\overline{X}_d$) for the same d according to the following formula:

$$Anomaly_{dy} = \frac{X_{dy} - \overline{X}_d}{\delta_d} \tag{5.1}$$

where δ_d is the standard deviation of fAPAR values in a specific 10 day period with reference to the historical time series. The anomaly is therefore defined for each pixel as the number of standard deviations the fAPAR for a specific 10 day period departs from the long-term average for that same period of the year. This measure eliminates the effect of seasonality, making the values comparable across the year, and highlights anomalous conditions with reference to the historical normal conditions for each particular 10 day period.

Given the relatively short historical record of MERIS fAPAR data available for anomaly detection, the MERIS fAPAR time series was extended back to 1997 with the fAPAR estimations obtained from the SeaWiFS sensor to obtain a more representative longer-term historical data set. The SeaWiFS fAPAR data were calculated using the SeaWiFS-based algorithm from the suite of JRC fAPAR algorithms described earlier, which was developed for comparability with MERIS data. This resulted in a 10+ year historical baseline of fAPAR information from which more meaningful anomalies could be calculated, compared to the shorter record available for MERIS data.

5.4 EVALUATION OF MERIS fAPAR AS A DROUGHT INDICATOR

5.4.1 Detection of Drought Events

In order to evaluate the ability of fAPAR for drought detection, we examined its behavior during past drought events and compared it to two other indices commonly used for drought monitoring: the satellite-derived NDVI and the Standardized Precipitation Index (SPI) (McKee et al., 1993, 1995). The SPI is a precipitation-based index that has been recommended for global drought monitoring by the World Meteorological Organization (WMO, 2009) and classifies droughts as "moderate" when SPI values are between −1 and −1.5, "severe" with SPI values ranging between −1.99 and −1.5, and "extreme" for SPI values less than −2.00 (McKee et al., 1993).

Figure 5.2 shows a fAPAR anomaly image for the 10 day period of August 11–20, 2003. During summer 2003, severe drought conditions were experienced across much of Europe. The calculation of the fAPAR anomalies allowed the detection of a large area of negative anomaly values (red areas) that extended across France, Germany, northern Italy, and much of Central Europe, reflecting the severe vegetation stress as compared to average conditions for that same period during the previous years. The fAPAR and fAPAR anomalies time series plotted in Figure 5.3 show vegetation conditions over a rainfed agricultural area of Lombardy (northern Italy) for 2003 and 2004. The negative fAPAR anomaly values for most of 2003 reflect the stressed vegetation conditions for this area, where yields were severely reduced because of the drought. In contrast, higher agricultural production was observed in 2004 (ISMEA, 2004), as reflected by the positive fAPAR anomaly values in the time series. Noise in the fAPAR time series, apparent in Figure 5.3, and in anomaly maps could be suppressed to some extent by application of a smoothing filter such as Savitsky–Golay (Savitzky and Golay, 1964).

Another example for Catalonia, Spain, is displayed in Figure 5.4, where fAPAR anomalies are plotted in comparison with time series of three other drought indicators operationally produced within EDO, including the SPI computed over 3 (SPI3) and 12 (SPI12) month intervals, as well as soil moisture anomalies derived from LISFLOOD hydrological model simulations (De Roo et al., 2000; Laguardia and Niemeyer, 2008). The fAPAR, fAPAR anomalies, and soil moisture estimations plotted in Figure 5.4 were produced on a 10 day time step and temporally averaged for each month in order to make them comparable with SPI for this analysis. The data plotted in Figure 5.4 represent conditions over a 4 year period for a pixel within a rainfed agricultural area of the Catalonia region in Spain. As mentioned in Section 5.2, this area was affected by prolonged drought during the years 2004–2008, which

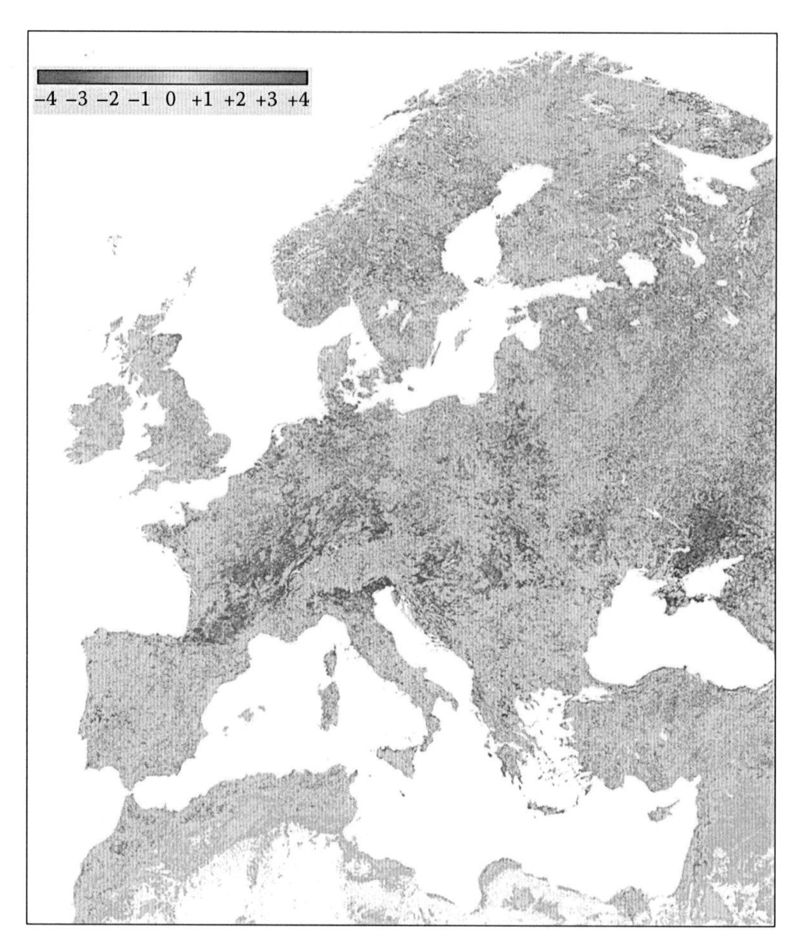

FIGURE 5.2 **(See color insert.)** fAPAR anomaly image for the August 11–20, 2003, period, showing considerably lower than average fAPAR values across much of Europe (red areas) due to severe vegetation water stress.

caused considerable economic damage. Different timescales of SPI calculations reflect the impact of drought on different water resources and therefore describe different types of droughts. Shorter timescales describe soil moisture availability variations and agricultural drought, while longer timescales describe the impact on groundwater, streamflow, and reservoirs (McKee et al., 1995). Figure 5.4 shows that fAPAR anomalies are in better agreement with the shorter SPI3 and soil moisture anomalies in identifying three drought periods (mid-2004 to mid-2005, mid-2006, and second half of 2007) with varying intensity ranging from severe to extreme conditions according to the SPI intensity scale (SPI3 less than 1.5). As noted by Rossi et al. (2008), fAPAR anomalies in some cases slightly lag the SPI3 time series because of an inherent lag in the response of vegetation to the precipitation anomalies that occurred in the previous 3 months described by SPI3. The longer SPI12 has less-pronounced troughs and remains negative for most of the displayed period,

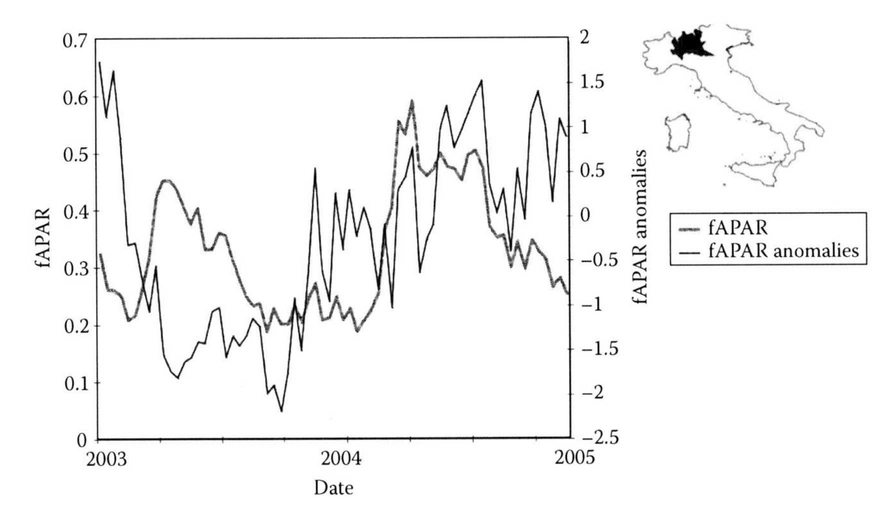

FIGURE 5.3 Time series of fAPAR and fAPAR anomalies during the 2003 drought in northern Italy.

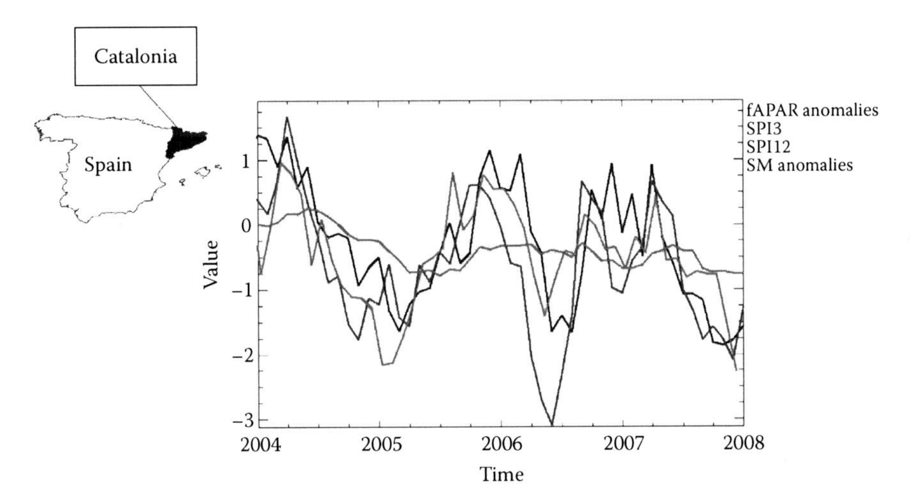

FIGURE 5.4 **(See color insert.)** Detection of a prolonged drought event in Catalonia (Spain) using fAPAR and soil moisture anomalies and SPI (3 and 12 month) products.

describing long-term, multiyear drought conditions. The temporal correspondence of the fAPAR anomaly time series with SPI3 and soil moisture anomalies suggests that this indicator is detecting vegetation state variations related to agricultural drought conditions.

5.4.2 Sensitivity to Antecedent Precipitation

The correlation between the monthly averages of the fAPAR anomalies and the SPI calculated at different timescales was also examined to evaluate the timing of vegetation response to precipitation anomalies. For this comparison, the Iberian

Peninsula was selected as the study area because water is the major limiting factor for vegetation development in this region (Roerink et al., 2003). The period under consideration was April 1998–October 2008. Following the work of Ji and Peters (2003), only the summer growing season months (between April and October) were considered because this is the time when impacts of droughts on vegetation are most severe. The SPI maps from the EDO are produced at a 5 km spatial resolution, and these were resampled in this experiment to the grid of the 1.2 km fAPAR anomaly data set. The statistical tests described later therefore compare MERIS-derived fAPAR data at their native resolution with SPI values characteristic of the surrounding 5 km pixel.

The temporal correlation between the fAPAR and SPI data sets was assessed at each pixel across the study area using the Pearson correlation coefficient, which is defined as follows:

$$R = \frac{\operatorname{cov}(x_1, x_2)}{\sqrt{\operatorname{cov}(x_1, x_1) \cdot \operatorname{cov}(x_2, x_2)}} \tag{5.2}$$

where x_1 and x_2 are contemporaneous data from the two data sets. The R coefficient values range between 1 (perfect positive correlation) and −1 (perfect negative correlation) with 0 indicating no correlation. The correlations were based on 77 samples (n = 7 months multiplied by 11 years). The minimum significant correlation for this sample size is 0.23 for $p = 0.05$ and 0.19 for $p = 0.1$.

Figure 5.5 shows the correlation maps between fAPAR anomalies and the 1, 3, and 6 month SPIs over the 11 year period for the study area. The three correlation images reveal that the fAPAR anomalies have a higher correlation with the 3 month SPI than with the shorter- and longer-term SPIs over most of the Iberian Peninsula. The improved correlation with SPI3 in comparison with SPI1 illustrates the cumulative impact that water deficits have on vegetation, generating a time lag in the response of plants to precipitation variations. Vegetation is therefore influenced not only by the recent precipitation but also by the rain received in previous months. Wang et al. (2003) found that NDVI response to precipitation showed a lag of 4–8 weeks. Since the monthly SPI (SPI1) considers precipitation from only the last month,

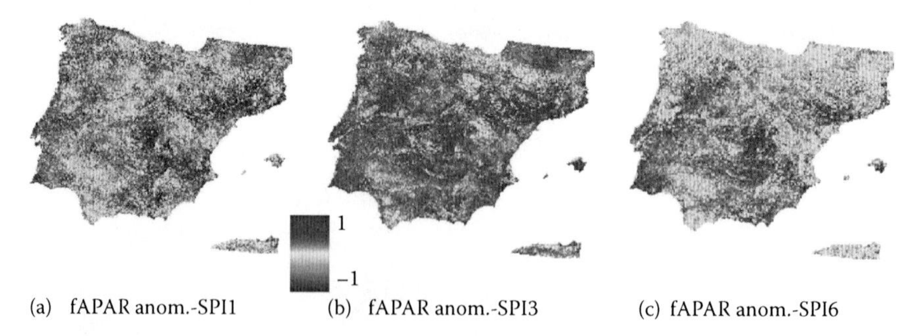

FIGURE 5.5 **(See color insert.)** Correlation of fAPAR anomalies with SPI1 (a), SPI3 (b), and SPI6 (c) over the Iberian Peninsula for the period April 1998 to October 2008, using 5 km SPI data from the EDO.

it does not match well with the anomalies in vegetation development monitored by fAPAR, which are better correlated with SPI3. Longer timescales ranging from 6 to 12 months (e.g., SPI6, SPI9, and SPI12) tend to have diminished temporal variance, as illustrated in Figure 5.5 by the 6 month SPI correlation map. Precipitation integrated over such long timescales becomes less coupled with the current state of vegetation. These results suggest that correlations between fAPAR anomalies and SPI peak for the SPI3 timescale, similar to the behavior found by Ji and Peters (2003) in their investigation of the NDVI-SPI relationship.

5.4.3 Comparison of fAPAR and NDVI Products for Drought Monitoring

The performance of the MERIS fAPAR for drought detection was compared with the widely used NDVI to evaluate whether an indicator such as the fAPAR obtained through a complex retrieval algorithm offers an added value compared to traditional empirical spectral VIs, which have been commonly applied for drought detection. Because the MERIS single-band data necessary for NDVI calculation are not available within the Level 3 time composite products delivered by ESA, a monthly 1 km NOAA AVHRR NDVI data set was used for comparison, generated using a maximum value time compositing technique and available within the JRC/Monitoring Agricultural Resources (MARS) Remote Sensing Database. The NDVI data were coregistered with the MERIS fAPAR data set and resampled at the same 1.2 km spatial resolution. NDVI anomalies were then calculated over the same 11 year period (September 1997–2008), following the methods used for the fAPAR anomalies described earlier.

The performance of these two indicators was evaluated by analyzing their correlation with SPI3, identified in Section 5.4.2 as the SPI interval having the strongest relationship with fAPAR anomalies. In order to compare the fAPAR and NDVI data to an SPI3 data set with a comparable spatial resolution, a gridded SPI3 data set at 2 km spatial resolution produced by the University of Valencia from a network of 3000 stations across Spain (Javier Garcia-Haro, personal communication) was used. These gridded data were resampled at 1.2 km spatial resolution and coregistered with the remote sensing data sets. Because this data set does not include SPI1 or SPI12 data, it was not used in fAPAR-SPI comparisons described in the previous section.

The national-level results presented in Figure 5.6 show that fAPAR anomalies have a positive correlation with SPI3 over a larger proportion of the study area (96% of total area) compared to the NDVI anomaly-SPI3 correlations (80% of total area). Yellow areas identify values of the correlation coefficient below the significance threshold 0.23 ($p = 0.05$). Overall, the fAPAR anomalies have a higher correlation with SPI3 than the NDVI anomalies over most of the study area.

Correlations between SPI3 and both the fAPAR and NDVI anomalies were also investigated by land cover class to understand how their relationship might vary for different vegetation types. Seven land cover types typical of Spain were analyzed, including rainfed cropland, cropland-grassland mosaic, irrigated cropland, needleleaf evergreen forest, broadleaf deciduous forest, rainfed tree crops,

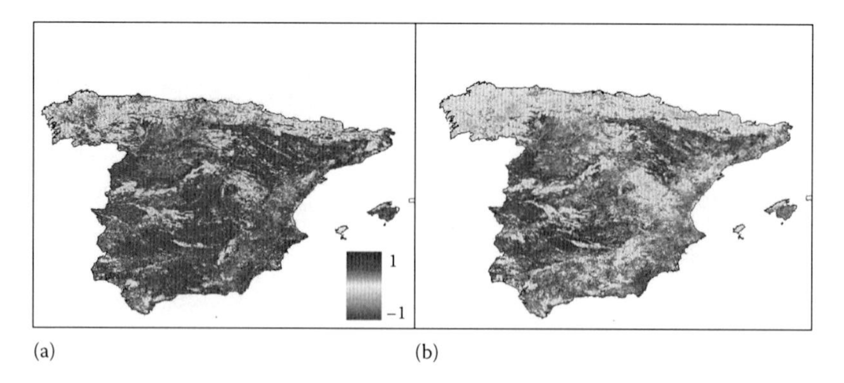

FIGURE 5.6 **(See color insert.)** Correlation between the fAPAR anomalies-SPI3 (a) and NDVI anomalies-SPI3 (b) over Spain, using 2 km SPI data from the University of Valencia.

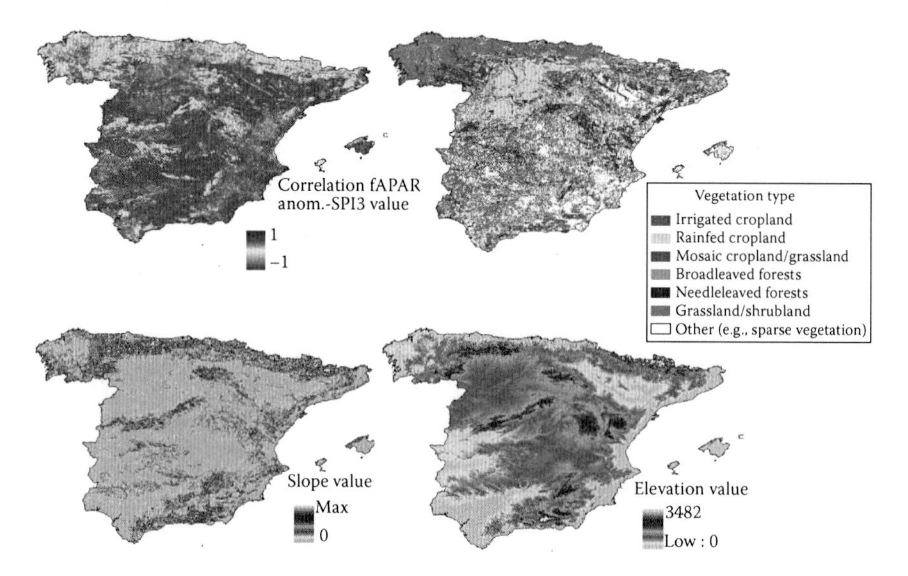

FIGURE 5.7 **(See color insert.)** From top left, clockwise: maps of MERIS fAPAR-SPI3 correlation, land cover, slope, and elevation.

and shrubland/grassland (see classification map in Figure 5.7). The ESA GlobCover land cover map (ESA GlobCover Project, 2008) was used to locate the geographic distribution of these land cover types across Spain. The original 300 m GlobCover map was resampled to a 1.2 km resolution, and the dominant land cover class was assigned to pixels when at least 90% of the area within their respective 1.2 km pixel footprint was classified to same land cover type in the original 300 m land cover map. Pixels covering heterogeneous land cover that did not have a single land cover type covering 90% of the area were not included in the analysis.

Table 5.2 shows, for each land cover type, the average and the maximum correlation of fAPAR and NDVI anomalies with SPI3. Rainfed cropland and natural

TABLE 5.2
Growing Season Correlations (April–October) between fAPAR/NDVI Anomalies and SPI2, Computed across Spain for Different Vegetation Types

	fAPAR Anomaly-SPI3		NDVI Anomaly-SPI3	
Land Cover	**Mean**	**Maximum**	**Mean**	**Maximum**
Rainfed cropland	0.42	0.78	0.24	0.59
Mosaic cropland-grassland	0.40	0.73	0.28	0.59
Irrigated cropland	0.14	0.76	0.10	0.59
Needleleaf evergreen forest	0.17	0.62	0.08	0.54
Broadleaf deciduous forest	0.15	0.70	0.07	0.53
Rainfed tree crops	0.13	0.63	0.11	0.57
Shrubland/grassland	0.37	0.73	0.22	0.60

herbaceous vegetation (grassland and shrubland) had the highest average correlations for both fAPAR and NDVI, while, for irrigated cropland, the correlation was lower and below both the significance thresholds cited earlier. The lower correlations over irrigated land might be expected because anomalous precipitation deficits have less effect on vegetation development than rainfed locations because of targeted water applications to maximize production. On average, correlations were below the significance thresholds for the two forest land cover types. Short-term precipitation variations described by SPI3 likely have less influence on the vegetation state of forests in comparison with herbaceous vegetation types because of deeper root systems, which allow trees to access water from deeper soil layers. For each land cover class, fAPAR anomalies were more highly correlated with precipitation anomalies expressed in the SPI3 than were the NDVI anomalies.

Note that this study cannot unequivocally determine the relative performance of fAPAR and NDVI as generic indicators of drought impact. We are using data from different data sets and sensors (MERIS and AVHRR) for the estimation of fAPAR and the calculation of NDVI, and differences in the instruments and in the data processing techniques may have contributed to some of the differences observed between the AVHRR- and MERIS-derived data sets in this study. However, the results suggest that the MERIS fAPAR Level 3 product is better correlated with precipitation deficits when compared to the AVHRR NDVI product over different vegetation types and over a wide range of terrain conditions. The MGVI algorithm, which is designed to reduce the sensitivity to geometrical effects such as terrain variations and varying viewing and illumination angles in the temporally composited MERIS data, may have an important role in this better performance. The advantage of using advanced algorithms for operational monitoring applications rather than empirical VIs was also highlighted by Gobron et al. (2007), who compared MERIS fAPAR performance for drought detection with an NDVI data set built using the same MERIS data and found that fAPAR has a much lower level of noise than does NDVI, offering an improved performance for drought detection.

5.4.4 Effects of Topography and Seasonality on fAPAR

Dependence of MERIS fAPAR anomaly-SPI3 correlation on slope and elevation (derived from a digital elevation model) (Vogt et al., 2007) was also analyzed in order to assess how these topographic features influence the ability of the fAPAR retrieval algorithm to detect water stress effects on the vegetation. Spain is characterized by a mountainous topography, with 24% of the country above 1000 m and 76% between 500 and 1000 m, and an average altitude of 660 m. Only 12% of mainland Spain has a slope less than 5%. Therefore, this domain provides a good testbed for evaluating influence of topography on response of vegetation-based drought indices to precipitation deficits. A visual comparison of correlation, slope, and elevation maps in Figure 5.7 suggests that both topographic features influence the correlation between the fAPAR anomalies and SPI3. Areas characterized by steeper slopes and high elevations (depicted in red and brown colors in the topographic maps) consistently have a lower fAPAR anomaly-SPI3 correlation than do the flatter, lower elevation areas. For example, there is a clear geographic correspondence between areas of lower fAPAR-SPI3 correlation and mountain systems such as the Sistema Central in central Spain, the Cordillera Cantabrica in the north, and the Pyrenees near the northeast border. However, comparing the topographic maps with the NDVI anomaly-SPI3 correlation map in Figure 5.6, the MERIS fAPAR anomaly seems to have an increased ability to characterize vegetation responses to drought stress over a wider range of slopes and altitudes.

Table 5.3 shows how the fAPAR-SPI3 correlation for three vegetation classes varied with elevation. Rainfed cropland was selected as the vegetation type where fAPAR performs better, and tends to be located in lower slope areas. Needleleaf forests are typically present in mountainous and steep areas, while grassland/shrubland is a much more geographically diffuse vegetation type that can be found across a wide range of topographic situations. The fAPAR-SPI3 correlation remained quite constant and decreased only at very high elevations. The lower correlations at high elevations signal a shift from an environment where moisture is the main factor limiting vegetation growth (typical of low elevations) to energy-limited, higher elevation environments where temperature and solar radiation are the main limiting factors. Karnieli et al. (2010) analyzed the limiting factors for vegetation development in

TABLE 5.3
Mean Correlation between fAPAR Anomalies and SPI3 at Different Elevations (above Sea Level)

Average Elevation (m)	Rainfed Cropland	Needleleaf Forest	Grassland/Shrubland
0–500	0.36	0.23	0.33
501–1000	0.43	0.20	0.42
1001–1500	0.50	0.16	0.37
1501–2000	0.42	0.15	0.30
2001–2500	N/A	0.14	0.27
>2500	N/A	N/A	0.24

TABLE 5.4
Mean Correlation between fAPAR Anomalies and SPI3 for Different Slope Classes

Average Slope within the 1.2 km Pixel (%)	Rainfed Cropland	Needleleaf Forest	Grassland/ Shrubland
0–10	0.42	0.20	0.38
11–20	0.32	0.14	0.36
21–30	0.25	0.13	0.33
31–40	0.25	0.13	0.26
41–50	0.03	0.10	0.16
>50	N/A	0.01	N/A

mountain environments and found that at high elevations water is relatively abundant, but low temperatures and clouds can reduce the energy available for plants, limiting their development. A sufficient amount of solar radiation is necessary for photosynthesis, while temperature determines respiration and photosynthetic rates and influences the available nutrients in the soil through its effect on the decomposition rate. In these areas, energy is the main limiting factor for vegetation development, and, therefore, vegetation state is less correlated with precipitation anomalies compared to lower elevations.

Table 5.4 shows the average correlations between fAPAR anomalies and SPI3 in different slope classes, for the same three vegetation classes. The results show that slope seems to contribute along with elevation to influencing the capability of fAPAR estimates to detect the effects of precipitation anomalies on vegetation. In pixels characterized by a steep average slope, the fAPAR anomalies are less correlated with precipitation anomalies. This may be the result of disturbance effects introduced by slope in the retrieval of data with satellite sensors. Although the MGVI algorithm is designed to reduce these disturbances, the fAPAR performance was poorer in complex mountain areas with a steep average slope than in plain areas with low average slope.

While the aforementioned comparisons combined all months of the summer growing season (April–September) into one correlation analysis, we can also examine how the strength of the fAPAR-SPI3 correlation varies across the growing season. Figure 5.8 shows scatter plots of fAPAR anomaly versus SPI3 for points selected from five rainfed agriculture sites across Spain, segregated by month ($n = 5$ sites × 11 years = 55 points per month). Care was taken to select pixels with homogeneous land cover in order to exclude variations introduced by different land use practices—therefore, correlations depicted in Figure 5.8 are higher than those listed in Table 5.2 which may have contained some mixed pixels. Although there is some variation in correlation strength, with higher correlations obtained in midseason, all correlations are significant at the $p = 0.30$ (April–May) to $p = 0.91$ (June, August) level as determined using the Fisher r-to-z transformation (Thoni, 1977). The correlations are relatively high and the slopes relatively uniform throughout the vegetation cycle. This suggests that the fAPAR anomaly can be a useful indicator of drought-related crop stress throughout most of the growing season.

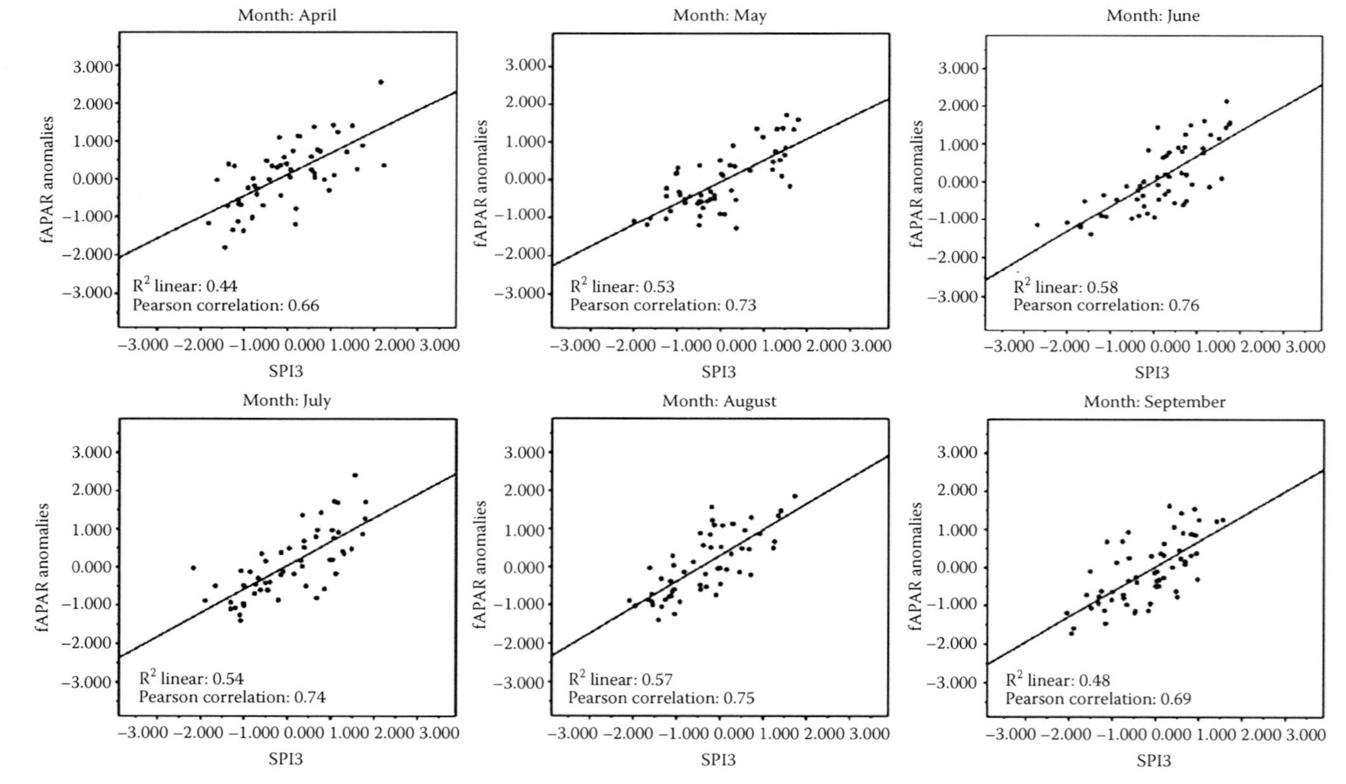

FIGURE 5.8 Monthly correlation between fAPAR anomalies and SPI3 for five selected rainfed cropland sites during the crop growth cycle in Spain ($n = 55$).

5.5 CONCLUSIONS

This chapter examined the potential of fAPAR retrievals using MERIS satellite data for drought monitoring in a case study over Spain, where water is the main limiting factor for vegetation growth over most of the country. The correlation between seasonal anomalies in MERIS fAPAR and the precipitation-based SPI was examined in order to evaluate the performance of fAPAR to detect vegetation response to drought. The results obtained with fAPAR were then compared with those obtained with an AVHRR-based NDVI data set to evaluate the relative performance of fAPAR with respect to a remote sensing product that has been commonly used for drought monitoring.

The results show that MERIS fAPAR anomalies have a positive correlation with SPI computed for different timescales; the highest correlation is with the 3-month SPI. These results confirm that the fAPAR is reflective of agricultural drought conditions caused by precipitation deficits over several weeks to a few months. When compared to the traditional satellite-based AVHRR NDVI product, fAPAR anomalies offered an increased sensitivity to drought conditions over a wider range of vegetation types and topographic characteristics. Unlike NDVI, fAPAR has a physical meaning, which provides the opportunity to better link the satellite indicator to the physiological processes of vegetation and to its response to drought stress. The MGVI algorithm used to produce the fAPAR estimations is explicitly designed to minimize the sensitivity to geometry variations and atmospheric effects and may have contributed to the improved monitoring performance of fAPAR compared to the simple NDVI. However, advanced complex algorithms also require longer processing times compared to simple indexes, and, consequently, product availability may be delayed. This can be a limitation that has to be taken into account for near-real time monitoring applications.

The role of contextual factors such as the vegetation type, elevation, and slope was also examined to determine if and how they influence the relationship between satellite-derived fAPAR estimation and moisture deficits, and to identify characteristic areas where fAPAR offers a greater potential for routine drought monitoring. Each factor was found to have some influence on the estimation of fAPAR. Vegetation type was the most important factor, with the highest correlations between fAPAR anomalies and precipitation anomalies observed for rainfed cropland and other herbaceous vegetation (i.e., grass and shrubs) and lower correlations over forests. A decrease in fAPAR correlation with SPI3 was observed at high elevations, which may be due to the shift from a moisture-limited environment to the energy-limited environment for vegetation development that occurs at higher elevation locations. Slope also affected the fAPAR performance, where fAPAR anomalies are less correlated with SPI3 in steeply sloped areas. This weaker relationship may be attributed to the difficulty of using satellite sensors to accurately retrieve spectral data for pixel locations in steep, complex mountainous areas. However, the correlation maps showed that the performance of fAPAR was less influenced by elevation and slope than NDVI.

Rainfed agriculture and grassland/shrublands are the vegetation types where fAPAR anomalies provide the best results in detecting the effects of droughts on vegetation. The accuracy is higher when monitoring areas characterized by elevations below 1500 m and moderate to low slopes. These environmental characteristics should be taken into consideration when interpreting and utilizing fAPAR information of a given area for drought monitoring. The fAPAR anomalies appear to have limited value for drought monitoring at high elevations, in very steep areas, and over forests, where less reliable information on drought conditions was acquired, as shown in this study. The relationship of fAPAR anomalies in forest areas with SPI calculated at timescales longer than 3 months (e.g., 9 and 12 months) has to be further examined to determine if a stronger relationship exists between fAPAR and longer-term precipitation deficits, which could yield interesting insight into the potential contribution of fAPAR to monitor longer-term effects of droughts over forests and other land cover types.

A study of the fAPAR-SPI3 correlation in five selected agricultural sites shows that variations during the different months of the summer growing season are minimal and the differences are not significant from a statistical point of view. The statistical results highlighted considerable temporal stability in the performance of fAPAR during the summer, when the impacts of agricultural droughts are the greatest.

The results presented in this chapter confirm that fAPAR is a useful indicator to detect water stress of plants and hence to monitor drought impacts on vegetation. Maps of MERIS fAPAR anomalies are already regularly produced within EDO for Europe, while other activities are ongoing in the JRC DESERT Action project to monitor droughts with fAPAR anomalies in other continents, such as Africa. Additional evaluation activities are foreseen in the future—for example, to verify how the performance of fAPAR varies in other parts of Europe with different climatic regimes. In fact, the Iberian Peninsula represented an ideal test area for this initial analysis because of its climate (most areas represent water-limited environment for vegetation development) and its diverse topography and land cover types. The comparison of MERIS-derived fAPAR data sets with other fAPAR products such as the MODIS-derived fAPAR data produced by NASA will also provide further insight into sensor and algorithm-induced differences that might exist among the various operational data sets that are currently available for drought monitoring applications.

ACKNOWLEDGMENTS

We thank Nadine Gobron and Monica Robustelli (JRC), Javier Garcia-Haro (University of Valencia), and the JRC DESERT Action. We thank also the JRC/MARS Remote Sensing Database, the ESA, and the ESA GlobCover Project led by MEDIAS-France for the GlobCover data set.

REFERENCES

ABUHRC. 2004. Disasters Bulletin 4. *Aon Benfield UCL Hazard Research Centre*. London, U.K.: University College of London.

Aussedat, O., M. Taberner, N. Gobron, and B. Pinty. 2007. *MERIS Level 3 Land Surface Aggregated Products Description*. Institute for Environment and Sustainability, EUR Report No. 22643 EN.

Baret, F., O. Hagolle, B. Geiger, P. Bicheron, B. Miras, M. Huc, B. Berthelot, F. Nino, M. Weiss, O. Samain, J.-L. Roujean, and M. Leroy. 2007. LAI, fAPAR and fCover CYCLOPES global products derived from VEGETATION: Part 1: Principles of the algorithm. *Remote Sensing of Environment* 110(3):275–286.

Brown, J.F., B.D. Wardlow, T. Tadesse, M.J. Hayes, and B.C. Reed. 2008. The Vegetation Drought Response Index (VegDRI): A new integrated approach for monitoring drought stress in vegetation. *GIScience and Remote Sensing* 45(1):16–46.

Camacho-de Coca, F., J. Garcia-Haro, J. Melia, and J.-L. Roujean. 2007. Prototyping algorithm for retrieving FAPAR using MSG data in the context of the LSA SAF project. *Geoscience and Remote Sensing Symposium, 2007. IGARSS 2007.* Boston, MA: IEEE International, pp. 1016–1020.

Chen, J.M. and J. Cihlar. 1999. *BOREAS RSS-07 LAI, Gap Fraction, and fPAR Data Set.* Oak Ridge National Laboratory Distributed Active Archive Center, Oak Ridge, TN.

CIA. 2003. *CIAINFORMA*. Rome, Italy: Confederazione Italiana Agricoltori.

Ciais, P., M. Reichstein, N. Viovy, A. Granier, J. Ogee, V. Allard, M. Aubinet, N. Buchmann, C. Bernhofer, A. Carrara, F. Chevallier, N. DeNoblet, A.D. Friend, P. Friedlingstein, T. Grunwald, B. Heinesch, P. Keronen, A. Knohl, G. Krinner, D. Loustau, G. Manca, G. Matteucci, F. Miglietta, J.M. Ourcival, D. Papale, K. Pilegaard, S. Rambal, G. Seufert, J.F. Soussana, M.J. Sanz, E.D. Schulze, T. Vesala, and R. Valentini. 2005. Europe-wide reduction in primary productivity caused by the heat and drought in 2003. *Nature* 437:529–533.

COPA-COGECA. 2003. *Assessment of the Impact of the Heat Wave and Drought of the Summer 2003 on Agriculture and Forestry*. COPA-COGECA, Bruxelles, Belgium.

De Roo, A.P.J., C.G. Wesseling, and W.P.A. Van Deursen. 2000. Physically based river basin modelling within a GIS: The LISFLOOD model. *Hydrological Processes* 14:1981–1992.

ESA GlobCover Project. 2008. *GlobCover Products Description Manual.* ESA/ESA Globcover Project, led by MEDIAS-Paris, France.

European Commission. 2007a. Addressing the challenge of water scarcity and droughts in the European Union. In *Communication from the Commission to the European Parliament and Council.* COM (2007) 414 Final, Commission of the European Communities, Luxembourg: Office for Official Publications of the European Communities. http://eur-lex.europa.eu/LexUriServ/LexUriServ.do?uri=COM:2007:0414:FIN:EN:PDF (accessed August 6, 2011).

European Commission. 2007b. *Water Scarcity and Droughts, In-Depth Assessment.* DG Environment, Brussels, Belgium. http://ec.europa.eu/environment/water/quantity/pdf/comm_droughts/2nd_int_report.pdf (accessed on December 13, 2011).

Gao, B.-C. 1996. NDWI—A normalized difference water index for remote sensing of vegetation liquid water from space. *Remote Sensing of Environment* 58:257–266.

GCOS. 2006. *Systematic Observation Requirements for Satellite-Based Products for Climate.* GCOS-107, UN Global Climate Observing System.

Ghulam, A., Q. Qin, and Z. Zhan. 2006. Designing of the perpendicular drought index. *Environmental Geology* 52(6):1045–1052.

Gobron, N., O. Aussedat, and B. Pinty. 2006b. *Moderate Resolution Imaging Spectroradiometer (MODIS)—JRC-FAPAR Algorithm Theoretical Basis Document.* EUR Report No. 22164 EN, Institute for Environment and Sustainability.

Gobron, N., O. Aussedat, B. Pinty, M. Taberner, and M. Verstraete. 2004. *Medium Resolution Imaging Spectrometer (MERIS—An Optimized FAPAR Algorithm—Theoretical Basis Document.* Revision 3.0. EUR Report No. 21386 EN, Institute for Environment and Sustainability.

Gobron, N., B. Pinty, O. Aussedat, J. Chen, W.B. Cohen, R. Fensholt, V. Gond, K.F. Hummerich, T. Lavergne, F. Mélin, J.L. Privette, I. Sandholt, M. Taberner, D.P. Turner, M. Verstraete, and J.-L. Widlowski. 2006a. Evaluation FAPAR products for different canopy radiation transfer regimes: Methodology and results using JRC products derived from SeaWiFS against ground-based estimations. *Journal of Geophysical Research* 111:D13110.

Gobron, N., B. Pinty, O. Aussedat, M. Taberner, O. Faber, F. Melin, T. Lavergne, M. Robustelli, and P. Snoeij. 2008. Uncertainty estimates for the FAPAR operational products derived from MERIS—Impact of top-of-atmosphere radiance uncertainties and validation with field data. *Remote Sensing of Environment* 112(4):1871–1883.

Gobron, N., B. Pinty, F. Mélin, M. Taberner, M. Verstraete, A. Belward, T. Lavergne, and J.-L. Widlowski. 2005. The state of vegetation in Europe following the 2003 drought. *International Journal Remote Sensing Letters* 26(9):2013–2020.

Gobron, N., B. Pinty, F. Mélin, M. Taberner, and M.M. Verstraete. 2002. *Sea Wide Field-of-View Sensor (SeaWiFS)-Level 2 Land Surface Products-Algorithm Theoretical Basis Document.* Institute for Environment and Sustainability, EUR Report No. 20144 EN, 23 p.

Gobron, N., B. Pinty, F. Mélin, M. Taberner, M. Verstraete, M. Robustelli, and J.-L. Widlowski. 2007. Evaluation of the MERIS/ENVISAT fAPAR Product. *Advances in Space Research* 39:105–115.

Gobron, N., B. Pinty, M.M. Verstraete, and J.-L. Widlowski. 2000. Advanced vegetation indices optimized for up-coming sensors: Design, performance, and applications. *IEEE Transactions on Geoscience and Remote Sensing* 38(6):2489–2505.

Gobron, N., B. Pinty, M. Verstraete, and Y. Govaerts. 1997. A semidiscrete model for the scattering of light by vegetation. *Journal of Geophysical Research-Atmospheres* 102:9431–9446.

Gobron, N., B. Pinty, M. Verstraete, and Y. Govaerts. 1999. The MERIS Global Vegetation Index (MGVI): Description and preliminary application. *International Journal of Remote Sensing* 20(9):1917–1927.

Gobron, N. and M. Verstraete. 2009. *fAPAR—Fraction of Absorbed Photosynthetically Active Radiation.* Essential Climate Variables Report, Version 8. Rome, Italy: Global Terrestrial Observing System.

Gonzalez Loyarte, M.M. and M. Menenti. 2008. Impact of rainfall anomalies on Fourier parameters of NDVI time series of northwestern Argentina. *International Journal of Remote Sensing* 29(4):1125–1152.

Govaerts, Y., M. Verstraete, B. Pinty, and N. Gobron. 1999. Designing optimal spectral indices: A feasibility and proof of concept study. *International Journal of Remote Sensing* 20:1853–1873.

Gu, Y., E. Hunt, B.D. Wardlow, J.B. Basara, J.F. Brown, and J.P. Verdin. 2008. Evaluation of MODIS NDVI and NDWI for vegetation drought monitoring using Oklahoma Mesonet soil moisture data. *Geophysical Research Letters* 35:doi:10.1029/2008GL035772.

ISMEA. 2004. Press Release, December 2, 2004. *Istituto di Servizi per il Mercato Agricolo Alimentare*, Rome, Italy.

Jamieson, P.D., R.J. Martin, G.S. Francis, and D.R. Wilson. 1995. Drought effects on biomass production and radiation-use efficiency in barley. *Field Crops Research* 43(2–3):77–86.

Jeyaseelan, A.T. 2004. Droughts & floods assessment and monitoring using remote sensing and GIS. In *Satellite Remote Sensing and GIS Applications in Agricultural Meteorology*, eds. M.V.K. Sivakumar, P.S. Roy, K. Harsen, and S.K. Saha, pp. 291–313. Geneva, Switzerland: World Meteorological Organization.

Ji, L. and A.J. Peters. 2003. Assessing vegetation response to drought in the northern Great Plains using vegetation and drought indices. *Remote Sensing of Environment* 87(1):85–98.

Karnieli, A., N. Agam, R.T. Pinker, M. Anderson, M.L. Imhoff, G.G. Gutman, N. Panov, and A. Goldberg. 2010. Use of NDVI and land surface temperature for drought assessment: Merits and limitations. *Journal of Climate* 23:618–633.

Knyazikhin, Y., J. Glassy, J.L. Privette, Y. Tian, A. Lotsch, Y. Zhang, Y. Wang, J.T. Morisette, P. Votava, R.B. Myneni, R.R. Nemani, and S.W. Running. 1999. MODIS leaf area index (LAI) and fraction of photosynthetically active radiation absorbed by vegetation (FPAR) product (MOD15) algorithm theoretical basis document. http://modis.gsfc.nasa.gov/data/atbd/atbd_mod15.pdf (accessed July 22, 2011).

Knyazikhin, Y., J.V. Martonchik, D.J. Diner, R.B. Myneni, M. Verstraete, B. Pinty, and N. Gobron. 1998. Estimation of vegetation canopy leaf area index and fraction of absorbed photosynthetically active radiation from atmosphere-corrected MISR data. *Journal of Geophysical Research* 103:32239–32256.

Kogan, F.N. 1995. Application of vegetation index and brightness temperature for drought detection. *Advances in Space Research* 11:91–100.

Laguardia, G. and S. Niemeyer. 2008. On the comparison between the LISFLOOD modelled and the ERS/SCAT derived soil moisture estimates. *Hydrology and Earth System Sciences Discussions* 5:1227–1265.

Legg, B.J., W. Day, D.W. Lawlor, and K.J. Parkinson. 1979. The effects of drought on barley growth: Models and measurements showing the relative importance of leaf area and photosynthetic rate. *Journal of Agricultural Science* 92:703–716.

Martin-Ortega, J. and A. Markandya. 2009. *The Costs of Drought: The Exceptional 2007–2008 Case of Barcelona*. BC3 Working Paper Series 2009-09. Bilbao, Spain: Basque Centre for Climate Change (BC3).

McCallum, I., W. Wagner, C. Schmullius, A. Shvidenko, M. Obersteiner, S. Fritz, and S. Nilsson. 2010. Comparison of four global FAPAR datasets over Northern Eurasia for the year 2000. *Remote Sensing of Environment* 114(5):941–949.

McKee, T.B, N.J. Doeskin, and J. Kleist. 1993. The relationship of drought frequency and duration to time scales. *Proceedings of the Eighth Conference on Applied Climatology*, pp. 179–184. Boston, MA: American Meteorological Society.

McKee, T.B, N.J. Doeskin, and J. Kleist. 1995. Drought monitoring with multiple time scales. *Proceeding of the Ninth Conference on Applied Climatology*, pp. 233–236. Boston, MA: American Meteorological Society.

Monteith, J.L. 1977. Climate and the efficiency of crop production in Britain. *Philosophical Transactions of the Royal Society Biological Sciences* 281:277–294.

Myneni, R.B. and D.L. Williams. 1994. On the relationship between fAPAR and NDVI. *Remote Sensing of Environment* 49:200–211.

Pinty, B., N. Gobron, F. Mélin, and M. Verstraete. 2002. *Time Composite Algorithm Theoretical Basis Document*. EUR Report No. 20150 EN. Brussels, Belgium: Institute for Environment and Sustainability.

Pinty, B., T. Lavergne, M. Vossbeck, T. Kaminski, O. Aussedat, R. Giering, N. Gobron, M. Taberner, M. Verstraete, and J.-L. Widlowski. 2007. Retrieving surface parameters for climate models from MODIS-MISR albedo products. *Journal of Geophysical Research* 112:D10116, doi: 10.1029/2006JD008105.

Plummer, S., O. Arino, W. Simon, and W. Steffen. 2006. Establishing an Earth observation product service for the terrestrial carbon community: The Globcarbon initiative. *Mitigation and Adaptation Strategies for Global Change* 11:97–111.

Rahman, H., B. Pinty, and M.M. Verstraete. 1993. Coupled surface-atmosphere reflectance (CSAR) model. 2. Semiempirical surface model usable with NOAA advanced very high resolution radiometer data. *Journal of Geophysical Research* 98(20):20791–20801.

Roerink, G.J., M. Menenti, W. Soepboer, and Z. Su. 2003. Assessment of climate impact on vegetation dynamics by using remote sensing. *Physics and Chemistry of the Earth* 28:103–109.

Rossi, S., C. Weissteiner, G. Laguardia, B. Kurnik, M. Robustelli, S. Niemeyer, and N. Gobron. 2008. Potential of MERIS fAPAR for drought detection. In *Proceedings of the Second MERIS/(A)ATSR User Workshop*, ESA SP-666, eds. H. Lacoste and L. Ouwehand. Frascati, Italy: ESA Communication Production Office.

Rouse, J.W., R.H. Haas, J.A. Schell, and D.W. Deering. 1974. Monitoring vegetation systems in the Great Plains with ERTS. In *Proceedings of the Third Earth Resources Technology Satellite-1 Symposium,* pp. 309–317. Washington, DC: NASA Scientific and Technical Information Office.

Savitzky, A. and M.J.E. Golay. 1964. Smoothing and differentiation of data by simplified least squares procedures. *Analytical Chemistry* 36(8):1627–1639.

Seixas, J., N. Carvalhais, C. Nunes, and A. Benali. 2009. Comparative analysis of MODIS-FAPAR and MERIS–MGVI datasets: Potential impacts on ecosystem modeling. *Remote Sensing of Environment* 113(12):2547–2559.

Song, X., G. Saito, M. Kodama, and H. Sawada. 2004. Early detection system of drought in East Asia using NDVI from NOAA AVHRR data. *International Journal of Remote Sensing* 25:3105–3111.

Su, Z., Y. Abreham, J. Wen, G. Roerink, Y. He, B. Gao, H. Boogaard, and C. van Diepen. 2003. Assessing relative soil moisture with remote sensing data: Theory, experimental validation, and application to drought monitoring over the North China Plain. *Physics and Chemistry of the Earth* 28:89–101.

Thoni, H. 1977. Testing the difference between two coefficients of correlation. *Biometrical Journal* 19(5):355–359.

UNEP. 2004. *Impacts of Summer 2003 Heat Wave in Europe*. Environment Alert Bulletin. Geneva, Switzerland: UNEP-DEWA/GRID-Europe.

Vermote, E., D. Tanré, J.-L. Deuzé, M. Herman, and J.J. Morcrette. 1997. Second simulation of the satellite signal in the solar spectrum: An overview. *IEEE Transactions on Geoscience and Remote Sensing* 35(3):675–686.

Verstraete, M.M., N. Gobron, O. Aussedat, M. Robustelli, B. Pinty, J.-L. Widlowski, and M. Taberner. 2008. An automatic procedure to identify key vegetation phenology events using the JRC-FAPAR products. *Advances in Space Research* 41:1773–1783.

Verstraete, M.M. and B. Pinty. 1996. Designing optimal spectral indices for remote sensing applications. *IEEE Transactions on Geoscience and Remote Sensing* 34:1254–1265.

Vogt, J., P. Soille, A.L. de Jager, E. Rimaviciute, W. Mehl, S. Foisneau, K. Bodis, J. Dusart, M.L. Paracchini, P. Haastrup, and C. Bamps. 2007. *A Pan-European River and Catchment Database*. JRC Reference Report 22920. Luxembourg: Office of the Official Publications of the European Communities. http://ccm.jrc.ec.europa.eu (accessed on December 13, 2011).

Wagner, W., G. Lemoine, and H. Rott. 1999. A method for estimating soil moisture from ERS scatterometer and soil data. *Remote Sensing of Environment* 70:191–207.

Wang, J., P.M. Rich, and K.P. Price. 2003. Temporal response of NDVI to precipitation and temperature in the central Great Plains, USA. *International Journal of Remote Sensing* 24:2345–3364.

Wang, L. and J.J. Qu. 2007. NMDI: A normalized multi-band drought index for monitoring soil and vegetation moisture with satellite remote sensing. *Geophysical Research Letters* 34:L20405, doi: 10.1029/2007GL031021.

WMO. 2009. *Lincoln Declaration on Drought Indices*. Geneva, Switzerland: World Meteorological Organization. http://www.wmo.int/pages/prog/wcp/agm/meetings/wies09/documents/Lincoln_Declaration_Drought_Indices.pdf (accessed on December 13, 2011).

Part II

Evapotranspiration

6 Remote Sensing of Evapotranspiration for Operational Drought Monitoring Using Principles of Water and Energy Balance

Gabriel B. Senay, Stefanie Bohms, and James P. Verdin

CONTENTS

6.1 INTRODUCTION

Evapotranspiration (ET) is an important component of the hydrologic budget because it reflects the exchange of mass and energy between the soil–water–vegetation system and the atmosphere. Prevailing weather conditions influence potential or reference ET through variables such as radiation, temperature, wind, and relativity humidity. In addition to these weather variables, actual ET (ET_a) is also affected by land cover type and condition, as well as soil moisture. The dependence of ET_a on land cover and soil moisture, and its direct relationship with carbon dioxide assimilation in plants, makes it an important variable for monitoring drought, crop yield, and biomass—a critical capability for decision makers interested in food security, grain markets, water allocation, and carbon sequestration (Bastiaanssen et al., 2005).

Because ET can be difficult to measure accurately, especially at large spatial scales, several different hydrologic modeling techniques have been developed to estimate ET_a using satellite remote sensing. In general, the ET modeling techniques can be grouped into two broad classes that include models based on surface energy balance (e.g., Bastiaanssen et al., 1998; Su et al., 2005; Allen et al., 2007; Anderson et al., 2007; Senay et al., 2007) and water balance (e.g., Allen et al., 1998, Senay, 2008) principles. While water balance models focus on tracking the pathways and magnitude of rainfall in the soil–vegetation system, most remote sensing energy balance models use land surface temperature (LST) as a primary constraint in partitioning radiant energy available at the surface between heat and water fluxes.

This chapter describes two ET models representing each of these approaches: the Vegetation ET (VegET) water balance model (Senay, 2008) and the Simplified Surface Energy Balance (SSEB) approach (Senay et al., 2007, 2011a), comparing their utility for operational drought monitoring and agrohydrologic applications. Both models use the concept of a reference ET (ET_o) to estimate the potential ET (ET_p) expected under unlimited water conditions, assuming an idealized reference crop with standardized bulk and aerodynamic resistance factors for vapor transport. The main difference between the two approaches is in the calculation of a correction factor accounting for soil moisture impacts on evaporation, estimating ET_a as a fraction of ET_o. VegET uses a vegetation water budgeting approach to track soil moisture changes, whereas the energy balance model uses spatial variations in LST.

Both models were designed for global operational applications and are therefore intentionally simplified in their representation of surface phenomena and modest in their input data requirements—based only on readily available remote sensing data. The simplified approaches facilitate real-time implementation in data-limited parts of the world, providing timely information for operational drought and food security analyses with minimal manual intervention and expert guidance.

6.2 MATERIALS AND METHODS

Here we provide a brief introduction to the VegET and SSEB ET modeling algorithms. The two approaches each have their own merits and limitations, and they can be used independently or in combination. The choice of model depends on the availability of data and the objective of the project. Both methods require an ET_o data set, which can be generated using meteorological data (net radiation, temperature, wind speed, relative humidity, and air pressure). In addition, the availability of rainfall and land surface phenology (LSP) data is critical for the VegET water balance model, while the SSEB energy balance approach requires LST information retrieved from thermal infrared satellite data. These differences in data inputs are important and define the applications and constraints that apply to each modeling approach. For example, the presence of cloud cover adversely affects the SSEB model because LST cannot be retrieved under cloudy conditions using thermal imaging. In contrast, the VegET model does not use thermal data and is less affected by cloud cover, which can be a significant advantage during the growing season in many parts of the world. On the other hand, VegET considers only rain-fed water inputs to the land surface system, whereas the LST inputs to SSEB provide diagnostic information about moisture inputs from all sources, including irrigation and shallow water tables. Another advantage of the SSEB approach is that it does not require precipitation data and thus is less prone to errors associated with the quality of the available precipitation data sets.

6.2.1 Data Requirements

To facilitate global applications, both the SSEB and VegET modeling systems have been designed to use readily available global remote sensing and weather data sets. Input data requirements by each model, and rationale thereof, are described in the following.

6.2.1.1 Precipitation Data

Precipitation is a key driver of the water balance VegET model. A combination of coarse (25 km for 1996–2004) and finer (5 km for 2005 to current) spatial resolution daily total rainfall data from the National Oceanic and Atmospheric Administration (NOAA) National Weather Service (NWS) (http://www.srh.noaa.gov/rfcshare/precip_about.php) is being used based on data availability. Both precipitation data sets yield comparable seasonal ET_a estimates from VegET (data not shown). Spatial resolution of ET_a output from VegET is not significantly limited by the input precipitation data set but rather by the scale of the LSP data used in the model. Furthermore, rainfall is relatively homogeneous at the subwatershed scale when aggregated over monthly or longer time scales.

6.2.1.2 Land Surface Phenology Data

As described in the following, LSP parameters used in VegET are defined using a time series of 1 km Normalized Difference Vegetation Index (NDVI) data derived from the NOAA Advanced Very High Resolution Radiometer (AVHRR) satellite imagery for the period of 1989–2004 (Eidenshink, 1990). These data sets have been normalized over multiple AVHRR instruments.

6.2.1.3 Soil Data

Soil water holding capacity, used in VegET, is derived from the State Soil Geographic Database (STATSGO) (http://www.ncgc.nrcs.usda.gov/products/datasets/statsgo/) for the United States, while data from the Food and Agriculture Organization (FAO) Digital Soils Map of the World are used for global applications.

6.2.1.4 Reference ET

Over the continental United States (CONUS), the ET_0 data used by both VegET and SSEB are produced at a daily time step as described by Senay et al. (2008) using the standardized Penman-Monteith equation (Allen et al., 1998). Global, six hourly weather data sets of net radiation, wind, relative humidity, and air temperature and pressure from the Global Data Assimilation System (GDAS) (Kanamitsu, 1989) are used to generate a global daily ET_0 at 1° spatial resolution.

6.2.1.5 Thermal Remote Sensing Data

The SSEB energy balance algorithm is mainly driven by LST derived from thermal band observations acquired by the Moderate Resolution Imaging Spectroradiometer (MODIS). Day-time, 8 day average LST tiles at 1 km resolution from the NASA Terra platform (MOD11A2), acquired from March 2000 to present, have been downloaded from the LP DAAC (Land Processes Data Active Archive Center) and reprojected and mosaicked using the MODIS reprojection tool. Although instantaneous LST data retrievals (e.g., from the MOD11_L2 swath product) are technically more appropriate for application of SSEB algorithms, the 8 day composite product is used here to reduce computational and data demands for operational global applications. Furthermore, use of the 8 day product reduces data gaps caused by cloud contamination. Ramifications of using the MODIS 8 day LST product are discussed further in Section 6.2.2.2.

6.2.2 Model Descriptions

6.2.2.1 Water Balance Model: VegET

The VegET approach is based on the most widely used water balance technique for operational crop performance monitoring: the Food and Agricultural Organization (FAO) algorithm for computing the crop Water Requirement Satisfaction Index (WRSI; FAO, 1986). The WRSI reflects the relative relationship (ratio/percent) between water supply (from rainfall and existing soil moisture) and demand (crop transpiration demand to meet its physiological needs) using observed data from the beginning of the crop season (planting) until the current date. WRSI is calculated as the ratio (or percentage) between the seasonal ET and the seasonal water requirement of the crop. The seasonal total water requirement is calculated as the ET_p adjusted by a crop coefficient (K_c), which varies by crop type and phenological stage. K_c generally varies between 0.3 and 1.2 for most cereal crops during the growing season (FAO, 1998).

The Famine Early Warning System Network (FEWS NET) demonstrated a regional implementation of the FAO WRSI over a modeling domain in southern Africa (Verdin and Klaver, 2002). Senay and Verdin (2003) further enhanced the geospatial model by introducing the concept of Maximum Allowable Depletion (MAD) and a soil water stress factor from irrigation engineering for better estimation of ET_a as a function of soil water content. The Senay and Verdin (2003) version of the model has been operational since 2000, with daily and 10 day outputs for Africa, Central America, and Afghanistan at 0.1° (~10 km) resolution. Graphics of model output are posted operationally at http://earlywarning.usgs.gov/fews/.

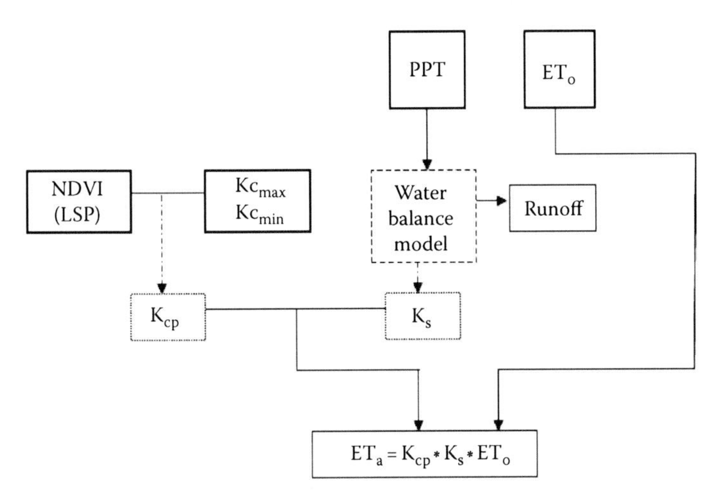

FIGURE 6.1 Simplified conceptual diagram of the VegET model. Major inputs are precipitation (PPT), reference ET (ET_o), and NDVI. Estimated parameters are a phenology-based crop coefficient (K_{cp}) and a soil-water stress factor (K_s). Model outputs are ET_a and runoff.

Building on the WRSI concept, the VegET modeling strategy (Senay, 2008) was recently developed for estimating ET_a in nonirrigated cropland and grassland environments as an enhancement to the U.S. Geological Survey (USGS)/FEWS NET crop water balance model (Senay and Verdin, 2003). VegET blends concepts from irrigation engineering with a remote sensing data stream to estimate ET_a quickly at low computational and data costs for sites anywhere in the world. Figure 6.1 shows a schematic representation of the VegET modeling framework.

A key innovation in the VegET model is the inclusion of the LSP parameter, which describes the seasonal progression of vegetation growth and development. The LSP allows the VegET model to be location (pixel)-specific, accommodating localized variations in vegetation growth patterns, as compared to the region-specific K_c function used in traditional agrohydrologic modeling. LSP can be observed by spaceborne sensors and is a key biophysical parameter that links the water and carbon cycles with anthropogenic activities, providing an important approach to change detection in terrestrial ecosystems (e.g., Goward et al., 1985; Reed et al., 1994; Tucker et al., 2001; de Beurs and Henebry, 2005). Integration of LSP information into a phenology-based crop coefficient (K_{cp}) is described later in this section.

VegET monitors soil water levels in the root zone through a daily (or longer time step) water balance algorithm and estimates ET_a in rain-fed cropland and grassland environments. Key input data to VegET are precipitation, ET_o, soil water holding capacity, and LSP. ET_a (in units of mm/day) is calculated as the product of the ET_o (mm/day), a soil moisture stress coefficient (K_s), and a phenology-based water-use coefficient (K_{cp}), as shown in Equation 6.1 (Senay, 2008):

$$ET_a = K_s * K_{cp} * ET_o \tag{6.1}$$

The K_s parameter is determined from a vegetation–soil–water balance model and has a value between 0 (dry soil) and 1 (moist soil). The water balance model works with a daily soil moisture accounting procedure over a soil bucket that is defined by the water holding capacity of the soil on a grid-cell basis. The LSP coefficient (K_{cp}) is comparable to the K_c widely used by agronomists (Allen et al., 1998) but includes an LSP dependence derived from remotely sensed time series of NDVI (Senay, 2008). K_{cp} represents both the spatial and temporal dynamics of the landscape water-use patterns on a grid-cell (or pixel) basis. The K_{cp} parameter is scaled between published crop coefficient minimum (Kc_{min}) and maximum (Kc_{max}) values based on current and climatological NDVI data:

$$K_{cp} = \frac{Kc_{max} - Kc_{min}}{NDVI_{max} - NDVI_{min}} * (NDVI_i - NDVI_o) \tag{6.2}$$

where

Kc_{max} is the maximum (mature) K_c value for a particular vegetation/crop type
Kc_{min} is the minimum (early stage) K_c value
$NDVI_{min}$ and $NDVI_{max}$ are the climatological minimum and maximum NDVI values in a year, respectively
$NDVI_i$ is the climatological NDVI value for a given period "i" (average weekly maximum value in this case)
$NDVI_o$ is the minimum reference NDVI value that is associated with the minimum K_c value

The calculation of $NDVI_o$ depends on the $NDVI_{max}$ specified at each pixel and is determined using one of two following cases:

Case I: If $NDVI_{max} >= 0.40$, then

$$NDVI_o = 0.3 \tag{6.3}$$

Case II: If $NDVI_{max} < 0.40$, then

$$NDVI_o = 0.33 * (NDVI_{max} - NDVI_{min}) + NDVI_{min} \tag{6.4}$$

Equations 6.3 and 6.4 were formulated to handle sparsely vegetated semiarid and arid regions differently from well-vegetated areas. Even a low maximum NDVI region will show a water-use phenology if it is rescaled differently in relation to its own minimum rather than the "global" minimum of NDVI = 0.3. Other researchers have used a different formulation to estimate K_c values or comparable coefficients from NDVI (e.g., Nagler et al., 2005; Groeneveld et al., 2007; Allen et al., 2011) for the same purpose of estimating ET_a.

A major assumption in the specification of K_{cp} is that there have been no major climate or land cover changes over the remote sensing data record to affect the water-use dynamics of a given individual modeling cell (or pixel) as represented by the LSP. This limits the utility of VegET (with the current setup) for monitoring highly managed landscapes such as irrigated agriculture and urban/rural fringe areas. However, with a modification of the water balance component of the model, current NDVI values are still capable of estimating ET from irrigated lands, as is demonstrated by Nagler et al. (2005).

For operational monitoring over the United States, the VegET model is run at 10 km spatial resolution (chosen to reduce computational time for a regional application) with operational products updated and posted daily at 7:00 pm (http://earlywarning.usgs.gov/usewem/swi.php). The operational products focus on the growing season period, defined as April 1–October 31.

6.2.2.2 Energy Balance Model: SSEB

Surface energy balance methods have been successfully applied by several researchers (Bastiaanssen et al., 1998; Su et al., 2005; Allen et al., 2007; Anderson et al., 2007) to estimate crop water use in irrigated areas and across the general landscape. The approach taken in these models requires solution of the energy balance equation at the land surface (Equation 6.5), computing the latent heat flux (ET_a converted into units of energy, W/m^2, as a residual):

$$\lambda E = R_n - G - H \tag{6.5}$$

where

λE is the latent heat flux (energy consumed by ET; W/m^2)
R_n is the net radiation at the surface (W/m^2)
G is the ground heat flux (energy stored in the soil and vegetation; W/m^2)
H is the sensible heat flux (energy used to heat the air; W/m^2)

Most thermal energy balance algorithms intended for operational ET monitoring have been explicitly designed to minimize sensitivity to errors in the absolute calibration and atmospheric correction of the LST data. Allen et al. (2007) describe a surface energy balance method that employs the *hot* (dry) and *cold* (wet) pixel approach of Bastiaanssen et al. (1998) in the SEBAL (Surface Energy Balance Algorithm for the Land) model, constraining ET_a estimates between reasonable bounds as defined at these end-member pixels. As such, these methods do not require absolute accuracy in LST but only relative accuracy in variability across the scene. For net radiation, SEBAL requires meteorological data on incoming radiation, along with the associated surface albedo and emissivity required to compute outgoing radiation. The ground heat flux is estimated using remote sensing estimates of surface temperature, albedo, and NDVI, while the sensible heat flux is estimated as a function of temperature gradient above the surface, surface roughness, and wind speed.

Although the full energy balance approach employed in SEBAL has been shown to give good results in many parts of the world (Bastiaanssen et al., 2005), well-trained operators are required to perform the selection of hot/cold end-member pixels, and input data requirements can be prohibitive, especially over large, data-sparse regions. As an alternative, the SSEB approach was developed at USGS Earth

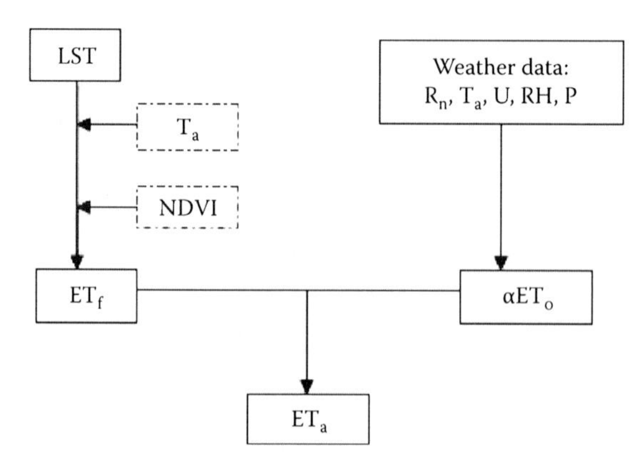

FIGURE 6.2 Schematic representation of the SSEB modeling setup. Suggested value for α is 1.2 when ET_o is based on clipped grass reference ET. R_n is net radiation, T_a is air temperature, U is wind speed, RH is relative humidity, and P is atmospheric pressure.

Resources Observation and Science (EROS) Center for operational application (Senay et al., 2007, 2011a).

The SSEB approach involves two basic steps (Figure 6.2). ET_a is computed as a product of the reference ET fraction (ET_f) and the reference ET (ET_o):

$$ET_a = ET_f * \alpha ET_o \tag{6.6}$$

where α is a multiplying factor that is generally set to 1.2 if ET_o is from the standardized clipped grass reference or 1.0 if an alfalfa-based reference ET is used. Local calibration using lysimeter data is recommended to accurately estimate α if the absolute magnitude of ET is critical for the study. For drought monitoring purposes, where anomalies with respect to "average" conditions are more important, the consistency of the method and data is more important than the absolute accuracy of ET (see Section 6.2.3).

The ET_f variable is the key to the SSEB approach since it captures the impact of soil moisture on ET_a, while ET_o determines the potential ET under nonlimiting water supply conditions. In the revised SSEB approach presented in this chapter, ET_f is calculated from the LST and air temperature data sets based on the assumptions that a *hot* pixel experiences little or no ET (Bastiaanssen et al., 1998; Allen et al., 2005), and a *cold* pixel represents maximum ET. An assumption is made in SSEB that ET can be scaled between these two end-point values of ET in proportion to the difference between LST and air temperature (T_a) measured at each pixel. Note that the method does not rely on absolute accuracy in either LST or T_a; however, it is required that the difference between the LST and T_a be relatively accurate across the study region. The main driver for the ET_f is the difference between LST and air temperature in relation to the same difference measured at the reference locations (hot and cold pixels). Across the LST scene, SSEB assumes that pixels with larger surface-to-air temperature differences have higher sensible heat (lower ET), while pixels with small (LST − T_a) have lower sensible heat (high ET). The inclusion of air temperature in the ET_f calculation in the revised SSEB approach is intended to facilitate continental application of the SSEB approach, reducing the need to select multiple end-member LST pairs across the continent in different climatic regions, which is a typical requirement for hot-cold pixel approaches.

The *hot* pixels are selected using an NDVI image as a guide to identify the locations of dry and nonvegetated (or sparsely vegetated) areas that exhibit very low NDVI values. Similarly, the *cold* pixels are selected from well-watered, healthy, and fully vegetated areas that have very high NDVI values. The ET fraction ($ET_{f,x}$) is calculated for each pixel "x" as

$$ET_{f,x} = \frac{dT_h - dT_x}{dT_h - dT_c} \tag{6.7}$$

where

dT_h is the difference between surface temperature (T_s) and air temperature (T_a) at the hot pixel

dT_c is the difference between T_s and T_a at the cold pixel

dT_x is the difference between T_s and T_a at a given pixel "x"

The method is sensitive to the selection of hot and cold pixels, and caution should be taken in selecting these reference points. The cold pixel can be a water body or well-watered dense vegetation, preferably with an NDVI value greater than or equal to 0.7. Since the energy balance partitioning of a water body is different from a land surface, a water body may be colder (if fed from snowmelt) or warmer (if fed from a geothermal source) than most dense vegetation, but this will only bring a systematic bias that can be corrected by checking against the LST from a well-watered vegetation in the same area and season. The main advantage of a water body is that it is generally available much of the year except the winter season of some regions when ET is low. This provides an advantage over the relatively short season of dense vegetation. However, it is important to remain consistent in the choice of the cold and hot pixels during the different parts of the year (i.e., if a cold pixel is chosen from a water body, it is advisable to select the same water body over time). The same principle applies to the hot pixel. In a large image scene, it is advisable to select the hot pixels from nonirrigated perpetually bare areas, with an NDVI value <0.2.

For this study over the CONUS, T_s is obtained from the MODIS 8 day LST product, while T_a is assigned from the monthly maximum air temperature (generally measured at around 1.5 m above ground level) from the PRISM (Parameter–Elevation Regressions on Independent Slopes Model; PRISM, 2011) data set, selecting the monthly interval closest to the 8 day time period corresponding to the LST data set. Eight-day ET_0 is computed from daily GDAS ET_0 output (Senay et al., 2008). The model is run at 8 day time increments over the period of record. In this chapter, only seasonal products from April through October are discussed because of their relevance for season-integrated drought monitoring. A temporally dynamic set of hot and cold pixels selected from representative locations (cold generally from the southeast United States [wetter area] and hot pixels [dry areas] in the western High Plains of the United States) has been used on the entire CONUS data set. It should be noted that although the hot and cold pixels are consistent in space, the LST values generally vary from season to season, so we prepare a unique set of hot and cold pixels for each period from the same region or location that meets the requirements. What is unique in this approach is the use of a single set of hot and cold pixels to scale across the CONUS for each 8 day period.

A number of simplifications regarding representation of land–atmosphere exchange are implicit in the SSEB algorithm, and these warrant some discussion. First, unlike SEBAL or METRIC and most thermal ET models, a full energy balance is not computed within SSEB. Rather, ET_f is scaled directly in inverse proportion to $T_s - T_a$, while other energy balance components are not assessed. This scaling neglects the effects of variable surface roughness and ground heat flux across the landscape on the surface energy balance. Also, the use of the 8 day LST composites can introduce errors into the methodology, because various pixels in the scene may be sampled on different days under different atmospheric and surface moisture conditions. Finally, local air temperature can be very different from T_a interpolated between station data (as in the PRISM data set), and this will add uncertainty to the ET estimates.

This SSEB method is experimental and requires further evaluation under a range of conditions; however, preliminary assessments are encouraging—particularly for long-term seasonal ET estimates. Gowda et al. (2009) evaluated the performance of the SSEB using lysimeter data in northwest Texas and found that it explained 84% of the lysimeter ET variation, with a mean bias of −0.6 mm/day, using pooled data sets from irrigated and rain-fed agricultural systems with corn and sorghum fields over a 2 year study (2006–2007). Recently, Senay et al. (2011b) evaluated the SSEB ET over the CONUS using an HUC-8 (Hydrologic Unit Code) level water balance approach. The annual differences between precipitation (P) and runoff (Q) at 1,399 HUC-8 level watersheds were compared to annual SSEB ET estimates with an r^2 of 0.90 and a mean bias of −67 mm or −11% of the difference between observed P and Q. The SSEB ET shows a general underestimation in the lower ET region (ET < 600 mm) compared to higher ET zones. More importantly, the high r^2 (0.90) demonstrates the precision and reliability of the approach in diverse ecosystems, especially when used as an anomaly product.

Because this method is intended for easy implementation for large-area monitoring by nonexperts, a simplified approach with minimal data requirements is desired. Additionally, the ET anomalies used for drought monitoring (see the following) are less sensitive to errors in the simplified modeling approach than are the absolute magnitudes of ET. In this context, SSEB can be considered as an index describing relative changes in ET over the satellite period of record.

6.2.2.3 Comparison of VegET and SSEB

Table 6.1 summarizes differences between the VegET and SSEB modeling approaches in terms of input and output data characteristics. Operational VegET output is currently produced over the CONUS on a daily basis, while SSEB-based ET_a for the CONUS is updated on an 8 day basis since the summer of 2011 (http://earlywarning.usgs.gov/usewem/eta_energy.php). Historical monthly SSEB ET outputs are currently available from 2000 to 2009 for the CONUS and are being validated using flux and water balance model outputs. In addition,

TABLE 6.1
Modeling and Data Characteristics of VegET and SSEB

	VegET	SSEB
Modeling approach	Water balance	Energy balance
Target monitored/output	ET_a, soil moisture, runoff	ET_a
Spatial resolution	Limited by LSP data MODIS: 250 m AVHRR: 1 km	Limited by thermal data (MODIS/AVHRR: 1 km) Landsat: ~100 m (local application)
Spatial extent	Global (potentially)	Global (potentially)
Frequency of product	Daily	8-day, daily is possible
Delay	1 day	About 2 weeks for MODIS
Period of record	Limited by rainfall data 1996–current: NexRad/ Station Blend 1979–current: GPCP (Global Precipitation Climatology Project)	Limited by thermal data AVHRR:1989–current MODIS: 2000–current
Web access	VegET model output is online at http://earlywarning.usgs.gov/usewem/swi.php	ET_a anomaly online Africa: http://earlywarning.usgs.gov/fews/africa/index.php CONUS: http://earlywarning.usgs.gov/usewem/eta_energy.php
Geographic projection	Latitude–longitude	Latitude–longitude
GIS environment	Yes	Yes
Description of product	Appropriate for rain-fed agriculture or grassland environments	Best applied to irrigated systems
Challenge/limitation for operational implementation	No major limitation is anticipated	Cloud cover and lack of climatic record

initial results for Africa and river basins in central Asia are showing promising results (UNEP, 2010; Senay et al., 2007). Applications for both approaches are presented later in this chapter.

6.2.3 ET_a Anomalies

For drought monitoring, indicators are typically formulated in terms of a monthly to seasonal anomaly, representing deviations of current conditions with respect to "normal" or "average" historical conditions for that time period. This is because anomalies (wetter or drier than usual) are easier to understand and measure than are absolute quantities (e.g., rainfall or ET in mm). Anomaly information is also more relevant for decision makers because it provides a historical context for how current conditions compare to conditions from previous years. In addition, impacts of model assumptions, formulation errors, and biases in input data are reduced in anomaly products. The main reason for this is that the statistical nature of anomaly calculation cancels multiplicative errors (e.g., due to model formulation and input data biases) that appear in both the numerator and denominator in Equation 6.8.

In this study, seasonal ET_a anomalies were calculated over the CONUS for both VegET and SSEB based on the median of the seasonal (April 1–October 31) total ET from 2000 to 2009 (data available years) as

$$ET_{a}\_ano = \frac{ET_{a}\_year}{ET_{a}\_median} * 100 \quad (6.8)$$

where

ET_{a}_ano is the ET_a anomaly for a given year in percent
ET_{a}_year is the seasonal ET_a total for a given year
ET_{a}_median is the median seasonal ET_a from 2000 to 2009

Although anomalies can be calculated at different time scales, this chapter focuses on seasonal time scales to highlight the utility of anomaly products for assessing agricultural drought impacts, which are generally felt at a seasonal time scale. An example of an international operational ET_a anomaly product for Africa using the SSEB model is presented later in the chapter.

6.3 VegET AND SSEB OUTPUT OVER THE CONUS

6.3.1 Cumulative ET_a

Seasonal cumulative ET maps for 2009 over the CONUS generated with the VegET and SSEB models are presented in Figure 6.3a and b, respectively. Generally, the two maps are comparable both in magnitude and spatial patterns in the predominantly rain-fed system of the eastern United States. Output from both models in this region exhibits high seasonal ET in excess of 500 mm, particularly in the southeast. More notable differences between models are observed in the western United States, for reasons that may vary by location. For example, the models predict significantly different fluxes in irrigated regions such as the Central Valley of California, where crops are expected to have high ET because of targeted water applications. The estimate from the VegET water balance model is low because irrigation water inputs are not accounted for in this modeling approach. In contrast, the contribution of irrigation is reflected in the MODIS LST data input into the SSEB model, resulting in higher, more representative ET values over these areas. There are also differences in the ET results for some areas of the northwest where vegetation with high ET may be benefiting from snowmelt/soil moisture/groundwater storage processes during the April–October growing season. This may be another example of impacts of nonrainfall-related moisture inputs that are captured by the SSEB model but not by the VegET model, which is driven by rainfall alone and does not account for snowmelt or runoff. Significant differences in the VegET and SSEB ET estimates over Minnesota and Wisconsin require further investigation. Extensive lakes, wetlands, and near-surface groundwater contributions to the evaporative flux may be contributing to the higher

ET fluxes predicted by SSEB in northern Minnesota, but in general, differences in this area may reflect regional surface properties and land-atmosphere couplings that are not properly accounted for in the simplified energy balance approach.

6.3.2 ET_a Anomalies

Seasonal ET anomaly maps for both VegET and SSEB models are presented in Figure 6.3c and d for 2009, respectively. The severe drought in south Texas and the southwest United States, in parts of Arizona and California, is clearly depicted in both maps, where below-average conditions (<50% normal ET) prevailed. These areas also compare well with the drought depiction by the U.S. Drought Monitor (USDM) of moderate to severe drought for much of the growing season (data not shown but available at http://drought.unl.edu/dm/archive.html). In contrast, above-average moisture conditions are indicated by both models for much of the High Plains region spanning parts of North Dakota to western Texas. Similar above-average moisture

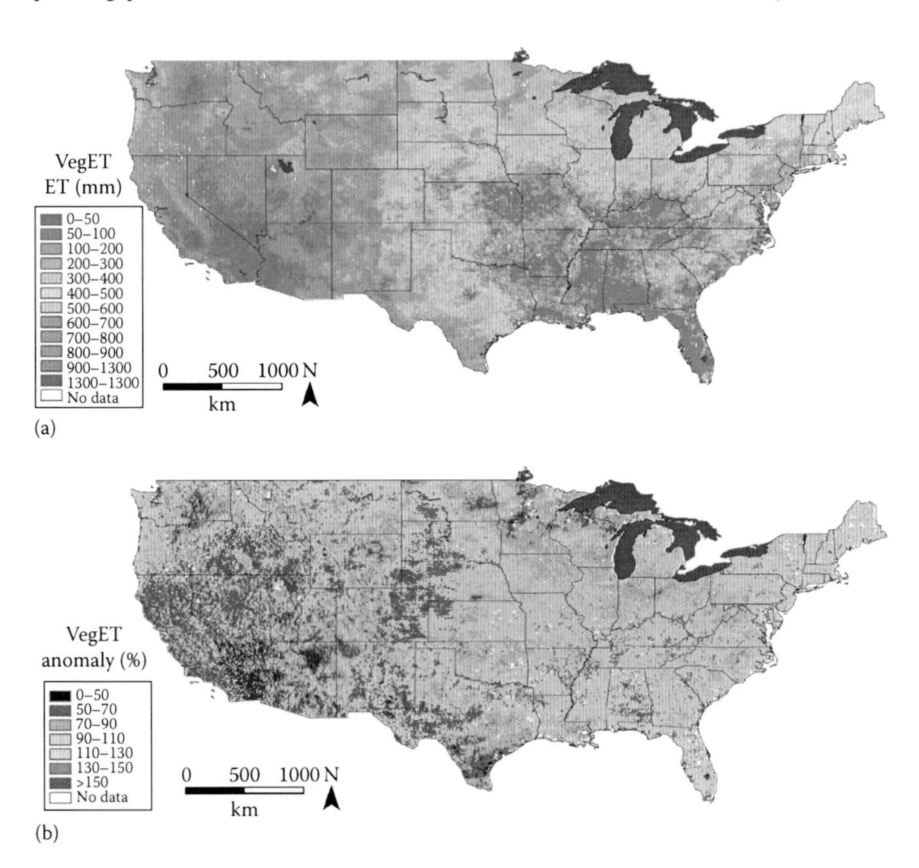

FIGURE 6.3 **(See color insert.)** Seasonal (April–October) total ET_a (mm) and ET anomalies (%) for CONUS in 2009: (a) seasonal VegET ET_a, (b) seasonal SSEB ET_a, (c) seasonal VegET ET_a anomaly, and (d) seasonal SSEB ET_a anomaly.

(continued)

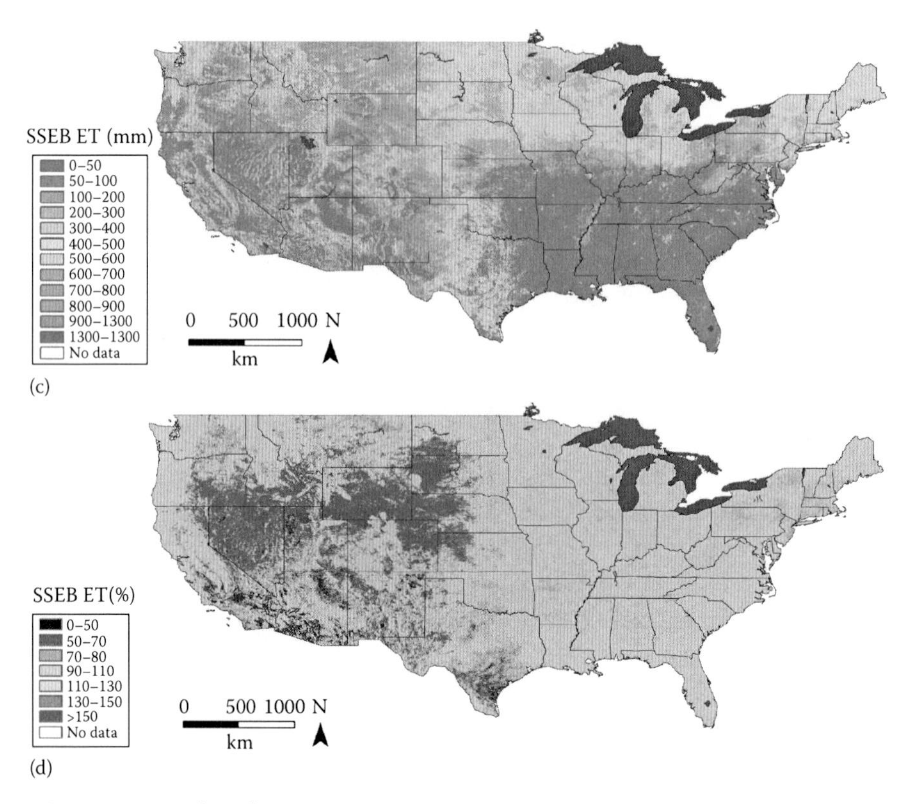

FIGURE 6.3 (continued)

conditions were also detected over much of the northern part of the semiarid western United States in both model results. Both of these results are consistent with the USDM, which assigned most of these areas a nondrought designation over the year.

Although there is a general agreement between the two maps, some regions exhibit significant discrepancies, including the areas in Minnesota and Wisconsin that were highlighted in the previous section. While SSEB indicates normal conditions, VegET suggests ET is below average from the viewpoint of rainfall distribution during the growing season. This is in agreement with the USDM, which classified the region as experiencing hydrological drought for much of the 2009 growing season. This suggests that the water balance and energy balance approaches may be responding differently to varying hydrologic processes that affect the timing, magnitude and severity of agricultural and hydrological droughts.

The obvious textural difference in the spatial patterns represented in the two maps results from differences in spatial resolution. The SSEB is modeled at 1 km while the VegET is produced at 10 km, but this should not affect results and conclusions made at a regional scale. These results illustrate the potential for both approaches to generate valuable ET information for operational drought monitoring, but more investigation is required to better understand the ET estimation differences between the two modeling approaches and determine how they can best be used as complementary data sources.

6.3.3 Case Study: Seasonal ET Time Series

To demonstrate the relative utility of the ET and anomaly products generated by the SSEB and VegET models, a county-based analysis was conducted for two selected counties with contrasting conditions in 2009. One county was located in a drought-affected part of south Texas (Duval County) and another from central Nebraska (Custer County), which had above-average rainfall over the growing season that year.

Figure 6.4a and b show monthly ET totals and anomalies, respectively, from the two models for the two counties. A closer look at Figure 6.4a shows that both models, as expected, predict higher ET for Custer County than for drought-stricken Duval

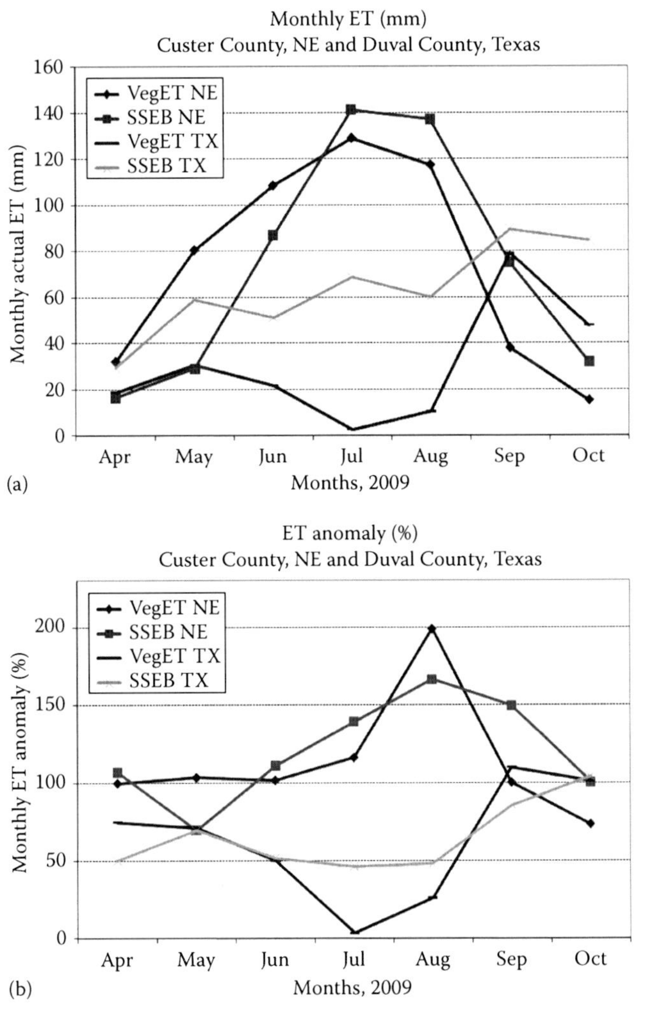

FIGURE 6.4 Monthly county-average ET (mm) and the corresponding anomalies (%) for two selected counties in the central United States in 2009 using VegET and SSEB models: (a) monthly totals for Custer County, Nebraska, and Duval County, Texas, and (b) monthly anomalies for Custer and Duval.

County. Better agreement between model outputs was obtained in Custer County, with a seasonal monthly average of 74 mm from both models. In contrast, the models gave significantly different results for Duval County, with VegET and SSEB estimating seasonal monthly averages of 30 and 63 mm, respectively.

This illustrates potential differences between the VegET and SSEB approaches to ET estimation and drought monitoring. According to a map of irrigated agricultural land area for the United States (Brown et al., 2009), agriculture in Custer County is generally under a predominantly rain-fed system, and rainfall moisture inputs are well represented in both modeling approaches. In comparison, Duval County appears to contain a higher fraction of irrigated land area in the irrigated agricultural land map. The increased ET due to applied irrigation water in Duval County would be captured by LST-based SSEB but is not accounted for in the VegET model. The largest difference between the VegET and SSEB ET curves for Duval County occur in June and July, which is generally a time of peak irrigation for most crops. By September, when irrigation is not as readily used and most of the ET is met by rainfall, the VegET and SSEB ET results were in better agreement (within 12%). This result suggests that a comparison of SSEB and VegET maps may be a valuable tool for identifying irrigated agricultural areas. Furthermore, in principle, the difference between the two approaches may be used to estimate the amount of irrigation that is applied (i.e., the SSEB ET would provide the total ET irrespective of the water source while the VegET ET_a can account for the amount of ET supplied by rainfall).

Although monthly ET totals in absolute terms are important for agrohydrologic analysis, we cannot infer from the plots in Figure 6.4a whether the counties are in a drought or how severe the moisture deficits might be. The monthly ET anomalies for both counties presented in Figure 6.4b are a more valuable tool for this application. The anomalies were calculated by comparing the monthly ET in 2009 to the historical 10 year median monthly ET values (2000–2009) for the same month. This plot shows that Duval County had below-average ET during 2009, reflecting the observed drought, while the ET for Custer County was above average for most of the season because of more favorable weather conditions. Furthermore, the anomalies from both models are in better agreement than are the monthly ET totals, which further illustrates the value of using ET anomaly information in drought detection. Despite the large discrepancies observed between total monthly ET from VegET and SSEB (30 mm vs. 63 mm) for Duval County, the seasonal monthly anomalies are 62% and 65% for VegET and SSEB, respectively (Figure 6.4b).

6.4 APPLICATIONS OF VegET AND SSEB FOR THE FAMINE EARLY WARNING SYSTEM NETWORK

The livelihood of most rural populations in sub-Saharan Africa is based on traditional rain-fed agriculture that is dependent on seasonal rainfall. Knowledge of crop water usage and soil moisture status that can be obtained through remotely sensed ET products provides valuable information for managing water resources and anticipating crop failure (Tadesse et al., 2008). The FEWS NET (http://earlywarning.usgs.gov/fews/)

has developed various tools that use readily available satellite-derived and model-assimilated weather data sets to monitor health and productivity of rain-fed agricultural areas. Based on ease of implementation and minimal input data requirements, the VegET and SSEB models are being integrated as operational monitoring tools within the FEWS NET system.

As noted earlier, the VegET model has its origins in the original FAO WRSI (FAO, 1986). The operational FEWS crop water balance model uses the same principles of FAO method in the calculations of the WRSI values based on region-specific crop calendars but parameterizes the calculation of ET as a function of soil moisture. VegET further improves the FEWS crop water balance model by introducing a location-specific crop water-use coefficient that is derived from remotely sensed data.

Although the Africa operational crop water balance model is still running with a prescribed crop calendar, a plan is underway to integrate the VegET parameterization, with a more objective vegetation calendar based on remotely sensed data that is specific to a location instead of a region. In light of this, initial work was done to apply the VegET model to estimate the Nile Basin water balance dynamics (Senay et al., 2009), highlighting the potential of the approach not only in drought monitoring but also for hydrologic studies. Figure 6.5a and b compare VegET ET_a estimates with mean annual precipitation over the basin, derived from satellite-based rainfall estimate (Xie and Arkin, 1997). As expected, high and low rainfall regions in Figure 6.5a show corresponding high and low ET, respectively, in Figure 6.5b as the result of rainfall and vegetation cover. Note that VegET does not capture the effects of intense irrigation that occurs along the Nile River, particularly at the Delta in Egypt where the Nile River empties into the Mediterranean Sea (extreme north). The diagnostic LST inputs to SSEB handle this better through the "total" ET estimation approach instead of the rainfall-driven water balance models (data not shown).

With the FEWS NET principle of reliance on a convergence of evidence, USGS/FEWS NET just launched an operational implementation of the SSEB ET modeling approach to complement the existing water balance products using a MODIS data stream for the entire African continent. An operational ET anomaly product has been produced and staged at the FEWS NET website (http://earlywarning.usgs.gov/fews/africa/index.php) since June 2011. The new SSEB products consist of monthly and cumulative ET anomalies at 1 km resolution. A sample product is shown in Figure 6.6, highlighting the severe drought (up to <50% of normal) in east Africa as a seasonal anomaly between January 1 and July 3, 2011 (most recent available data). The product shows an above-average ET (>110% of normal) in parts of southern Africa and normal conditions (ranging between 90% and 100% of normal) in much of Africa, including the irrigated basin of the Nile River Delta. Irrigated areas tend to show normal conditions from year to year since irrigation application is not affected by the year-to-year variability of rainfall as long as the water is sourced from large reservoirs, as is the case for the Nile River Delta, which is regulated by the Aswan High Dam/Lake Nasser.

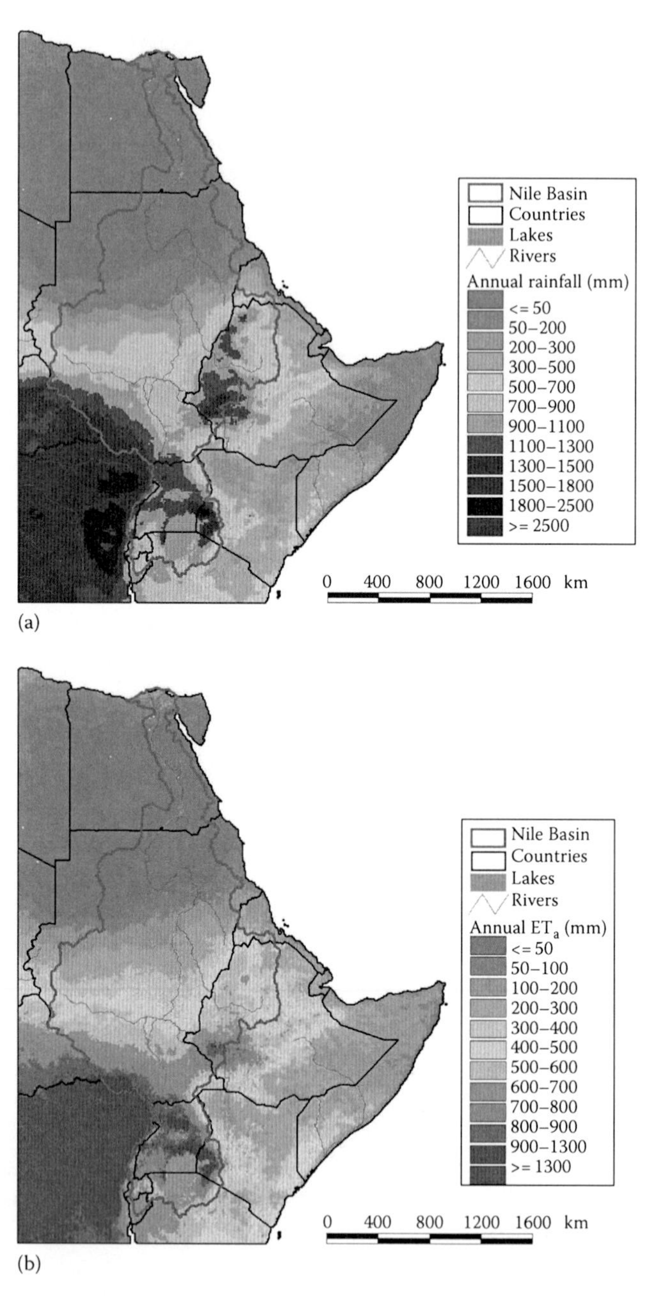

FIGURE 6.5 **(See color insert.)** Spatial distribution of satellite-derived annual rainfall in northeastern Africa (median of 2001–2007) (a) and annual ET_a from the VegET model (median from the same period as the rainfall) (b).

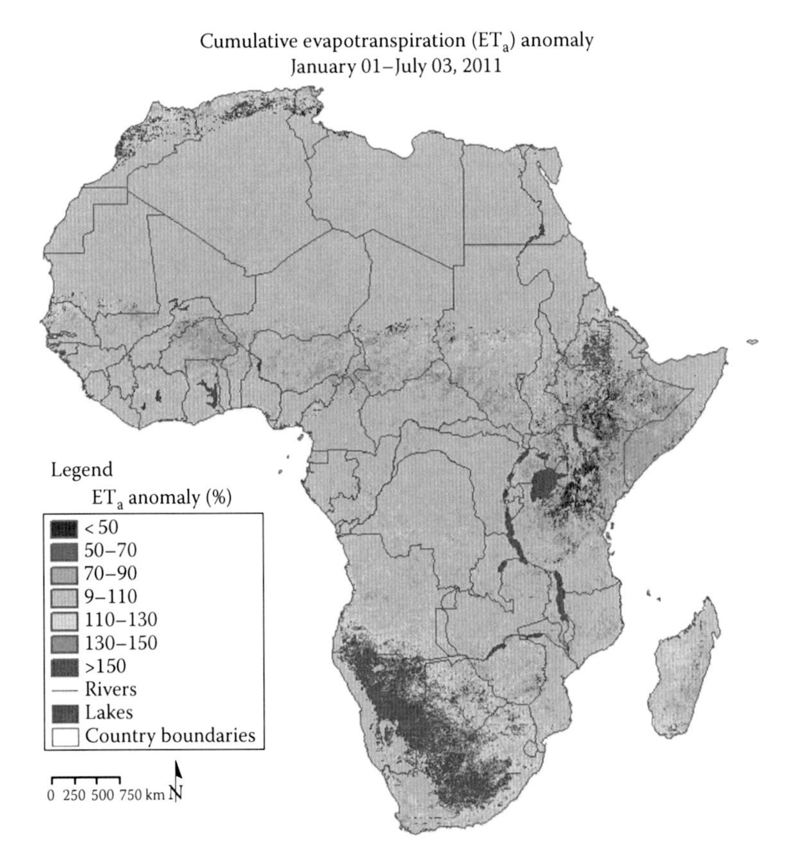

FIGURE 6.6 **(See color insert.)** Africa-wide seasonal anomaly of ET_a from the SSEB model output for 2011 as of July 3, 2011 (January 1–July 3). SSEB ET anomaly is operationally processed and posted on a FEWS NET website regularly on an 8 day time step.

6.5 CONCLUSIONS

The main objective of this chapter was to demonstrate the use of remotely sensed data in simplified energy and water balance modeling approaches to estimating ET for drought monitoring and agrohydrologic applications. Both VegET and SSEB models were able to capture the general spatial patterns of seasonal ET over much of the CONUS. However, notable differences were observed between their seasonal ET totals, particularly over locations where water sources other than rainfall (e.g., irrigation, snowmelt runoff, and subsurface irrigation from high water tables) are available to vegetation. The anomaly maps proved to be more useful in detecting year-to-year changes than seasonal ET totals, which are prone to errors associated with data and model assumptions and simplifications.

Although both approaches can provide comparable results for drought monitoring using the anomaly products, each may have unique advantages in some specific applications and locations. For example, the water balance approach (rainfall based) provides more information on temporal soil moisture variability and runoff as a by-product of the VegET model, which is useful for other hydrologic applications. In comparison, the SSEB ET modeling is more useful for quantifying ET from non-rain-fed systems such as irrigation and groundwater-fed vegetation systems since SSEB ET estimates ET regardless of the water source. However, some of the differences exhibited between VegET and SSEB require further investigation to fully understand the processes involved and determine synergistic applications.

This study highlights that simplified modeling techniques and parameterization that use readily available global satellite data and model-assimilated weather data sets can be implemented effectively in an operational setup for timely assessment of drought hazards and monitoring agrohydrologic conditions in data-poor parts of the world. Recently, FEWS NET has implemented an operational setup of the SSEB over Africa for agricultural monitoring in Africa using the MODIS data stream as part of the convergence of evidence principle advocated by FEWS NET. Because of the global nature of the input data sets to both the SSEB and VegET models, there are opportunities to expand these products in different parts of the world. Field data are required to validate and calibrate these models before using the products in an absolute sense for water balance applications. However, the models can produce reliable anomaly products that can be used for drought detection.

ACKNOWLEDGMENT

Any use of trade, firm, or product names is for descriptive purposes only and does not imply endorsement by the U.S. Government.

REFERENCES

Allen, R.G., L.S. Pereira, T.A. Howell, and M.E. Jensen. 2011. Evapotranspiration information reporting: 1. Factors governing measurement accuracy. *Agricultural Water Management* 98:899–920.

Allen, R.G., L. Pereira, D. Raes, and M. Smith. 1998. *Crop Evapotranspiration.* Food and Agriculture Organization (FAO) Publication 56. Rome, Italy: FAO of the United Nations.

Allen, R.G., M. Tasumi, A. Morse, and R. Trezza. 2005. A Landsat-based energy balance and evapotranspiration model in Western US water rights regulation and planning. *Irrigation and Drainage Systems* 19:251–268.

Allen, R.G., M. Tasumi, A. Morse, R. Trezza, W. Kramber, I. Lorite, and C.W. Robison. 2007. Satellite-based energy balance for mapping evapotranspiration with internalized calibration (METRIC) – Applications. *Journal of Irrigation and Drainage Engineering* 133:395–406.

Anderson, M.C., J.M. Norman, J.R. Mecikalski, J.A. Otkin, and W.P. Kustas. 2007. A climatological study of evapotranspiration and moisture stress across the continental United States based on thermal remote sensing: 1. Model formulation. *Journal of Geophysical Research* 112:D10117, doi:10.1029/2006JD007506.

Bastiaanssen, W.G.M., M. Menenti, R.A. Feddes, and A.A.M. Holtslag. 1998. A remote sensing surface energy balance algorithm for land (SEBAL): 1. Formulation. *Journal of Hydrology* 212(213):213–229.

Bastiaanssen, W.G.M., E.J.M. Noordman, H. Pelgrum, G. Davids, B.P. Thoreson, and R.G. Allen. 2005. SEBAL model with remotely sensed data to improve water-resources management under actual field conditions. *Journal of Irrigation and Drainage Engineering* 131:85–93.

Brown, J.F., S. Maxwell, and M.S. Pervez. 2009. Mapping irrigated lands across the United States using MODIS satellite imagery. In *Remote Sensing of Global Croplands for Food Security*. eds. P.S. Thenkabail, J.G. Lyon, H. Turral, and C.M. Biradar, pp. 177–198. Boca Raton, Florida: Taylor & Francis.

de Beurs, K.M. and G.M. Henebry. 2005. A statistical framework for the analysis of long image time series. *International Journal of Remote Sensing* 26(8):1551–1573.

Eidenshink, J.C. 1990. The 1990 conterminous US AVHRR data set. *Photogrammetric Engineering and Remote Sensing* 58:809–913.

FAO. 1986. *Yield Response Function*. FAO Irrigation and Drainage Paper 3. Rome: Food and Agricultural Organization of the United Nations.

FAO. 1998. *Crop Evapotranspiration: Guideline for Comparing Crop Water Requirements*. FAO Irrigation and Drainage Paper 56. Rome: Food and Agricultural Organization of the United Nations.

Goward, S.N., C.J. Tucker, and D.G. Dye. 1985. North American vegetation patterns observed with the NOAA-7 advanced very high resolution radiometer. *Plant Ecology* 64(1):3–14.

Gowda, P.H., G.B. Senay, T.A. Howell, and T.H. Marek. 2009. Lysimetric evaluation of simplified surface energy balance approach in the Texas High Plains. *Applied Engineering in Agriculture* 25:665–669.

Groeneveld, D.P., W. Baugh, J. Sanderson, and D. Cooper. 2007. Annual groundwater evapotranspiration mapped from single satellite scenes. *Journal of Hydrology* 344:146–156.

Kanamitsu, M. 1989. Description of the NMC global data assimilation and forecast system. *Weather and Forecasting* 4:335–342.

Nagler, P., R. Scott, C. Westenburg, J. Cleverly, E. Glenn, and A. Huete. 2005. Evapotranspiration on western U.S. rivers estimated using the Enhanced Vegetation Index from MODIS and data from eddy covariance and Bowen ratio flux towers. *Remote Sensing of Environment* 97:337–351.

PRISM. 2011. *PRISM Products Matrix*. PRISM Climate Group, Oregon State University, Corvallis, Oregon. http://www.prism.oregonstate.edu/products/matrix.phtml?vartype=tmax&view=maps (accessed on December 12, 2011).

Reed, B.C., J.F. Brown, D. VanderZee, T.R. Loveland, J.W. Merchant, and D.O. Ohlen. 1994. Measuring phenological variability from satellite imagery. *Journal of Vegetation Science* 5(5):703–714.

Senay, G.B. 2008. Modeling landscape evapotranspiration by integrating land surface phenology and a water balance algorithm. *Algorithms* 1:52–68.

Senay, G.B., K.O. Asante, and G.A. Artan. 2009. Water balance dynamics in the Nile Basin. *Hydrological Processes* 23:3675–3681.

Senay, G.B., M.E. Budde, and J.P. Verdin. 2011a. Enhancing the Simplified Surface Energy Balance (SSEB) approach for estimating landscape ET: Validation with the METRIC model. *Agricultural Water Management* 98:606–618.

Senay, G.B., M. Budde, J.P. Verdin, and A.M. Melesse. 2007. A coupled remote sensing and simplified surface energy balance approach to estimate actual evapotranspiration from irrigated fields. *Sensors* 7:979–1000.

Senay, G.B., S. Leake, P.L. Nagler, G. Artan, J. Dickinson, J.T. Cordova, and E.P. Glenn. 2011b. Estimating basin scale evapotranspiration (ET) by water balance and remote sensing methods. *Hydrological Processes* (in press). DoI:10.1002/hyp.8379.

Senay, G.B. and J. Verdin. 2003. Characterization of yield reduction in Ethiopia using a GIS-based crop water balance model. *Canadian Journal of Remote Sensing* 29(6):687–692.

Senay, G.B., J.P. Verdin, R. Lietzow, and A.M. Melesse. 2008. Global daily reference evapotranspiration modeling and evaluation. *Journal of American Water Resources Association* 44:969–979.

Su, H., M.F. McCabe, E.F. Wood, Z. Su, and J. Prueger. 2005. Modeling evapotranspiration during SMACEX: Comparing two approaches for local and regional scale prediction. *Journal of Hydrometeorology* 6(6):910–922.

Tadesse, T., M. Haile, G. Senay, B.D. Wardlow, and C.L. Knutson. 2008. The need for integration of drought monitoring tools for proactive food security management in sub-Saharan Africa. *Natural Resources Forum* 32(4):265–279.

Tucker, C.J., D.A. Slayback, J.E. Pinzon, S.O. Los, R.B. Myneni, and M.G. Taylor. 2001. Higher northern latitude normalized difference vegetation index and growing season trends from 1982 to 1999. *International Journal of Biometeorology* 45:184–190.

UNEP. 2010. *Africa Water Atlas*. Division of Early Warning and Assessment (DEWA). United Nations Environmental Programme (UNEP). Nairobi, Kenya.

Verdin, J. and R. Klaver. 2002. Grid cell based crop water accounting for the famine early warning system. *Hydrological Processes* 16:1617–1630.

Xie, P. and P.A. Arkin. 1997. A 17-year monthly analysis based on gauge observations, satellite estimates, and numerical model outputs. *Bulletin of the American Meteorological Society* 78(11):2539–2558.

7 Thermal-Based Evaporative Stress Index for Monitoring Surface Moisture Depletion

Martha C. Anderson, Christopher R. Hain, Brian D. Wardlow, Agustin Pimstein, John R. Mecikalski, and William P. Kustas

CONTENTS

7.1 INTRODUCTION

The standard suite of indicators currently used in operational drought monitoring reflects anomalous conditions in several major components of the hydrologic budget—representing deficits in precipitation, soil moisture content, runoff, surface and groundwater storage, snowpack, and streamflow. In principle, it is useful to have a diversity of indices because drought can assume many forms (meteorological, agricultural, hydrological, and socioeconomic), over broad ranges in timescale (weeks to years), and with varied impacts of interest to different stakeholder groups. Farmers, for example, may be principally interested in soil moisture deficits, river forecasters will focus on streamflow fluctuations, and water managers will be concerned with longer-term stability in municipal water supply and reservoir levels. Only recently has actual evapotranspiration (ET) been considered as a primary indicator of drought conditions (e.g., Anderson et al., 2007b; Labedzki and Kanecka-Geszke, 2009; Li et al., 2005; Mo et al., 2010). ET is a valuable drought indicator because it reflects not only moisture availability but also the rate at which water is being consumed. Because transpiration (T) and carbon uptake by vegetation are tightly coupled through stomatal exchange, ET anomalies are indicative of vegetation health and growing conditions. In addition, the importance of so-called flash droughts is becoming increasingly evident, where hot, dry, and windy atmospheric conditions can lead to unusually rapid soil moisture depletion and, in some cases, devastating crop failure. Such events cannot be easily identified using local precipitation anomalies but should have a detectable ET signature.

In general, techniques for mapping ET can be classified into either prognostic or diagnostic modeling approaches. Prognostic approaches, like the land-surface models (LSMs) implemented within the National Land Data Assimilation System (NLDAS; Mitchell et al., 2004), use spatially distributed observations of precipitation as input to compute the full water balance at every model grid cell, considering soil texture and moisture holding capacity, runoff, and the local rate of infiltration and drainage. Such models are extremely useful in drought monitoring because they can generate self-consistent anomaly indicators relating to each component of the surface water budget. However, accurate assessment of critical model inputs can be challenging, particularly for large-scale applications. Real-time precipitation analyses of reasonable quality are available over most of the United States (e.g., McEnery et al., 2005), but many parts of world lack sufficiently dense radar and rain-gauge networks to generate comparable data sets. Satellite-derived global precipitation products provide improved spatial coverage over these data-limited areas (Huffman et al., 2007; Joyce et al., 2004) but are known to exhibit seasonally and spatially dependent biases (Villarini et al., 2009; Zeweldi and Gebremichael, 2009) and often rely on gauge correction to produce realistic rainfall amounts. Soil moisture storage, runoff, and infiltration are strongly determined by the assumed soil properties, which are typically linked to soil type and may have high uncertainties. Biased specifications of precipitation rates and soil hydraulic properties can introduce significant cumulative biases into prognostic water budget estimates (Schaake et al., 2004), with potentially deleterious effects on climatologies derived for drought monitoring. Furthermore, LSM predictions will not reflect non-precipitation-related moisture inputs to the local land-surface system (e.g., irrigation or influence of shallow groundwater) unless these inputs are explicitly modeled.

In contrast, diagnostic ET mapping techniques typically require significantly less a priori knowledge of antecedent moisture inputs and subsurface conditions. These methods use remote-sensing measurements of key land-surface state variables to "diagnose" the current surface moisture status. Because evaporation cools surfaces, the land-surface temperature (LST) state conveys valuable proxy information regarding soil moisture and is commonly used in diagnostic estimates of

ET (Courault et al., 2005; Kalma et al., 2008; Norman et al., 1995b). The observed LST implicitly reflects all moisture inputs to the land-surface system, both known (precipitation) and unknown (e.g., natural or anthropogenic groundwater extraction), and, therefore, these inputs do not need to be explicitly specified as in the prognostic modeling approach. Diagnostic ET models therefore provide information about actual water consumption, which may exceed local short-term moisture inputs when water is mined or manually transferred between basins. This information is a valuable supplement to the "natural consumption" estimates typically conveyed by prognostic water balance models.

In this chapter, we discuss a new drought index based on diagnostic remote sensing of ET. The Evaporative Stress Index (ESI) represents temporal anomalies in the ratio of actual ET to potential ET (PET), derived from satellite imagery collected in the thermal infrared (TIR) atmospheric window channel (10–12 μm). TIR imagery is used to compute LST, which serves as a boundary condition on the surface energy balance, including the evaporative flux. The modeling system used here is multiscale, running over continental scales using TIR imagery acquired with the Geostationary Operational Environmental Satellites (GOES) combined with shortwave information about vegetation cover fraction to diagnose evaporative fluxes at 5–10 km spatial resolution using the Atmosphere-Land Exchange Inverse Model (ALEXI; Anderson et al., 2007b). Higher resolution assessments of surface moisture stress can be obtained through spatial disaggregation (DisALEXI; Norman et al., 2003) using TIR data from polar orbiting systems such as Landsat, the Advanced Spaceborne Thermal Emission and Reflection Radiometer (ASTER), and the Moderate Resolution Imaging Spectroradiometer (MODIS), which image at resolutions ranging between 100 and 1000 m. Because the ESI does not use rainfall data as input, it provides an independent check on precipitation-based drought indicators and may be more robust in regions with minimal ground-based meteorological infrastructure. The remotely sensed ET fields have the advantage that they inherently include non-precipitation-related moisture signals that need to be modeled a priori in prognostic LSM schemes.

Here we compare the ESI with standard precipitation-based drought indices and with drought classifications recorded over the continental United States (CONUS) in retrospective U.S. Drought Monitor (USDM; Svoboda et al., 2002) maps from 2000 to 2009. The goal of this analysis is to establish the level of similarity between ET- and precipitation-based indices and to identify new and unique information regarding drought conditions that only diagnostic ET estimates can provide.

7.2 TIR-BASED MODELING OF EVAPOTRANSPIRATION

The ALEXI surface energy balance model (Anderson et al., 1997, 2007b,c; Mecikalski et al., 1999) was specifically designed to minimize the need for ancillary meteorological data while maintaining a physically realistic representation of land-atmosphere exchange over a wide range of vegetation cover conditions. It is one of few diagnostic LSMs designed explicitly to exploit the high temporal resolution afforded by geostationary satellites like GOES, which is ideal for operational large-area applications such as drought monitoring.

Surface energy balance models estimate ET by partitioning the energy available at the land surface ($RN - G$, where RN is net radiation and G is the soil heat conduction flux, in W m^{-2}) into turbulent fluxes of sensible and latent heating (H and λE, respectively, in W m^{-2}):

$$RN - G = H + \lambda E \tag{7.1}$$

where,

λ is the latent heat of vaporization (J kg^{-1})
E is ET (kg s^{-1} m^{-2} or mm s^{-1})

Surface temperature is a valuable metric for constraining λE because varying soil moisture conditions yield a distinctive thermal signature. Moisture deficiencies in the root zone lead to vegetation stress and elevated canopy temperatures, while depletion of water from the soil surface layer causes the soil component of the scene to heat up rapidly.

The land-surface representation in the ALEXI model is based on the series version of the two-source energy balance (TSEB) model of Norman et al. (1995a) (see also Kustas and Norman, 1999, 2000), which partitions the composite surface radiometric temperature, T_{RAD}, into characteristic soil and canopy temperatures, T_S and T_C, based on the local vegetation cover fraction apparent at the thermal sensor view angle, $f(\theta)$:

$$T_{RAD}(\theta) \approx \left(f(\theta)T_C^{\,4} + [1 - f(\theta)]T_S^{\,4}\right)^{1/4} \tag{7.2}$$

(see schematic in Figure 7.1). For a canopy with a spherical leaf angle distribution and leaf area index (LAI), $f(\theta)$ can be approximated as

$$f(\theta) = 1 - \exp\left(\frac{-0.5\Omega(\theta)\,LAI}{\cos\theta}\right) \tag{7.3}$$

where $\Omega(\theta)$ is a view angle–dependent clumping factor, currently assigned by vegetation class (Anderson et al., 2005). With information about T_{RAD}, LAI, and radiative forcing, the TSEB evaluates the soil (subscript "s") and the canopy (subscript "c") energy budgets separately, computing system and component fluxes of net radiation ($RN = RN_C + RN_S$), sensible and latent heat ($H = H_C + H_S$ and $\lambda E = \lambda E_C + \lambda E_S$), and soil heat conduction (G). Importantly, because angular effects are incorporated into the decomposition of T_{RAD}, the TSEB can accommodate thermal data acquired at off-nadir viewing angles and can therefore be applied to geostationary satellite images.

The TSEB has a built-in mechanism for detecting thermal signatures of vegetation stress. A modified Priestley–Taylor relationship (PT; Priestley and Taylor, 1972), applied to the divergence of net radiation within the canopy (RN_C), provides an initial estimate of canopy transpiration (λE_C), while the soil evaporation rate (λE_S) is computed as a residual to the system energy budget. If the vegetation is stressed and transpiring at significantly less than the potential rate, the PT equation will overestimate

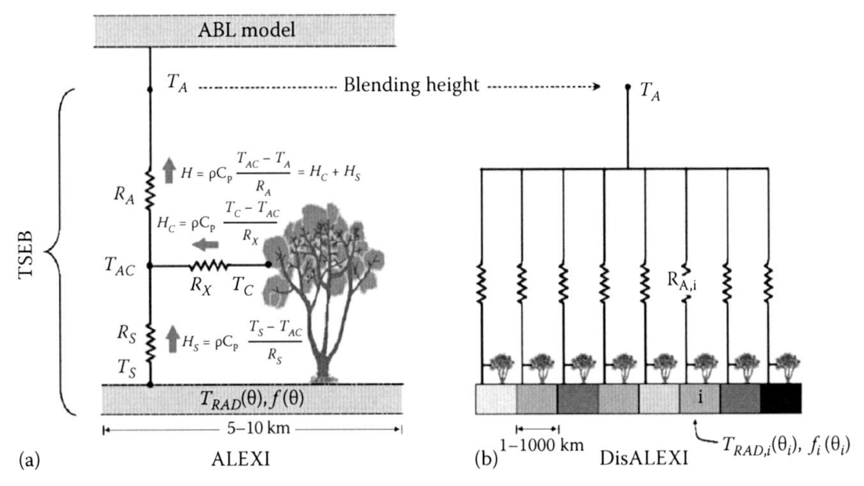

FIGURE 7.1 Schematic diagram representing the ALEXI (a) and DisALEXI (b) modeling schemes, highlighting fluxes of sensible heat (H) from the soil and canopy (subscripts "s" and "c") along gradients in temperature (T), and regulated by transport resistances R_A (aerodynamic), R_x (bulk leaf boundary layer), and R_S (soil surface boundary layer). DisALEXI uses the air temperature predicted by ALEXI near the blending height (T_A) to disaggregate 10km ALEXI fluxes, given vegetation cover ($f(\theta)$) and directional surface radiometric temperature ($T_{RAD}(\theta)$) information derived from high-resolution remote-sensing imagery at look angle θ.

λE_C and the residual λE_S will become negative. Condensation onto the soil is unlikely during midday on clear days, and, therefore, $\lambda E_S < 0$ is considered a signature of system stress. Under such circumstances, the PT coefficient is iteratively throttled back until $\lambda E_S \sim 0$ (expected under dry conditions). Both λE_C and λE_S will then be some fraction of the PET rates associated with the canopy and soil. This approach therefore opens the potential for surface and root zone moisture pool assessment and, thus, concomitant tracking of both meteorological and agricultural drought.

For regional-scale applications, the TSEB has been coupled with an atmospheric boundary layer (ABL) model to internally simulate land–atmosphere feedback on near-surface air temperature (T_A in Figure 7.1). In the ALEXI model, the TSEB is applied at two times during the morning ABL growth phase (~1 h after sunrise and before local noon) using radiometric temperature data obtained from a geostationary platform like GOES at spatial resolutions of 5–10km. Energy closure over this interval is provided by a simple slab model of ABL development (McNaughton and Spriggs, 1986), which relates the rise in air temperature in the mixed layer to the time-integrated influx of sensible heat from the land surface. As a result of this configuration, ALEXI uses only time-differential temperature signals, thereby minimizing flux errors due to absolute sensor calibration, as well as atmospheric and emissivity corrections (Kustas et al., 2001). The primary radiometric signal is the morning surface temperature rise, while the ABL model component uses only the general slope (lapse rate) of the atmospheric temperature profile (Anderson et al., 1997), which is more reliably analyzed from synoptic radiosonde data than is the absolute temperature reference.

Anderson et al. (2007a) summarize ALEXI validation experiments employing a spatial flux disaggregation technique (DisALEXI; Norman et al., 2003), which uses higher resolution TIR imagery presently only available from aircraft or polar orbiting systems like Landsat, ASTER, or MODIS to downscale the GOES-based flux estimates (10 km resolution) to the flux measurement footprint scale (on the order of 100–1000 m; see Figure 7.1). Typical root-mean-square deviations in comparison with tower flux measurements (30 min averages) of H and λE are 35–40 W m^{-2} (15% of the mean observed flux) over a range of vegetation cover types and climatic conditions, while errors at daily time steps are typically 10%. Disaggregation also facilitates high spatial resolution assessment of moisture flux and stress conditions, supporting a broader range of local-scale decision-making activities than can be serviced with the coarser-resolution ALEXI results and providing detailed insights into landscape-level ET dynamics. However, high-resolution ET mapping is constrained in temporal resolution by the overpass frequency of the polar orbiting satellite.

A complete ALEXI processing infrastructure has been developed to automatically ingest and preprocess all required input data, to execute the model, and to postprocess model output for visual display and use in other applications. The model currently runs daily on a 10 km resolution grid covering the CONUS, and to date, model input/output from this framework has been archived for the period 2000 to present. Snow-covered regions are currently masked using the 24 km resolution Daily Northern Hemisphere Snow and Ice Analysis product distributed through the National Snow and Ice Data Center (NSIDC; http://nsidc.org/data/docs/noaa/g02156_ims_snow_ice_analysis/index.html).

7.3 EVAPORATIVE STRESS INDEX

In this chapter, we explore applications of remotely sensed *ET* to drought monitoring. In particular, we examine information conveyed by anomalies in the ratio of actual to *PET*:

$$f_{PET} = \frac{ET}{PET} \tag{7.4}$$

determined under clear-sky conditions. In this analysis, *ET* and *PET* are instantaneous estimates at shortly before local noon, retrieved using ALEXI. Equation 7.4 follows from earlier work using TIR band data in agricultural applications, as reviewed by Moran (2003), where f_{PET} has been used as a tool for crop stress detection and irrigation scheduling. Limiting the assessment to clear-sky conditions separates variability in ET due to soil moisture from impacts of varying cloud climatology. In addition, TIR-based LST retrieval can be accomplished only through clear skies. Division by PET serves to normalize out some degree of variability in ET due to seasonal variations in available energy and vegetation cover amount, further refining the focus on the soil moisture signal. Standardized anomalies in f_{PET} will be referred to as the ESI.

Because the ET values used to compute the ESI are dependent on clear-sky conditions, only a portion of the ALEXI modeling domain can be filled on any given day.

On average, pixels in 75% of the CONUS domain are executed at least once every 6 days, and 95% are updated at least every 20 days. Therefore, temporal compositing of clear-sky f_{PET} values is required to fill in the full model domain. Compositing also serves to reduce effects of noise in the ET retrievals, primarily arising from incomplete cloud clearing in the LST inputs to ALEXI. ESI composites can be generated for multiple intervals to reflect different timescales of drought, analogous to the production of the Standardized Precipitation Indices. Here we examine composites computed over a moving window of 1, 2, and 3 months and a 6 month window defining the nominal growing season average for April through September. Composites are computed as an unweighted average of all index values that passed cloud-screening tests over the interval in question:

$$\langle v(w, y, i, j)\rangle = \frac{1}{nc}\sum_{n=1}^{nc} v(n, y, i, j) \quad (7.5)$$

where $\langle v(w, y, i, j)\rangle$ is the composite for week w, year y, and i, j grid location; $v(n, y, i, j)$ is the value on day n; and nc is the number of clear days during the compositing interval.

To highlight differences in moisture conditions between years, drought indices are typically presented as anomalies or percentiles with respect to multiyear average fields determined over some period of record. Standardized anomalies in f_{PET} were computed over the period 2000–2009 and are expressed as a pseudo z-score, normalized to a mean of 0 and a standard deviation of 1. Fields describing "normal" (mean) conditions and temporal standard deviations at each pixel were computed for each compositing interval. Then standardized anomalies were computed as

$$\Delta\langle v(w, y, i, j)\rangle = \frac{\langle v(w, y, i, j)\rangle - \frac{1}{ny}\sum_{y=1}^{ny}\langle v(w, y, i, j)\rangle}{\sigma(w, i, j)} \quad (7.6)$$

where the second term in the numerator defines the normal field, averaged over all years ny, and the denominator is the standard deviation, also computed over all years.

In this notation, ESI-X is defined as $\Delta\langle f_{PET}\rangle$ computed for an X-month f_{PET} composite. Like most other drought indices, this formulation generates negative values when conditions are drier than normal and positive values for wetter than normal conditions. Implicit in the application of Equation 7.6 to ALEXI f_{PET} is the assumption that these quantities are normally distributed in time at every i, j location in the CONUS grid during 2000–2009. In this case, values of ESI less than −2 represent dry conditions exceeding 2σ, which should occur 2% of the time. At present, there are not enough years in the ALEXI archive (10 points) to warrant fitting of a non-normal distribution, but such adjustments may be applied as the archive's length of record is extended.

7.4 PRECIPITATION-BASED DROUGHT METRICS

To better understand the behavior the new ET index, Anderson et al. (2011a) conducted an intercomparison with more commonly used drought indices based on precipitation observations. The suite of precipitation indices considered in the intercomparison is listed in Table 7.1 and described briefly in the following.

7.4.1 Palmer Indices

The Palmer indices examined here include the Palmer Drought Severity Index (PDSI; Palmer, 1965), the Palmer Modified Drought Severity Index (PMDI; Heddinghaus and Sabol, 1991), the short-term (monthly timescale) Palmer Z Index, and the longer-term Palmer Hydrological Drought Index (PHDI). The principle advantages of the Palmer indices are a long period of record and a long history of usage, which has fostered familiarity within the drought community. However, the Palmer indices do have some specific limitations, which are reviewed in detail by Alley (1984) and Karl (1983). These limitations relate to lack of spatial and temporal standardization, along with simplistic treatment of evaporative losses and soil moisture storage.

Palmer-Z, PDSI, PDMI, and PHDI data sets are distributed by the National Climatic Data Center (NCDC) at the climate division level and on a monthly time step from 1895 to present (http://www1.ncdc.noaa.gov/pub/data/cirs/). These products are based on rain gauge and air temperature data that have been spatially averaged at the climate division scale (Guttman and Quayle, 1996). For this study, these Palmer data sets were regridded to the 10 km ALEXI grid for the study period, maintaining a constant index value over each of the climate division polygons.

7.4.2 Standardized Precipitation Index

Issues with PDSI and variants thereof inspired the formation of the SPI (McKee et al., 1993, 1995), which uses observed precipitation as its only input. Precipitation data at a given location are converted into probabilities based on a local long-term climatology.

TABLE 7.1
Drought Indicators Included in the Intercomparison Study

Index	Acronym	Type
U.S. Drought Monitor	USDM	Multi-index synthesis
Evaporative Stress Index (X-month composite)	ESI-X	Remote sensing of f_{PET}
Standardized Precipitation Index (X-month)	SPI-X	Precipitation
Palmer Z Index	Z	Precipitation + storage
Palmer Drought Severity Index	PDSI	Precipitation + storage
Palmer Modified Drought Index	PMDI	Precipitation + storage
Palmer Hydrological Drought Index	PHDI	Precipitation + storage

The probabilities are then standardized such that a value of 0 indicates the median precipitation amount (in comparison with the climatology), which is measured at that pixel over the time interval in question (Edwards and McKee, 1997). The SPI can be computed for multiple timescales (typically ranging from 2 to 52 weeks) to monitor both short- and long-term drought conditions. Because the SPI is based only on precipitation data, a long period of record spanning many decades can be constructed. A major disadvantage of the SPI (and the Palmer indices) for mapping applications is that high-quality gridded precipitation data are not available at high spatial resolution for most parts of the world. In addition, lack of a temperature component means there is no accounting for rate of atmospheric consumption through evaporation.

SPI data are distributed by NCDC at the climate division level and on a monthly time step from 1895 to present (http://www1.ncdc.noaa.gov/pub/data/cirs/). These products are based on gauge data spatially averaged to the climate division scale. In this analysis, we have evaluated 1, 2, 3, and 6 month SPI products (referred to as SPI-1, SPI-2, SPI-3, and SPI-6, respectively). Longer SPI products (e.g., 9 and 12 months) extend beyond the annual growing season extent of the current ESI archive and will be assessed in a future study when the archive has been expanded to year round. The SPI data sets were regridded to the 10 km ALEXI grid, maintaining a constant index value over each of the climate division polygons.

7.4.3 U.S. Drought Monitor

Through expert analysis, authors of the weekly USDM subjectively integrate information from many drought indicators, including the Palmer indices and the SPI, other hydrological parameters such as streamflow and groundwater, and local reports from state climatologists and observers across the country (Svoboda et al., 2002). Archived USDM data are distributed by the National Drought Mitigation Center (NDMC) at http://drought.unl.edu/dm/ in both tabular and vector GIS formats. USDM data were downloaded in table form; these data report the percent area of each USDM drought class by calendar date at the county level.

County polygons were used to assign a USDM value for each date to each pixel in the 10 km ALEXI grid. All pixels contained within a given county polygon were assigned the same value, corresponding to the most severe drought class observed over at least 33% of the county. For computational purposes, the drought classes were mapped to numerical values, with "no drought" assigned a value of −1, D0 = 0 (abnormally dry), D1 = 1 (moderate drought), D2 = 2 (severe drought), D3 = 3 (extreme drought), and D4 = 4 (exceptional drought). For example, if a particular county was classified as 38% D1, and 0% D2–4, the pixels in that county were assigned a value of 1.

The USDM is unique among the drought indicators examined here in that it includes information at multiple drought timescales, as well as some socioeconomic and management/policy considerations. Because it is a subjective assessment, it should not be considered an absolute metric of "truth" in drought monitoring. Still, it is useful to assess spatiotemporal correlations between the USDM and various indicators used in its construction. This process gives us insight as to how new indices like the ESI can be most effectively used to inform production of the USDM.

7.4.4 Standardized Anomalies and Seasonal Composites

The Palmer and SPI data used here were normalized by the NCDC to the period 1931–1990. The period of record for the ESI (2000–2009) is considerably shorter, with average climatic conditions that are not necessarily representative of the normalization periods for the other indices. Therefore, the terms *wetter* and *drier* may convey different meaning for the ESI than for the PDSI and SPI. To improve comparability between the indices evaluated here, anomalies for each precipitation-based index included in the intercomparison, and for the USDM drought classes, were recomputed over the period 2000–2009 using Equation 7.6, analogous to the ESI formulation. Recomputation of anomalies with respect to the same period of record significantly improved spatial agreement between indices.

7.5 DROUGHT INDEX INTERCOMPARISON

7.5.1 Seasonal Anomalies

Annual standardized anomalies in several drought indicators are compared in Figure 7.2, computed from 6 month composites (April–September) over the 2000–2009 growing seasons. The metrics displayed include anomalies in USDM drought classifications, the ESI, and three standard precipitation-based drought indices (Palmer-Z, SPI-3, and PDSI), which were selected to exemplify a range in timescales and modeling approaches. A visual intercomparison is useful in determining what new utility a satellite-based index like ESI can contribute to the effort of drought monitoring in the United States by placing its performance within the context of more familiar drought metrics. It also highlights the difficulty in forming a single objective synthesis of information conveyed by a diversity of drought indices.

In examining Figure 7.2, a few caveats should be considered. First, the USDM is not necessarily independent of the Palmer and SPI indices, as these are commonly used in the delineation of USDM drought classifications. In contrast, the ESI constitutes a completely independent assessment of surface moisture status because it does not use precipitation data as input and it is not currently used in the construction of the USDM. Second, USDM drought classes incorporate information relevant to different kinds of drought over varying timescales, and we cannot expect a single indicator to agree perfectly with the USDM. For example, some drought features in the USDM may indicate increased human demand for water (e.g., due to urban expansion) rather than natural hydrologic deficits, yet such impacts will not be conveyed in the ESI, SPI, or Palmer indices. Socioeconomic droughts cannot be easily identified solely by using remote sensing or meteorological data.

In Figure 7.2, drought features apparent in the USDM anomalies are generally reflected in one or more of the other indices, depending on the type and timescale of the drought event. An exception is the multiyear hydrological drought in the western United States in 2004, which is not well delineated by any of the indices included in the intercomparison. Such events should become evident in longer-term ESI composites (e.g., 12–24 months) once a year-round archive has been developed. In other years, the ESI successfully reproduces patterns evident in the precipitation indices, indicating the value of the LST signal as a surface moisture proxy. For example,

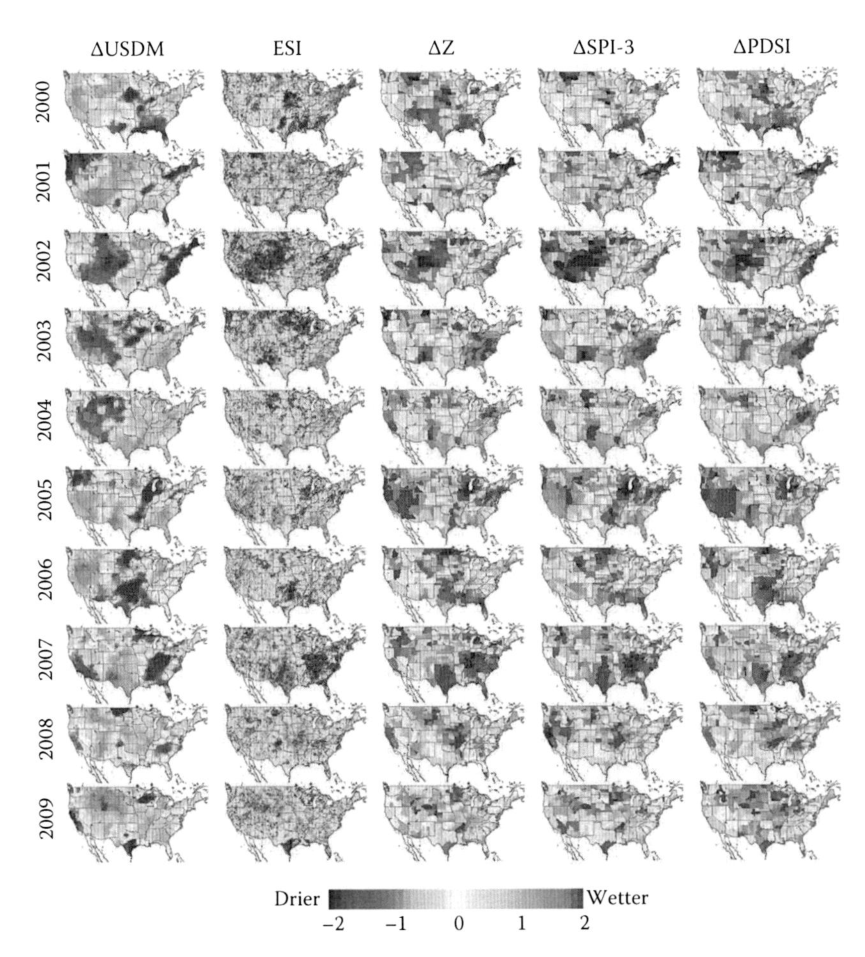

FIGURE 7.2 **(See color insert.)** Seasonal (26 week) anomalies in USDM, ESI, Z, SPI-3, and PDSI for 2000–2009. All indices are presented as z-scores or standard deviations from mean values determined over the 2000–2009 period.

the thermal band inputs to ALEXI capture the major drought events occurring in 2002 and 2007, even in the eastern United States, where there is dense vegetation cover during the middle of the growing season and little direct exposure of the dry soil surface.

Figure 7.3 looks in greater detail at the drought of 2007 that ravaged much of the southeastern United States (particularly in Alabama, Georgia, and the Carolinas), leading to low streamflows, depleted water supplies, and significant agricultural losses. This is a part of the CONUS where standard soil moisture retrievals based on passive microwave (MW) remote sensing tend to lose sensitivity because of strong attenuation of the soil signal by water contained in the dense forest canopy, as demonstrated in Figure 7.3c. In the thermal-derived ESI, however, the moisture deficit signal is strong—vegetation stress and soil moisture depletion in the surface skin contribute to elevated canopy and soil components of the composite surface radiometric

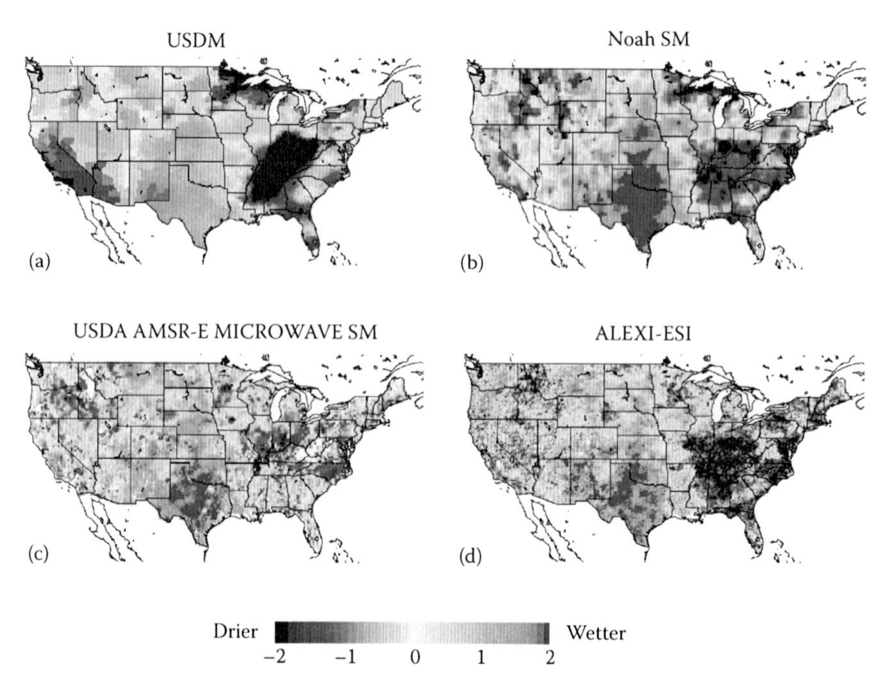

FIGURE 7.3 **(See color insert.)** Standardized anomalies for the 2007 growing season (April–September) in (a) the USDM drought classes, (b) soil moisture predicted by the Noah LSM, (c) USDA AMSR-E passive MW soil moisture retrieval, and (d) ALEXI-ESI.

temperature. The ESI reproduces patterns in soil moisture predicted by the Noah LSM (part of the Land Data Assimilation System [LDAS] modeling suite; Mitchell et al., 2004), with the advantage of requiring no antecedent precipitation information or soil texture data.

7.5.2 Monthly Comparisons

Using shorter timescale ESI composites, we can examine how quickly the ESI responds to changing moisture conditions. Figure 7.4 looks at delineations of drought conditions at monthly time steps in 1 and 3 month ESI composites over the southeastern United States during the drought of 2007, in comparison with anomalies in the Palmer Z index and the USDM. Temporal variability in the ESI-1 shows good general correspondence with monthly rainfall amounts evident in the Z index. For example, heavy rains in July led to short-term increases in ET, followed by reintensified evaporative stress in August in response to anomalously low rainfall that month. Spatial patterns in monthly rainfall expressed in the Z-index maps are also reproduced with reasonable fidelity in the monthly ESI-1 maps.

In contrast, the longer-term ESI-3 better follows the monthly evolution in the USDM drought classifications, which are relatively conservative and typically do not change at the county level by more than one drought class between weekly reports. Anderson et al. (2011a) found that temporal correlations between the ESI and anomalies in the longer-term drought indicators included in the intercomparison

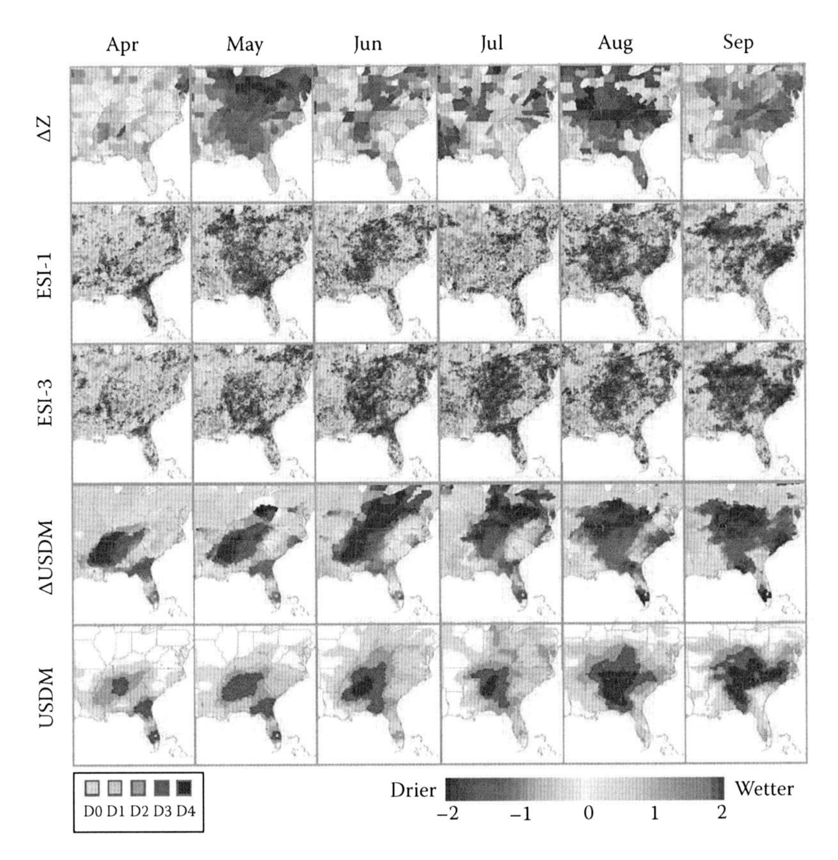

FIGURE 7.4 **(See color insert.)** Maps of monthly drought indicators during April–September 2007, focusing on the severe drought event that occurred in the southeastern United States.

(USDM PMDI, PHDI, PDSI, and SPI-6) improve with increasing ESI compositing interval, reaching a plateau at approximately 2–3 months. Again, good spatial correspondence is observed between patterns in monthly ESI-3 and ΔUSDM, including formation of a hot spot over the Carolinas and Virginia in September, along with a pocket of reduced stress over southeast Georgia.

7.5.3 Spatiotemporal Correlations between Indices

Anderson et al. (2011a) examined spatial and temporal correlations between monthly ESI and several drought indicators, including the USDM, and these analyses are extended here. Table 7.2 lists spatial correlation coefficients computed between pairs of monthly drought index maps, averaged over the growing season months (April to September) for all years included in the intercomparison (2000–2009). These statistics help show which indicators most closely resemble other indicators in terms of monthly spatial patterns. All indicators were aggregated to the climate division scale prior to correlation computation. Temporal correlation coefficients between monthly indicators, computed at each pixel and averaged over the modeling domain, yield

TABLE 7.2
Spatial Correlation Coefficient (Pearson's r) Computed between Pairs of Monthly Composites (April–September) of Several Drought Indicators, Averaged over All Years (2000–2009)

	ESI-1	ESI-2	ESI-3	ΔZ	ΔSPI-1	ΔSPI-2	ΔSPI-3	ΔSPI-6	ΔPDSI	ΔPMDI	ΔPHDI
ΔUSDM	0.485	0.526	0.522	0.417	0.290	0.437	0.506	0.600	0.669	0.710	0.706
ESI-1		0.819	0.722	0.498	0.402	0.516	0.519	0.484	0.508	0.531	0.484
ESI-2			0.899	0.398	0.271	0.479	0.536	0.532	0.545	0.577	0.546
ESI-3				0.364	0.239	0.404	0.513	0.544	0.550	0.585	0.565
ΔZ					0.900	0.741	0.658	0.538	0.619	0.620	0.512
ΔSPI-1						0.694	0.563	0.405	0.444	0.437	0.335
ΔSPI-2							0.813	0.583	0.574	0.592	0.486
ΔSPI-3								0.730	0.653	0.689	0.597
ΔSPI-6									0.752	0.807	0.768
ΔPDSI										0.933	0.897
ΔPMDI											0.956

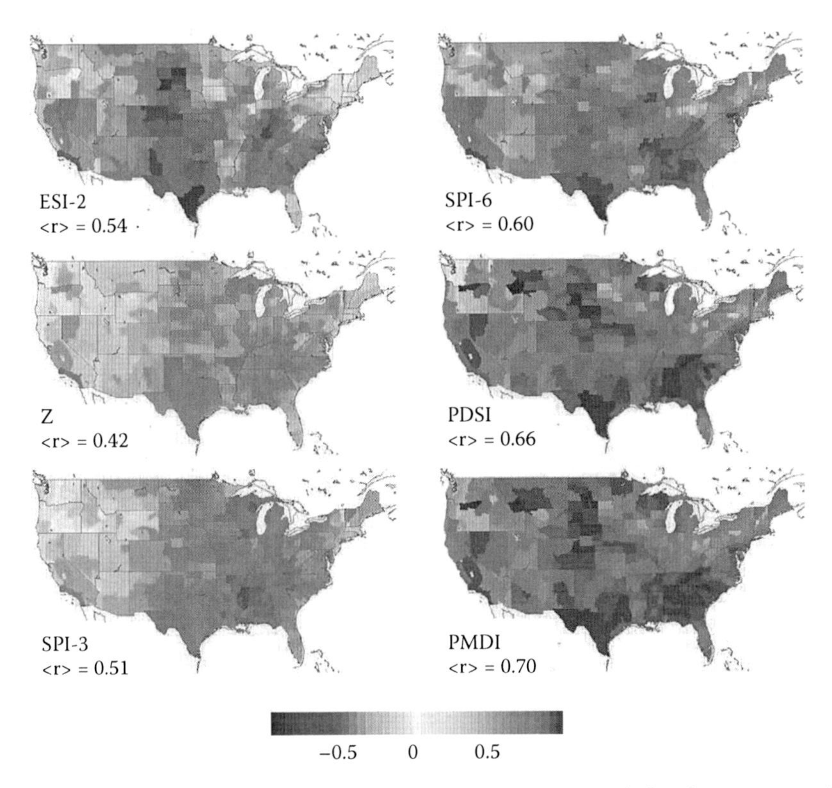

FIGURE 7.5 **(See color insert.)** Coefficient of temporal correlation between monthly maps of USDM anomalies and other drought indices included in the intercomparison for 2000–2009. Domain-averaged values of correlation coefficient are indicated as <r>. To compute correlations, the ESI has been aggregated up to the climate division scale—the native resolution of the precipitation-based indices.

similar results (Anderson et al., 2011a). Maps of temporal correlation between the USDM and several drought indices are shown in Figure 7.5.

In terms of monthly spatial patterns over the growing season, the USDM is most strongly correlated with the PMDI (<r> = 0.71; Table 7.2). Spatial agreement with the USDM improves as index timescale increases (e.g., from SPI-1 to SPI-6). The modifications to the PDSI embedded within the PMDI formulation result in improved agreement with USDM anomalies (and with most other indices), suggesting that these modifications have had a positive impact by improving spatial comparability across the CONUS domain. It must be remembered, however, that the precipitation indices in Table 7.2 (Palmer and SPI) are used in the construction of the USDM and, therefore, are not independent estimators of drought conditions. This will lead to enhanced correlations with USDM drought classes.

In contrast, the ESI can be considered a truly independent indicator within the context of this intercomparison. Both ESI-2 and -3 maps yield similar spatial correlation (<r> = 0.52) with the USDM, ranking between those of SPI-3 and SPI-6 (0.51 and 0.60, respectively). This suggests that ET, as a physical process, integrates over a longer time period than the equivalent precipitation interval—that is, it retains some memory of moisture conditions prior to the composite interval. In terms of pixel-based temporal correlation, Anderson et al. (2011a) demonstrated that the shorter-term precipitation indices (Z and SPI-1 to SPI-3) show a weakly negative correlation with USDM rankings in the northwestern United States (see Figure 7.5), which may be related to hydrologic delays in snowpack-forming regions (Shukla and Wood, 2008). In addition, the 1, 2, and 3 month SPIs show weaker correlations with the USDM in the southwestern United States. Wu et al. (2007) argue that distributions of the 3 month SPI are highly skewed in arid climates, peaking strongly in the no-rain case and causing SPI-3 to underpredict the severity and frequency of drought events. The 6 month SPI has a more normal distribution and better represents the observed drought occurrence frequency. This is consistent with the results of Anderson et al. (2011a) shown in Figure 7.5, which indicate that SPI-6 is more highly correlated with the USDM in the western United States than SPI-3.

The ESI does not exhibit the strongly degraded performance in the western United States seen in SPI products of comparable timescale (Figure 7.5). Drought rankings by the ESI and the USDM are most highly correlated in time over the Great Plains and in the southern United States. These are areas where LST and indicators of vegetation fraction, like the Normalized Difference Vegetation Index (NDVI), tend to be anticorrelated through much of the April–September period (see Figure 7.4 in Karnieli et al., 2010), indicating moisture-limiting (as opposed to energy-limiting) vegetation growth conditions. ET will be most sensitive to changing subsurface moisture conditions under these conditions. Of the indices considered here, the ESI is most similar to the PMDI, with agreement improving with ESI compositing interval (Table 7.2). We anticipate that ESI agreement with USDM and PMDI will further improve when snow processes have been incorporated into ALEXI and longer timescale ESI moving composite intervals (e.g., 6 and 12 months) become more robust (see Section 7.6).

7.6 LIMITATIONS AND FUTURE WORK

Operational execution of the ALEXI model is being transitioned to the National Environmental Satellite, Data, and Information Service (NESDIS) within NOAA in support of the monthly North American Drought Briefing generated by the Climate Prediction Center (http://www.cpc.noaa.gov/products/Drought/). As part of this transition, model preprocessing infrastructure is being reconfigured to use standard NOAA data sources. Hourly skin temperature and insolation will be obtained from the NESDIS GOES Surface and Insolation Product (GSIP; www.star.nesdis.noaa.gov/smcd/opdb/goes/gcip/html/gsip_home.html). Downwelling long-wave radiation at the earth's surface (needed to compute net radiation) is also provided within the GSIP product suite. Ultimately, the ESI archive can be extended back to 1979 using GOES imagery archived at the NCDC through the International Satellite Cloud Climatology Project (ISCCP) B1 Data Rescue project. Meteorological inputs to ALEXI (primarily wind speed and lapse rate) will be extracted from the Regional Climate Data Assimilation System (R-CDAS)—the real-time continuation of the North American Regional Reanalysis (NARR) performed by the National Centers for Environmental Prediction (NCEP) Environmental Modeling Center (EMC) (Mesinger et al., 2006). While ESI evaluation is currently limited to snow-free periods coincident with the growing season for most of the CONUS (approximately April–October), a snow energy balance modeling component in TSEB, adapted from the work of Kongoli and Bland (2000), is under development to provide year-round coverage. Improved treatment of winter ET processes will allow assessment of longer-term ESI composites (e.g., 6–12 months), which may better characterize impacts of hydrological droughts.

The spatial domain of ALEXI-ESI application is also undergoing expansion. As part of the transition to NESDIS, the domain will expand to include North and South America (approximately −60° to +60° latitude) using GOES data. Other domains are being established over southern Europe, the Middle East, and the African continent using land-surface products from the European Meteosat Second Generation (MSG) satellites (Anderson et al., 2011b). A longer-term goal of global ESI coverage (excluding the poles) can be obtained with the current international system of geostationary satellites, as archived at hourly time steps and 5 km resolution by the Geoland2 project under the European GMES (Global Monitoring for Environment and Security) initiative (Lacaze et al., 2010), or using the ISCCP B1 data set at 3 hourly temporal and 10 km spatial resolution (Knapp, 2008).

Work is underway to improve the spatial resolution of ESI products over targeted regions (e.g., the U.S. Corn Belt), employing a multisensor fusion strategy to map daily ET and f_{PET} at 30 m resolution. Using DisALEXI, daily ALEXI fields at 10 km can be downscaled to 1 km using TIR data from MODIS (available every 1–2 days) and to 30 m using Landsat TIR imagery (60–120 m native resolution and 8–16 day revisit depending on the number of satellites concurrently in orbit) that has been improved to the spatial resolution of the shortwave bands (30 m) using a thermal sharpening technique (Agam et al., 2008; Kustas et al., 2003). Finally, a new Spatial and Temporal Adaptive Reflectance Model (STARFM; Gao et al., 2006) can be used to merge the MODIS and Landsat-scale ET evaluations, generating daily predicted fields at the Landsat scale (Anderson et al., 2011b). In this way, we make full use of all available TIR data in interpolating surface moisture conditions between infrequent Landsat overpasses. The 30 m resolution will enable sub-field-scale sampling, leading to more robust assessments of vegetation condition in agricultural landscapes and other ecosystems with small-scale heterogeneity (Figure 7.6).

One of the major limitations of TIR-based indices is the inability to collect thermal images of the land surface through cloud cover. This can severely limit update capacity in perpetually cloudy regions of the globe, such as in the Intertropical Convergence Zone. Temporal sampling can be improved by incorporating moisture information during cloudy periods retrieved using MW remote sensing—for example, using the Advanced Microwave Scanning Radiometer-Earth Observing System (AMSR-E) and instruments on the upcoming Soil Moisture Active Passive (SMAP) mission. Hain (2010) showed that joint assimilation of TIR f_{PET} (from ALEXI) and MW soil moisture into the Noah LSM in NLDAS provides better soil moisture estimates than either retrieval method (TIR or MW) does in isolation. The two retrievals are quite complementary: TIR provides relatively high resolution (60 m–10 km) and low temporal resolution (due to cloud cover) retrievals over a wide

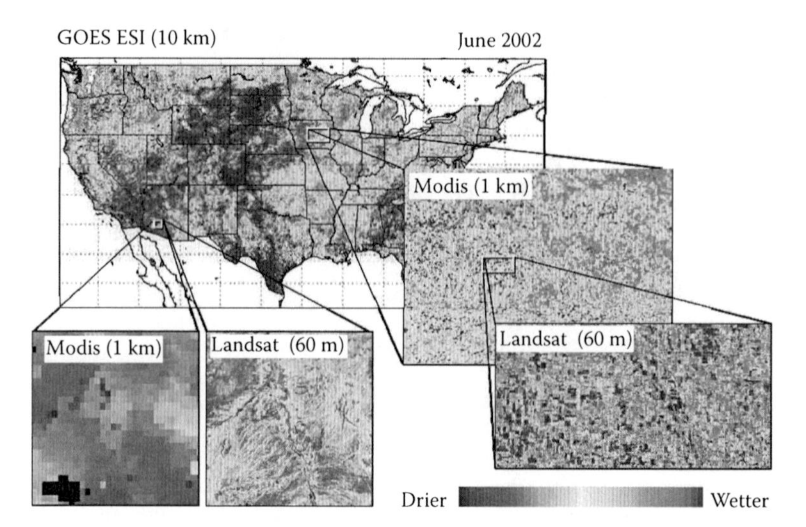

FIGURE 7.6 **(See color insert.)** Comparison of spatial information content provided by ESI fields generated from GOES, MODIS, and Landsat TIR imagery. In this figure, red indicates drier conditions and green indicates wetter conditions.

range of vegetation cover fraction, while MW provides relatively low spatial resolution (10–50 km) and high temporal resolution (can penetrate through clouds), but accurate retrievals are only possible over areas with sparse vegetation. Furthermore, MW retrievals are sensitive to soil moisture only in the shallow soil surface layer (0–5 cm), while TIR provides information about soil moisture conditions integrated over the full root zone, reflected in the observed canopy temperature. The added value of TIR assimilation over MW alone is most significant in areas of moderate to dense vegetation cover (greater than 60%, characterizing much of the eastern United States), where MW retrievals have little sensitivity to soil moisture at any depth due to absorption/emission by water contained within the vegetative canopy. Joint assimilation of both TIR f_{PET} and MW soil moisture into a prognostic LSM would serve to maximize both spatial and temporal sampling of surface moisture conditions, and provide additional hydrologic information such as runoff, streamflow, and groundwater recharge.

7.7 CONCLUSIONS

An intercomparison was conducted between drought indices based on remotely sensed ET, ground observations of rainfall, and drought classifications reported in the USDM during 2000–2009. The ESI, which represents anomalies in the ratio of actual to PET (f_{PET}) derived from thermal remote sensing, was demonstrated to provide useful information for routine monitoring of drought conditions at continental scales without requiring any knowledge of antecedent precipitation. Spatial distributions in ESI correlate with patterns in traditional precipitation-based drought indices, responding to rainfall events at monthly time steps. Therefore, this new drought index appears to be a valuable complement to standard precipitation indices.

ET and f_{PET} products from the LSM suite in NLDAS are also available over the CONUS and are employed as constituents of the NCEP North American Drought Briefing. The main benefit of the ALEXI approach is that it requires no prior information regarding antecedent precipitation and nonprecipitation moisture inputs or about soil moisture storage capacity—the current surface moisture status is deduced directly from the remotely sensed radiometric temperature signal. Thus, thermal remote-sensing models provide independent moisture information that can be useful for validating or updating spatially distributed soil moisture variables in prognostic land-surface water balance models and forecasts.

Because precipitation is not used in the construction of the ESI, this index will have utility for real-time monitoring in regions with sparse rainfall data or significant delays in meteorological reporting. Remote sensing provides a more complete sampling of land-surface conditions than do point-based surface measurement networks and, therefore, better facilitates upscaling to larger scales (Anderson et al., 2007a). High-resolution remote-sensing precipitation products exist (e.g., Joyce et al., 2004), but validation studies show that although these are reasonably good at predicting the occurrence of precipitation, they are less accurate at estimating precipitation amount (Zeweldi and Gebremichael, 2009), which is critical for assessing drought impact. Using multisensor data fusion, integrating TIR-based ET data from geostationary and polar orbiting systems, the ESI can potentially be generated at sub-field-scale resolutions.

The thermal remote-sensing inputs to the ESI can also identify signatures of moisture inputs that are not collocated with precipitation events (e.g., irrigation and wicking from shallow water tables), which may mitigate the impact of local rainfall deficits. Significant horizontal redistribution of precipitation will also be reflected in ESI. These additional moisture sources will not be represented in LSMs without prior knowledge of their existence. The ESI therefore conveys information about *actual* stress rather than the potential for stress. It reflects a different component of the hydrologic cycle that has typically not been considered in drought monitoring, focusing on water use rather than water supply. Inclusion of an ET-based index into the existing drought index suite may provide valuable information about "rapid-onset" events, a situation where intense hot, dry, and windy conditions lead to rapid water loss and the potential for catastrophic crop yield loss, especially when crops are newly emerged and immature. Such events have caused great economic damage in the United States and other parts of the world but are difficult to detect and explain using standard meteorological drought indices.

Future analyses will include comparisons with drought indices based on soil moisture and surface runoff estimates from the NLDAS LSM system, which are currently used in the North American Drought Briefing generated by the NOAA Climate Prediction Center (http://www.cpc.noaa.gov/products/Drought/), and with other remote-sensing drought indices, such as the Vegetation Heath Index (VHI; Kogan, 1997) and the Vegetation Drought Response Index (VegDRI; Brown et al., 2008), and indicators from the Simplified Surface Energy Balance (SSEB) and Vegetation ET (VegET) approaches described by Senay et al. (2012; Chapter 6). The domain of ALEXI application is under expansion to provide global coverage between −60° and 60° latitude using the international system of geostationary satellites. Finally, work is under way to incorporate a snow module in ALEXI so that it can be applied year-round as the archive is expanded retrospectively to the beginning of the GOES era (into the late 1970s) and forward in real time.

ACKNOWLEDGMENTS

Support for this work was provided by funding from the NOAA Climate Program Office under grant GC09-236. USDA is an equal opportunity provider and employer.

REFERENCES

Agam, N., W.P. Kustas, M.C. Anderson, F. Li, and P.D. Colaizzi. 2008. Utility of thermal image sharpening for monitoring field-scale evapotranspiration over rainfed and irrigated agricultural regions. *Geophysical Research Letters* 35, doi:10.1029/2007GL032195.

Alley, W.M. 1984. The Palmer Drought Severity Index: Limitations and assumptions. *Journal of Climate and Applied Meteorology* 23:1100–1109.

Anderson, M.C., C.R. Hain, B. Wardlow, J.R. Mecikalski, and W.P. Kustas. 2011a. Evaluation of a drought index based on thermal remote sensing of evapotranspiration over the continental U.S. *Journal of Climate* 24:2025–2044.

Anderson, M.C., W.P. Kustas, and J.M. Norman. 2007a. Upscaling flux observations from local to continental scales using thermal remote sensing. *Agronomy Journal* 99:240–254.

Anderson, M.C., W.P. Kustas, J.M. Norman, C.R. Hain, J.R. Mecikalski, L. Schultz, M.P. Gonzalez-Dugo, C. Cammalleri, G. d'Urso, A. Pimstein, and F. Gao. 2011b. Mapping daily evapotranspiration at field to continental scales using geostationary and polar orbiting satellite imagery. *Hydrology and Earth System Sciences* 15:223–239.

Anderson, M.C., J.M. Norman, G.R. Diak, W.P. Kustas, and J.R. Mecikalski. 1997. A two-source time-integrated model for estimating surface fluxes using thermal infrared remote sensing. *Remote Sensing of Environment* 60:195–216.

Anderson, M.C., J.M. Norman, J.R. Mecikalski, J.P. Otkin, and W.P. Kustas. 2007b. A climatological study of evapotranspiration and moisture stress across the continental U.S. based on thermal remote sensing: II. Surface moisture climatology. *Journal of Geophysical Research* 112:D11112, doi:11110.11029/12006JD007507.

Anderson, M.C., J.M. Norman, J.R. Mecikalski, J.P. Otkin, and W.P. Kustas. 2007c. A climatological study of evapotranspiration and moisture stress across the continental U.S. based on thermal remote sensing: I. Model formulation. *Journal of Geophysical Research* 112: D10117, doi:10110.11029/12006JD007506.

Anderson, M.C., J.M. Norman, W.P. Kustas, F. Li, J.H. Prueger, and J.M. Mecikalski. 2005. Effects of vegetation clumping on two-source model estimates of surface energy fluxes from an agricultural landscape during SMACEX. *Journal of Hydrometeorology* 6:892–909.

Brown, J.F., B.D. Wardlow, T. Tadesse, M.J. Hayes, and B.C. Reed. 2008. The Vegetation Drought Response Index (VegDRI): A new integrated approach for monitoring drought stress in vegetation. *GIScience and Remote Sensing* 45:16–46.

Courault, D., B. Seguin, and A. Olioso. 2005. Review on estimation of evapotranspiration from remote sensing data: From empirical to numerical modeling approaches. *Irrigation and Drainage Systems* 19:223–239.

Edwards, D.C. and T.B. McKee. 1997. Characteristics of 20th century drought in the United States at multiple time scales. Climatology Report Number 97-2, Colorado State University, Fort Collins, CO.

Gao, F., J. Masek, M. Schwaller, and F.G. Hall. 2006. On the blending of the Landsat and MODIS surface reflectance: Predicting daily Landsat surface reflectance. *IEEE Transactions on Geoscience and Remote Sensing* 44:2207–2218.

Guttman, N.B. and R.G. Quayle. 1996. A historical perspective of U.S. climate divisions. *Bulletin of the American Meteorological Society* 77:293–303.

Hain, C.R. 2010. Developing a dual assimilation approach for thermal infrared and passive microwave soil moisture retrievals. PhD thesis, University of Alabama, Huntsville, AL.

Heddinghaus, T.R. and P. Sabol. 1991. A review of the Palmer Drought Severity Index and where do we go from here? *Proceedings of the 7th Conference on Applied Climatology*, American Meteorological Society, Boston, MA.

Huffman, G.J., R.F. Adler, D.T. Bolving, G. Gu, E.J. Nelkin, Y. Hong, D.B. Wolff, K.P. Bowman, and E.F. Stocker. 2007. The TRMM Multisatellite Precipitation Analysis (TMPA): Quasi-global, multiyear, combined-sensor precipitation estimates at fine scales. *Journal of Hydrometeorology* 8:38–55.

Joyce, R.J., J.E. Janowiak, P.A. Arkin, and P. Xie. 2004. CMORPH: A method that produces global precipitation estimates from passive microwave and infrared data at high spatial and temporal resolution. *Journal of Hydrometeorology* 5:487–503.

Kalma, J.D., T.R. McVicar, and M.F. McCabe. 2008. Estimating land surface evaporation: A review of methods using remotely sensing surface temperature data. *Surveys in Geophysics* 29(4–5):421–469.

Karl, T.R. 1983. Some spatial characteristics of drought duration in the United States. *Journal of Climate and Applied Meteorology* 22:1356–1366.

Karnieli, A., N. Agam, R.T. Pinker, M.C. Anderson, M.L. Imhoff, G.G. Gutman, N. Panov, and A. Goldberg. 2010. Use of NDVI and land surface temperature for drought assessment: Merits and limitations. *Journal of Climate* 23:618–633.

Knapp, K.R. 2008. Scientific data stewardship of International Satellite Cloud Climatology Project B1 geostationary observations. *Journal of Applied Remote Sensing* 2:023548.

Kogan, F.N. 1997. Global drought watch from space. *Bulletin of the American Meteorological Society* 78:621–636.

Kongoli, C.E. and W.L. Bland. 2000. Long-term snow depth simulations using a modified atmosphere-land exchange model. *Agricultural and Forest Meteorology* 104:273–287.

Kustas, W.P. and J.M. Norman. 1999. Evaluation of soil and vegetation heat flux predictions using a simple two-source model with radiometric temperatures for partial canopy cover. *Agricultural and Forest Meteorology* 94:13–29.

Kustas, W.P. and J.M. Norman. 2000. A two-source energy balance approach using directional radiometric temperature observations for sparse canopy covered surfaces. *Agronomy Journal* 92:847–854.

Kustas, W.P., G.R. Diak, and J.M. Norman. 2001. Time difference methods for monitoring regional scale heat fluxes with remote sensing. *Land Surface Hydrology, Meteorology, and Climate: Observations and Modeling* 3:15–29.

Kustas, W.P., J.M. Norman, M.C. Anderson, and A.N. French. 2003. Estimating subpixel surface temperatures and energy fluxes from the vegetation index–radiometric temperature relationship. *Remote Sensing of Environment* 85:429–440.

Labedzki, L. and E. Kanecka-Geszke. 2009. Standardized evapotranspiration as an agricultural drought index. *Irrigation and Drainage* 58:607–616.

Lacaze, R., G. Balsamo, F. Baret, A. Bradley, J.-C. Calvet, F. Camacho, R. D'Andrimont, S.C. Freitas, H. Makhmara, V. Naeimi, P. Pacholczyk, H. Poilvé, B. Smets, K. Tansey, I.F. Trigo, W. Wagner, and M. Weiss. 2010. Geoland2—Towards an operational GMES land monitoring core service; First results of the biogeophysical parameter core mapping service. In *ISPRS TC VII Symposium—100 Years ISPRS*, Vienna, Austria, July 5–7, 2010, eds. W. Wagner and B. Székely, IAPRS, Vol. XXXVIII, Part 7B, pp. 354–359.

Li, H., Y. Lei, L. Zheng, and R. Mao. 2005. Calculating regional drought indices using evapotranspiration (ET) distribution derived from Landsat7 ETM+ data. In *Remote Sensing and Modeling of Ecosystems for Sustainability II* (Proceedings Volume 5884), eds. W. Gao and D.R. Shaw. Bellingham, Washington, DC: SPIE.

McEnery, J., J. Ingram, Q. Duan, T. Adams, and L. Anderson. 2005. NOAA's Advanced Hydrologic Prediction Service. *Bulletin of the American Meteorological Society* 86:375–385.

McKee, T.B., N.J. Doesken, and J. Kleist. 1993. The relationship of drought frequency and duration to time scales. *Preprints, 8th Conference on Applied Climatology*, Anaheim, CA.

McKee, T.B., N.J. Doesken, and J. Kleist. 1995. Drought monitoring with multiple time scales. *Preprints, 9th Conference on Applied Climatology*, Dallas, TX.

McNaughton, K.G. and T.W. Spriggs. 1986. A mixed-layer model for regional evaporation. *Boundary-Layer Meteorology* 74:262–288.

Mecikalski, J.M., G.R. Diak, M.C. Anderson, and J.M. Norman. 1999. Estimating fluxes on continental scales using remotely-sensed data in an atmosphere-land exchange model. *Journal of Applied Meteorology* 38:1352–1369.

Mesinger, F., G. DiMego, E. Kalnay, K. Mitchell, P.C. Shafran, W. Ebisuzaki, D. Jovic, J. Woollen, E. Rogers, E.H. Berbery, M.B. Ek, Y. Fan, R. Grumbine, W. Higgins, H. Li, Y. Lin, G. Manikin, D. Parrish, and W. Shi. 2006. North American regional reanalysis. *Bulletin of the American Meteorological Society* 87:343–360.

Mitchell, K.E., D. Lohmann, P.R. Houser, E.F. Wood, J.C. Schaake, A. Robock, B.A. Cosgrove, J. Sheffield, Q. Duan, L. Luo, R.W. Higgins, R.T. Pinker, J.D. Tarpley, D.P. Lettenmaier, C.H. Marshall, J.K. Entin, M. Pan, W. Shi, V. Koren, J. Meng, B.H. Ramsay, and A.A. Bailey. 2004. The multi-institution North American Land Data Assimilation System (NLDAS): Utilizing multiple GCIP products and partners in a continental distributed hydrological modeling system. *Journal of Geophysical Research* 190:D07S90, doi:10.1029/2003JD003823.

Mo, K.C., L.N. Long, Y. Xia, S.K. Yang, J.E. Schemm, and M.B. Ek. 2010. Drought indices based on the Climate Forecast System Reanalysis and ensemble NLDAS. *Journal of Hydrometeorology* 12:181–205. doi: 10.1175/2010JHM1310.1171.

Moran, M.S. 2003. Thermal infrared measurement as an indicator of plant ecosystem health. In *Thermal Remote Sensing in Land Surface Processes*, eds. D.A. Quattrochi and J. Luvall, 257–282. Boca Raton, FL: CRC Press, Taylor & Francis Group.

Norman, J.M., M.C. Anderson, W.P. Kustas, A.N. French, J.R. Mecikalski, R.D. Torn, G.R. Diak, T.J. Schmugge, and B.C.W. Tanner. 2003. Remote sensing of surface energy fluxes at 10^1-m pixel resolutions. *Water Resources Research* 39(8), doi:10.1029/2002WR001775.

Norman, J.M., M. Divakarla, and N.S. Goel. 1995b. Algorithms for extracting information from remote thermal-IR observations of the earth's surface. *Remote Sensing of Environment* 51:157–168.

Norman, J.M., W.P. Kustas, and K.S. Humes. 1995a. A two-source approach for estimating soil and vegetation energy fluxes from observations of directional radiometric surface temperature. *Agricultural and Forest Meteorology* 77:263–293.

Palmer, W.C. 1965. Meteorological drought. Research Paper 45, U.S. Department of Commerce Weather Bureau, Washington, DC.

Priestley, C.H.B. and R.J. Taylor. 1972. On the assessment of surface heat flux and evaporation using large-scale parameters. *Monthly Weather Review* 100:81–92.

Schaake, J.C., Q. Duan, V. Koren, K.E. Mitchell, P.R. Houser, E.F. Wood, A. Robock, D.P. Lettenmaier, D. Lohmann, B. Cosgrove, J. Sheffield, L. Luo, R.W. Higgins, R.T. Pinker, and J.D. Tarpley. 2004. An intercomparison of soil moisture fields in the North American Land Data Assimilation System (NLDAS). *Journal of Geophysical Research* 109, doi:10.1029/2002JD00309.

Senay, G.B., S. Bohms, and J.P. Verdin. 2012. Remote sensing of evapotranspiration for operational drought monitoring using principles of water and energy balance. In *Remote Sensing of Drought: Innovative Monitoring Approaches,* eds. B.D. Wardlow, M.C. Anderson, and J.P. Verdin. Boca Raton, FL: CRC Press.

Shukla, S. and A.W. Wood. 2008. Use of a standardized runoff index for characterizing hydrologic drought. *Geophysical Research Letters* 35, doi:10.1029/2007GL032487.

Svoboda, M., D. LeComte, M. Hayes, R. Heim, K. Gleason, J. Angel, B. Rippey, R. Tinker, M. Palecki, D. Stooksbury, D. Miskus, and S. Stephens. 2002. The U.S. Drought Monitor. *Bulletin of the American Meteorological Society* 83:1181–1190.

Villarini, G., W.F. Krajewski, and J.A. Smith. 2009. New paradigm for statistical validation of satellite precipitation estimates: Application to a large sample of the TMPS 0.25° 3-hourly estimates over Oklahoma. *Journal of Geophysical Research* 114:D12106, doi:12110.11029/12008JD011475.

Wu, H., M.D. Svoboda, M.J. Hayes, D.A. Wilhite, and F. Wen. 2007. Appropriate application of the Standardized Precipitation Index in arid locations and dry seasons. *International Journal of Climatology* 27:65–79.

Zeweldi, D.A. and M. Gebremichael. 2009. Evaluation of CMORPH Precipitation Products at fine space-time scales. *Journal of Hydrometeorology* 10:300–307.

8 Agricultural Drought Monitoring in Kenya Using Evapotranspiration Derived from Remote Sensing and Reanalysis Data

Michael T. Marshall, Christopher Funk, and Joel Michaelsen

CONTENTS

8.1 INTRODUCTION

More than half of the people in sub-Saharan Africa live on less than US$ 1.25 per day, and nearly 30% do not receive sufficient nourishment to maintain daily health (UN, 2009a). These figures are expected to rise as a result of the recent global financial crisis that has led to an increase in food prices. Food for Peace (FFP), the program that administers more than 85% of U.S. international food aid, recently reported that the seven largest recipient countries of food aid worldwide are in sub-Saharan Africa (FFP, 2010). In Kenya, the fifth largest recipient of food aid from FFP and a country highly dependent on rainfed agriculture, below-average precipitation in 2009 led to a 20% reduction in maize production and a 100% increase in domestic maize prices (FEWS NET, 2009). Given these sorts of climatic shocks, it is imperative that mitigation strategies be developed for sub-Saharan Africa and other regions of the developing world to improve the international and national response to impending food crises. Crop monitoring is an important tool used by national agricultural offices and other stakeholders to inform food security analyses and agricultural drought mitigation. Remote sensing and surface reanalysis data facilitate efficient and cost-effective approaches to measuring determinants of agricultural drought. In this chapter, we explore how remotely sensed estimates of actual evapotranspiration (ET_a) can be integrated with surface reanalysis data to augment agricultural drought monitoring systems.

Although water availability is important throughout every stage of crop development, from germination to harvest, crops are most sensitive to moisture deficits during the reproductive stages (Shanahan and Nielsen, 1987). A study that analyzed maize, for example, showed that a 1% decline in seasonal ET_a led to an average loss of 1.5% in crop yield, whereas water stress in the same proportion concentrated during the reproductive phases led to a 2.6% decline in crop yield (Stegman, 1982). Agricultural drought can therefore be defined as inadequate soil water availability, particularly during the reproductive phase, caused by low precipitation, insufficient water-holding capacity in the root zone of the soil, and/or high atmospheric water demand (potential evapotranspiration, ET_p), which results in a reduction in crop yield. Agricultural droughts differ in timescale and impact from shorter-term meteorological droughts, which are characterized by negative precipitation anomalies on the order of days to weeks, and the longer-term negative runoff and water storage anomalies that characterize hydrological drought (Dracup et al., 1980).

Deficits in ET_a are a direct measure of crop stress and can be integrated into agricultural drought monitoring systems. In response to large soil moisture deficits, for example, plants close their stomata, thereby downregulating water and CO_2 exchange with the atmosphere (Jones, 1992). This protects plants from excessive water loss and cavitation and reduces ET_a, which is the quantity of moisture that is lost to the atmosphere via wet canopy and soil evaporation and transpiration. There are several options for integrating ET_a estimates into agricultural drought monitoring. One possibility is using ET_a anomalies derived from a suite of observed, reanalysis, and remote sensing data to provide a simple means for assessing crop stress, although this approach would have limited use in the most agriculturally intensive regions of the developing world where the variability in soil moisture (and therefore variability in ET_a) is low. A more sophisticated application would involve a crop model that uses ET_a explicitly to determine impact of moisture anomalies on start of season, length of growing period, crop yield, and water availability. The major disadvantage of this approach in data-sparse regions is that the models require detailed meteorological data derived from a dense network of weather stations. Remote sensing and surface reanalysis meteorological models, however, have made this approach more feasible in recent decades.

In this chapter, an ET_a-based crop index that represents a balance between simple and complex monitoring tools is presented and applied in sub-Saharan Africa to illustrate its utility in monitoring agricultural drought. Given the scarcity of observational data in the region, an ET_a model is used to derive the index. The model is a hybrid that combines components of two models that were parameterized with freely available remote sensing and surface reanalysis meteorological data. The ET_a model was evaluated for major biomes throughout sub-Saharan Africa using eddy covariance flux tower data in Marshall et al. (2011). This chapter builds on the previous work by extending model performance using district-level crop yield data. An ET_a-based crop index derived from the model is then applied to food security analysis. The index is compared qualitatively against the Standardized Precipitation Index (SPI), which is a common index used to monitor agricultural droughts in this region of Africa. The chapter concludes with recommendations on how to integrate the ET_a-based index into a regular monitoring scheme and a brief, general discussion regarding the future direction of agricultural drought monitoring in sub-Saharan Africa.

8.2 REVIEW OF INDICES FOR CROP MONITORING

8.2.1 Climate-Based Indices

McGuire and Palmer (1957) were the first to develop an index specifically for agricultural drought monitoring. The index was simply the ratio of atmospheric water demand (ET_p) to supply (precipitation + soil moisture). The resulting Moisture Adequacy Index (MAI) is a relative measure of the moisture available to support normal plant functions. Crops are assumed to be under stress when water demand is much greater than supply. Since the development of the MAI, several other conceptually simple indices have been developed that are better able to capture the spatio-temporal complexities of an agricultural drought. Heim (2002) provides a thorough review of agricultural drought indices commonly used in the United States and globally.

The most notable achievement in crop monitoring occurred shortly after the development of MAI with the application of a simple water balance model to evaluate meteorological droughts (Palmer, 1965). The Palmer Drought Severity Index (PDSI) measures the precipitation required to balance the changes in moisture inputs (precipitation and soil recharge) with outputs (ET_a and runoff). A two-layer soil profile is used to estimate runoff and soil moisture recharge from the difference between precipitation and ET_a, which is calculated as the excess ET_p above precipitation. PDSI is typically standardized using a climatologically driven weighting function to yield the Moisture Anomaly Index (PDSI-z). PDSI-z responds to moisture conditions over short periods of time, so it is more appropriately used for identifying meteorological and agricultural droughts (Karl, 1986). In addition, PDSI-z is a standardized index, making its application across regions more appropriate than PDSI for meteorological and agricultural droughts. Palmer (1965) also developed the Crop Moisture Index (CMI), which uses PDSI constrained by dry and wet ranges to identify agricultural droughts. There is overlap between the dry and wet ranges, often leading to contradictions in drought onset and cessation. Alley (1984) gives a thorough description of PDSI and CMI, including sources of error and major limitations when applied to agricultural drought.

The strong dependence of PDSI on rainfall has contributed to the development of precipitation-based indices, the most notable of which is the SPI (McKee et al., 1993). Although the SPI has been used primarily to monitor meteorological droughts, it is scalable and has been applied successfully to agricultural drought monitoring as well (Keyantash and Dracup, 2002). The index is a standardized representation of the deviation of rainfall probabilities from the long-term mean. These probabilities are typically approximated using a gamma or Pearson III distribution (Guttman, 1999). In contrast with the PDSI-z, the SPI requires fewer potentially uncertain inputs, it can be estimated at various temporal scales, and its physical meaning is easier to interpret.

8.2.2 Satellite-Based Indices

Satellite observations of vegetation conditions have provided another basis for crop monitoring. The Normalized Difference Vegetation Index (NDVI) (Rouse et al., 1974), derived from remotely sensed visible and near-infrared reflectance, is sensitive to the photosynthetic capacity of plants and typically lags precipitation by 1–2 months in sub-Saharan Africa (Nicholson et al., 1990). NDVI has been used to characterize trends in vegetation (Philippon et al., 2007) and interannual variability in precipitation (Anyamba and Eastman, 1996). A common approach to monitoring crop stress using NDVI is to combine it with remotely sensed temperature, yielding a composite index such as the Vegetation Health Index (VHI) (Kogan, 1995). The inclusion of temperature conveys additional information on drought conditions (e.g., soil dryness, plant stress) not captured by NDVI alone. VHI assumes that NDVI and surface temperature are inversely related and is therefore useful where soil moisture is a constraint on plant health. In canopies found commonly in tropical Africa, plants are light limited. In these cases, temperature and NDVI are directly related, because plants prefer to photosynthesize at warmer temperatures when moisture is not a limiting factor (Karnieli et al., 2006). The uncertainty in the NDVI–temperature relationship for different climatic zones and land cover types therefore makes regional drought monitoring difficult with VHI.

Most recently, process-based models of ET_a have been used to develop indices for agricultural drought monitoring. Unlike precipitation, ET_a takes into account additional factors contributing to crop stress, including vegetation type and phenology, antecedent soil moisture conditions, and soil properties (Narasimhan and Srinivasan, 2005). In addition, ET_a-based indices are more sensitive than precipitation-based indices to gradual changes in soil moisture and crop stress that occur during agricultural droughts. An ET_a-based index used extensively in sub-Saharan Africa is the Water Requirement Satisfaction Index (WRSI), first developed by the United Nations (UN) Food and Agriculture Organization (FAO) (Doorenbos and Pruitt, 1977). WRSI is the ratio of ET_a to ET_0, the reference evapotranspiration adjusted by an empirically derived coefficient that accounts for crop type and its specific growth stage. ET_0 is derived for a wet grass using the Monteith (1965) equation driven by surface net radiation, temperature, and vapor pressure. WRSI uses a simple bucket model, with no soil profile, making ET_a essentially a residual of ET_p and precipitation (Senay and Verdin, 2003). Given the complexities of modeling ET_a in data-sparse regions, the major disadvantages of the model are that start season and length of growing period are estimated from NDVI, the required climate data tend to have very coarse spatial resolution, and ET_p is derived from an empirical relationship between the crop coefficient and ET_0. In reality, this coefficient varies dramatically across land cover types, as well as intra- and interseasonally, making its application over broad areas and time frames difficult (Allen, 2000). Anderson et al. (2007) developed the Evaporative Stress Index (ESI) for the United States, which is conceptually similar to WRSI, but uses an energy balance approach to estimate ET_a. Unlike bucket models, the energy balance approach avoids assumptions of the soil profile and requires vegetation and temperature inputs, which are more certain than remotely sensed precipitation data in sub-Saharan Africa. The approach, however, does require land-surface temperature (LST) derived from thermal satellite data acquired under clear-sky conditions, which is limited by periods of persistent low-level stratus cloud cover that tends to obscure conditions during the dry season in the most food insecure (subtropical) region of Africa.

8.3 DESCRIPTION OF THE EVAPOTRANSPIRATION MODEL

8.3.1 Theoretical Background

The use of temporally continuous surface reanalysis meteorological data can enhance temporally coarse resolution remote sensing data used to develop ET_a-driven indices for crop monitoring applications where dense climate data networks are unavailable. Surface reanalysis meteorological data are developed from the synthesis of remote sensing, global circulation models (GCMs), and meteorological station data using a suite of sophisticated assimilation techniques. Recently, Marshall et al. (2010) evaluated two fundamentally different approaches to estimating ET_a, both driven by the Global Land Data Assimilation System (GLDAS) surface reanalysis data set (Rodell et al., 2004). One approach was driven primarily by remotely sensed vegetation, and the other approach was driven primarily by surface reanalysis precipitation. The former is described in detail in Fisher et al. (2008) and will be identified as the Fisher model for this chapter. The Fisher model uses a series of vegetation and soil moisture factors to constrain ET_p. The constraints are intentionally simplified to facilitate easy application over large data-poor areas. The second approach, called the National Centers for Environmental Prediction, Oregon State University, Air Force, and Hydrologic Research Lab (Noah) Land Surface Model, is described in detail in Chen et al. (2007). Both approaches compute ET_a as the sum of transpiration (ET_c), wet canopy evaporation (ET_i), and bare soil evaporation (ET_s). Marshall et al. (2010) demonstrated that ET_c from the Fisher model, which included a dynamic vegetation component, performed particularly well with reanalysis data for energy-limited (humid) sites, while Noah more accurately captured the soil and wet canopy components in moisture-limited (dry) sites. This led to the integration of ET_c of the Fisher model with ET_s and ET_i from the Noah model to provide a more robust ET estimation over a range of environmental conditions. The components of the model are described in detail later.

In the hybrid ET_a formulation, the transpiration component of the Fisher model uses the Priestley and Taylor (1972) equilibrium equation for ET_p and retains the Priestley–Taylor advection coefficient ($\alpha = 1.26$). ET_c is estimated by constraining Priestley–Taylor ET_p (the term in square brackets in Equation 8.1) by five coefficients analogous to the Jarvis and McNaughton (1986) decoupling coefficient (Ω):

$$ET_c = f_c f_g f_t f_m (1 - f_{wet}) \left[\frac{\alpha \Delta}{\Delta + \gamma} R_N \right] \tag{8.1}$$

where f_c, f_g, f_t, f_m, and f_{wet} are total vegetation fraction, photosynthetically active canopy fraction, air temperature constraint, plant moisture constraint, and relative surface wetness, respectively. The constraints for ET_c are defined for daylight hours only, when the contribution of ET to the atmospheric water balance is significantly larger than at nighttime. At dekadal or monthly time steps, which are generally used for agricultural drought monitoring in sub-Saharan Africa, soil heat flux is negligible and therefore has been omitted from the equation. ET_p is defined by three terms: slope of the saturation-to-vapor pressure curve (Δ), psychometric constant (γ), and daytime net radiation (R_N) in mass units. The vegetation cover fraction f_c is used to constrain R_N in terms of the canopy portion of ET_a and is calculated as a linear function of NDVI. Early studies showed that NDVI is highly correlated with f_c calculated from Beer's law (Sellers, 1987). The active green vegetation cover fraction (f_g) is computed as the ratio of f_{APAR} to f_c, where f_{APAR} is the fraction of absorbed photosynthetically active radiation, estimated using a linear function of the Enhanced Vegetation Index (EVI) (Huete et al., 2002). Two different vegetation indices were used in these calculations, because EVI is sensitive to the chlorophyll content (photosynthetically active portion) of the canopy, while NDVI is more effective in capturing the total biomass of the canopy (Huete et al., 2002). The temperature constraint f_t assumes that photosynthesis will increase with air temperature until an optimal temperature is achieved (June et al., 2004), and once the temperature becomes higher than the optimal temperature, there is a decline in photosynthesis at a proportional rate. Optimal temperature is defined as the temperature at which the ratio of f_{APAR} and available radiation to the vapor pressure deficit (VPD) is maximum. It ranges from near 0°C in the Arctic to 35°C in subtropical deserts (Potter et al., 1993). The plant moisture constraint f_m is defined as the relative change in light absorptance (f_{APAR}/maximum f_{APAR}), assuming that plant moisture stress (moisture availability) varies with the amount of light a plant absorbs. When a plant is moisture stressed, it will close its stomata (reduce photosynthesis) to prevent cavitation. Maximum f_{APAR} is determined over the available time series using the equation for f_{APAR} (function of EVI) defined earlier. The factor f_{wet} gives the probability that the surface is wet, defined as a function of near-surface relative humidity. The constraint assumes that during daylight hours, when relative humidity is at 100%, the surface is completely wetted and will evaporate moisture from the wet canopy and bare soil portions to meet atmospheric demand. In this case, ET_c approaches 0. On the other hand, when relative humidity approaches 0%, the surface is completely dry, and the contribution from the wet canopy and soil components to ET_a is 0. This constraint is an integral component of the wet canopy and soil evaporation components of the Fisher model (not shown here), which the Marshall et al. (2010) study revealed was particularly sensitive to specific humidity, a highly uncertain surface reanalysis variable (Kalnay et al., 1996). The model parameterization further amplified these errors.

The Noah model, first described by Chen et al. (1996), takes a water balance approach to estimating ET_a. The soil evaporation component of the model was based on the work of Mahfouf and Noilhan (1991), who compared several approaches to estimating soil evaporation and found that the preferred method constrains ET_p, as defined in Penman (1948), by a moisture availability parameter (β). This parameter is driven by soil moisture using the following equation:

$$\beta = \left(\frac{\Theta_1 - \Theta_w}{\Theta_{ref} - \Theta_w} \right)^f \tag{8.2}$$

where

Θ_1 is the soil moisture in the top soil layer
Θ_w is the wilting point
Θ_{ref} is the field capacity of the soil
f is a constant (Betts et al., 1997)

The wet canopy component of the Noah model, which tends to represent the smallest contribution to total ET_a, is computed as ET_p constrained by the amount of precipitation intercepted by the canopy (ratio of intercepted canopy water content to maximum canopy water capacity). The transpiration component of the Noah model (not shown here) is driven primarily by the total vegetation fraction, which is derived from NDVI climatological means (Hogue et al., 2005). This tends to make the model insensitive to large phenological changes that are typical in semiarid regions found throughout most of sub-Saharan Africa.

The hybrid model uses ET_c from the Fisher model and $ET_{i,s}$ from the Noah LSM, and was shown to perform as well or better than the individual Fisher and Noah models when compared to eddy covariance flux tower data collected in areas representing major climate zones and land cover types in sub-Saharan Africa (Marshall et al., 2010). The major benefit of the hybrid model over the Noah LSM was the use of a Fisher-based ET_c component driven by dynamic vegetation, which tended to better capture the seasonality of ET_a at drier sites. The hybrid model performed best in humid areas, conditions representative of a significant portion of tropical Africa. The performance of the hybrid model at the humid sites was attributed to the use of Priestley–Taylor (equilibrium) ET_p in the computation of ET_c, as implemented in the Fisher model. In the equilibrium case, advection is implicitly accounted for by α. ET_p is therefore driven primarily by R_N, so the Priestley–Taylor formulation performs better for energy-limited (dense) vegetation and more poorly for moisture-limited (sparse) vegetation (Mu et al., 2007), because surface resistance rapidly increases as soil dries out due to advection, decreasing α (Agam et al., 2010). Although constraints in the Fisher model account for changes in soil moisture, the relatively poor performance of the model in semiarid regions suggested that the model could be improved with the use of Penman ET_p, where advection is handled explicitly; however, further analysis revealed that both formulations gave similar results at all the sites. The model used in this chapter, therefore, retains the Priestley–Taylor approach, under the assumption that at large (satellite) scales, advection is driven primarily by the growth of the convective boundary layer, which in turn is driven by R_N (Raupach, 2000).

8.3.2 Data Handling and Processing

The canopy component of the Fisher model was run at a daily time step and requires five inputs: NDVI, EVI, maximum air temperature, minimum relative humidity, and R_N. Daily maximum temperature and relative humidity were used instead of averages because the relationship between ET_a and these variables is strongest during midday when convection is high. The climate data used to drive the ET_c (Fisher) component of the hybrid model were derived from 3-hourly 0.25° (~25 km) resolution GLDAS climate data (Rodell et al., 2004). The GLDAS data set is a synthesis of various reanalysis, remote sensing, and ground sources, including National Oceanic and Atmospheric Administration Global Data Assimilation System (NOAA/GDAS) atmospheric fields, Climate Prediction Center (CPC) Merged Analysis of Precipitation fields (CMAP), and observation-driven shortwave and longwave radiation using the Air Force Weather Agency's AGRicultural METeorological modeling system. These data are available from 2001 to the present, while only the NOAA/GDAS reanalysis is available for 2000. Saturation vapor pressure was computed using the Allen et al. (1998) temperature function and combined with specific humidity to estimate relative humidity. Global vegetation fields available at 0.05° (~5 km) spatial resolution from reflectance data acquired by the Moderate Resolution Imaging Spectroradiometer (MODIS) on board the Earth Observing System (EOS)-Terra platform (Huete et al., 2002) were subset for sub-Saharan Africa. The vegetation index products consist of 16-day composites to reduce noise due to atmospheric effects. Postprocessing included a piecewise weighted least squares filter (Swets et al., 1999), which further reduces atmospheric effects that can degrade the signal. The $LE_{i,s}$ (Noah) components of the hybrid model were downloaded and processed in a similar fashion to the GLDAS forcing data (Rodell et al., 2004). The model was run over the African domain considered in this study for 2000–2009.

8.4 MODEL CALIBRATION AND VALIDATION

8.4.1 Study Area

The Republic of Kenya, located approximately between 5° 7′ N and 4° 39′ S longitude, is part of the Greater Horn region of Africa along the Indian Ocean (SEDAC, 2005). Kenya has a surface area of 579,617 km^2 comprising 8 provinces that are subdivided into 47 districts (Figure 8.1a). According to 2009 figures, the country's population was more than 30 million people, with nearly 22% of Kenyans living in urban centers (UN, 2009b). Farming is the primary livelihood of more than 75% of the population, conducted either on subsistence plots in marginal farming areas or on large plantations in the more arable areas (Uwechue, 1996), with less than 4%

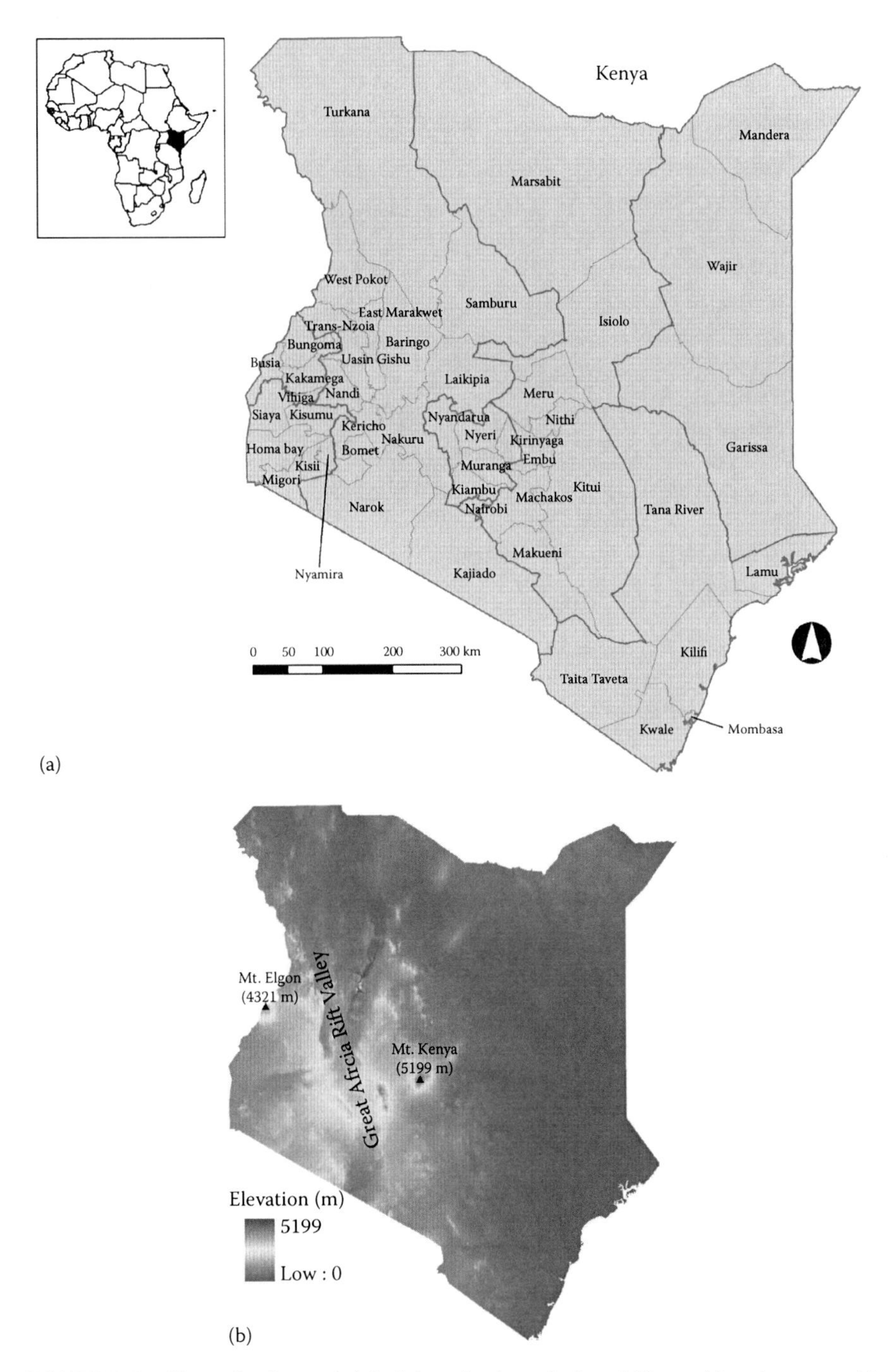

FIGURE 8.1 **(See color insert.)** Administrative boundaries of Kenya (a) and topographic map of Kenya (b). Forty-seven districts span eight provinces outlined in dark brown on the administrative map. The western and eastern highlands of Kenya are divided by the Great African Rift Valley shown on the topographic map.

of people being pastoralists. Districts to the northwest and east are pastoral, transitioning to mixed agriculture and pastoral areas to the south, such as West Pokot, Baringo, and Laikipia districts (FEWS NET, 2010). The recent increase in drought frequency and intensity and conspiring factors such as poor trade infrastructure, poverty, lack of government intervention, HIV/AIDS, failed adoption of drought-tolerant crops, and lack of grazing resources make these districts particularly food insecure. The most arable land is found in a high population density corridor consisting of Meru and Nithi districts to the east and Western Province to the west of the Great African Rift Valley (Figure 8.1b). These districts include major portions of the valley (Bomet, Nakuru, Kericho, Trans-Nzoia, and Uasin Gishu) and are characterized largely by cereal and dairy farming, while maize, the primary food staple, is the major crop grown outside these districts.

Kenya can be divided into five broad climatic zones: coastal, eastern/north, eastern/south, eastern/central, and western rift valley (DSK, 2003). The rainfall pattern in Kenya is typically bimodal, with a short rainy season (October–December) driven by convergence and the southward migration of the Intertropical Convergence Zone and a long rainy season (March–May) driven by southeasterly trades and the Indian Monsoon in January–February. Figure 8.2a shows the average date of long rain onset in Kenya. The onset of rains for the vast majority of the country occurs in March, with the largest monthly totals occurring in May for the coastal areas and 1 month earlier in the central and northern areas of the country. More than 85% of crops (primarily staples) are planted during the month of onset. The western and eastern highlands ascend from the Great African Rift Valley. Orographic uplift enhances rainfall in the highlands, which receive the largest amount of rainfall (>1000 mm per annum) in the country. The driest parts of the country (<250 mm per annum) are in the lowlands of northern Kenya. Rainfall in the western rift valley is strongly influenced by Lake Victoria and orographic uplift in the western highlands. The development of deep convection and cumulous clouds brings rain throughout the year, with surges during the long rainy season and the lowest amounts in January–February. Given the extended long rainy season, crops are staggered, and the harvest season in the western rift valley (October–January) is much longer than the harvest season for the remainder of the country (July–August).

8.4.2 Evapotranspiration and Crop Yield

Water loss through transpiration is a consequence of carbon uptake, and, therefore, ET_a often correlates well to fluctuations in plant biomass in areas of dense vegetation. Prolonged climate-related stress, as discussed earlier, results in lower carbon uptake and plant biomass over the growing season, thus reducing annual crop production and yield. Maize is the staple food throughout most of Kenya and is grown primarily during the long rainy season, so production and planted area statistics from this season of the year were deemed appropriate to evaluate the ET_a hybrid model. The agricultural ministry conducts an exhaustive field campaign that measures cropped area (ha) before harvest and estimates crop production by counting bags of a particular crop at harvest (Freund, 2005). These point measurements are then extrapolated to each district, yielding a district-level estimate of crop area, production, and yield. For this study, crop production statistics at the district level from 2000 to 2004 were

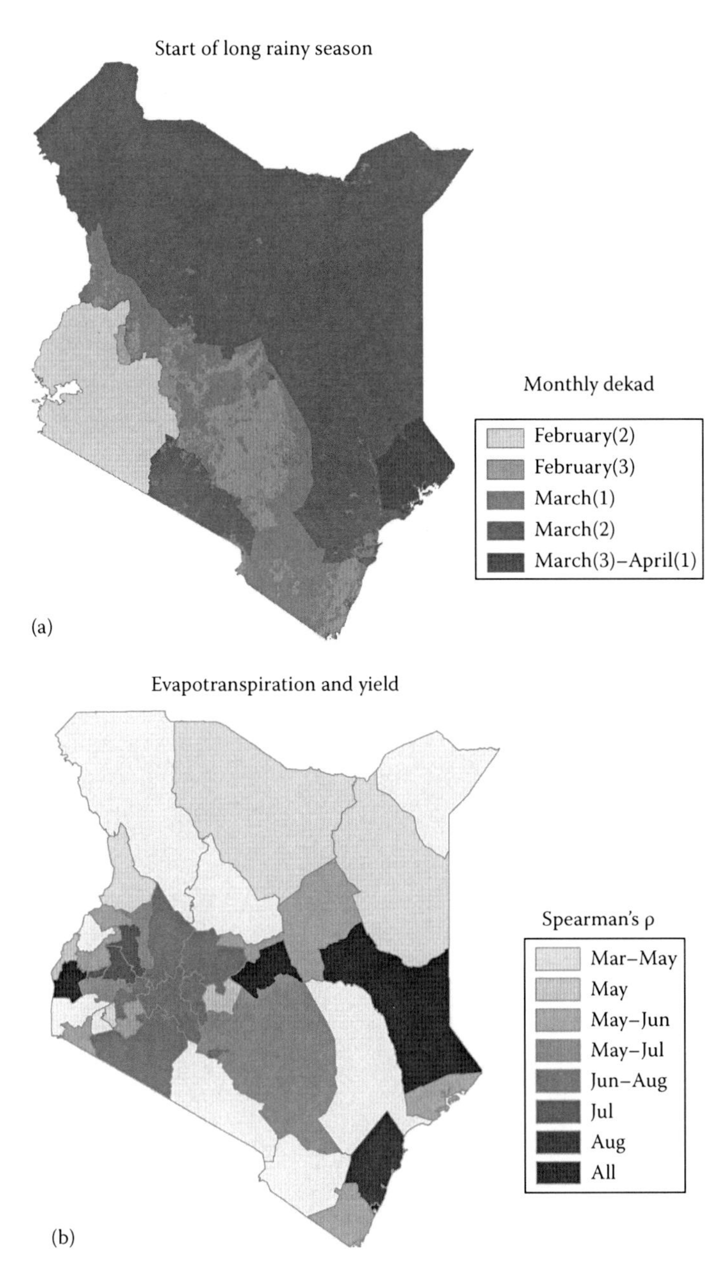

FIGURE 8.2 Mean date of long rain onset identified by the dekad (1–3) for each month, with yellow shading indicating cropped area (a) and months showing high ($\rho \geq 0.7$) correlations between ET_a and district-level maize yield (b). Districts in white showed no correlation between ET_a and yield in any month, while districts in brown showed correlations outside the growing season.

acquired from Kenya's Department of Resource Surveys and Remote Sensing of the Ministry of Planning and National Development. The hybrid (Fisher ET_c + Noah $ET_{i,s}$) model results were compared against the Noah ET_a and ET_c results from the Fisher model. ET_c from the Fisher model was also included in the evaluation because it was expected that transpiration, a direct measure of moisture availability in the root zone, would track crop stress well on its own.

Statistics from 70 districts were aggregated to 47 districts to match a vector file that contained pre-2001 district-level administrative boundaries. The vector file was used to obtain monthly areal statistics from the 0.05° hybrid (Fisher ET_c + Noah $ET_{i,s}$) and 0.25° Noah gridded data sets. ET_a totals (mm) averaged over each of the 47 districts for each month in the extended long rainy season (March–August) were used in the comparison. Monthly anomalies were computed from the 5 year (2000–2004) means. The Spearman rank correlation coefficient (ρ) was used to identify months where ET_a anomalies correlated well with maize yield. This resulted in only five data points (i.e., one per year in study period) for each month of analysis. A $\rho \geq 0.7$ (confidence = 90%) threshold was used to discern between strong and weak correlations. The Spearman rank correlation is a nonparametric technique used to detect monotonic trends (Sprent and Smeeton, 2007) and is essentially the Pearson correlation coefficient (R) for ranked data where ties are accounted for by taking the arithmetic average of the ranks. The Spearman rank correlation was chosen for several reasons: (1) production data are highly uncertain and often contain several outliers that tend to limit the ability of parametric techniques; (2) crop metrics, such as seasonal average NDVI, tend to do well at discriminating between "wet" and "dry" years when crops are limited by rainfall, but poorly differentiate "good" and "very good" production years when relatively small differences in production are a result of nonclimatic factors (Funk and Verdin, 2009); and (3) it is appropriate for small sample sizes.

The map in Figure 8.2b demonstrates the regional seasonality in the relationship between ET_a and yield, as indicated by the consensus of the two models on the correlation of yield to ET_a. For each district, the months in which monthly ET_a anomalies correlate well ($\rho \geq 0.7$) with yield from the hybrid model are identified in the figure. The months are nearly identical to those from the Noah model, but the correlations were typically lower for the latter. It is important to note that the correlations between ET_a and maize yield reflect only climate-driven variability. It is therefore assumed that several nonclimatic factors, such as labor input and soil tilth, play a lesser role in controlling maize yield. Districts highlighted in white showed poor correlations across all the months, while districts in brown showed correlations outside the expected growing season. In 36 of the 47 districts (77%), the ET_a anomalies from the hybrid model were better correlated with yield than were Noah model ET_a anomalies, as assessed on a month-by-month basis. For four of the wettest districts (Bungoma, Kisii, Taita Taveta, and Vihigia), ET_a anomalies from the Noah model were marginally better correlated with yield, at a low level of confidence (<50%). Of the five remaining districts, four showed higher correlations between yield and Noah ET_a than the hybrid model, but these correlations occurred during noncritical (outside the growing season) months of the year. The only district in which the Noah model provided significantly higher correlations during critical growing season months was in another wet district (Siaya), but the cause (or causes) of this result in this district is (are) unclear.

The variability in ET_a and maize yield correlations and inconsistencies with the expected growing season are highest in Western Kenya. Possible causes for this include poor crop reporting or the importance of nonclimatic factors. The most probable cause, however, deals with the climate and planting regime of this region. The western districts of Kenya are characterized by fairly consistent rainfall and a staggered planting regime throughout the year, meaning the variability in ET_a is low. Areas of low variability tend to transpire at or close to the atmospheric demand because they are not moisture limited. This could explain why some districts showed high correlations between ET_a and maize yield for some of the districts and low to no correlations in other districts. For the majority of the crop-producing districts in Kenya, ET_a appears to correlate with the grain-filling (reproductive) period of the seasonal calendar. The Rift Valley districts have a prolonged rainy season (February–September) and a relatively late harvest (October–February), so grain filling (June–August) is later than other districts. In the central and eastern highlands, the rainy season is shorter (March–June), and the harvest (July–September) is much earlier than the Rift Valley districts, so grain filling occurs earlier (May–July). Pastoral districts to the north have a shorter growing and later rainy season than the central and eastern highlands, so the highest correlation between ET_a and maize yield is earlier, again reflecting the grain-filling period of the seasonal calendar. Districts along the southeast coast showed weak or no correlations between ET_a and maize yield, and this is most likely due to the low variability in maize yield. These districts grow crops primarily during the short rainy season (October–January).

Districts highlighted with a graduated grayscale scheme in Figure 8.3 were selected for further analysis because they represent the most intensely cultivated and populated districts of the country. These districts showed significant ($\rho \geq 0.7$) correlations between maize yield and modeled ET_a averaged over May–July, the critical grain-filling period of the seasonal calendar. The Spearman rank correlation between the hybrid and Noah model ET_a and district-level yield in this focus area for the May–July season are shown in Figure 8.3a and b, respectively. Modeled ET_a anomalies in these months tended to show the strongest correlations with maize yield and therefore may be the most optimal midseason predictors. In general, both models provided similar correlations, with slightly stronger predictive values from the hybrid model except in the Kitui District. Noah correlations were significantly lower in the East Marakwet and Nyeri districts, most likely because of poor observed rainfall representation and reanalysis data used to drive Noah ET_a. The apparent 2–3 month lag between optimal ET_a (strongest ET_a-yield relationship) and rainfall onset is consistent with previous studies that showed a 2–3 month lagged relationship between rainfall onset and peak NDVI (Nicholson et al., 1990). This lag accounts for the vegetation response to accumulated soil moisture, which is most critical during the grain-filling period.

To further study the predictive capacity of midseason ET_a for end-of-season yield, May–July ET_a anomalies computed from monthly averages from each district polygon highlighted in Figure 8.3 were compared with yield anomalies in Figure 8.4. Here we examine anomalies in ET_a from both models and ET_c (canopy component only)

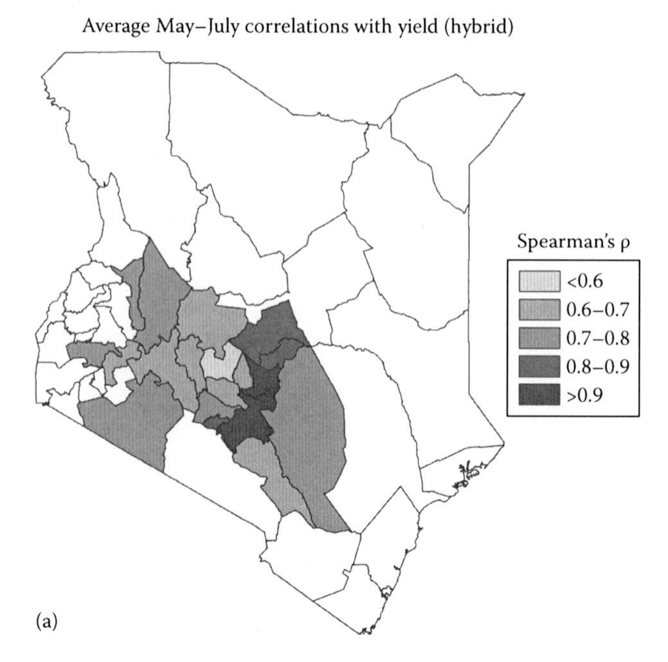

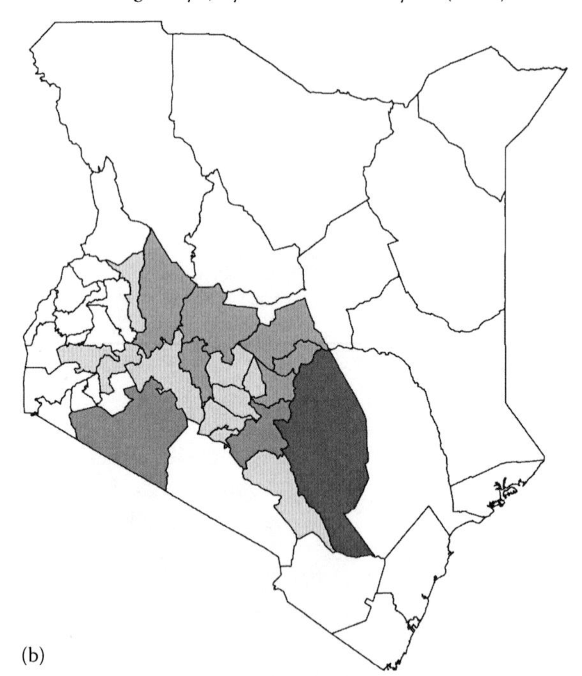

FIGURE 8.3 Spearman rank correlations for select districts between May–July average ET_a and district-level crop yield (2000–2004) for the combined (Fisher ET_c + Noah $ET_{i,s}$) model (a) and Noah LSM (b). White indicates areas where correlations were 0.

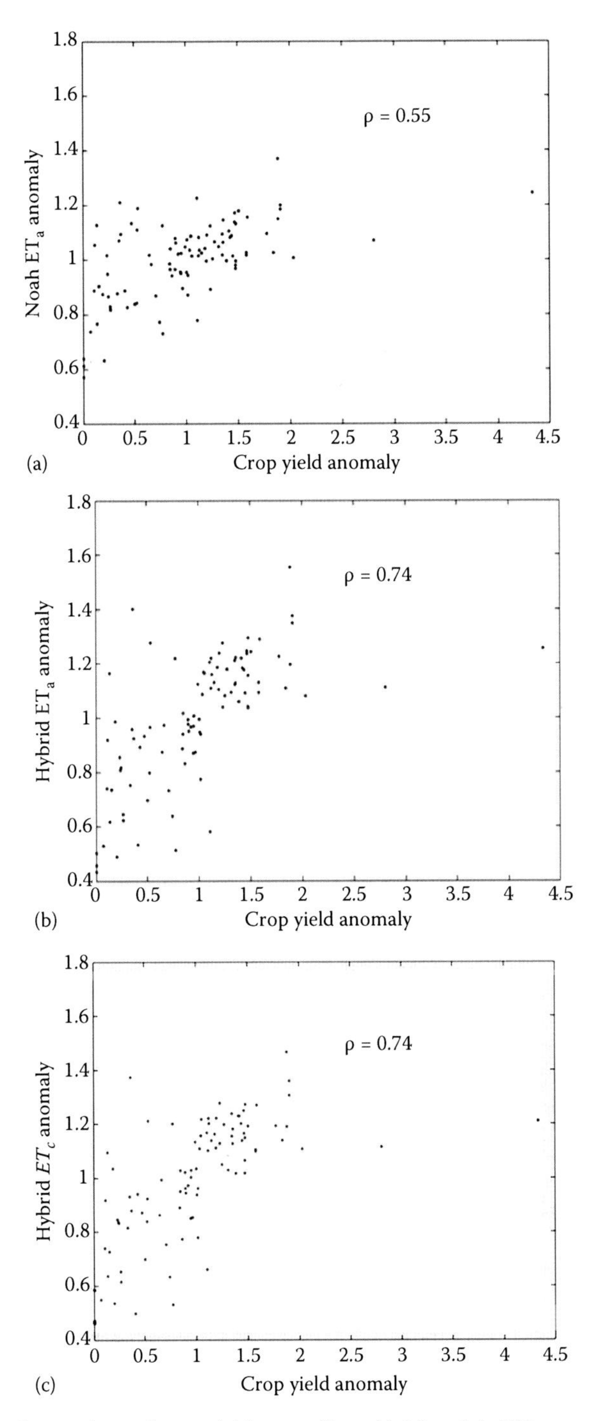

FIGURE 8.4 Comparison of crop yield anomalies with May–July ET_a anomalies using ET_a from Noah (a), ET_a from the hybrid model (b), and ET_c from the Fisher model ($N = 95$) (c).

from the Fisher model, to determine the relative predictive power of ET_c in isolation. The use of 19 districts and 5 years of yield data gave a total of 95 data points for analysis. The confidence levels were not computed, because each sample is not truly independent (i.e., wet and dry years will tend to cluster, because neighboring districts will exhibit similar climate and growth response). The correlation between Noah ET_a anomalies and yield ($\rho = 0.55$, $R = 0.57$) was significantly lower than the hybrid model ($\rho = 0.74$, $R = 0.64$) or the canopy component of the Fisher model ($\rho = 0.74$, $R = 0.65$). However, the transpiration component correlations were similar to those from the hybrid model ET_a. From an operational standpoint, ET_c is a preferred indicator because it requires fewer inputs to compute. ET_a from the hybrid model and Fisher ET_c each explained ~55% of the variance in maize yield, which is within the acceptable range for crop monitoring applications related to food security.

8.5 ET-BASED DROUGHT INDEX

Given that ET_c from the Fisher model appears to have utility in predicting end-of-season yield variability, ET_c was used in this section to develop a drought monitoring index analogous to the ESI introduced by Anderson et al. (2007), given by $1 - ET_a/ET_p$. As with WRSI, ET is constrained by the atmospheric water demand. An ESI value approaching 1 indicates very low ET_a, reflecting low soil moisture and associated crop stress conditions. As ESI approaches 0, ET_a approaches ET_p, which occurs when soil moisture is high and the crop is not constrained by a loss of water to the atmosphere. Here we formulate a modified version of the ESI focusing on crop stress and agricultural drought monitoring using the ratio of ET_c to canopy ET_p. Substituting the Fisher transpiration component into this relationship yields the following index:

$$ESI_c = 1 - f_g f_t f_m \left(1 - f_{wet}\right) \tag{8.3}$$

The index eliminates the need for the estimation of R_N used in ET_p. The ESI_c requires only estimates of NDVI and EVI derived from remote sensing and air temperature and relative humidity derived from surface reanalysis meteorological data. This formulation neglects evaporation from the canopy and soil surface, which is reasonable given that crop stress is directly proportional to the amount of water available in the root zone. Marshall et al. (2010) demonstrated that the Fisher model for ET_c is driven primarily by EVI, and, therefore, this index essentially tracks vegetation anomalies as modified by the air temperature and humidity response functions (f_t and f_m). Unlike VHI, which uses radiometric surface temperature as a proxy for soil dryness, this index uses surface reanalysis meteorological data (temperature, pressure, and specific humidity) to simulate soil dryness.

Seasonal (May–July) average ESI_c was computed over Kenya from 2000 to 2009 at 0.05° resolution using the data sources described in Section 8.3. To account for differences in ESI_c across crop types and climatic zones and to improve comparability with other statistical drought indices, the ESI_c data set was normalized by the mean and variance to form a Z-score. Z-scores standardize a sample distribution to a normal distribution. A Z-score of negative one represents 1σ below the normal mean (0), while a Z-score of positive one represents 1σ above the normal mean. Values ranged

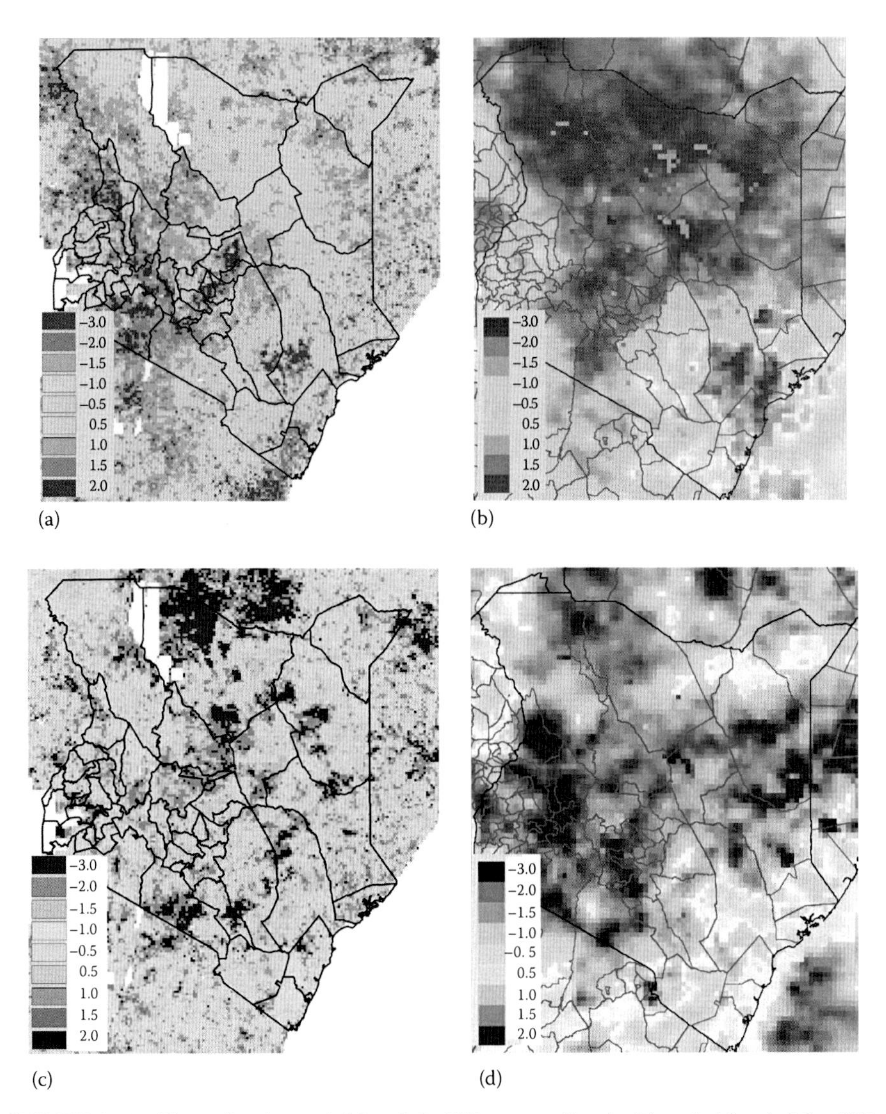

FIGURE 8.5 **(See color insert.)** May–July ESIc anomalies (a, b) and March–May SPI, (c, d) for Kenya in 2000 and 2003, respectively. Values are expressed as Z-scores of ESIc and the gamma probability of rainfall. Areas in white indicate missing data/bad values.

between -2.5σ and $+2.5\sigma$, where negative values indicate wet anomalies and positive values indicate dry anomalies.

In Figure 8.5, ESI_c seasonal anomalies for Kenya are compared with SPI seasonal anomalies using visualizations created with the Early Warning Explorer (EWX: http://earlywarning.usgs.gov/fews/) interface recently developed by the Climate Hazards Group at the University of California, Santa Barbara, to inform food security researchers and policy makers. EWX is a web-based, desktop-like application for exploring geospatial as well as time series data related to famine early warning activities. The EWX enables food security researchers and policy makers to view related data sets side-by-side, and has many advanced features found in traditional GIS applications. The SPI was computed from 0.1° (~10 km) resolution dekadal Rainfall Estimates (RFE) 2.0 data, and normalized to the period 2000–2009. RFE 2.0 rainfall combines a suite of data sources using the methodology described in Xie and Arkin (1997), using satellite data from the Meteosat geostationary satellites, the Advanced Microwave Sounding Unit (AMSU), the Special Sensor Microwave/Imager (SSM/I) infrared data, and meteorological data from approximately 1000 stations throughout sub-Saharan Africa that are part of the World Meteorological Organization's Global Telecommunication System. The SPI shown in Figure 8.5 was computed for the long rainy season (March–May) to accommodate for the time lag between precipitation and vegetation green-up. SPI ranges from -3σ to $+3\sigma$, where positive and negative scores indicate dry and wet anomalies, respectively.

Figure 8.5 shows SPI and ESI_c for an extremely dry year (2000) and wet year (2003). ESI_c and SPI show general agreement in the major crop-producing districts. In 2000, strong positive (dry) ESI_c anomalies and negative (dry) SPI anomalies can be seen in Nakuru, Narok, Nyandarua, Meru, Nithi, Embu, Muranga, Kiambu, and Nairobi. The negative (wet) ESI_c anomalies observed in Nyeri cover an area that includes Mount Kenya (elevation = 5199 m), where most of the precipitation is the result of orographic lifting. As a result, this area receives rainfall throughout most of the year and is not representative of other parts of the district. SPI is at a coarser spatial resolution than ESI_c and is not well represented by meteorological stations in this area, which may contribute to the observed bias in SPI. Strong wet anomalies for both ESI_c and SPI can be seen in southeastern districts along the coast. Normal conditions in the west of the country along Lake Victoria are well represented by both data sets as well. Food security reports corroborate these relatively localized patterns in ESI_c, which reflects the normal long rains confined to localized areas in the west, along the coast, and around Mount Kenya (Nyeri district) (FEWS NET, 2000). Food security reports in 2000 indicate high food insecurity in districts with chronic food shortages that include Turkana, Samburu, Marsabit, and Wajir with more sporadic insecurity in the districts of Isiolo, Garissa, and Mandera. The extremity of deficits in the north and east is less pronounced in ESI_c than SPI. This is most likely due to a low vegetation signal (i.e., low canopy cover) in these arid areas. In Marshall et al. (2010), the hybrid ET_a model showed its weakest performance at the most arid eddy covariance flux tower sites, and this was attributed in part to the strong dependence of the model on EVI and the low EVI signal in these areas. The coarse resolution of the EVI data and dominance of bare soil in these areas could make EVI insensitive to phenological changes in drier areas. Other contributing factors could include poor specific humidity reanalysis data or soil moisture formulation. In 2003, wet anomalies can be seen for both indices in the major crop-producing districts. Neutral conditions and wet anomalies for both indices in the north and along the coast can be seen as well. The moderate dry anomaly in Uasin Gishu, West Pokot, and Trans-Nzoia districts that is not indicated by SPI may be a result of the delayed onset of rains in this area, which caused below-normal food production (FEWS NET, 2003).

The ESI_c and SPI were also compared for 2009 (Figure 8.6) because this was a unique year climatically in the 10 year time series, with a delayed onset of the long rains throughout much of central, eastern, and coastal Kenya. This delay combined with 3 years of successive drought made 2009 particularly challenging for farmers. In the Kitui district (south-central Kenya), for example, maize prices increased by more

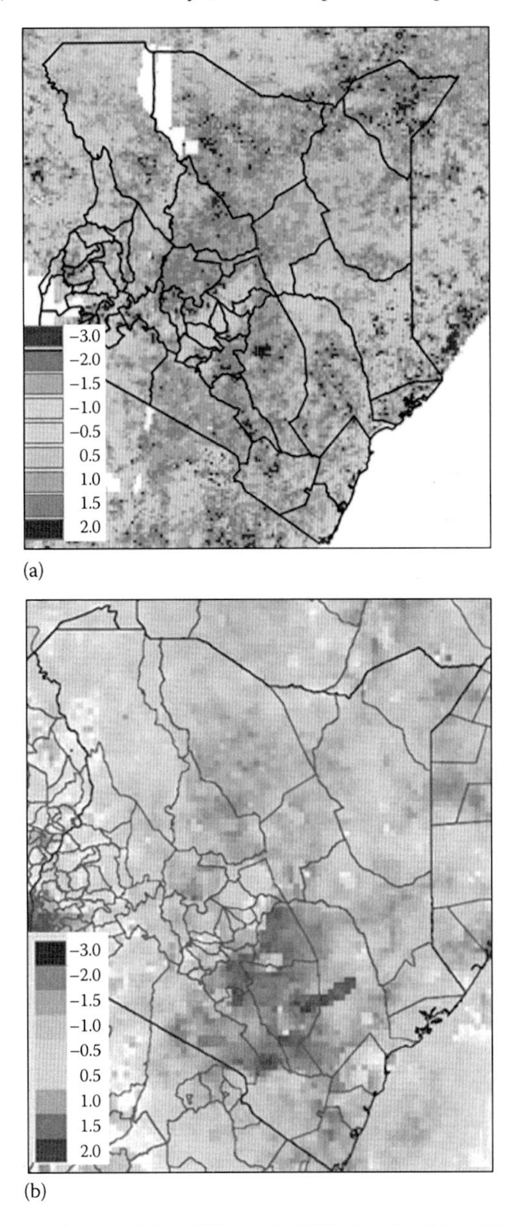

FIGURE 8.6 (See color insert.) Map of Kenya in 2009 showing May–July ESI_c anomalies (a), March–May SPI (b), and

(continued)

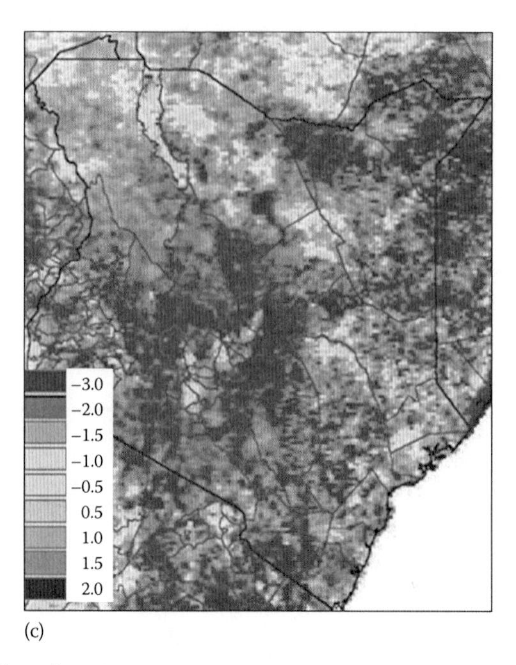

FIGURE 8.6 (continued) **(See color insert.)** May–July MODIS LST anomalies (c).

than 200% in response to crop failure (FEWS NET, 2009). Figure 8.6 compares ESI_c and SPI along with anomalies in the MODIS land surface temperature (LST) product (Wan et al., 2004) computed for May–July. LST was included in the analysis, because it can be readily viewed with SPI on the EWX webpage, and because it provides proxy remote sensing information about the current surface moisture status. The LST product used on the EWX webpage is derived from the MODIS thermal sensor on board the Aqua platform, resampled to 0.05° spatial resolution globally at a daily time step. ESI_c provides a good representation of the range of food security conditions that existed in 2009. In that year, many districts in the west and center of the country with normal or slightly above-normal ESI_c and SPI conditions (−0.5 to 0.0) were moderately food secure with localized areas of high food insecurity (FEWS NET, 2009). Marginal lands, such as those found in Baringo, Kitui, Machakos, and Makueni districts, were highly food insecure, as reflected by the large positive ESI_c anomalies. Pronounced dry anomalies in the SPI appear in the Laikipia and Nyandarua districts, but, overall, the SPI fails to capture the severity of drought over the full region of insecurity, which may be partially explained by the LST anomaly map in Figure 8.6c. In 2009, temperatures were abnormally warm, which, when combined with late rains, would increase the evaporative moisture loss rate and enhance crop stress. These higher temperatures and vegetation stress are reflected over much of Kenya in the MODIS LST anomaly map, but they would not have been as clearly represented in the SPI, which is solely dependent on precipitation inputs and does not incorporate temperature into its calculation. As a result, both climatic drivers impacting the rate of water consumption and overall condition of crops in 2009 will be captured in ET_a-based indices, but not in precipitation-based indices like SPI.

8.6 DISCUSSION AND FUTURE WORK

The ESI_c index can be implemented in several ways to inform food security analysts and policy makers. One option would be to include it in an internet-based tool like EWX, which allows side-by-side user comparisons with other indices to give a more complete interpretation of agricultural droughts. EWX is stand-alone and interactive, allowing stakeholders at multiple levels to develop their own mitigation plans. The results of the analysis could be disseminated to the larger stakeholder community via weekly, seasonal, and annual reports. The African Dissemination Service is another online resource that includes a more comprehensive collection of data than EWX, but it is not interactive and requires additional software that may not be available to all stakeholders.

The poor representation of ESI_c in the arid areas in the north and east of Kenya could be improved to better represent the dry and sparsely vegetated conditions there. ESI_c is driven primarily by anomalies in vegetation indices and secondarily to surface temperature. Comparisons with other empirical indices that use vegetation and temperature as drivers, such as the VHI, will help to determine whether the more physical ESI_c formulization and inclusion of a formal moisture constraint produces any significant improvements in moisture-limited areas. ESI_c values in this study generally exhibited a skewed distribution over the period 2000–2009. Therefore, using Z-scores to represent anomalies may not properly stress the importance of extreme events, as was witnessed in the north and east of Kenya in 2000. The gamma distribution works particularly well for standardizing rainfall in Africa and should be evaluated, along with other distributions to improve interpretation of ESI_c. Given the stronger SPI signal in arid areas and ESI_c signal in the primary agricultural areas, a combined index that uses standardized ESI_c and SPI could potentially be a powerful tool for crop stress monitoring. Given the lagged relationship between rainfall and vegetation response, a simple autoregressive model of rainfall and vegetation could be used to reformulate the ESI_c index in terms of precipitation and temperature. The index could then be back-casted to measure historical trends in crop stress and projected to determine crop stress probabilities under future Intergovernmental Panel on Climate Change (IPCC) climate scenarios.

The ET_a model hybrid could also be used to improve yield estimate crop models, defining start and end of season at a higher resolution than data sets currently used in sub-Saharan Africa. This would improve the crop phase adjusted ET_p used in crop models, such as WRSI. ET_a is a leading component of the surface energy and water balance. Improved estimates of ET_a would be of benefit in applying crop models for monitoring crop stress.

8.7 CONCLUSIONS

Precipitation, soil moisture, and vegetation-based indices have historically been used to monitor crop stress. ET is more intimately connected to moisture available to crops under stress than is precipitation or volumetric soil moisture, but it has not been used extensively in the past because of poor parameterization and a general lack of calibration and validation data. In this chapter, we presented a simple low-cost model for estimating ET_a that combines remote sensing and surface reanalysis meteorological data. The model was evaluated against an energy and water balance approach to estimating ET_a using maize yield data in Kenya. The analysis revealed that the canopy component of the ET_a model performed as well as the hybrid model in areas of dense vegetation cover, given its sensitivity to root zone moisture.

The canopy component of the ET_a was used to parameterize a new vegetation–temperature–based crop stress index. Unlike other vegetation–temperature indices, the new index includes a moisture constraint on crop health and incorporates a more realistic parameterization of vegetation with the inclusion of EVI and temperature by including an optimal temperature-response function. However, like other vegetation-based drought indices, the coarse resolution of satellite imagery used to drive the model fails to adequately capture the vegetation signal (and crop stress) in arid regions. For the major crop-producing areas of Kenya, however, the analysis demonstrated that the ET_c-based index is an effective crop monitoring tool using yield data and food security reports. The index, which is partially derived from remote sensing, is seen as an improvement on remotely sensed vegetation and temperature indices, because the formulization includes more general concepts on light and soil moisture limitation and surface reanalysis improves the temporal resolution of model inputs. Precipitation-based indices, which are currently used in sub-Saharan Africa, are a good indicator of moisture supply, but do not account for atmospheric moisture demand (ET_p), which can be critical to maintaining plant health, especially during the reproductive phase. Atmospheric demand is directly proportional to surface temperature. Combining an index of moisture supply with a temperature component that captures the demand side would help to characterize not only onset conditions but the reproductive phase of crop growth as well. A compound index that includes ET_c and precipitation would therefore give a more complete picture of crop development.

The use of ET_a-driven models in agricultural studies is still in its infancy in sub-Saharan Africa. Dense eddy covariance flux tower and lysimeter data sets collected in selected areas of sub-Saharan Africa combined with upscaling techniques provide a test bed to further refine the ET_a model and crop stress index and apply them to an array of questions related to agricultural development.

ACKNOWLEDGMENTS

This research was supported by the U.S. Agency for International Development Famine Early Warning System Network, under U.S. Geological Survey Cooperative Agreement #G09AC00001, and the National Aeronautics and Space Administration, under Precipitation Science Grant #NNX07AG266. Special thanks to Gideon Galu, who provided the agricultural statistics from the government of Kenya.

REFERENCES

Agam, N., W.P. Kustas, M.C. Anderson, J.M. Norman, P.D. Colaizzi, T.A. Howell, J.H. Prueger, T.P. Meyers, and T.B. Wilson. 2010. Application of the Priestley-Taylor approach in a two-source surface energy balance model. *Journal of Hydrometeorology* 95:185–198.

Allen, R.G. 2000. Using the FAO-56 dual crop coefficient method over an irrigated region as part of an evapotranspiration intercomparison study. *Journal of Hydrology* 229:27–41.

Allen, R.G., L.S. Pereira, D. Raes, and M. Smith. 1998. *Crop Evapotranspiration—Guidelines for Computing Crop Water Requirements*. Rome, Italy: Food and Agriculture Organization of the United Nations.

Alley, W.M. 1984. The Palmer drought severity index: Limitations and assumptions. *Journal of Climate and Applied Meteorology* 23:1100–1109.

Anderson, M.C., J.M. Norman, J.R. Mecikalski, J.A. Otkin, and W.P. Kustas. 2007. A climatological study of evapotranspiration and moisture stress across the continental United States based on thermal remote sensing: 1. Model formulation. *Journal of Geophysical Research* 112:D10117.

Anyamba, A. and J.R. Eastman. 1996. Interannual variability of NDVI over Africa and its relation to El Niño/Southern Oscillation. *International Journal of Remote Sensing* 17:2533–2548.

Betts, A.K., F. Chen, E. Mitchell, and Z.I. Janjic. 1997. Assessment of the land surface and boundary layer models in two operational versions of the NCEP Eta Model using FIFE data. *Monthly Weather Review* 125:2896–2916.

Chen, F., K.W. Manning, M.A. LeMone, S.B. Trier, J.G. Alfieri, R. Roberts, M. Tewari, D. Niyogi, T.W. Horst, S.P. Oncley, J.B. Basara, and P.D. Blanken. 2007. Description and evaluation of the characteristics of the NCAR high-resolution land data assimilation system. *Journal of Applied Meteorology and Climatology* 46:694–713.

Chen, F., K. Mitchell, J. Schaake, Y. Xue, H. Pan, V. Koren, Q.Y. Duan, M. Ek, and A. Betts. 1996. Modeling of land surface evaporation by four schemes and comparison with FIFE observations. *Journal of Geophysical Research* 101:2896–2916.

Doorenbos, J. and W.O. Pruitt. 1977. Crop water requirements. FAO irrigation and drainage paper 24, Rome, Italy: United Nation Food and Agriculture Organization.

Dracup, J.A., K.S. Lee, and E.G. Paulson, Jr. 1980. On the definition of droughts. *Water Resources Research* 16:297–302.

DSK. 2003. *National Atlas of Kenya*. Nairobi, Kenya: Survey of Kenya.

FEWS NET. 2000. *Kenya Food Security Update: July 2000*. Washington, DC: United States Agency for International Development.

FEWS NET. 2003. *Kenya Food Security Update—September 5, 2003*. Washington, DC: United States Agency for International Development.

FEWS NET. 2009. *Kenya Food Security Outlook July to December 2009*. Washington, DC: United States Agency for International Development.

FEWS NET. 2010. *Kenya: Food Security Framework*. Washington, DC: United States Agency for International Development.

FFP. 2010. *Fact Sheet: Office of Food for Peace 2009 Statistics*. Washington, DC: United States Agency for International Development.

Fisher, J.B., K.P. Tu, and D.D. Baldocchi. 2008. Global estimates of the land-atmosphere water flux based on monthly AVHRR and ISLSCP-II data, validated at 16 FLUXNET sites. *Remote Sensing of Environment* 112:901–919.

Freund, J.T. 2005. Estimating crop production in Kenya: A multi-temporal remote sensing approach. Master's thesis, Santa Barbara, CA: University of California-Santa Barbara.

Funk, C. and J. Verdin. 2009. Real-time decision support systems: The famine early warning system network. In *Satellite Rainfall Applications for Surface Hydrology*, eds. M. Gebremichael and F. Hossain, p. 327. New York: Springer.

Guttman, N.B. 1999. Accepting the standardized precipitation index: A calculation algorithm. *Journal of the American Water Resources Association* 35:311–322.

Heim, R.R. 2002. A review of twentieth-century drought indices used in the United States. *Bulletin of the American Meteorological Society* 83:1149–1165.

Hogue, T.S., L. Bastidas, H. Gupta, S. Sorooshian, K. Mitchell, and W. Emmerich. 2005. Evaluation and transferability of the Noah land surface model in semiarid environments. *Journal of Hydrometeorology* 6:68–84.

Huete, A., K. Didan, T. Miuru, E.P. Rodriguez, X. Gao, and L.G. Ferreira. 2002. Overview of the radiometric and biophysical performance of the MODIS vegetation indices. *Remote Sensing of Environment* 83:195–213.

Jarvis, P. and K. McNaughton. 1986. Stomatal control of transpiration: Scaling up from leaf to region. *Advances in Ecological Research* 15:1–50.

Jones, H.G. 1992. *Plants and Microclimate: A Quantitative Approach to Environmental Plant Physiology*. Cambridge, MA: Cambridge University Press.

June, T., J.R. Evans, and G.D. Farquhar. 2004. A simple new equation for the reversible temperature dependence of photosynthetic electron transport: A study on soybean leaf. *Functional Plant Biology* 31:275–283.

Kalnay, E., M. Kanamitsu, R. Kistler, W. Collins, D. Deaven, L. Gandin, M. Iredell, S. Saha, G. White, J. Wollen, Y. Zhu, A. Keetmaa, R. Reynolds, M. Chelliah, W. Ebisuzaki, W. Higgins, J. Janowiak, K.C. Mo, C. Ropelewski, J. Wang, R. Jenne, and D. Joseph. 1996. The NCEP/NCAR 40-year reanalysis project. *Bulletin of the American Meteorological Society* 77:437–471.

Karl, T.R. 1986. The sensitivity of the Palmer drought severity index and Palmer's Z-Index to their calibration coefficients including potential evapotranspiration. *Journal of Climate and Applied Meteorology* 25:77–86.

Karnieli, A., M. Bayasgalan, Y. Bayarjargal, N. Agam, S. Khudulmur, and C.J. Tucker. 2006. Comments on the use of the vegetation health index over Mongolia. *International Journal of Remote Sensing* 27:2017–2024.

Keyantash, J. and J.A. Dracup. 2002. The quantification of drought: An evaluation of drought indices. *Bulletin of the American Meteorological Society* 83:1167–1180.

Kogan, F.N. 1995. Application of vegetation index and brightness temperature for drought detection. *Advances in Space Research* 15:91–100.

Mahfouf, J.F. and J. Noilhan. 1991. Comparative study of various formulations of evaporations from bare soil using in situ data. *Journal of Applied Meteorology* 30:1354–1365.

Marshall, M., K. Tu, C. Funk, J. Michaelsen, P. Williams, C. Williams, J. Ardö, B. Marie, B. Cappelaere, A. de Grandcourt, A. Nickless, Y. Nouvellon, R. Scholes, and W. Kutsch. 2011. Combining surface reanalysis and remote sensing data for monitoring evapotranspiration (in press).

McGuire, J.K. and W.C. Palmer. 1957. The 1957 drought in the eastern United States. *Monthly Weather Review* 85:305–314.

McKee, T.B., N.J. Doesken, and J. Kleist. 1993. The relationship of drought frequency and duration to time scales. *Eighth Conference on Applied Climatology*, American Meteorological Society, Anaheim, CA, January 17–22.

Monteith, J.L. 1965. Evaporation and environment. *Symposia of the Society for Experimental Biology* 19:205–234.

Mu, Q., M. Zhao, F.A. Heinsch, M. Liu, H. Tian, and S.W. Running. 2007. Evaluating water stress controls on primary production in biogeochemical and remote sensing based models. *Journal of Geophysical Research* 12:01010.01029/02006JG000179.

Narasimhan, B. and R. Srinivasan. 2005. Development and evaluation of soil moisture deficit index (SMDI) and evapotranspiration deficit index (ETDI) for agricultural drought monitoring. *Agricultural and Forest Meteorology* 133:69–88.

Nicholson, S.E., M.L. Davenport, and A.R. Malo. 1990. A comparison of the vegetation response to rainfall in the Sahel and East Africa, using normalized difference vegetation index from NOAA AVHRR. *Climatic Change* 17:209–241.

Palmer, W.C. 1965. *Meteorological Drought*. Research Paper No. 45, Washington, DC: U.S. Department of Commerce Weather Bureau.

Penman, H.L. 1948. Natural evaporation from open water, bare soil and grass. *Proceedings of the Royal Society of London. Series A, Mathematical and Physical Sciences* 193:120–145.

Philippon, N., L. Jarlan, N. Martiny, P. Camberlin, and E. Mougin. 2007. Characterization of the interannual and intraseasonal variability of West African vegetation between 1982 and 2002 by means of NOAA AVHRR NDVI data. *Journal of Climate* 20:1202–1218.
Potter, C.S., J.T. Randerson, C.B. Field, P.A. Matson, P.M. Vitousek, H.A. Mooney, and S.A. Klooster. 1993. Terrestrial ecosystem production: A process model based on global satellite and surface data. *Global Biogeochemical Cycles* 7:811–841.
Priestley, C.H.B. and R.J. Taylor. 1972. On the assessment of surface heat flux and evaporation using large-scale parameters. *Monthly Weather Review* 100:81–92.
Raupach, M.R. 2000. Equilibrium evaporation and the convective boundary layer. *Boundary-Layer Meteorology* 96:107–142.
Rodell, M., P.R. Houser, U. Jambor, J. Gottschalck, K. Mitchell, C.-J. Meng, K. Arsenault, B. Cosgrove, J. Radakovich, M. Bosilovich, J.K. Entin, J.P. Walker, D. Lohmann, and D. Toll. 2004. The global land data assimilation system. *Bulletin of the American Meteorological Society* 85:381–394.
Rouse, J.W. Jr., R.H. Haas, D.W. Deering, and J.A. Schell. 1974. Monitoring the vernal advancement of retrogradation of natural vegetation. Greenbelt, MD: ASA/GSFC.
SEDAC. 2005. Gridded Population of the World, version 3 (GPWv3). Center for International Earth Science Information Network, Columbia University and Centro Internacional de Agricultura Tropical. http://sedac.ciesin.columbia.edu/gpw/ (accessed July 8, 2011).
Sellers, P.J. 1987. Canopy reflectance, photosynthesis, and transpiration, II. The role of biophysics in the linearity of their interdependence. *Remote Sensing of Environment* 21:143–183.
Senay, G.B. and J. Verdin. 2003. Characterization of yield reduction in Ethiopia using a GIS-based crop water balance model. *Canadian Journal of Remote Sensing* 29:687–692.
Shanahan, J.F. and D.C. Nielsen. 1987. Influence of growth retardants (anti-gibberellins) on corn vegetative growth, water use, and grain yield under different levels of water stress. *Agronomy Journal* 79:103–109.
Sprent, P. and N.C. Smeeton. 2007. *Applied Nonparametric Statistical Methods*. New York: Chapman and Hall/CRC.
Stegman, E.C. 1982. Corn grain yield as influenced by timing of evapotranspiration deficits. *Irrigation Science* 3:75–87.
Swets, D.L., B.C. Reed, J.R. Rowland, and S.E. Marko. 1999. A weighted least-squares approach to temporal smoothing of NDVI. *Proceedings of the 1999 ASPRS Annual Conference*, Portland, OR, May 17–21.
UN. 2009a. *The Millennium Development Goals Report*. New York: United Nations.
UN. 2009b. World urbanization prospects: The 2009 revision population database. Population Division of the Department of Economic and Social Affairs of the United Nations Secretariat, http://esa.un.org/wup2009/unup/index.asp?panel=1 (accessed on November 15, 2009).
Uwechue, R. (ed.). 1996. *Africa Today*. London, U.K.: Africa Books Limited.
Wan, Z., Y. Zhang, Q. Zhang, and Z.-L. Li. 2004. Quality assessment and validation of the MODIS global land surface temperature. *International Journal of Remote Sensing* 25:261–274.
Xie, P. and P.A. Arkin. 1997. Global precipitation: A 17-year monthly analysis based on gauge observations, satellite estimates, and numerical model outputs. *Bulletin of the American Meteorological Society* 78:2539–2558.

Part III

Soil Moisture/Groundwater

9 Microwave Remote Sensing of Soil Moisture *Science and Applications*

Son V. Nghiem, Brian D. Wardlow, David Allured, Mark D. Svoboda, Doug LeComte, Matthew Rosencrans, Steven K. Chan, and Gregory Neumann

CONTENTS

9.1 INTRODUCTION

Soil moisture is a fundamental link between global water and carbon cycles and has major applications in predicting natural hazards such as droughts and floods (National Research Council, 2007). From precipitation data, soil wetness can be estimated by hydrological land-surface models. In the United States, preliminary precipitation data are based on measurements gathered from many active stations nationwide each month, and it takes 3–4 months to assemble final, quality-controlled data. In the western United States, some climate divisions may have no stations reporting in a particular month or may lack first- or second-order stations, and significant blockages by mountains limit the capability of precipitation measurement by surface rain radars (Maddox et al., 2002).

Soil moisture can also be measured directly, using data from networks like the Oklahoma Mesonet System (Illston et al., 2004) and the Soil Climate Analysis Network (SCAN) (USDA, 2009a). However, measurements from such networks are generally too sparse for most applications and are of varying accuracy. Soil moisture observations have been added to the SNOTEL network (USDA, 2009b), but fully calibrated data are not yet available routinely. Given the limited number of stations collecting point-based, in situ data, this information may not be representative of regional soil moisture conditions.

Soil moisture measurements over a large spatial extent (areal data rather than point data) with few or no missing gaps are crucial for characterizing the land surface water distribution from regional to continental scales. Recognizing the importance of soil moisture as a key variable for drought monitoring, satellite microwave remote sensing soil moisture retrievals using both passive and active sensors hold the potential to begin to fill this informational void in the United States and elsewhere.

Passive microwave radiometers, such as the Scanning Multichannel Microwave Radiometer (SMMR), Special Sensor Microwave/Imager (SSM/I), Tropical Rainfall Measuring Mission (TRMM) Microwave Imager (TMI), Advanced Microwave Scanning Radiometer on the Earth Observing System (AMSR-E), and Soil Moisture and Ocean Salinity sensor (SMOS), measure the natural emission of microwave energy from the land surface, which is used to derive soil moisture using various algorithms (Wang, 1985; Owe et al., 1988; Kerr and Njoku, 1990; Teng et al., 1993; van de Griend and Owe, 1994; Engman, 1995; Jackson, 1997; Kerr et al., 2001; Njoku et al., 2003). These passive radiometers operate at microwave frequencies from L to Ka bands with additional higher frequencies for other applications.

In contrast, active sensors, including synthetic aperture radar (SAR) and scatterometers, transmit signals to a targeted surface area and measure the scattering return. Many approaches have been used to estimate soil moisture from data sets acquired by SARs including Seasat, Spaceborne Imaging Radar-C (SIR-C), European Remote Sensing (ERS), RADARSAT, Environmental Satellite (Envisat), and Advanced Land Observing Satellite (ALOS) (Blanchard and Chang, 1983; Cognard et al., 1995; Dubois et al., 1995; Loew et al., 2006; Shrivastava et al., 2009; Takada et al., 2009), and by scatterometers such as ERS and QuikSCAT (QSCAT) (Wagner et al., 1999; Nghiem et al., 2000; Wagner and Scipal, 2000). In this chapter, we review the science principle of active and passive remote sensing of soil moisture and then illustrate results from AMSR-E and QSCAT for drought applications.

9.2 MICROWAVE REMOTE SENSING SCIENCE

The principle of microwave remote sensing of soil moisture is based on the sensitivity of soil permittivity to the amount of liquid water. The permittivity of a medium, like moist soil, characterizes electromagnetic wave propagation and attenuation in the medium. Both brightness temperature (BT) (measured by a radiometer) and backscatter (measured by a radar) are dependent on the soil permittivity. Empirical models have been developed in order to relate volumetric content (m_v) for different soil types to the dielectric constant (the permittivity of a medium relative to that of free space) at microwave frequencies between 1.4 and 18 GHz (Dobson et al., 1985; Hallikainen et al., 1985).

Although in situ measurements of soil dielectric constant can be made with a probe (Jackson, 1990), satellite remote sensors do not directly provide soil dielectric measurements. Instead, these sensors acquire BT or backscatter signatures, which are dependent on soil dielectric properties and thus soil moisture. Such a relationship enables the inversion of soil moisture from BT or backscatter data, but it can be complicated by vegetation cover, surface roughness, rainfall, and anthropogenic effects (e.g., radio frequency interference [RFI]), which have different impacts on the accuracy of soil moisture retrieval at different microwave frequencies.

9.2.1 Passive Remote Sensing

The retrieval of soil moisture from BT has been studied by many researchers (see summary by Njoku et al., 2003) and is reviewed briefly here. For an isothermal vegetated soil surface with physical temperature T_s, BT (T_{bp}) can be expressed as follows:

$$T_{bp} = T_s \left\{ e_{sp} \exp(-\tau_c) + (1-\omega_p)\left[1-\exp(-\tau_c)\right]\left[1 + r_{sp}\exp(-\tau_c)\right]\right\} \quad (9.1)$$

where the soil emissivity is $e_{sp} = 1 - r_{sp}$ for soil reflectivity r_{sp}, which is influenced by soil moisture through the effect of moisture on the soil dielectric constant. In Equation 9.1, τ_c and ω_p are the vegetation opacity and the vegetation single scattering albedo, respectively. Multiple scattering in the vegetation layer is neglected, and a quasi-specular soil surface and no reflection at the air–vegetation boundary are assumed in Equation 9.1. Vegetation opacity and multiple scattering have less effect at lower microwave frequencies. The effective emitting depth is controlled by the near-surface moisture profile and is smaller for higher microwave frequencies and for wetter soils. Although microwaves can only sense soil moisture in the top soil layer (in millimeters to decimeters, depending on frequencies), there is a correlation to soil moisture in deeper soil at night when the soil moisture and temperature profiles are more uniform.

For a fixed viewing angle, an empirical formulation has been found useful for relating the reflectivity of a rough soil surface, r_{sp}, to that of the equivalent smooth surface, r_{oq} (Wang and Choudhury, 1981; Wang, 1983), which is expressed as follows:

$$r_{sp} = \left[(1-Q)r_{op} + Qr_{oq}\right]\exp(-h) \quad (9.2)$$

where

p and q represent either of the orthogonal polarization states (vertical, v, or horizontal, h)

Q and h are roughness parameters

Q may be approximated as zero at low frequencies (e.g., L and C bands). The separation of soil moisture and roughness effects through Equation 9.2 is not precise, and the parameter h has a residual moisture dependence (Li et al., 2000; Wigneron et al., 2001).

To normalize the surface temperature (T_s) dependence in Equation 9.1, the polarization ratio (PR) is obtained by

$$PR = \frac{T_{bv} - T_{bh}}{T_{bv} + T_{bh}} \tag{9.3}$$

which is suitable for multichannel data taken at the same incidence angle (Kerr and Njoku, 1990). At large incidence angles (e.g., >50°), the difference between the vertically and horizontally polarized BTs for bare soils is large, giving rise to a significant PR signal. However, the observation path length through the vegetation becomes longer at large incidence angles, increasing vegetation attenuation and thus decreasing sensitivity to soil moisture.

While Equations 9.1 through 9.3 form a general theoretical basis for soil moisture retrieval from passive microwave data, several approaches have been developed for different satellite data sets using different methods to correct for effects of soil type, roughness, vegetation, and surface temperature (Njoku et al., 2003). Nevertheless, further advances are needed for various nonisothermal conditions and multiple interactions between the soil surface and vegetation cover at different growth stages. For data from the AMSR-E on the EOS Aqua satellite, the soil moisture retrieval utilizes primarily the frequency channels of 10.7 and 18.7 GHz to consider effects of atmospheric and vegetative attenuation and to minimize the requirement for ancillary data inputs. The TMI has 10.7 and 19.3 GHz channels, which can be used to obtain PR for soil moisture applications with a better consistency at the lower frequency (Njoku et al., 2003). Further details of the retrieval can be found in the literature (Njoku and Li, 1999; Njoku et al., 2003; Njoku, 2004).

9.2.2 Active Remote Sensing

In active remote sensing, soil moisture can be derived from backscatter measured by an SAR at a high spatial resolution with a limited spatial and infrequent repeat coverage and by a scatterometer at a low spatial resolution with a large areal and frequent coverage. Many theoretical models have been developed to characterize backscatter signatures of vegetated soil. Here, a scattering model based on the analytic vector wave theory (Nghiem et al., 1993a) together with a practical formulation is reviewed.

Backscatter σ_0 from moist soil with vegetation cover is determined from an ensemble average of the correlation of scattered field components E as follows:

$$\left\langle \bar{E}_{0s}(\bar{r}) \cdot \bar{E}_{0s}^{*}(\bar{r}) \right\rangle = \sum_{i,j,k,l,m}^{x,y,z} k_0^4 \int_0^{\pi} d\psi_f \int_0^{2\pi} d\phi_f \, p(\psi_f, \phi_f) \int_{V_1} d\bar{r}_1 \int_{V_1} d\bar{r}_1^{0} C_{\xi 1 jklm}\left(\bar{r}_1, \bar{r}_1^{0}; \psi_f, \phi_f\right)$$
$$\times \left[\left\langle G_{01ij}(\bar{r}, \bar{r}_1) \right\rangle \left\langle F_{1k}(\bar{r}_1) \right\rangle \right] \cdot \left[\left\langle G_{01il}\left(\bar{r}, \bar{r}_1^{0}\right) \right\rangle \left\langle F_{1m}\left(\bar{r}_1^{0}\right) \right\rangle \right]^{*} \tag{9.4}$$

where subscript 0 represents the air space above the vegetation, and subscript 1 indicates the vegetation cover occupying volume V_1 over the soil surface. The dyadic Green's function G and the mean field F are obtained as described by Nghiem et al. (1990). The correlation function C characterizes the vegetation scatterers having different size, shape, and orientation angle ψ_f in elevation and ϕ_f in azimuth. For the vegetation canopy, the effective permittivity is calculated under the strong permittivity fluctuation theory, which accounts for wave attenuation including scattering and absorption loss (Nghiem et al., 1993a). The analytic vector wave theory accounts for fully polarimetric scattering, preserves the phase information, and includes multiple reflection and transmission interactions of upgoing and downgoing electromagnetic waves with the soil surface. The solution conveys soil moisture information, because soil transmissivity and reflectivity are controlled by the soil dielectric constant as a function of volumetric soil moisture.

Rough surface scattering can be included in the contribution to the total backscatter. The small-scale roughness of the soil surface is described with a standard deviation height and a slope. When a large-scale roughness also exists, the overall roughness is accounted for by a joint probability density function for both roughness scales (Nghiem et al., 1995). The vegetation volume scattering and soil surface scattering are assumed to be uncorrelated because of independent statistical representations of vegetation scatterers (e.g., leaves, twigs, and branches) and soil surface roughness. As a result, the total backscatter is a sum of the vegetation volume backscatter and soil surface backscatter. In the layer scattering configuration, such as a vegetation layer over a rough soil surface, contributions from the rough surface scattering are considered with wave interactions, differential propagation delay, and wave attenuation in the vegetation layer (Nghiem et al., 1995), which can be effectively anisotropic when vegetation scatters have a preferential directional structure (e.g., planophile, plagiophile, erectophile, or extremophile orientation distribution) (Nghiem et al., 1993b).

The backscatter from a rough soil surface depends strongly on the soil dielectric constant and the transmissivity and reflectivity because of wave interactions with the soil boundary. Thus, the surface scattering also contains a soil moisture signature in addition to the soil moisture information in the interactive volume scattering components. However, a dense vegetation canopy can have a large imaginary part in its effective permittivity, which attenuates both the soil surface scattering and soil interactions in the volume scattering, and consequently masks the soil moisture signature. Specific mathematical details of the volume and surface scattering in layered media can be found in earlier publications by Nghiem et al. (1990, 1993a,b, 1995).

Although the formulation mentioned earlier provides physical insights and a theoretical basis for active remote sensing of soil moisture, in practice, it is not possible to set up a soil moisture inversion method strictly based on theoretical modeling of electromagnetic scattering because of the complexities of natural environments in different climate regimes. The alternative is a simple empirical linear equation that relates backscatter σ_0 to volumetric soil moisture m_v as

$$\sigma_0 = am_v + b \tag{9.5}$$

where coefficients a and b are dependent on incidence angle, polarization, vegetation conditions, soil type, surface variation, and climate regime (Mo et al., 1984, Prevot et al., 1993, Shrivastava et al., 1997; Shoshany et al., 2000; Hutchinson, 2003). Particularly for Ku-band backscatter data from the SeaWinds scatterometer aboard the QSCAT satellite, the bias term b in Equation 9.5 contains a signature of seasonal vegetation change, while changes in volumetric soil moisture m_v from rainwater are detectable in backscatter variations in a timescale consistent with the initial impulse increase in wetness from the precipitation input throughout the subsequent drying process (Nghiem et al., 2005). Thus, soil moisture change (SMC) can be directly inferred from Equation 9.5 using the temporal backscatter-change method, which removes most of the background bias.

9.2.3 Passive and Active Blending

As presented in Sections 9.2.1 and 9.2.2, passive and active sensors measure different parameters: passive BT and active radar backscatter, each of which has different sensitivities to soil moisture and vegetation cover. This section explains how blending of passive and active can better represent the overall state of soil moisture on land surface compared to the separate use of each data type.

In the ideal theoretical case of smooth bare soil ($\tau_c = 0$), Equation 9.1 dictates that the BT is directly proportional to the emissivity e_{sp}, which is determined by soil dielectric constant and is thereby most sensitive to soil moisture. In comparison, there is no active radar backscatter because there is no vegetation ($V_1 = 0$ in Equation 9.4 without vegetation) and no rough surface; hence, the soil moisture is not measurable by a radar for bare soil without any roughness. For real surfaces, surface roughness and/or vegetation cover will exist and will affect the sensitivity to soil moisture differently in passive (Njoku et al., 2003) versus active data (Nghiem et al., 1993a) until the vegetation cover becomes sufficiently dense to start masking the soil effects.

As an illustration of passive and active blending, a correlation analysis was conducted comparing satellite-based remote sensing signatures to in situ soil moisture and vegetation measurements at a U.S. Department of Agriculture (USDA) Natural Resources Conservation Service (NRCS) SCAN site in Lonoke, Arkansas (91.867°W and 34.833°N) (USDA, 2009a). The vegetation cover in the Lonoke area is primarily agricultural crops, including soybeans, rice, and wheat (Njoku et al., 2003). More than 1 year (1999–2000) of TMI passive microwave data at 10.7 and 19.3 GHz were analyzed, centered within 25 km of the Lonoke SCAN site. Results showed a wide range of sensitivity in the response of instantaneous *PR* (obtained at each local overpass time) and the transient SMC after rain events. The variance between measurements and linear fit values of daily *PR* (10.7 GHz) versus the contemporaneous daily m_v became so large at larger soil moisture values that *PR* varied by a factor of 3 at $m_v = 34\%$, while a transient soil moisture can change 6%–34% for the same *PR* value around 0.017. This is consistent with the findings by Njoku et al. (2003), which indicate that transient soil moisture events are not effectively captured by TMI data.

In contrast, seasonal trends in TMI *PR* (90 day running average) are well correlated with seasonal soil moisture (90 day running average) measured at a depth

TABLE 9.1
Correlation Results between Seasonal TMI *PR* and Seasonal SCAN Volumetric Soil Moisture m_v at 5 cm Depth from Linear Regression Analysis in the Form of $PR = \alpha \cdot m_v + \beta$ with a Correlation Coefficient ρ

	10.7 GHz			19.3 GHz		
	α	β	ρ	α	β	ρ
Fall–winter	0.00109	0.00766	0.977	0.000931	0.00587	0.953
Spring–summer	0.00131	−0.00108	0.988	0.000960	−0.000908	0.946
All year	0.00124	0.00235	0.936	0.000894	0.00320	0.792

of 5 cm at the Lonoke SCAN site (Table 9.1). Plots of seasonal data for contemporaneous SCAN m_v and TMI *PR* at both frequencies reveal a hysteresis behavior (Figure 9.1). Theoretically, Equation 9.1 suggests that the hysteresis is caused by attenuation effects on the passive microwave signatures under different vegetation conditions during different seasons. In Figure 9.1, the linear fit for all data in the entire year is used as a reference for each frequency. The *PR* generally lies above the annual linear fit during the fall and winter seasons of 1999 with vegetation cover decreasing in early fall, reaching a minimum in winter, and then increasing toward the spring equinox in 2000. In contrast, *PR* is mostly below the annual linear fit during spring and summer as the vegetation cover increases to a peak in summer and then slightly decreases toward the fall equinox. Vegetation attenuation effects cause the hysteresis in seasonal *PR* versus m_v observed at both frequencies with less severe impacts at the lower frequency evidenced by the smaller spread at 10.7 GHz around the best-fit line in Figure 9.1.

For the active microwave analysis, time-series QSCAT data were extracted within 25 km around the same SCAN site in the same manner as for the TMI data. In contrast to the passive microwave case, daily QSCAT backscatter change correlates well with contemporaneous SMC from rainwater. Daily QSCAT data capture 91% of the rain events recorded at the Lonoke SCAN site in 1999–2000. To illustrate the high correlation of QSCAT backscatter to transient soil moisture, a regression analysis using daily observations was performed for the period of October 4 to November 19, 1999, when two major rain events occurred over the SCAN site. With the linear formulation in the inverted form of Equation 9.5 such that $m_v = a' \cdot \sigma_0 + b'$ for backscatter σ_0 in dB, m_v in percent, and $a' = 8.9\%/\text{dB}$ and $b' = 111.1\%$, a high correlation coefficient of 0.91 and a small standard deviation of 3.7% were found for backscatter at the horizontal polarization (σ_{0HH}). This indicates that both the initial impulse of soil moisture increase from rain and the subsequent soil moisture decrease in the ensuing drying process were well represented. For backscatter at the vertical polarization (σ_{0VV}), the result is similar, with SMC of 8.4% for a dB change in σ_{0VV}, and thus the backscatter at the vertical polarization is slightly less sensitive than the horizontal polarization to transient soil moisture. This is consistent with Equation 9.4, where the dyadic Green's function includes soil reflection, which is stronger at the horizontal polarization than

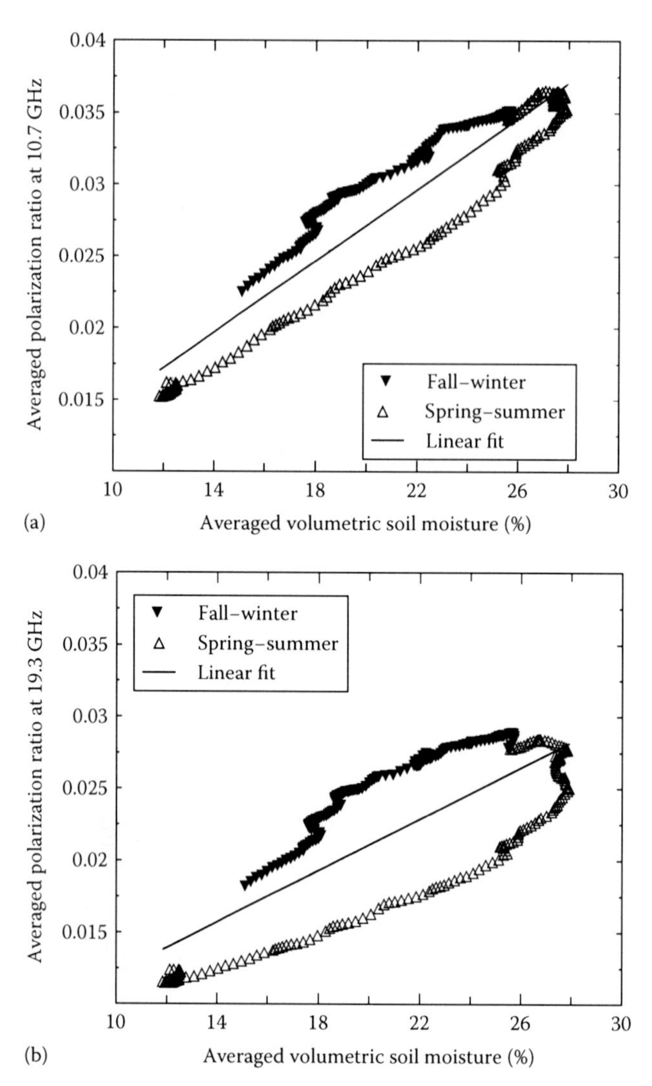

FIGURE 9.1 Seasonal TMI *PR* at (a) 10.7 GHz and (b) 19.3 GHz versus seasonal SCAN soil moisture at 5 cm depth in an agricultural area at Lonoke, Arkansas. All data are contemporaneous (collocated in time) and are 90 day running averages.

the vertical polarization. Also, the incidence angle at 54° for σ_{0VV} is larger than 46° for σ_{0HH}, which means that σ_{0VV} suffers from higher attenuation effects because of the longer path length in the vegetation cover. Nevertheless, QSCAT data can identify sufficiently heavy rainfall events even at peak vegetation conditions when the backscatter increases above the seasonal level of the background backscatter. As a result, QSCAT has the capability to identify transient SMC. This illustrates the complementary information that can be estimated from the combination of active and passive microwave data, together capturing both transient and seasonal trends in soil moisture content.

Seasonal trends in active backscatter data primarily convey information about vegetation. To demonstrate this, Normalized Difference Vegetation Index (NDVI) data representing seasonal vegetation change (Justice et al. 1985; Verdin et al., 1999; Zhang et al., 2010) were compared with seasonal QSCAT backscatter data (90 day running average). Advanced Very High Resolution Radiometer (AVHRR) NDVI data from the National Oceanic and Atmospheric Administration (NOAA) AVHRR were averaged within 25 km around the SCAN site so that the spatial scale of AVHRR NDVI data was compatible with the QSCAT data. A high correlation coefficient of 0.946 was found between the NDVI and linear σ_{0VV} and a slightly lower correlation coefficient of 0.864 between the NDVI and linear σ_{0HH}. Therefore, seasonal QSCAT backscatter can be used to characterize seasonal vegetation change regardless of cloud cover, which is transparent to QSCAT at the Ku-band frequency of 13.4 GHz. This is consistent with earlier results on the relation of Ku-band backscatter with NDVI (Moran et al., 1997), green leaf area index (Moran et al., 1998), and aboveground biomass (Nghiem, 2001).

Seasonal running averaged QSCAT σ_{0VV} (more sensitive to seasonal vegetation change compared to σ_{0HH}) and TMI *PR* at 10.7 GHz (more sensitive to seasonal SMC compared to 19.3 GHz data) were compared over the fall–winter season and spring–summer season. The hysteresis behavior is clearly observed in the curve of σ_{0VV} versus *PR* (Figure 9.2). In fall and winter, *PR* is below the annual linear fit, corresponding to less vegetation cover as compared to spring–summer *PR* above the linear fit with more vegetation cover. The lower vegetation cover indicated by lower backscatter in fall and winter supports the fact that *PR* is above the annual linear fit in the *PR*-m_v hysteresis (Figure 9.1) for less vegetation attenuation effects on *PR*, and vice versa for spring and summer. We observe that the vegetation peak seen in σ_{0VV} occurs in summer after the seasonal soil moisture reaches the maximum seen in *PR*

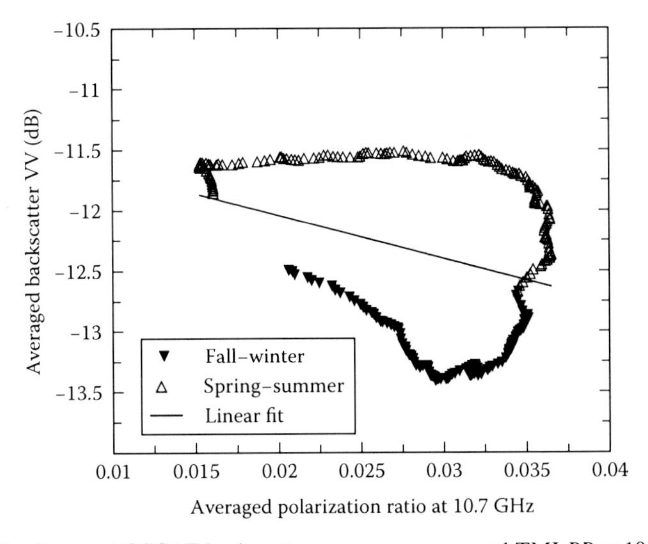

FIGURE 9.2 Seasonal QSCAT backscatter σ_{0VV} versus seasonal TMI *PR* at 10.7 GHz in an agricultural area within 25 km around Lonoke, Arkansas. All data are contemporaneous (collocated in time) and are 90 day running averages.

in spring. This analysis shows that independent information on seasonal vegetation change in active backscatter data can explain vegetation cover effects on passive microwave signatures.

9.3 DROUGHT APPLICATIONS

9.3.1 Drought Monitoring Issues

For hydrological and agricultural drought assessment and monitoring, water on the land surface and in the soil are both relevant, and thus soil moisture data must play a key role. Nevertheless, the current in-situ station network is inadequate, and soil moisture measurements are too sparse for effective use or are nonexistent in many areas (NIDIS, 2007).

For county-level monitoring, which is an important goal of the National Integrated Drought Information System (NIDIS) (Western Governors' Association, 2004; NIDIS 2006, 2007), the National Weather Service (NWS) has determined that an effective Cooperative Observer Network would require a minimum spatial density of one observing site per 1000 km^2 across the country or a separation of about 24–32 km (NIDIS, 2007). The location of each in-situ sensor must be carefully selected such that the measured soil moisture is representative of the surrounding area. Furthermore, consistency and persistency in data collections are important in terms of data quality and data availability across different agencies and across different states.

9.3.2 Uses of Satellite Data

In view of the aforementioned issues in drought monitoring, recent efforts have enabled certain uses of soil moisture measurements derived from satellite remote sensing data for enhancing drought monitoring systems (Nghiem et al., 2010). Several specific results are presented in this section to illustrate various uses of satellite data with different temporal and spatial scales.

9.3.2.1 Temporal Data at Local Scale

Temporal QSCAT observations combined with in-situ station measurements are used to illustrate how satellite data can help to enhance drought monitoring capabilities. Figure 9.3 presents results at the NCDC Global Summary of the Day (GSOD) (NCDC, 2010a) Station 727760 in Great Falls, Montana (47.467°N, 111.383°W). Time series of QSCAT data together with in situ measurements around this station are constructed with the Special Satellite-Station Processor (SSSP) (Nghiem et al., 2003). Daily QSCAT data at horizontal polarization (more sensitive to soil moisture than vertical polarization) were selected with centroids located within 25 km around Station 727760 and from ascending orbits (~6 a.m. local overpass) that are better correlated with soil moisture than data from descending orbits.

QSCAT σ_{0HH} data around Great Falls (top panel in Figure 9.3) clearly identify rain events before and after the dry period between July 9 and September 5, 2000, when very little rain fell. In August 2000, the long-term Palmer Drought Index for the

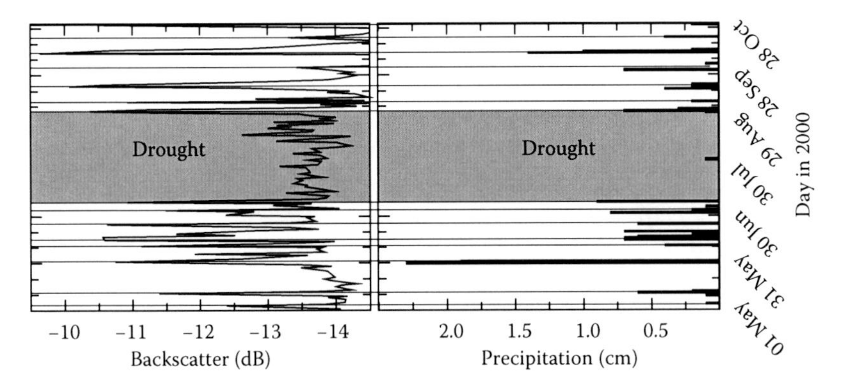

FIGURE 9.3 Measurements around the NCDC GSOD Station 727760 (47.467°N 111.383°W) at Great Falls in Montana. Left panel is QSCAT σ_{0HH} within 25 km around the station, and right panel is precipitation from station rain gauge. Thin horizontal lines align rain events to backscatter impulses. Gray shaded area defines the period between July 9 and September 5, 2000, when there was very little rainfall.

region around Great Falls was −4 or below, indicating long-term, extreme drought conditions. USDA issued Natural Disaster Determinations for drought for the entire state of Montana in 2000, when severe and persistent drought caused significant losses to agriculture and other sectors (Resource Management Services, 2004). The summer drought period observed by QSCAT is validated by the lack of rain in in situ precipitation data (right panel in Figure 9.3) for the same period, when several heat waves occurred. Both before and after this midsummer drought period, QSCAT detected a number of significant rain events that increased backscatter by about 3 dB, which is equivalent to a 26.8% increase in volumetric soil moisture (per the Lonoke rating value of $a' = 8.921\%/\text{dB}$). Thus, water from these rain events reached the land surface and significantly increased the moisture in soil. In contrast, rain gauge precipitation (RGP) data corresponding to these significant rain events inconsistently and disparately ranged across one order of magnitude from low values (<0.2 cm) to high values (>2.0 cm).

9.3.2.2 Spatial Data at Regional Scale

Satellite microwave remote sensing data, such as AMSR-E or QSCAT, can be used to monitor drought and water resources at regional to global scales. Both have swath widths of 1400 km or larger (Tsai et al., 2000; Njoku et al., 2003), which allow a near-daily global coverage and as many as two data acquisitions per day at high latitudes. Several attributes related to water can be obtained from microwave satellite data for drought monitoring. Examples of these information products derived from AMSR-E and QSCAT data are provided for 2009 over the state of Texas, when much of the state was afflicted by drought (Texas Water Development Board, 2007).

A relevant attribute for water resource and drought assessments is precipitation frequency, which quantifies the recurrence of rain events in a given period (González and Valdés, 2004). Instead of apparent precipitation frequency (APF), derived from in situ rain gauge data or surface rain radar data, a different measure of effective precipitation frequency (EPF) can be derived from satellite scatterometer data. EPF accounts for rainwater that effectively reaches the land surface and increases soil moisture, as opposed to APF, which may have problems with apparent precipitation, virga, or inconsistency in gauge data collection. For applications to QSCAT data, EPF = $100(N_W/N_C)$ is defined as the percentage of the number of wet days (N_W) when the soil moisture increase is ≥5% in the topsoil layer (5 cm) such that the corresponding backscatter increase is ≥0.56 dB for σ_{0HH} or ≥0.60 dB for σ_{0VV} above the background level, over the total number of satellite coverage days (N_C) excluding days when satellite data were missing in a given period.

EPF was retrieved from QSCAT data across Texas from June 1 to August 31, 2009 (left panel in Figure 9.4). In summer 2009, exceptional drought occurred over much of south-central Texas, as shown in the U.S. Drought Monitor (USDM) maps from June to August 2009 (right panels in Figure 9.4). By August 2009, extreme and exceptional drought conditions (D3 and D4, respectively) remained persistent across south-central Texas, where the topsoil conditions were very dry and river levels were near historic lows (NCDC, 2009). Consistent with these drought conditions, QSCAT EPF showed few to no rain events across most of southern Texas (black to magenta areas, left panel of Figure 9.4). In contrast, soil in part of the Texas Panhandle was shown to be wetted by several rain events during this time period (light blue to green and yellow areas, left panel of Figure 9.4), which is reflected by the change of conditions in the USDM maps, which showed most of the area classified as abnormally dry (D0) in June had improved to no drought by late August 2009.

Although EPF carries information on wet precipitation frequency or how often the land surface becomes wet because of rainwater, daily SMC from QSCAT data

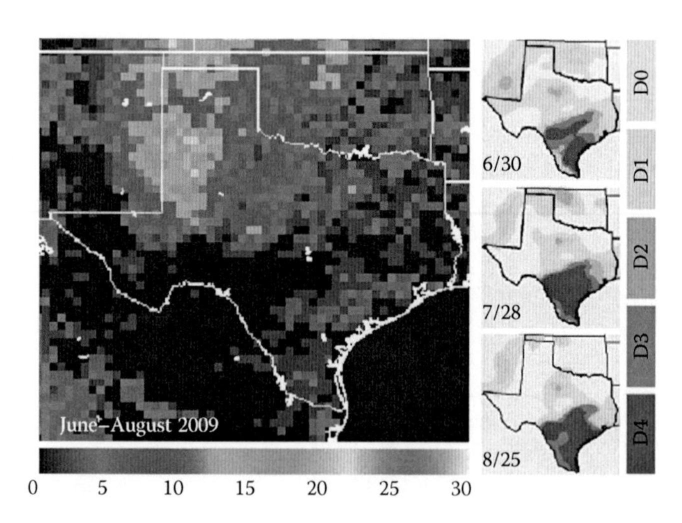

FIGURE 9.4 **(See color insert.)** Effective precipitation frequency (%) measured by QSCAT for the period June–August 2009 (left panel) and drought levels from D0–D4 from the USDM for weeks ending on the marked dates in 2009 (right panels). The USDM drought levels include D0 for abnormally dry, D1 for moderate drought, D2 for severe drought, D3 for extreme drought, and D4 for exceptional drought. (Ref. Svoboda et al. 2002.)

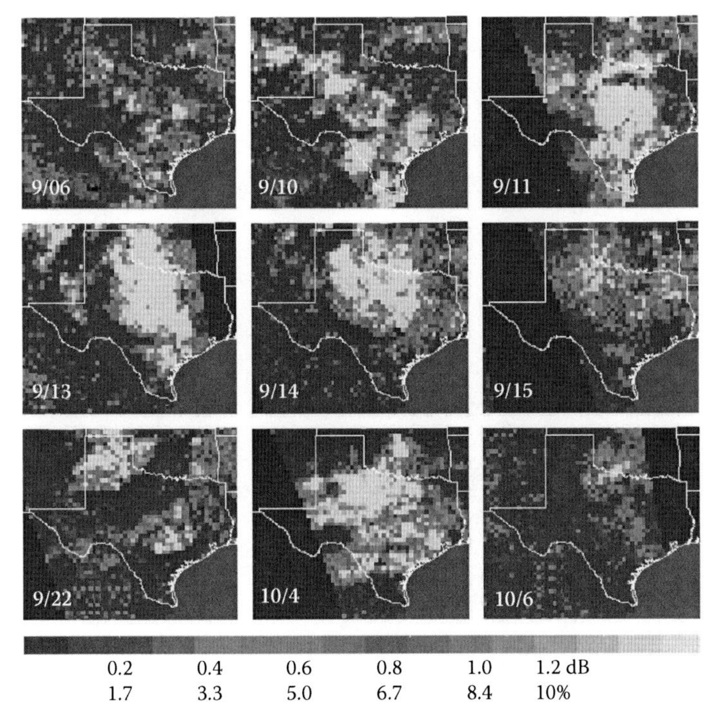

FIGURE 9.5 **(See color insert.)** SMC measured by QSCAT with the vertical polarization along ascending orbits in September to early October 2009. The color scale represents backscatter change in dB and volumetric SMC in % with the Lonoke rating.

represents the quantitative change in soil moisture or the amount (intensity) of rainwater that accumulates on land surface each day. Therefore, SMC is an attribute relevant to monitoring hydrological drought because it is related to water on land rather than raindrops in the atmosphere (a meteorological parameter). Hydrological drought is associated with shortfalls on surface or subsurface water supply whereas meteorological drought is related to deficiencies of precipitation (Wilhite and Glantz, 1985). SMC is also appropriate for early warning of agricultural drought (drought that has agricultural impacts) because SMC represents the source of rainwater that can infiltrate into the root zone after a rain event.

Figure 9.5 presents maps of selected daily SMC compared to the semimonthly average over Texas from early September to early October 2009. Intense SMC (in yellow), which reflects large increases in soil moisture, occurred across large areas of central Texas on September 10, 11, 13, and 14 and October 4. The SMC results on these days are consistent with torrential rainfall events reported across central and south Texas (up to 20 in. of total rainfall recorded in some locations) in September 2009, causing flash flooding (NWS, 2009). With this new water input, drought conditions in central and south Texas significantly improved by early October 2009 (as shown in the USDM map for October 6 in Figure 9.6).

Complementary to the transient change observed in the QSCAT daily SMC, AMSR-E passive microwave data provide good measurements of seasonal soil

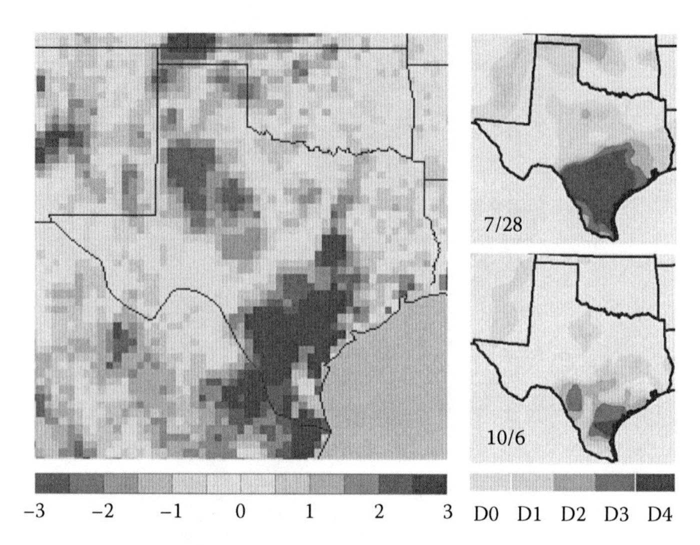

FIGURE 9.6 **(See color insert.)** Difference of AMSR-E monthly averaged soil moisture in % of m_v (September 7 to October 6, 2009) and m_v (June 29 to July 28, 2009) showing seasonal SMC (left panel), and drought condition change between USDM drought maps in July and in September 2009 (right panels).

moisture (as discussed in Section 9.2.3). Figure 9.6 (left panel) shows the difference in seasonal soil moisture between the June–July and September–October periods in 2009. AMSR-E seasonal soil moisture results reveal a large region of increased soil moisture in south-central Texas (blue areas). This corresponds to the marked improvement in drought conditions in September compared to those in July 2009 as depicted on the USDM drought maps (right panels in Figure 9.6). In contrast, an area in western Texas had a substantial reduction in soil moisture by the September–October period (red–brown areas in the left panel of Figure 9.6) compared to more moist conditions in the June–July period. This area had a larger EPF observed by QSCAT in the earlier months, as seen in the left panel of Figure 9.4 for June–August 2009.

The independent attributes derived from different remote sensing data sets (QSCAT and AMSR-E) are consistent with the changes in true drought conditions that occurred over Texas in 2009 and provide complementary perspectives for drought assessments. The improvement in drought conditions classified in the USDM map on October 6, 2009, (lower right panel in Figure 9.6) reflects the recent transient wetting events observed in daily SMC from QSCAT (e.g., SMC map for October 4, 2009, in Figure 9.5) and the seasonal SMC observed by AMSR-E (Figure 9.6). These results demonstrate the capability and consistency of different microwave-based parameters to depict the state of soil moisture, as well as its transient and seasonal changes from local to regional scales.

9.3.2.3 Spatial Data at Continental Scale

A major advantage of satellite data is its large spatial coverage at continental to global scales compared to local, surface in situ measurements from station networks.

Here, the pattern of SMC as observed by QSCAT and AMSR-E satellites is examined across the contiguous United States (CONUS) and compared to rainfall patterns from the regional multisensor precipitation analysis assembled into national maps of Stage-4 daily precipitation (SDP) available from the National Mosaic and Multi-Sensor Quantitative Precipitation Estimation algorithm (NMQ, 2009). This comparison is to identify the similarities as well as the differences between precipitation data and soil moisture data, which are relevant for monitoring different drought types (SDP for meteorological drought versus SMC for early warning of agricultural drought).

The large swaths of measurements by QSCAT and AMSR-E provide near-daily coverage over the CONUS. However, data gaps exist, and a full coverage of the entire CONUS is not possible every day, especially when ascending- and descending-orbit data are used separately. Figure 9.7 shows daily SMC maps in May 2009 from QSCAT ascending-orbit data (Figure 9.7a) at about 6 a.m. local overpass time and from AMSR-E descending-orbit data (Figure 9.7b) at about 1:30 a.m. local overpass time.

Overall, the patterns of daily SMC from QSCAT and AMSR-E are similar. Both reveal precipitation water on land surface in the Midwest and the Great Lakes states extending toward the northeastern United States, whereas most of the western United States was dry. An extensive wet region is observed across Kansas and Nebraska in both SMC maps (marked by the circles in Figure 9.7a and b). Interestingly, a well-defined dry area is detected by both QSCAT and AMSR-E just south of Lake Michigan along the Illinois–Indiana border. However, some discrepancies exist between the AMSR-E and QSCAT SMC results. First, the volumetric SMC observed by QSCAT can be more than 10% in various areas (yellow areas in Figure 9.7a), where AMSR-E SMC barely exceeds 5% (blue areas, Figure 9.7b). For example, the region east of Lake Ontario in New York had a large positive SMC (wet) in the QSCAT map while the AMSR-E SMC showed a slightly negative value (dry). These differences are not surprising given the better sensitivity of QSCAT data to transient SMC, as discussed earlier in Section 9.2.3.

In the case of the discrepancy between QSCAT and AMSR-E SMC in New York, it could be hypothesized that the difference was due to the different observation times of the two instruments (6 a.m. for QSCAT and 1:30 a.m. for AMSR-E). However, SDP maps indicate significant rainfall on May 27 continuing to May 28, 2009, in New York (Figure 9.7c and d). The lower sensitivity in AMSR-E data to transient SMC is likely the cause of the differences in the SMC results. The SDP map on May 28 (Figure 9.7c) also shows a large-scale overall pattern similar to the SMC observed by QSCAT and by AMSR-E (to a lesser degree) with band of heavier rainfall across the upper Midwest and Great Lakes region extending into the northeastern states. However, the SDP map on May 28 indicates no precipitation in Kansas and Nebraska where both QSCAT and AMSR-E detected rainwater on land surface resulting from the intense rainfall on the previous day (see the region marked by circles in Figure 9.7). This case illustrates that SMC can represent the rainwater accumulated from preceding strong precipitation events with the water still remaining in the top soil for some period of time after the rain events. As such, SMC is

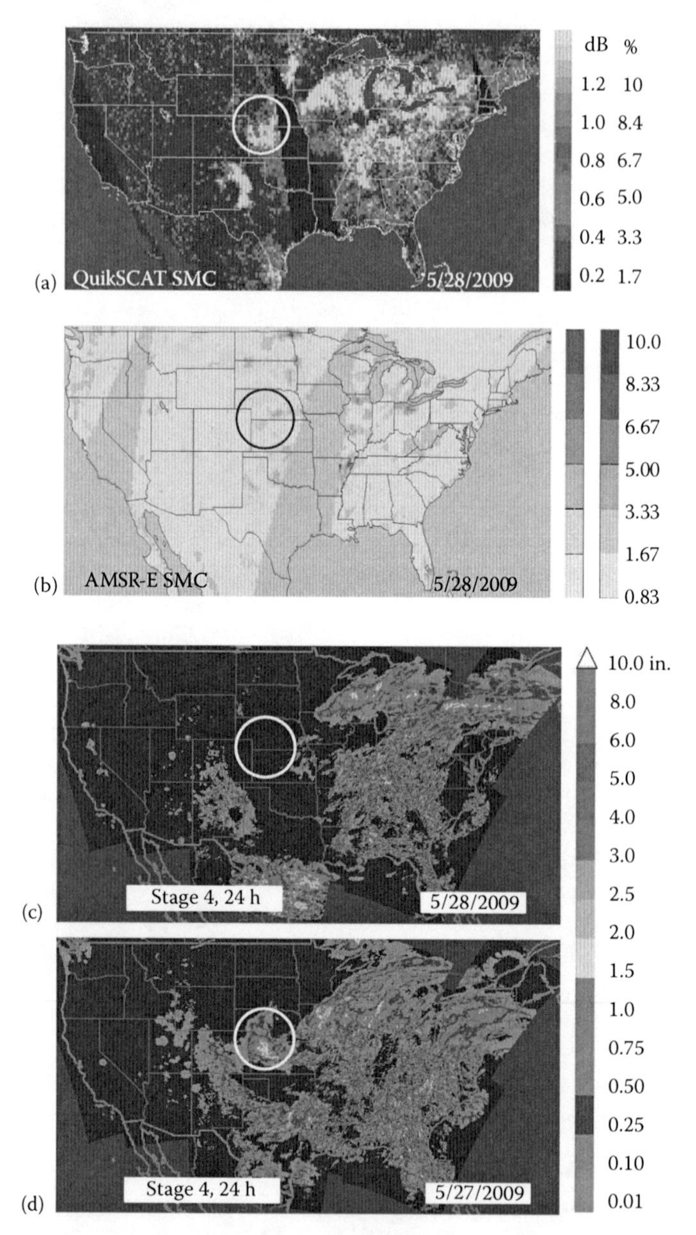

FIGURE 9.7 **(See color insert.)** SMC on May 28, 2009, compared to the 2-week average between May 14 and 28, 2009, observed by (a) QSCAT SMC represented by backscatter change in dB and by volumetric moisture change in % from the Lonoke rating, and (b) AMSR-E by volumetric moisture change in % with yellow brown for drier and cyan blue for wetter conditions. The SMC maps are compared with Stage-4 24 h precipitation measurements (NMQ, 2009) at 12:00 UTC in inches for (c) May 28, 2009, and (d) and May 27, 2009.

also an indicator of the intensity or amount of rainwater on land surface (vis-à-vis rainwater as raindrops in the atmosphere) in terms of the SMC duration.

There are other discrepancies between SMC and SDP. For example, in New Mexico, SDP observed extensive precipitation across the state while SMC from both QSCAT and AMSR-E found wetness only in some areas of the state (such as in northeastern New Mexico). This difference suggests either the rainwater did not fully reach the land surface (virga problem) or SDP has uncertainties in surface radar data (AP problem). Similarly, the SDP pattern was much more widespread compared to the SMC pattern (Figure 9.7) in adjoining Texas, where AP problems can cause significant difficulties in precipitation mapping (Story, 2009). These observations suggest that SMC, pertaining to land surface conditions, is more relevant to hydrological and agricultural drought monitoring, while SDP as a parameter for precipitation rate is useful for meteorological drought monitoring. In the case of light rains, the small amount of rainwater that may reach the land surface can be evaporated before the next orbit pass, and thus SMC may not capture the wetness from light rains evaporated in a short time.

9.3.2.4 Soil Moisture Products for Drought Monitoring and Forecasting

In an operational environment, science results need to be transitioned into data and image products with appropriate formats and protocols that can be rapidly and easily used by drought experts, such as the USDM authors. Here, examples of various SMC products produced for the USDM are presented and compared with other traditional drought products to identify their advantages and limitations.

Three SMC attributes, including daily SMC, weekly maximum SMC, and weekly mean SMC, are produced and a USDM-defined color palette applied to classify the various levels of change. Because the USDM is an operational tool, the SMC data are updated weekly on Monday to be in sync with other updated products and analyses used to create the weekly USDM on Tuesday. The overall SMC processing system allows the flexibility in making SMC products with different time periods for various purposes, including the Monday-updated SMC for USDM operational assessment and 5–8 day SMC products for comparison and benchmarking with different NOAA precipitation maps.

Figure 9.8 presents an example of 8 day mean and 8 day maximum SMC maps, which are compared with the RGP product, representative of precipitation from October 14, 2008, and the ensuing 7 days. The RGP product is produced by NOAA's Climate Prediction Center (CPC) from several quality-controlled surface weather measurement data sources, including the Automated Surface Observing Systems (ASOS) and cooperative observers. Approximately 7000 daily in situ rain gauge observations are included in the making of the RGP product (Higgins et al., 2000). RGP maps are made with different time periods from 5 to 8 days for drought assessment. In this example, the full 8 day RGP product is compared with the corresponding 8 day SMC mean and maximum SMC maps (Figure 9.8a and b). In the maps, yellow to brown represent drier conditions and green to blue represent wetter conditions, which are shown with the corresponding USDM D-level contour lines for October 14, 2008.

The mean SMC map (Figure 9.8a) reveals a significant soil moisture increase (light blue area) extending in western Kansas, as well as a noticeable increase in soil moisture (green areas) in the Texas Panhandle, southeastern Texas, central Oklahoma, eastern New Mexico, and eastern Montana. Significant drying also appears across several states in the upper Midwest (Minnesota, Nebraska, and North and South Dakota). In comparison, the maximum SMC map (Figure 9.9b) indicates

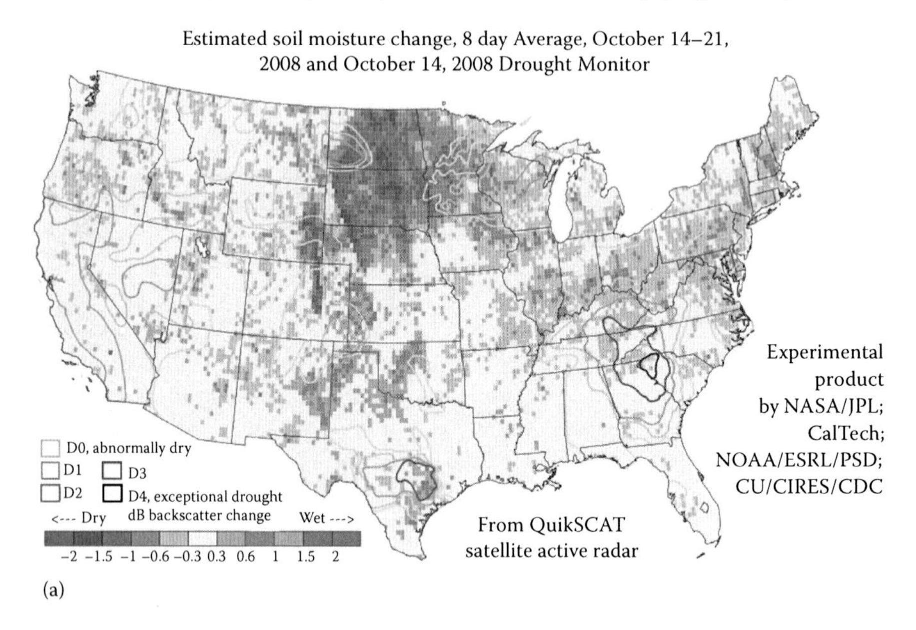

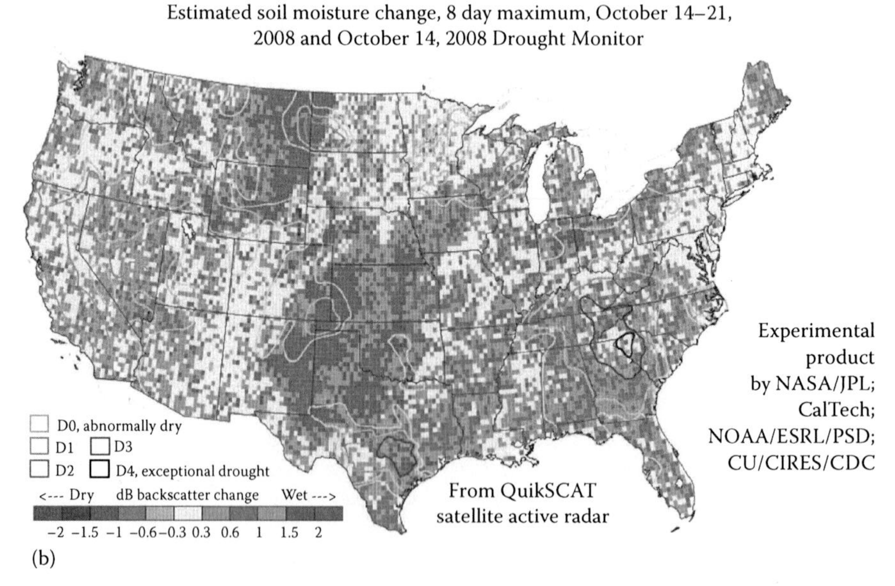

FIGURE 9.8 (See color insert.) Comparison of QSCAT SMC with RGP for the period of October 14, 2008, and the ensuing 7 days: (a) mean SMC, (b) max SMC, and

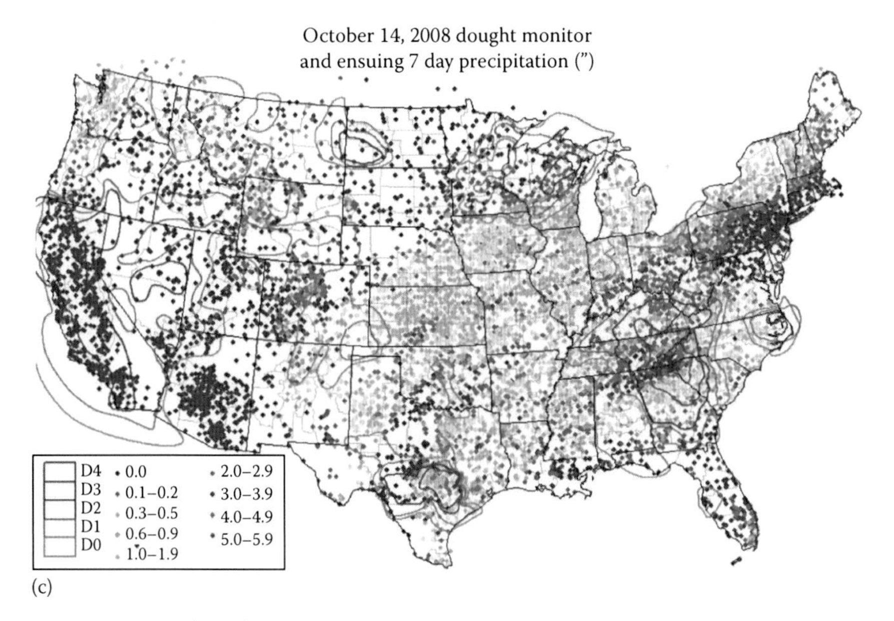

FIGURE 9.8 (continued) **(See color insert.)** (c) RGP used in making USDM maps.

a much more intensive soil moisture increase over extensive regions (blue to magenta areas) because it represents the peak soil moisture increase detected at any time in the 8 day period. The maximum SMC corresponds to the largest value of rainwater detectable on the land surface on any given day of the period including the remnant rainwater from previous days. In the maximum SMC map, a caveat is that low SMC values (gray and light green) are noisy because of spatial variability and limited accuracy in satellite data.

While the maximum SMC corresponds to the peak water accumulation on land surface, the mean SMC provides an assessment of the persistence of rainwater in soil, because the greater the number of days when a significant amount of soil moisture increase occurs, the larger the mean SMC value for that given time period. Therefore, it is possible to have a large peak SMC due to an intensive single-day rain event over an area (e.g., blue area between Indiana and Ohio in Figure 9.8b) where the maximum SMC value is high but the mean SMC is low because no rainwater accumulated on the other days during the 8 day period. Since the persistence of SMC (i.e., how long rainwater accumulates and remains in soil) depends on factors such as soil type, infiltration rate, and runoff processes, the mean SMC carries information that is relevant for hydrological and agricultural drought monitoring. The mean and the maximum SMC carry different information, and both can contribute to drought assessments.

For benchmarking, the traditional RGP product used in USDM is included in Figure 9.8c to compare with the mean and maximum SMC products. A comparison of the RGP and SMC maps clearly points to the different characteristics of these measurements: RGP consists of point data at separate rain gauge station locations, while the SMC maps are composed of 25 km pixels that provide continuous

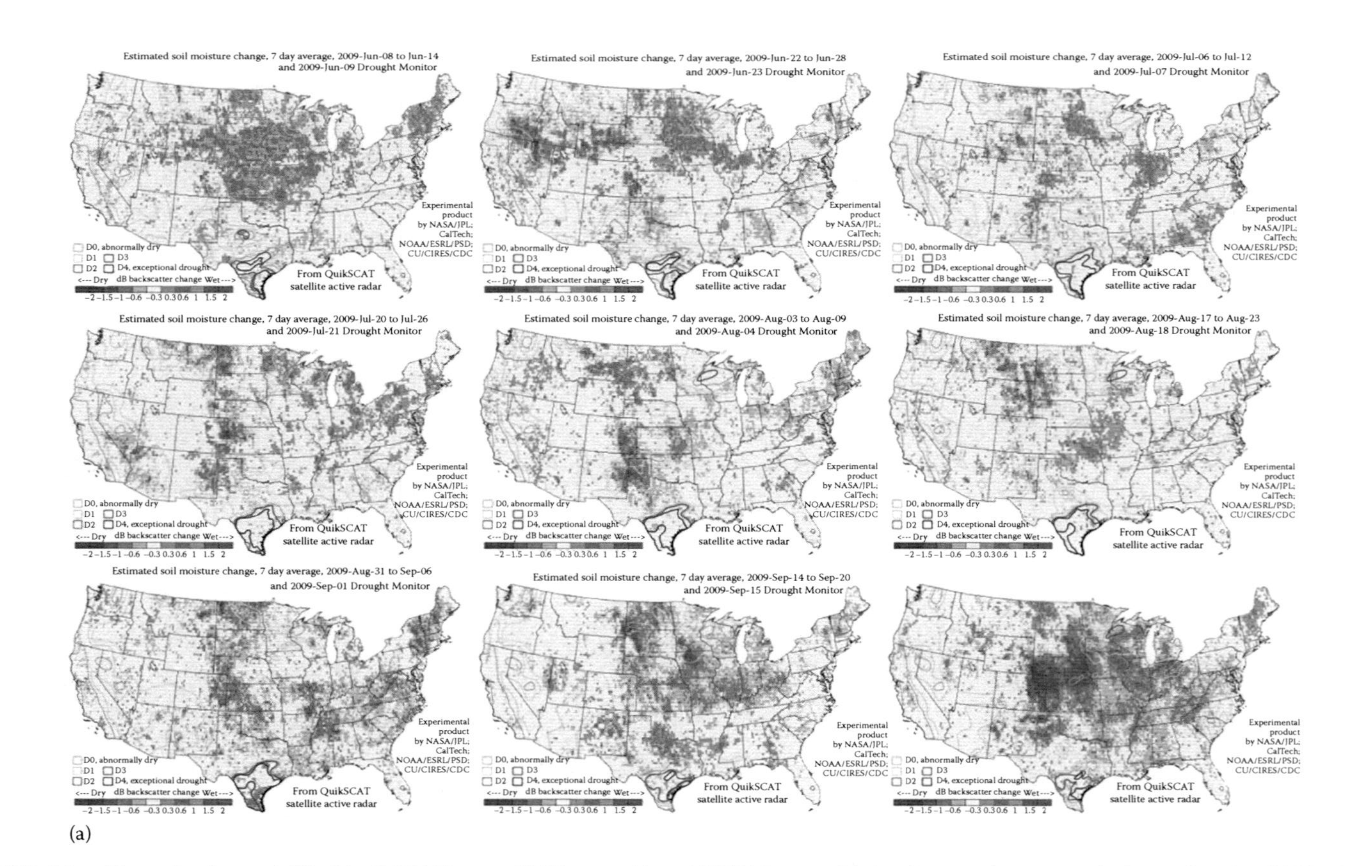

FIGURE 9.9 **(See color insert.)** Weekly QSCAT mean SMC maps (a) and USDM maps (b) for the growing season in June–October 2009 (skipping a map once every other week).

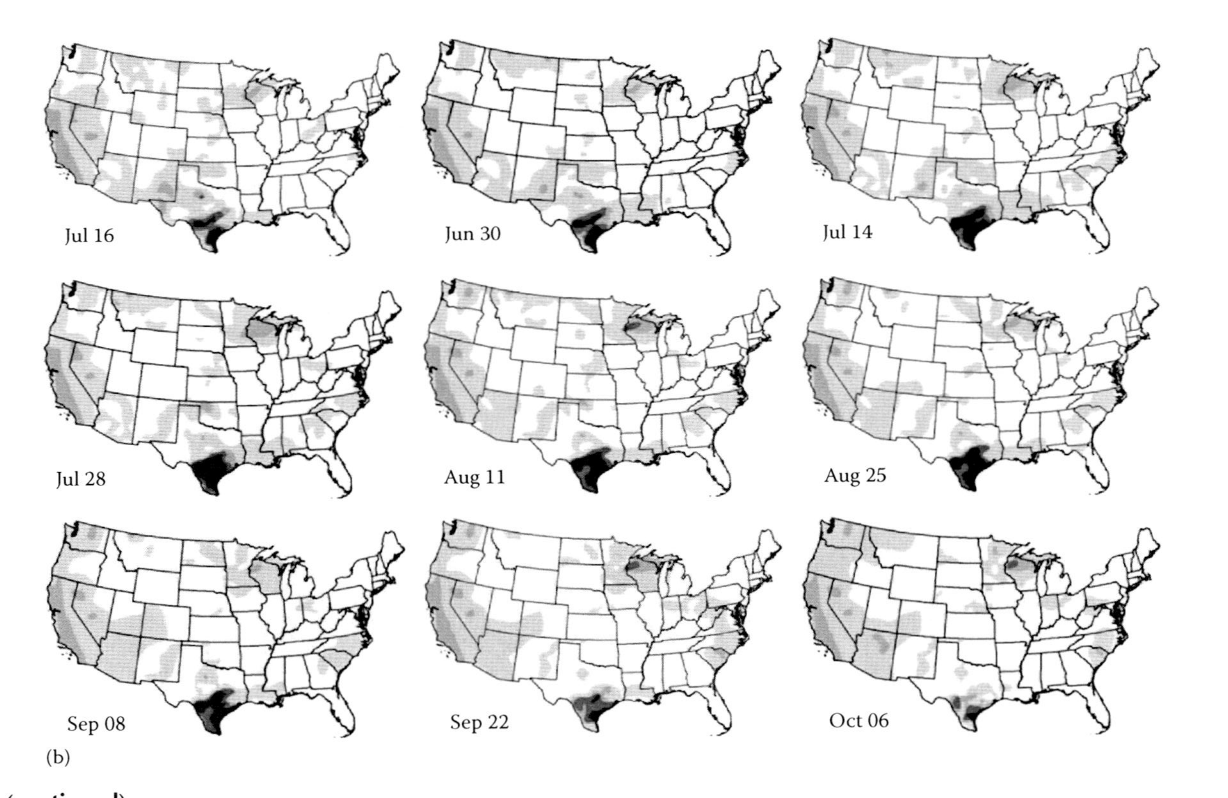

FIGURE 9.9 (continued)

spatial coverage of measurements across the CONUS. The vastly improved spatial coverage at an appropriate resolution of the satellite observations is important for resolving the county-level drought condition, which is currently lacking for most inputs into the USDM and is a goal of NIDIS. The average county size for the CONUS is approximately 50 km in linear scale (~2500 km^2 in area) as estimated from census data (U.S. Census Bureau, 2005). Thus, to resolve information at the 50 km county scale, a spatial scale of 25 km is required according the Nyquist sampling theorem, which is satisfied by the SMC data. However, RGP can provide more frequent hourly data, whereas the SMC is only available two times per day at most. Although the SMC temporal scale is suitable for the weekly USDM, better temporal coverage can improve the overall result, especially in the tropics, where current satellite data gaps are the largest because of the divergence of satellite swaths at lower latitudes.

Although the spatial patterns in both RGP and maximum SMC maps (Figure 9.8b and c) agree in general over the areas of extensive precipitation discussed earlier, a large area of discrepancy exists over Montana, Wyoming, and part of North Dakota. This discrepancy is primarily due to the SMC having a memory of any precipitation water as long as it remains on land surface at the time of the satellite measurement (as in the cases presented in Figure 9.7 with a comparison of SMC to precipitation data) as opposed to the instantaneous and temporally discrete rain gauge measurements at specific station location. These results point out a key advantage of SMC in "memorizing" the rainwater staying on land surface integrated up to the time of measurement, allowing less frequent SMC measurements to capture the state of soil moisture. In contrast, satellite precipitation measurements need to be very frequent to capture the amount of rainwater falling through the atmosphere at the exact discrete time of each precipitation event.

Regarding the mean SMC (Figure 9.8a), a high value requires sufficient rainwater to accumulate on land surface over a significant duration during the period under consideration. The mean SMC represents both the quantity and persistence of new precipitation water in soil, which is more relevant to drought monitoring than both maximum SMC and RGP. For example, maximum SMC and RGP identify a precipitation pattern in eastern Nebraska and Iowa; however, the same region appears dry in the mean SMC. This indicates that the transient rainwater may not have been sufficient to sustain the presence of soil moisture over a significant fraction of the 8 day period to have an overall impact on soil moisture condition over the given period.

For the 2009 growing season (i.e., June to early October), Figure 9.9 presents a comparison of QSCAT SMC and USDM results across the CONUS. There is an overall consistency between the two sets of results on a regional scale. This is observed as a monitoring process in detecting the frequency, intensity, and extent of SMC rather than an isolated examination of the spatial pattern in each map at a given time. For example, throughout the 2009 growing season, not much water from precipitation was detected on land surface, as seen in the SMC maps in the West and the Southwest, where USDM maps consistently show either no improvement (e.g., California) or worsening drought conditions (e.g., Arizona). For south Texas, no significant wet events occurred in the first part of the growing season, which is reflected by the severe to extreme drought conditions in the USDM, while the rainfall events in August and September that improved the drought conditions (as shown in the USDM maps in September and early October) are represented by the positive SMC in September.

In the Midwest, Figure 9.9 reveals extensive SMC in South Dakota, southern Minnesota, eastern Nebraska, and western Iowa in June and July 2009. During this same time period, USDM results consistently indicate some improvement in South Dakota, Nebraska, and Iowa (primarily change from D0 to no drought classification), and USDM maps suggest drought levels in Minnesota remained the same or became slightly worse. In early June 2009, SDP results showed an extensive rain pattern over the Midwest, including Minnesota, which suggests that rainwater was still present on the land surface as detected in the SMC (top left panel in Figure 9.9). Since the mean SMC represents a persistent amount of rainwater on land, SMC inherently reflects information about temperature, wind, insolation, and other parameters that affect soil wetness. Therefore, SMC may supplement information in synergy with other parameters currently used in the USDM to enhance the results.

SMC as measured by satellite can benefit not only drought monitoring but also drought forecasting. Skillful forecasts of drought or soil moisture would have significant uses for agriculture and hydrology (water planning). Recognizing the importance of seasonal forecasts of drought, NOAA CPC has been issuing such forecasts since March 2000. These forecasts are designed to indicate whether existing droughts will persist or improve and whether a new drought will form. An important first step in creating an improved forecast would be better knowledge of existing conditions. The SMC, as shown in Figures 9.8 and 9.9, is an appropriate parameter to contribute to a more accurate depiction of near-surface moisture supplies.

9.4 SUMMARY AND CONCLUDING REMARKS

Soil moisture derived from active and passive microwave remote sensing data can be used to enhance drought monitoring capabilities as summarized in the following:

1. SMC from active scatterometer data can characterize transient changes including the intensity, frequency, and extent of rainwater that actually reaches and accumulates on land surface, whereas passive radiometer data are for seasonal soil moisture. Together, soil moisture measurements from active and passive satellite data represent the state of dryness or wetness pertaining to land surface and are thus relevant to both hydrological and agricultural drought, as opposed to precipitation data (such as the specific precipitation index), which is more relevant to meteorological drought (WMO, 2009).
2. Satellite soil moisture measurements from scatterometer and radiometer data have continuous coverage across large geographic areas, whereas in situ RGP data or soil moisture networks such as SCAN consist of a relatively sparse spatial distribution of point data from networks with varying station densities and different data quality standards. Also, in the weekly

time scale for drought monitoring by the USDM and NIDIS, both QSCAT and AMSR-E data have a full coverage of the CONUS.

3. Satellite areal data represent the condition over the whole pixel size as opposed to a single local site at each in situ station. Moreover, QSCAT data with a 25 km resolution satisfy the Nyquist requirement to resolve the county-level drought condition, which is currently lacking for most inputs into the USDM and is an important goal of NIDIS.

Nevertheless, satellite measurements also have limitations, which necessitate using them in combination with in situ observations to more fully characterize soil moisture conditions. For example, in situ measurements can be obtained many times in a day (e.g., hourly measurements), whereas a satellite sensor typically collects data one or two times per day depending on the latitude. In addition, many in situ stations have a longer observational time series compared to satellite data. In particular, AMSR-E data have been collected since 2002 (starting in October 2011, the AMSR-E instrument was no longer operational), and QSCAT data were obtained over a decade (1999–2009), while many rain gauge stations were established several decades ago. Furthermore, there are differences in the characteristics of attributes measured by in situ gauges and by satellite sensors as presented in the benchmark study in the previous section, which should be combined to better represent various drought conditions.

For in situ data to be more useful, in situ measurements should characterize the conditions as far as possible beyond the local site. Here, satellite data can assist in the assessment of the extent beyond which local measurements are valid. For example, soil moisture data from SCAN can be compared or correlated in time (across months, seasons, or years) to satellite soil moisture signatures collected over areas with different radii away from the in situ station, to determine whether and how far the different measurements are correlated. The larger the radius at which satellite data and in situ measurements are well correlated, the larger the extent to which in situ data are representative. This is valuable in the selection of station locations for long-term maintenance so that the surface data are valid over the largest area as possible (not just in a close proximity of each station), thereby minimizing the number of stations required to monitor a certain region such as a county (in view of the county-scale goal of NIDIS).

Assimilation systems (Mitchell et al., 2004; Kumar et al., 2008) can be used to integrate various ground measurements and satellite observations within an ensemble framework of community land-surface models. This modeling approach allows data with various time scales and spatial coverage to be incorporated in a systematic manner. Because in situ networks are changing and improving and new satellite data and products are being developed, land data assimilation systems need to continuously evolve to provide enhanced products for drought monitoring. Furthermore, new measurements can allow better cross-verifications and validations among different models used in the land data assimilation systems in an effort to produce accurate, high-quality products.

Drought is a common climatic phenomenon throughout the world and a global problem requiring international efforts for drought assessment, forecasting, and mitigation. In this regard, satellite data from different nations can contribute to this overall goal. The QSCAT antenna ceased to spin in November 2009 after its continuous operation since July 1999. Meanwhile, the Indian Space Agency successfully launched another scatterometer similar to QSCAT aboard the Oceansat-2 Satellite (OSCAT) (Jayaraman et al., 1999) in September 2009. These satellite events, together with a scatterometer data agreement signed among the different nations, highlight the importance of international collaborations in improving global satellite coverage for drought monitoring.

As of December 2011, QSCAT is still measuring valid backscatter data along narrow tracks at a fixed azimuth, which are valuable for validating OSCAT measurements. Once verified with QSCAT, OSCAT can continue the QSCAT time series of SMC. A long-term SMC record is important for assessing drought conditions within the climatic historical perspective. In August 2011, China launched the Haiyan-2 (HY-2) satellite carrying another scatterometer (Dong et al., 2004). In addition, the development of another advanced satellite scatterometer is being studied in the United States, stemming from the recommendation of the NRC Decadal Survey (National Research Council, 2007). The current European SMOS mission (Kerr et al., 2010) and the proposed NASA Soil Moisture Active and Passive (SMAP) mission (Entekhabi et al., 2010) could potentially provide global soil moisture measurements critical for drought monitoring. Collectively, these successive satellite missions would provide multidecadal data important for addressing the nonstationarity issue in climate change. Moreover, long-term data are necessary for developing a probabilistic standardized index approach with multiple time scales of soil moisture variability that could be used for drought monitoring.

Experiences in using microwave satellite data to enhance the USDM will be valuable in improving international drought monitoring systems, such as the North American Drought Monitor (NADM) covering Canada, Mexico, and the United States (NCDC, 2010b). In developing countries that lack in situ or surface measurement networks, the role of satellite data for drought monitoring becomes increasingly important because products such as AMSR-E soil moisture and QSCAT SMC can be retrieved globally and fill informational gaps that are currently pervasive. Such products can enhance global drought monitoring, for which the Global Earth Observation System of Systems (GEOSS) (Lautenbacher, 2006) will be crucial as an overall integrator.

Long-term satellite-based moisture records are also important for developing climatologies used in forecasting drought conditions. Given the limited skill of seasonal forecasts of temperature and precipitation, drought forecasters place considerable weight on projecting current conditions forward based on what has happened in the past. In this regard, careful attention should be paid to the issue of nonstationarity due to significant changes in regional climatic trends in recent years. Improved knowledge of short-term moisture trends can contribute positively to drought forecasts. Although there is no guarantee that short-term trends will persist, forecasters need to know whether and how fast moisture conditions are deteriorating. Such trends serve to flag situations that require additional analysis. Once the SMC products are obtained for a suitable number of years to capture contemporary changes, forecasters may gain new knowledge of the probabilities that soil moisture conditions will likely improve or deteriorate during the following season. One of the goals of drought forecasting is to formulate the forecasts in terms of probabilities to provide a more accurate portrayal of confidence levels, and the statistics of historical soil moisture conditions can contribute to this effort. In short, improved knowledge of initial moisture conditions, short-term trends, and climatology have the potential to enhance the skill of current and future drought forecasts globally as well as in the United States.

ACKNOWLEDGMENTS

The research carried out at the Jet Propulsion Laboratory, California Institute of Technology, was supported by the National Aeronautics and Space Administration (NASA) Water Resources Area of the NASA Applied Sciences Program. Thanks to Robert Rabin from the NOAA National Severe Storms Laboratory for his direction in acquiring Stage-4 daily precipitation products for comparison with SMC patterns.

REFERENCES

Blanchard, B.J. and A.T.C. Chang. 1983. Estimation of soil-moisture from SEASAT-SAR data. *Water Resources Bulletin* 19(5):803–810.

Cognard, A.L., C. Loumagne, M. Normand, P. Olivier, C. Ottle, D. Vidalmadjar, S. Louahala, and A. Vidal. 1995. Evaluation of the ERS-1 synthetic aperture radar capacity to estimate surface soil-moisture—2-Year results over the Naizin watershed. *Water Resources Research* 31(4):975–982.

Dobson, M.C., F.T. Ulaby, M.T. Hallikainen, and M.A. Elrayes. 1985. Microwave dielectric behavior of wet soil, Part II: Dielectric mixing models. *IEEE Transactions on Geoscience and Remote Sensing* GE-23(1):35–46.

Dong, X., K. Xu, H. Liu, and J. Jiang. 2004. The radar altimeter and scatterometer of China's HY-2 satellite. *Proceedings of IEEE International Geoscience and Remote Sensing Symposium* 3:1703–1706.

Dubois, P., J.J. van Zyl, and T. Engman. 1995. Measuring soil moisture with imaging radars. *IEEE Transactions on Geoscience and Remote Sensing* 33(4):916–926.

Engman, E.T. 1995. Recent advances in remote-sensing in hydrology. *Reviews of Geophysics* 33(Part 2, Supplement S):967–975.

Entekhabi, D., E.G. Njoku, P.E. O'Neill, K.H. Kellogg, W.T. Crow, W.N. Edelstein, J.K. Entin, S.D. Goodman, T.J. Jackson, J. Johnson, J. Kimball, J.R. Piepmeier, R.D. Koster, N. Martin, K.C. McDonald, M. Moghaddam, S. Moran, R. Reichle, J.C. Shi, M.W. Spencer, S.W. Thurman, L. Tsang, Leung, and J.J. Van Zyl. 2010. The soil moisture active passive (SMAP) mission. *Proceedings of the IEEE* 98(5):704–716.

González, J. and J.B. Valdés. 2004. The mean frequency of recurrence of in-time-multidimensional events for drought analyses. *Natural Hazards and Earth System Science* 4(1):17–28.

van de Griend, A.A. and M. Owe. 1994. Microwave vegetation optical depth and inverse modelling of soil emissivity using Nimbus/SMMR satellite observations. *Meteorology and Atmospheric Physics* 54:225–239.

Hallikainen, M.T., F.T. Ulaby, M.C. Dobson, M.A. El-Rayes, and L.-K. Wu. 1985. Microwave dielectric behavior of wet soil, Part I: Empirical models and experimental observations. *IEEE Transactions on Geoscience and Remote Sensing* GE-23(1):25–34.

Higgins, R.W., W. Shi, E. Yarosh, and R. Joyce. 2000. Improved United States precipitation quality control system and analysis. *NCEP/Climate Prediction Center ATLAS No. 7*, U.S. Department of Commerce, National Weather Service, NOAA. http://www.cpc.noaa.gov/products/outreach/research_papers/ncep_cpc_atlas/7/ (last accessed on December 15, 2011).

Hutchinson, J.M.S. 2003. Estimating near-surface soil moisture using active microwave satellite imagery and optical sensor inputs. *Transactions of the American Society of Agricultural and Biological Engineers (ASABE)* 46(2):225–236.

Illston, B.G., J.B. Basara, and K.C. Crawford. 2004. Seasonal to interannual variations of soil moisture measured in Oklahoma. *International Journal of Climatology* 24:1883–1896.

Jackson, T.J. 1990. Laboratory evaluation of a field-portable dielectric soil-moisture probe. *IEEE Transactions on Geoscience and Remote Sensing* 28(2):241–245.

Jackson, T.J. 1997. Soil moisture estimation using SSM/I satellite data over a grassland region. *Water Resources Research* 33:1475–1484.

Jayaraman, V., V.S. Hedge, M. Rao, and H.H. Gowda. 1999. Future earth observation missions for oceanographic applications: Indian perspectives. *Acta Astronautica* 44(7–12):667–674.

Justice, C.O., J.R.G. Townshend, B.N. Holben, and C.J. Tucker. 1985. Analysis of the phenology of global vegetation using meteorological satellite data. *International Journal of Remote Sensing* 6(8):1271–1318.

Kerr, Y.H. and E.G. Njoku. 1990. A semiempirical model for interpreting microwave emission from semiarid land surfaces as seen from space. *IEEE Transactions on Geoscience and Remote Sensing* 28:384–393.

Kerr, Y.H., P. Waldteufel, J.P. Wigneron, S. Delwart, F. Cabot, J. Boutin, M.J. Escorihuela, J. Font, N. Reul, C. Gruhier, S.E. Juglea, M.R. Drinkwater, A. Hahne, M. Martin-Neira, and S. Mecklenburg. 2010. The SMOS mission: New tool for monitoring key elements of the global water cycle. *Proceedings of the IEEE* 98(5):666–687.

Kerr, Y.H., P. Waldteufel, J.-P. Wigneron, J.-M. Martinuzzi, J. Font, and M. Berger. 2001. Soil moisture retrieval from space: The soil moisture and ocean salinity (SMOS) mission. *IEEE Transactions on Geoscience and Remote Sensing* 39(8):1729–1735.

Kumar, S.V., R.H. Reichle, C.D. Peters-Lidard, R.D. Koster, X.W. Zhan, W.T. Crow, J.B. Eylander, and P.R. Houser. 2008. A land surface data assimilation framework using the land information system: Description and applications. *Advances in Water Resources* 31(11):1419–1432.

Lautenbacher, C.C. 2006. The Global Earth Observation System of Systems: Science serving society. *Space Policy* 22(1):8–11.

Li, Q., L. Tsang, J. Shi, and C.H. Chan. 2000. Application of physics-based two-grid method and sparse matrix canonical grid method for numerical simulations of emissivities of soils with rough surfaces at microwave frequencies. *IEEE Transactions on Geoscience and Remote Sensing* 38:1635–1643.

Loew, A., R. Ludwig, and W. Mauser. 2006. Derivation of surface soil moisture from ENVISAT ASAR wide swath and image mode data in agricultural areas. *IEEE Transactions on Geoscience and Remote Sensing* 44(4):889–899.

Maddox, R.A., J. Zhang, J.J. Gourley, and K.W. Howard. 2002. Weather radar coverage over the contiguous United States. *Weather and Forecasting* 17:927–934.

Mitchell, K.E., D. Lohmann, P.R. Houser, E.F. Wood, J.C. Schaake, A. Robock, B.A. Cosgrove, J. Sheffield, Q.Y. Duan, L.F. Luo, R.W. Higgins, R.T. Pinker, J.D. Tarpley, D.P. Lettenmaier, C.H. Marshall, J.K. Entin, M. Pan, W. Shi, V. Koren, J. Meng, B.H. Ramsay, and A.A. Bailey. 2004. The multi-institution North American Land Data Assimilation System (NLDAS): Utilizing multiple GCIP products and partners in a continental distributed hydrological modeling system. *Journal of Geophysical Research* 109(D7):D07S90, doi:10.1029/2003JD003823.

Mo, T., T.J. Schmugge, and T.J. Jackson. 1984. Calculations of radar backscattering coefficient of vegetation covered soils. *Remote Sensing of Environment* 15:119–133.

Moran, M.S., A. Vidal, D. Troufleau, Y. Inoue, and T.A. Mitchell. 1998. Ku- and C-band SAR for discriminating agricultural crop and soil conditions. *IEEE Transactions of Geoscience and Remote Sensing* 36:265–272.

Moran, M.S., V. Vidal, D. Troufleau, J. Qi, T.R. Clarke, P.J. Pinter, Jr., T.A. Mitchell, Y. Inoue, and C.M.U. Neale. 1997. Combining multifrequency microwave and optical data for crop management. *Remote Sensing of Environment* 61:96–109.

National Research Council. 2007. *Earth Science and Applications from Space: National Imperatives for the Next Decade and Beyond*. Washington, DC: The National Academies Press.

NCDC. 2009. *State of the Climate—Drought, August 2009*. National Climatic Data Center, NESDIS, NOAA, U.S. Department of Commerce. http://www.ncdc.noaa.gov/sotc/?report=drought& year=2009&month=8 (last accessed on December 15, 2011).

NCDC. 2010a. Global surface summary of the day—GSOD. National Climatic Data Center, NESDIS, NOAA, U.S. Department of Commerce. http://www.data.gov/geodata/g600037/ (last accessed on December 15, 2011).

NCDC. 2010b. North American drought monitor. National Climatic Data Center, NESDIS, NOAA, U.S. Department of Commerce. http://www.ncdc.noaa.gov/temp-and-precip/drought/nadm/index.html (last accessed on December 15, 2011).

Nghiem, S.V. 2001. Advanced scatterometry for geophysical remote sensing. In *Jet Propulsion Laboratory (JPL) Document D-23048*. Pasadena, CA: Jet Propulsion Laboratory, California Institute of Technology.

Nghiem, S.V., M. Borgeaud, J.A. Kong, and R.T. Shin. 1990. Polarimetric remote sensing of geophysical media with layer random medium model. In *Progress in Electromagnetics Research—Polarimetric Remote Sensing*. Vol. 3, ed. J.A. Kong, pp. 1–73. New York: Elsevier.

Nghiem, S.V., R. Kwok, S.H. Yueh, and M.R. Drinkwater. 1995. Polarimetric signatures of sea ice, 1, Theoretical model. *Journal of Geophysical Research* 100(C7):13665–13679.

Nghiem, S.V., T. Le Toan, J.A. Kong, H.C. Han, and M. Borgeaud. 1993a. Layer model with random spheroidal scatterers for remote sensing of vegetation canopy. *Journal of Electromagnetic Waves and Applications* 7(1):49–76.

Nghiem, S.V., E.G. Njoku, G.R. Brakenridge, and Y. Kim. 2005. Land surface water cycles observed with satellite sensors. *Proceedings of the 19th Conference on Hydrology and 16th Conference on Climate Variability and Change, 85th American Meteorological Society Meeting*. San Diego, CA, January 9–13.

Nghiem, S.V., E.G. Njoku, J.J. Van Zyl, and Y. Kim. 2003. Global energy and water cycle—Soil moisture variability pattern over continental extent observed with active and passive satellite data. In *Jet Propulsion Laboratory (JPL) Document. D-26225*. Pasadena, CA: Jet Propulsion Laboratory, California Institute of Technology.

Nghiem, S.V., J.J. Van Zyl, W.-Y. Tsai, and G. Neumann. 2000. Potential application of scatterometry to large-scale soil moisture monitoring. In *Jet Propulsion Laboratory (JPL) Document D-19523*. Pasadena, CA: Jet Propulsion Laboratory, California Institute of Technology.

Nghiem, S.V., J. Verdin, M. Svoboda, D. Allured, J. Brown, B. Liebmann, G. Neumann, E. Engman, and D. Toll. 2010. Improved drought monitoring with NASA satellite data. *Environmental and Water Resources Institute (EWRI) Currents* 12(3):7.

Nghiem, S.V., S.H. Yueh, R. Kwok, and D.T. Nguyen. 1993b. Polarimetric remote sensing of geophysical medium structures. *Radio Science* 28(6):1111–1130.

NIDIS. 2006. National integrated drought information system act of 2006. Public Law 109–430. *U.S. Statutes at Large* 120(2006):2918. http://www.gpo.gov/fdsys/pkg/PLAW-109publ430/pdf/PLAW-109publ430.pdf (last accessed on December 15, 2011).

NIDIS. 2007. The national integrated drought information system implementation plan—A pathway for national resilience. http://www.drought.gov/pdf/NIDIS-IPFinal-June07.pdf (last accessed on December 15, 2011).

Njoku, E. 2004. *AMSR-E/Aqua L2B Surface Soil Moisture, Ancillary Parms, & QC EASE-Grids* V002. Boulder, Co: National Snow and Ice Data Center. http://nsidc.org/data/docs/daac/ae_land_l2b_soil_moisture.gd.html (last accessed on December 15, 2011).

Njoku, E.G., T.J. Jackson, V. Lakshmi, T.K. Chan, and S.V. Nghiem. 2003. Soil moisture retrieval from AMSR-E. *IEEE Transactions on Geoscience and Remote Sensing* 41(2):215–229.

Njoku, E.G. and L. Li. 1999. Retrieval of land surface parameters using passive microwave measurements at 6–18 GHz. *IEEE Transactions on Geoscience and Remote Sensing* 37:79–93.

NMQ. 2009. Stage-4 24hr QPE accumulation. National Mosaic & Multi-Sensor QPE (NMQ). OK: NOAA National Severe Storms Laboratory, Norman. http://nmq.ou.edu/ (last accessed on December 15, 2011).

NWS. 2009. September 2009 weather in review. NOAA National Weather Service (NWS), Southern Region Headquarters. http://www.srh.noaa.gov/images/ewx/wxevent/sep2009.pdf (last accessed on December 15, 2011).

Owe, M., A. Chang, and R.E. Golus. 1988. Estimating surface soil moisture from satellite microwave measurements and a satellite-derived vegetation index. *Remote Sensing of Environment* 24:131–345.

Prevot, M., M. Dechambre, O. Taconet, D. Vidal-Madjar, M. Normand, and S. Galle. 1993. Estimating the characteristics of vegetation canopies with airborne radar measurements. *International Journal of Remote Sensing* 14:2803–2818.

Resource Management Services. 2004. State of montana multi-hazard mitigation plan and statewide hazard assessment. Fort Harrison, MT: Department of Military Affairs, Disaster and Emergency Services.

Shrivastava, H.S., P. Patel, Y. Sharma, and R.R. Navalgund. 2009. Large-area soil moisture estimation using multi-incidence-angle RADARSAT-1 SAR data. *IEEE Transactions on Geoscience and Remote Sensing* 47(8):2528–2535.

Shrivastava, S.K., N. Yograjan, V. Jayaraman, P.P.N. Rao, and M.G. Chandrasekhar. 1997. On the relationship between ERS-1 SAR/backscatter and surface/sub-surface soil moisture variations in vertisols. *Acta Astronautica* 40(10):693–699.

Shoshany, M., T. Svoray, P.J. Curran, G.M. Foody, and A. Perevolotsky. 2000. The relationship between ERS-2 SAR backscatter and soil moisture: Generalization from a humid to semi-arid transect. *International Journal of Remote Sensing* 21(11):2337–2343.

Story, G. 2009. The difficulty of achieving good precipitation estimates for use in real-time drought monitoring. Sixth U.S. Drought Monitor Forum, Austin, TX, October 7–8.

Svoboda, M.D., M.J. Hayes, and D.A. Wilhite. 2001. The role of integrated drought monitoring in drought mitigation planning. *Annals of Arid Zone* 40(1):1–11.

Svoboda, M., D. LeComte, M. Hayes, R. Heim, K. Gleason, J. Angel, B. Rippey, R. Tinker, M. Palecki, D. Stooksbury, D. Miskus, and S. Stephens. 2002. The drought monitor. *Bulletin of the American Meteorological Society* 83(8):1181–1190.

Takada, M., Y. Mishima, and S. Natsume. 2009. Estimation of soil surface properties in peatland using ALSO/PALSAR. *Landscape and Ecological Engineering* 5(1):45–58.

Teng, W.L., J.R. Wang, and P.C. Doriaswamy. 1993. Relationship between satellite microwave radiometric data, antecedent precipitation index, and regional soil moisture. *International Journal of Remote Sensing* 14:2483–2500.

Texas Water Development Board. 2007. Highlights of the 2007 state water plan. *Water for Texas 2007*, Document No. GP-8-1.

Tsai, W.-Y., S.V. Nghiem, J.N. Huddleston, M.W. Spencer, B.W. Stiles, and R.D. West. 2000. Polarimetric scatterometry: A promising technique for improving ocean surface wind measurements. *IEEE Transactions on Geoscience and Remote Sensing* 38:1903–1921.

U.S. Census Bureau. 2005. County and county equivalent areas. http://www.census.gov/geo/www/cob/co_metadata.html (last accessed on December 15, 2011).

USDA. 2009a. SCAN—soil climate analysis network. SCAN Brochure, Natural Resources Conservation Service, National Water and Climate Center National Soil Survey Center, Lincoln, NE. http://www.wcc.nrcs.usda.gov/scan/SCAN-brochure.pdf (last accessed on December 15, 2011).

USDA. 2009b. SNOTEL and snow survey and water supply forecasting, SNOTEL Brochure, NWCC Rev3/09, Natural Resources Conservation Service, National Water and Climate Center (NWCC), Portland, OR. http://www.wcc.nrcs.usda.gov/snotel/SNOTEL-brochure.pdf (last accessed on December 15, 2011).

Verdin, J., C. Funk, R. Klaver, and D. Roberts. 1999. Exploring the correlation between Southern Africa NDVI and Pacific sea surface temperatures: Results for the 1998 maize growing season. *International Journal of Remote Sensing* 20(10):2117–2124.

Wagner, W., J. Noll, M. Borgeaud, and H. Rott. 1999. Monitoring soil moisture over the Canadian prairies with the ERS scatterometer. *IEEE Transactions on Geoscience and Remote Sensing* 37(1):206–216.

Wagner, W. and K. Scipal. 2000. Large-scale soil moisture mapping in western Africa using the ERS scatterometer. *IEEE Transactions on Geoscience and Remote Sensing* 38(4, Part 2): 1777–1782.

Wang, J.R. 1983. Passive microwave sensing of soil moisture content: The effects of soil bulk density and surface roughness. *Remote Sensing of Environment* 13:329–344.

Wang, J.R. 1985. Effect of vegetation on soil moisture sensing observed from orbiting microwave radiometers. *Remote Sensing of Environment* 17:141–151.

Wang, J.R. and B.J. Choudhury. 1981. Remote sensing of soil moisture content over bare field at 1.4 GHz frequency. *Journal of Geophysical Research* 86:5277–5282.

Western Governors' Association. 2004. *Creating a Drought Early Warning System for the 21st Century, The National Integrated Drought Information System*. Denver, CO: Western Governors' Association. http://www.westgov.org/wga/publicat/nidis.pdf (last accessed on December 15, 2011).

Wigneron, J.-P., L. Laguerre, and Y.H. Kerr. 2001. A simple parameterization of the L-band microwave emission from rough agricultural soils. *IEEE Transactions on Geoscience and Remote Sensing* 39:1697–1707.

Wilhite, D.A. and M.H. Glantz. 1985. Understanding the drought phenomenon: The role of definitions. *Water International* 10:111–120.

WMO (World Meteorological Organization). 2009. Experts agree on a universal drought index to cope with climate risks. Press release, World Meteorological Organization, WMO No.-872, United Nations, Geneva, Switzerland. http://www.wmo.int/pages/mediacentre/press_releases/pr_872_en.html (last accessed on December 15, 2011).

Zhang, X.Y., M. Goldberg, D. Tarpley, M.A. Friedl, J. Morisette, F. Kogan, and Y.Y. Yu. 2010. Drought-induced vegetation stress in southwestern North America. *Environment Research Letters* 5(2):024008, doi:10.1088/1748-9326/5/2/024008.

10 North American Land Data Assimilation System

A Framework for Merging Model and Satellite Data for Improved Drought Monitoring

Justin Sheffield, Youlong Xia, Lifeng Luo, Eric F. Wood, Michael Ek, and Kenneth E. Mitchell

CONTENTS

10.1 INTRODUCTION

Drought is a pervasive natural climate hazard that has widespread impacts on human activity and the environment. In the United States, droughts are billion-dollar disasters, comparable to hurricanes and tropical storms and with greater economic impacts than extratropical storms, wildfires, blizzards, and ice storms combined (NCDC, 2009). Reduction of the impacts and increased preparedness for drought requires the use and improvement of monitoring and prediction tools. These tools are reliant on the availability of spatially extensive and accurate data for representing the occurrence and characteristics (such as duration and severity) of drought and their related forcing mechanisms. It is increasingly recognized that the utility of drought data is highly dependent on the application (e.g., agricultural monitoring versus water resource management) and time (e.g., short- versus long-term dryness) and space (e.g., local versus national) scales involved. A comprehensive set of drought indices that considers all components of the hydrological–ecological–human system is necessary. Because of the dearth of near-real-time in situ hydrologic data collected over large regions, modeled data are often useful surrogates, especially when combined with observations from remote sensing and in situ sources.

This chapter provides an overview of drought-related activities associated with the North American Land Data Assimilation System (NLDAS), which purports to provide an incremental step toward improved drought monitoring and forecasting. The NLDAS was originally conceived to improve short-term weather forecasting by providing better land surface initial conditions for operational weather forecast models. This reflects increased recognition of the role of land surface water and energy states, such as surface temperature, soil moisture, and snowpack, to atmospheric processes via feedbacks through the coupling of the water and energy cycles. Phase I of the NLDAS (NLDAS-1; Mitchell et al., 2004) made tremendous progress toward developing an operational system that gave high-resolution land hydrologic products in near real time. The system consists of multiple land surface models (LSMs) that are driven by an observation-based meteorological data set both in real time and retrospectively. This work resulted in a series of scientific papers that evaluated the retrospective data (meteorology and model output) in terms of their ability to reflect observations of the water and energy cycles and the uncertainties in the simulations as measured by the spread among individual models (Pan et al., 2003; Robock et al., 2003; Sheffield et al., 2003; Lohmann et al., 2004; Mitchell et al., 2004; Schaake et al., 2004). These evaluations led to the implementation of significant improvements to the LSMs in the form of new model physics and adjustments to parameter values and to the methods and input meteorological data (Xia et al., 2012). The system has since expanded in scope to include model intercomparison studies, real-time monitoring, and hydrologic prediction and has inspired other activities such as high-resolution land surface modeling and global land data assimilation systems (e.g., the Global Land Data Assimilation System [GLDAS], Rodell et al., 2004; the Land Information System [LIS], Kumar et al., 2006).

The second phase of the project (NLDAS-2) extended the original concept (improved weather forecasting) in recognition of the value of the NLDAS products to the wider scientific community and stakeholders interested in hydrologic processes and data. A key part of this impetus was to provide water-related data to support water resources management, energy demand assessment, agricultural monitoring, fire risk assessment, drought monitoring, and flood prediction. In particular, the National Centers for Environmental Prediction (NCEP) Environmental Modeling Center (EMC) has collaborated with partners at the National Oceanic and Atmospheric Administration (NOAA) Climate Program Office (CPO) Climate Prediction Program of the Americas (CPPA) to develop an NLDAS drought monitoring and forecast system, which is the focus of this chapter. This system will provide a single stream of data that can be used for drought monitoring in support of the National Integrated Drought Information System (NIDIS; http://www.drought.gov), which is a federally mandated initiative to provide improved and consistent national drought information. Key characteristics of the NLDAS-2 drought products supporting this activity include improved reliability in model output demonstrated through rigorous LSM intercomparisons; the ability to detect the onset, extent, and duration of major drought events; the capability to perform long-term simulations so that robust climatologies can be calculated for meaningful anomaly detection; and rapid updating in near real time. To this end, the NLDAS-2 has focused on reducing the differences in calculated values among models and improving the representation of measured land fluxes and states to improve the reliability of the results. The project has also evaluated model depiction of major historic drought events over multiple decades to establish the consistency in the information and develop a reliable climatology, and demonstrated its use in an operational setting at the EMC. In addition, the system has been enhanced through the implementation of a seasonal forecasting component that has benefited from the improved land surface states that are essential to seasonal hydrologic prediction (Li et al., 2009).

This chapter describes the development and application of NLDAS-2 products for drought monitoring and seasonal forecasting, as well as future challenges to improving the system. First, an overview of the long-term retrospective simulations in terms of their depiction of drought and highlights of some of the major drought events over the United States in the past 30 years are presented. Second, we explore how the models differ in their depiction of drought at various temporal and spatial scales, which relates to the reliability of the predictions. An overview of the real-time drought monitor and seasonal forecast systems is then provided with recent examples. Finally, we discuss the potential for augmenting the system with expanded use of remote sensing data. In particular, we assess the utility of remote sensing–based estimates of soil moisture for drought monitoring and discuss how they might be used in a model-based drought assessment such as the NLDAS to provide better predictions, for example, in regions with sparse precipitation measurement networks.

10.2 NLDAS APPROACH TO DROUGHT MONITORING

10.2.1 Overview of NLDAS-2

NLDAS-2 is a core project of the NCEP EMC funded by NOAA's CPPA with collaboration from several groups, including the National Aeronautics and Space Administration (NASA) Goddard Space Flight Center (GSFC), Princeton University, the National Weather Service (NWS) Office of Hydrological Development (OHD), the University of Washington, and NCEP's CPC. The system comprises three parts: a 28 year (1979–2007) retrospective simulation component that forms a climatology against which current conditions can be assessed, a real-time monitoring component that updates hydrologic fields daily, and a forecast component that makes seasonal forecasts on a monthly basis using ensemble (probabilistic) forecast techniques with lead times up to 6 months. Each of these components makes use of remote sensing data that are combined with ground observations and atmospheric model data to provide input data and boundary conditions for the LSMs. The remote sensing data include precipitation data, surface shortwave radiation, and vegetation spatial distribution and characteristics.

10.2.2 NLDAS-2 Models

The system incorporates four LSMs: the Noah, Mosaic, Variable Infiltration Capacity (VIC), and Sacramento (SAC) models. The Noah model was developed as the land component in the NOAA/NCEP mesoscale Eta model (Betts et al., 1997; Chen et al., 1997; Ek et al., 2003) and is the land model in the Weather Research and Forecasting (WRF) regional atmospheric model and the NOAA/NCEP coupled Climate Forecast System (CFS) and Global Forecast System (GFS) for short-term and medium-term weather forecasting, respectively. The Mosaic model was developed for use in the NASA global climate model (Koster and Suarez, 1994, 1996). The VIC model is a macroscale, semidistributed hydrologic model (Liang et al., 1994; Wood et al., 1997) that was developed at the University of Washington and Princeton University. The SAC model was developed as a lumped conceptual hydrologic model (Burnash et al., 1973), calibrated for small catchments and used operationally at NWS River Forecast Centers (RFC). It is run in a semidistributed mode for NLDAS. These models simulate the coupled water and energy cycles at the earth's surface at varying degrees of complexity. However, the SAC model only simulates the water cycle. Each model has unique attributes that reflect the origin of their development either as a hydrologic model or a Soil Vegetation Atmosphere Transfer (SVAT) model, intended to serve as the land component in atmospheric and climate models. For example, the VIC and Mosaic models use a unique tiling scheme to represent the heterogeneity of vegetation within a model grid cell. The Noah and SAC models use a single dominant vegetation type for each grid cell.

Each of these models is run over a common spatial domain on a regular 1/8th degree (~12 km) grid that covers the conterminous United States, northern Mexico, and southern Canada (125°–67°W and 25°–53°N). They share a common land mask, underlying elevation, hourly input surface meteorological forcing, soil texture, vegetation classes and distribution, streamflow network, streamflow routing model, and input and output file format. Common hourly output fields from each model include surface state variables such as soil moisture, soil temperature, snow water equivalent, surface fluxes (e.g., latent, sensible, and ground heat flux), and runoff. Although all models use common maps of vegetation and soil classes, they retain their unique soil and vegetation parameter values such as root depth and density, different soil column layering (number and thickness of layers), and different seasonal cycles of vegetation characteristics. The vegetation classification was derived from the University of Maryland Advanced Very High Resolution Radiometer (AVHRR)-based data set of Hansen et al. (2000). All models specify the seasonality of vegetation on a climatological basis using AVHRR-based data. For Mosaic, SAC, and VIC, values of Leaf Area Index (LAI) and its seasonal cycle are derived from AVHRR-based Normalized Difference Vegetation Index (NDVI) data (Myneni et al., 1997). Noah uses the global, monthly 5 year climatology of the green vegetation fraction (GVF) derived by Gutman and Ignatov (1998) from AVHRR-based NDVI. Runoff and baseflow from each model are routed using a common river routing scheme (Lohmann et al., 2004) to produce streamflow at selected gauging points for comparison to measurements and analysis of flow characteristics.

10.2.3 Meteorological Forcings and Retrospective (1979–2007) Simulation

The atmospheric forcings that drive the models are derived from a combination of data from atmospheric model reanalysis, ground measurements, and satellite remote sensing. The underlying data set comes from North American Regional Reanalysis (NARR; Mesinger et al., 2006) products with a 32 km spatial and 3 h temporal resolution. The data set includes 2 m (above the ground surface) air temperature, 2 m specific humidity, 10 m wind speed, surface pressure, precipitation, incoming solar radiation, and incoming longwave radiation. These are interpolated to the 1/8th degree (~12 km) spatial resolution and 1 h temporal resolution of the NLDAS grid, accounting for changes in elevation and solar angle based on methods developed in NLDAS-1 (Cosgrove et al., 2003). Although the NARR improves on previous global reanalyzes in terms of its depiction of near-surface meteorology, especially through the assimilation of gauge precipitation data, some biases still remain, and for some variables (i.e., precipitation and downward surface shortwave radiation), observational data are used instead of the NARR. Precipitation is anchored to the CPC unified gauge-based precipitation analysis with orographic enhancements derived from Parameter-elevation Regressions on Independent Slopes Model data (PRISM; Daly et al., 1994), with NARR precipitation data used in parts of Canada and Mexico where gauge density is low. The daily data are disaggregated to hourly time steps using ground-based Doppler radar data and remote sensing data from the NOAA CPC Morphing Technique (CMORPH) (Joyce et al., 2004). For shortwave radiation, a large bias in the NARR was removed by scaling it to match the remote sensing–based product of Pinker et al. (2003), which uses data from NOAA Geostationary Operational Environmental Satellites (GOES). Details of the NLDAS-2 forcings are given at http://www.emc.ncep.noaa.gov/mmb/nldas/LDAS8th/forcing/forcing_narr.shtml. The four models were run retrospectively for the period 1979–2007. Each model simulation was initialized from a "spin-up" simulation run for 1979–1995 so that the moisture and temperature states were brought to equilibrium.

10.2.4 Drought Indices

As the full terrestrial water and energy cycles (except for the SAC model) are represented in the NLDAS, it is possible to depict drought in terms of any one or combination of hydrologic components such as precipitation, streamflow, and soil moisture. The NLDAS-2 real-time monitor provides a range of drought indices (see Section 10.4), including daily, weekly, and monthly anomalies, as well as percentiles of various hydrologic fields (soil moisture, snow water equivalent, total runoff, streamflow, evaporation, precipitation) output from the four LSMs on their common 1/8th degree grid.

Here we focus on soil moisture, given its role as an aggregator in the hydrologic system, reflecting precipitation and snowmelt inputs and the loss of water from the system via evapotranspiration, runoff, and drainage. Soil moisture forms the basis for improved short-term and seasonal weather prediction, through exertion of control over water and energy exchange with the atmosphere. Soil moisture fields can also be used for seasonal hydrologic prediction by providing antecedent states that are crucial to flood prediction, as well as future drought emergence. In this chapter, drought is shown in terms of monthly mean soil moisture percentiles, which normalize the data with respect to climatological values for each month at a 1/8th model grid cell resolution. This approach has been successfully used in model-based drought studies for the United States (Sheffield et al., 2004; Andreadis et al., 2005; Wang et al., 2009) and globally (Sheffield and Wood, 2007, 2008a; Wang et al., 2010) and in the assessment of drought under projected future climates (Sheffield and Wood, 2008b). The 20th percentile was chosen as the threshold for drought, which has been used in previous studies (Andreadis et al., 2005; Sheffield et al., 2009a), as well as in operational systems such as the U.S. and North American Drought Monitors (Lawrimore et al., 2002; Svoboda et al., 2002).

10.3 RESULTS FROM RETROSPECTIVE SIMULATIONS

We begin by providing an overview of historic drought as represented by the retrospective simulations from the four models. An essential element of drought monitoring is the background climatology to which current and future conditions can be compared and dry anomalies detected within an extended historical context. The use of multiple models helps to quantify the uncertainties due to model physics and parameterizations, but we also calculate a multi-model ensemble (MME) average that represents a best estimate of drought conditions given these model uncertainties. The multi-model average is calculated by averaging the percentiles of the four models and then recalculating percentile values with respect to the multi-model average, as the model averaging will tend to reduce extreme values (Wang et al., 2009, 2010).

Figure 10.1 shows soil moisture percentiles averaged over the continental United States and regionally for the multi-model average (calculated from the four NLDAS-2 LSMs). The regions are based on the NWS RFC regional delineations shown in Figure 10.2. In general, the spread among the models was very small, especially when averaged over the whole United States and for drier regions in the West (e.g., California–Nevada RFC). Despite the spin-up of the model states, there are still noticeable differences at the beginning of the time series (i.e., the first year, 1979), particularly for the drier western regions, where the model states used for the initial conditions may not have reached equilibrium. Although averaging over the United States smoothes the temporal variations in soil moisture time series, some

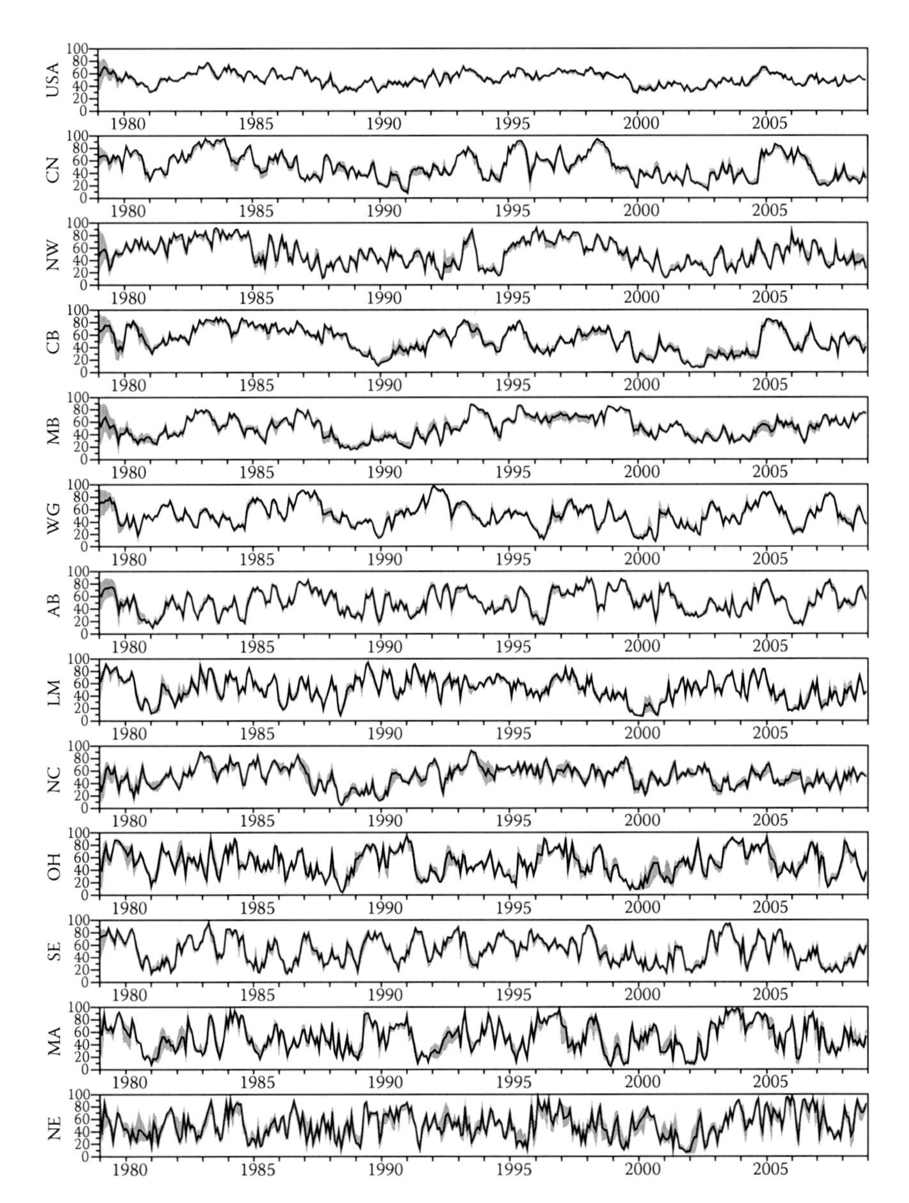

FIGURE 10.1 A 30 year time series (1979–2008) of soil moisture percentiles averaged over the continental United States and NWS RFC regions for the multi-model mean (calculated from the four NLDAS-2 LSMs). The gray shading represents the range in the models. The regions are ordered from west (upper panels) to east (lower panels).

periods in the early 1980s, late 1980s, and early 2000s are noticeably drier than normal, with large areas of drought within the regions during these periods (shown in Figure 10.3) based on a 20th percentile soil moisture threshold for drought. The peak areal percentage of the CONUS that is in drought for the MME is 54% in June 1988. Regionally, the more humid eastern regions (bottom graphs) exhibited

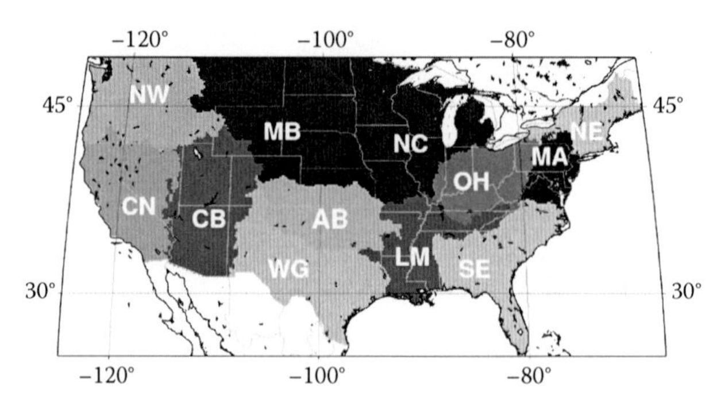

FIGURE 10.2 Map of the 12 continental U.S. RFC regions: Northwest (NW), Missouri Basin (MB), North Central (NC), Ohio Basin (OH), Northeast (NE), Middle Atlantic (MA), California–Nevada (CN), Colorado Basin (CB), Arkansas-Red Basin (AB), West Gulf (WG), Lower Mississippi (LM), and Southeast (SE).

greater variability than the drier western regions (upper graphs) and also had a larger spread among the model results. The largest differences among the models generally occurred during wet periods following a drought (e.g., during 2000 in the Ohio Basin (OHRFC) and during 2003–2004 in the Colorado Basin (CBRFC); Figure 10.2), which indicates that the differences are primarily derived from how each individual model partitions precipitation and propagates hydrologic anomalies through the system rather than how the models depict drying.

Figure 10.4 shows the spatial extent and severity of four major drought events in the United States (1988, 1996, 2002, and 2007) as simulated by the models and the MME. The 1988 drought spanned the central United States and northern Great Plains (Lawford, 1992; Trenberth and Branstator, 1992) and had the largest economic impacts of any drought or natural hazard in the United States, totaling ~$39 billion in losses (Riebsame et al., 1991) (only surpassed by Hurricane Katrina in total economic impacts), mainly because of its geographic extent over regions of high agricultural intensity (e.g., U.S. Corn Belt) and population density (e.g., eastern United States). This broad extent is well captured by the four models, as shown in Figure 10.4. The 1996 drought over Texas and parts of the Southwest resulted in estimated losses of $6 billion in Texas alone (Wilhite, 2000). Again, the four models capture the broad pattern of the drought, but the differences between models can be large. For example, in northwestern Texas, the SAC model does not depict drought conditions, in contrast to the other models. The 2002 drought was part of a long-term drought in the western United States that had persisted since about 1999 but reached its peak areal extent during the summer of 2002, covering about 45% of the country, as indicated in the top time series graph in Figure 10.3. This is especially apparent for the Colorado Basin (CBRFC), where drought conditions covered more than 90% of the region, as defined by modeled soil moisture results in Figure 10.4. This drought was driven by near record low precipitation (Lawrimore and Stephens, 2003) and caused severe water resources impacts, with record low levels in the Lake Powell reservoir, increased wildfire hazards, and tree die-off directly attributed to drought and an associated insect outbreak (Betancourt, 2003). The more recent

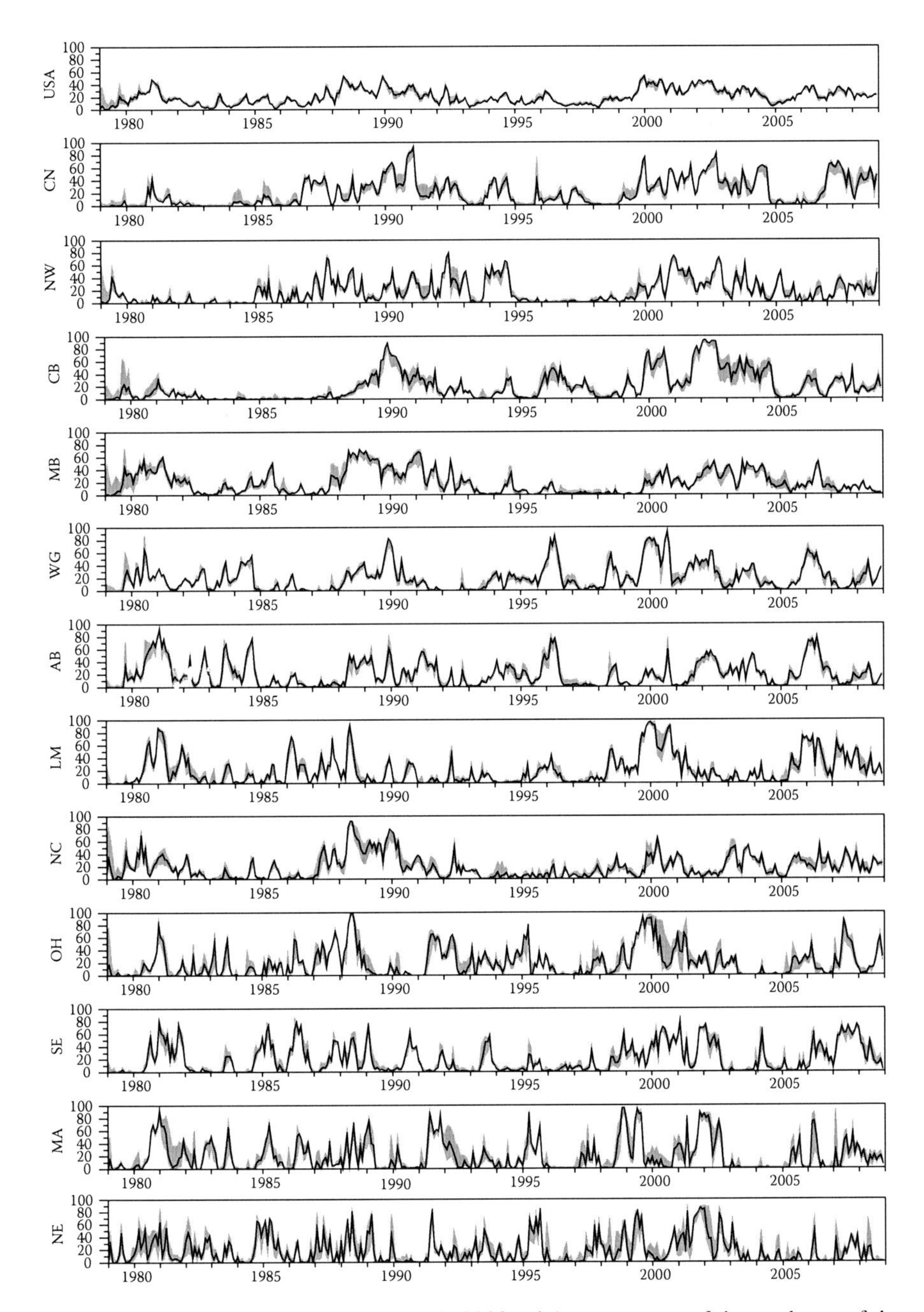

FIGURE 10.3 A 30 year time series (1979–2008) of the percentage of the total area of the continental United States and individual NWS RFC regions detected to be in drought from the soil moisture percentile information (drought defined as percentile <20%) for the multi-model mean (calculated from the four NLDAS-2 LSMs). The gray shading represents the range across the models. The regions are ordered from west (upper panels) to east (lower panels).

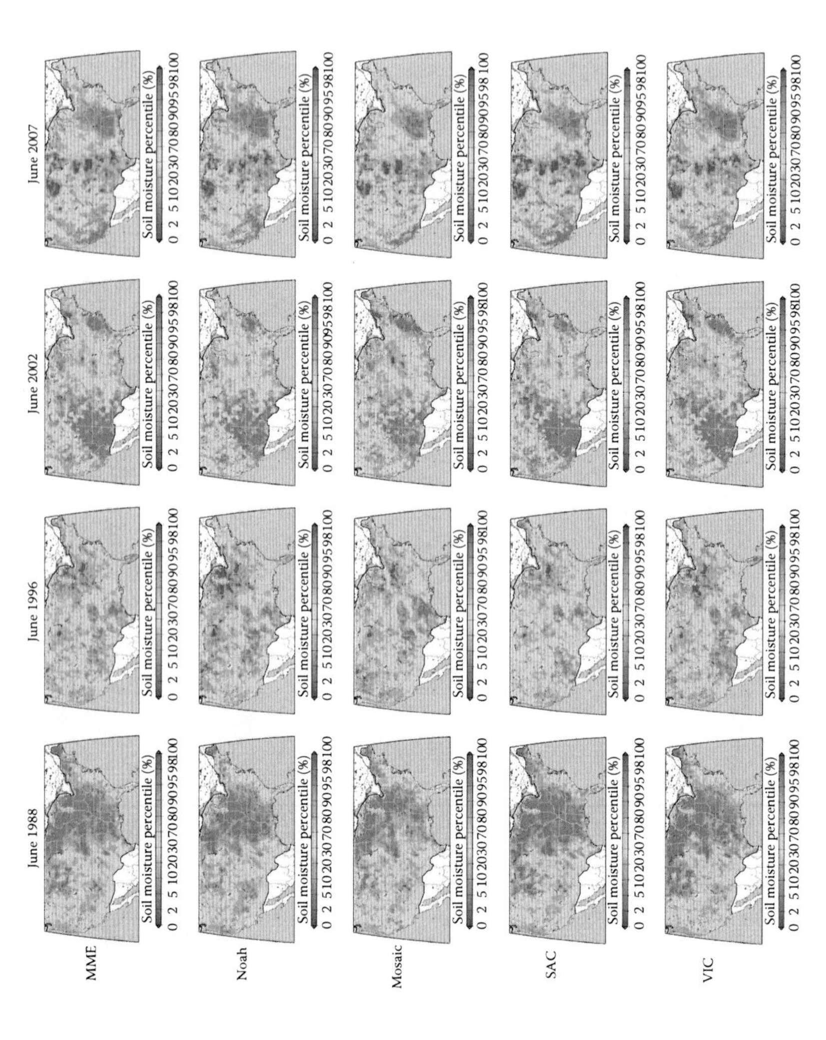

FIGURE 10.4 **(See color insert.)** Snapshots of four major drought events from June soil moisture percentiles from the MME and the four models. Columns are (1) 1988, (2) 1996, (3) 2002, and (4) 2007.

2007 drought in the southeast United States reduced reservoir levels to record lows, affecting water supplies to the city of Atlanta and exacerbating interstate conflicts on water allocations. The model results in Figure 10.4 indicate that this drought covered the southern states to the east of the Mississippi River, while drought also affected California and other western states.

The four models produced remarkably consistent depictions of the peak extent of these large-scale drought events, indicating that their rank correlations are high. They also agree well on the location and magnitude of wet regions. Some level of agreement is to be expected among the models given the commonality in the meteorological forcings and underlying land surface characteristics, as well as the use of percentiles for quantifying drought severity. Nevertheless, the level of intermodel consistency is encouraging, and the classified drought patterns appear to coincide with the known extents and impacts of these major events as discussed earlier.

However, as we look at local scales and the overall statistics of soil moisture deficits, the differences between the models become more apparent. Figure 10.5 shows the statistics of the duration and frequency of soil moisture deficits on a grid cell basis for the four models. As before, deficits are defined as soil moisture below the 20th percentile lasting for one or more months and are referred to as a run (Yevjevich, 1972). This is a somewhat loose definition of drought, but it reveals the differences in the timescales of soil moisture variation between the models. The SAC model has the highest number of deficits of any duration; this in part is driven by the higher frequency of short-term runs (1–3 months) and lower frequency of long-term runs (>12 months), and is reflected by the lower mean duration. In contrast, the Mosaic model has the lowest total number of runs, lowest number of short-term runs, and highest number of long-term runs. The VIC and Noah models have similar statistics, although the Noah tends to have longer duration runs and a sharper delineation between regions, with high short-term frequencies in the east and northwest and high long-term (>12 months) frequencies in the west. These differences are related to the variability of soil moisture in each of the models, which can vary considerably (Schaake et al., 2004) across models and climate regimes because of differences in model parameterizations of soil water movement and the relationships to soil water drainage and infiltration. Although the models agree well at large scales, there are significant differences at the shorter time steps and in the local-scale spatial patterns, which affect how the depicted drought develops, persists, and recovers, with ramifications for regional- and local-scale monitoring. This highlights the value of an MME for reducing the impacts of model-specific parameterizations.

10.4 NLDAS DROUGHT APPLICATIONS AND OPERATIONAL ASPECTS

One rationale for running NLDAS-2 is to support operational drought monitoring and seasonal drought forecasting. To this end, the NLDAS is now producing real-time information and future predictions of the hydrologic cycle across the United States, including drought monitoring products and seasonal drought forecasts. The NLDAS-2 drought monitor provides daily updates at 1–2 days behind real time, and the seasonal forecast system makes predictions every

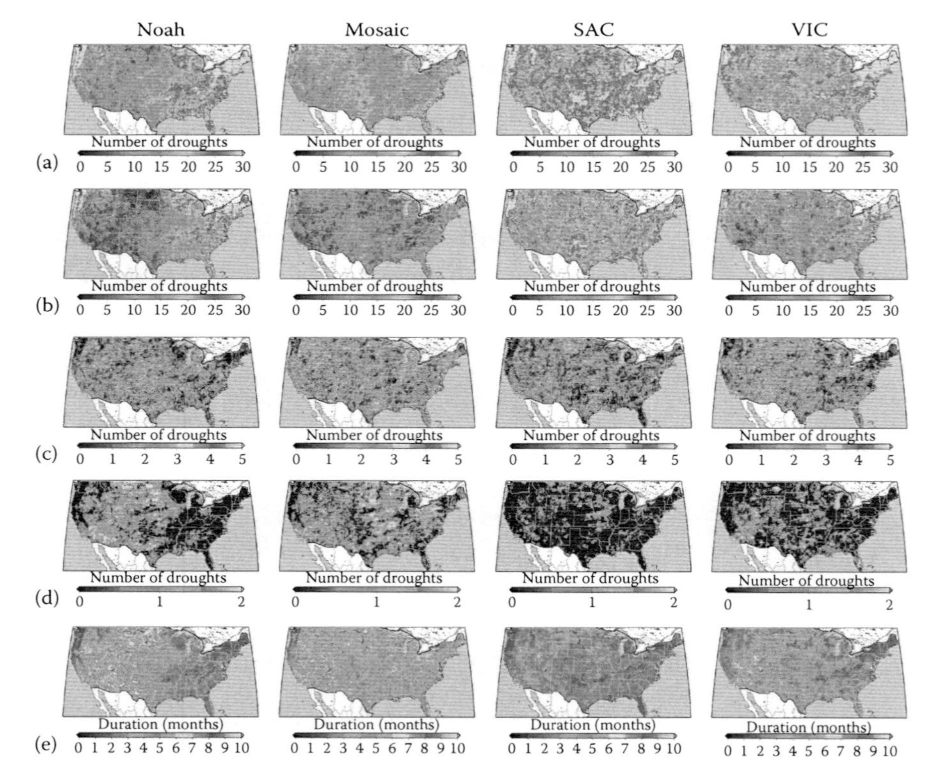

FIGURE 10.5 (See color insert.) Statistics of drought duration and frequency for the four LSMs for 1979–2008 calculated from monthly soil moisture percentiles. A drought is defined at each grid cell when the soil moisture percentile drops below 20%. (a) Total number of droughts, (b) number of short-term (1–3 month duration) droughts, (c) number of medium-term (7–12 month duration) droughts, (d) number of long-term (>12 month duration) droughts, and (e) the mean drought duration.

month out to 9 months into the future. These products are provided to the community at http://www.emc.ncep.noaa.gov/mmb/nldas/. The near-real-time forcing data and output from the models are available in hourly time step in binary compressed GRIB2 format files and can be accessed from the NCEP EMC via ftp (ftp://hydro1.sci.gsfc.nasa.gov/data/s4pa/NLDAS) and from the GES DISC GrADS Data Server (http://hydro1.gsfc.nasa.gov/dods/).

10.4.1 NLDAS Real-Time Drought Monitoring

The NLDAS experimental drought monitor is based on near-real-time output of soil moisture and other hydrologic variables from the four LSMs, thus providing an MME estimate of current drought conditions across the United States. The anomalies and percentiles are based on a 28 year climatology (1980–2007). Two separate climatology files are used, one for the calculation of anomalies and the other for the calculation of percentiles. Anomalies are calculated by comparing the current soil moisture values to mean values for the same time of year over each grid cell. Percentiles are based on a 5 day moving window of soil moisture values. This acts to smooth out the soil moisture record and removes any high frequency variations (or noise) in the data. Weekly analyses for each grid cell are computed by comparing the past 7 days to the corresponding period in the percentile climatology. Taking day 1 of the week as an example, hourly soil moisture values from this day are averaged together to form a single daily value. This value is then ranked against the soil moisture values from each day of the 5 day window surrounding day 1 of the corresponding week in the percentile climatology. This same process is then repeated for days 2–7 of the week, with each day of the week contributing equally to the overall ranking value. Monthly (30 day) percentile analyses are computed in a similar fashion. Figure 10.6 shows an example of the real-time monitor for the MME from December 2010 for precipitation, evapotranspiration, runoff, streamflow, soil moisture, and snow water equivalent. Comparison of the anomalies shows drier than normal conditions in the Gulf Coast states, with particularly dry conditions in east Texas and Louisiana, yet there are striking differences between the different hydrologic components, highlighting the differences in how anomalies propagate through the system. Collectively, these differences would be expected, given that different components of the hydrologic system respond at varying temporal intervals to moisture deficits ranging from days to weeks for precipitation, ET, and soil moisture to weeks to months for streamflow. As a result, each of these NLDAS-2 products has the potential to provide useful information about specific hydrologic components that could enhance hydrological drought monitoring.

10.4.2 NLDAS Seasonal Forecasting

The experimental seasonal hydrologic forecast systems of Luo and Wood (2008) and Wood and Lettenmaier (2006) have been combined and applied to the NLDAS suite of models and data products to form the NLDAS-2 seasonal hydrologic

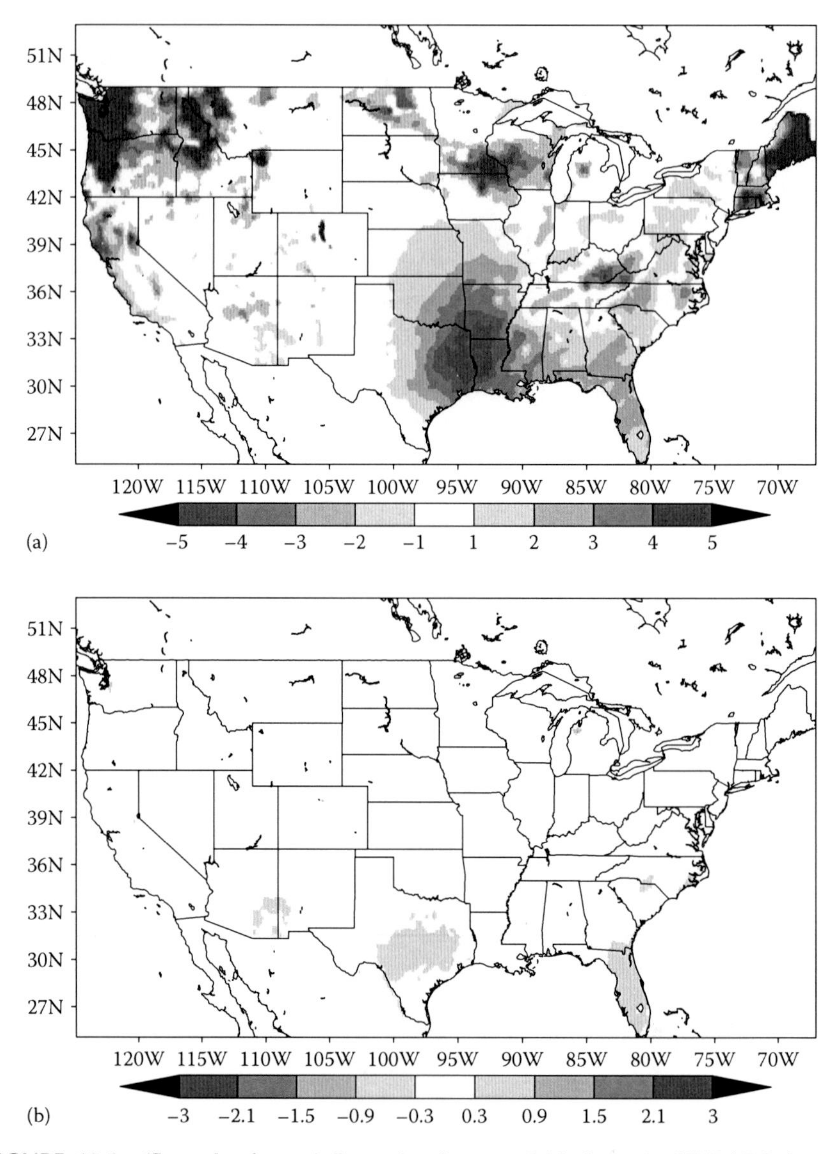

FIGURE 10.6 (See color insert.) Example of output fields from the NLDAS-2 drought monitor (http://www.emc.ncep.noaa.gov/mmb/nldas/drought/), showing anomaly data for the week ending on December 16, 2010, for (a) precipitation and multi-model averages of (b) evapotranspiration,

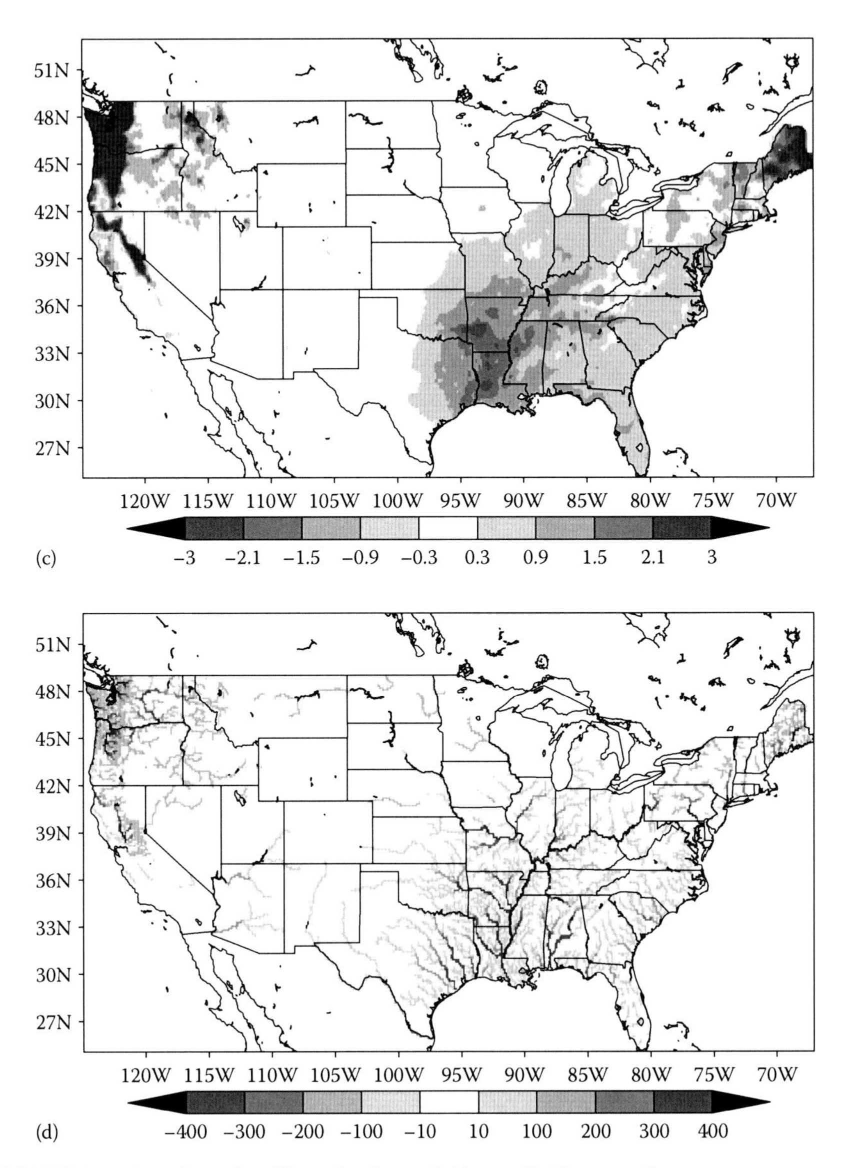

FIGURE 10.6 (continued) **(See color insert.)** (c) runoff, (d) streamflow,

(continued)

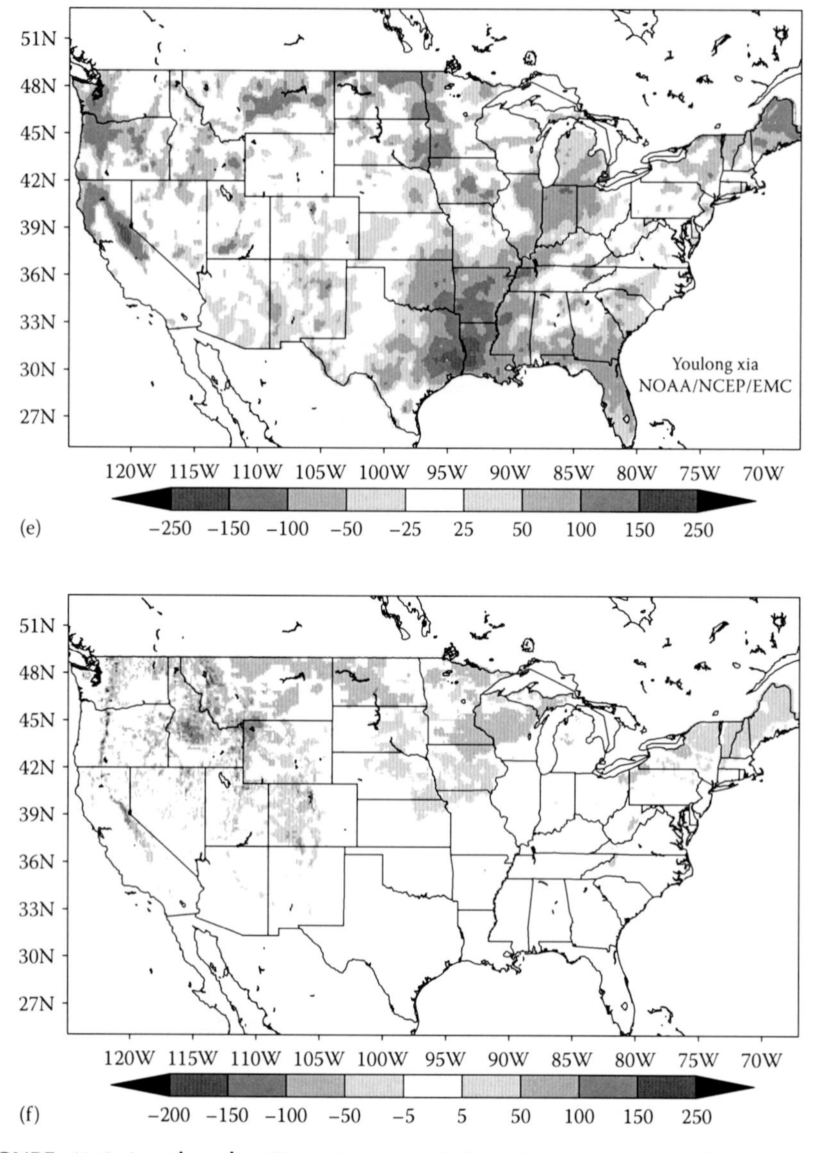

FIGURE 10.6 (continued) **(See color insert.)** (e) soil moisture, and (f) snow water equivalent.

forecast system. Currently, only the VIC model is used in the experimental forecast, with plans to incorporate the other models in the future. Forecasts are produced monthly at the beginning of each month using the initial condition that is closest to the first day of the month. Three forecast approaches are implemented that depend on the source of the climate forecast data from either statistical or dynamical forecasts. The climate data are then downscaled to 1/8th degree, bias corrected, and used to drive the VIC model to produce predictions of hydrology and drought conditions. The three forecast approaches are

1. The CFS-based forecast (Saha et al., 2006) uses seasonal forecasts from the NCEP CFS dynamical model. The set of CFS forecasts from the previous month are combined to form the model forecast distribution, which is then merged with historic observations using a Bayesian approach developed by Luo et al. (2003) and Luo and Wood (2008).
2. The CPC outlook–based forecast (http://www.cpc.ncep.noaa.gov/products/forecasts/) is based on expert merging of statistical and dynamical (including CFS) forecasts and is generally comparable in skill to the CFS-only forecasts. It uses the seasonal outlook of probability of exceedance (POE) released by the NCEP CPC during the previous month as the forecast distribution for each of the 102 U.S. climate divisions. These distributions are applied to all NLDAS grid boxes within the climate division.
3. The Extended Streamflow Prediction (ESP) (Day, 1985) method is based on resampling of the historic record, and therefore, its skill will generally not exceed that of the other two methods. The ESP method uses 20 randomly selected historical atmospheric forcing time series as possible realizations of future conditions. To be comparable and practical, both CFS and CPC outlook–based approaches also generate 20 ensemble members.

Figure 10.7 shows forecasts made in March 2010 using the three methods. Each column is the 6 month forecast from one forecast approach, and each row is the specific forecast for each individual month, showing the probability of drought persisting at lead times of 1–6 months, where drought is defined by monthly average soil moisture percentiles and a 20th percentile threshold. Given that the NLDAS-2 seasonal forecast system is an ensemble forecast system, the drought forecast includes a forecast anomaly, forecast percentile, and forecast probability analysis. The anomaly and percentile of the ensemble mean or median are used as a single-valued deterministic forecast. When interpreting the ensemble forecast in a probabilistic fashion, the probability of drought (when soil moisture is below 20th percentile) is derived from the ensemble. More details of the system and development history can be found on the Princeton Seasonal Hydrological Forecast System website (http://hydrology.princeton.edu/forecast).

The drought forecast system has been evaluated with respect to hindcasts of the 1988 drought and also tested in real time for the 2007 drought in the southeast United States. Figure 10.8 shows the drought forecast made on January 1, 2007, for the subsequent 3 months compared to the drought conditions estimated later from the observation-forced monitoring that represents our best estimate of true conditions. In this case, the system was able to forecast the development of drought conditions in California (with low uncertainty as represented by the ensemble spread) and in the southeast (with larger uncertainty in the magnitude and location of the drought center) up to 3 months in advance. In general, the system has demonstrated significant skill in the first 2 months of a forecast and shows marginal skill out to 4–6 months (Luo and Wood, 2008).

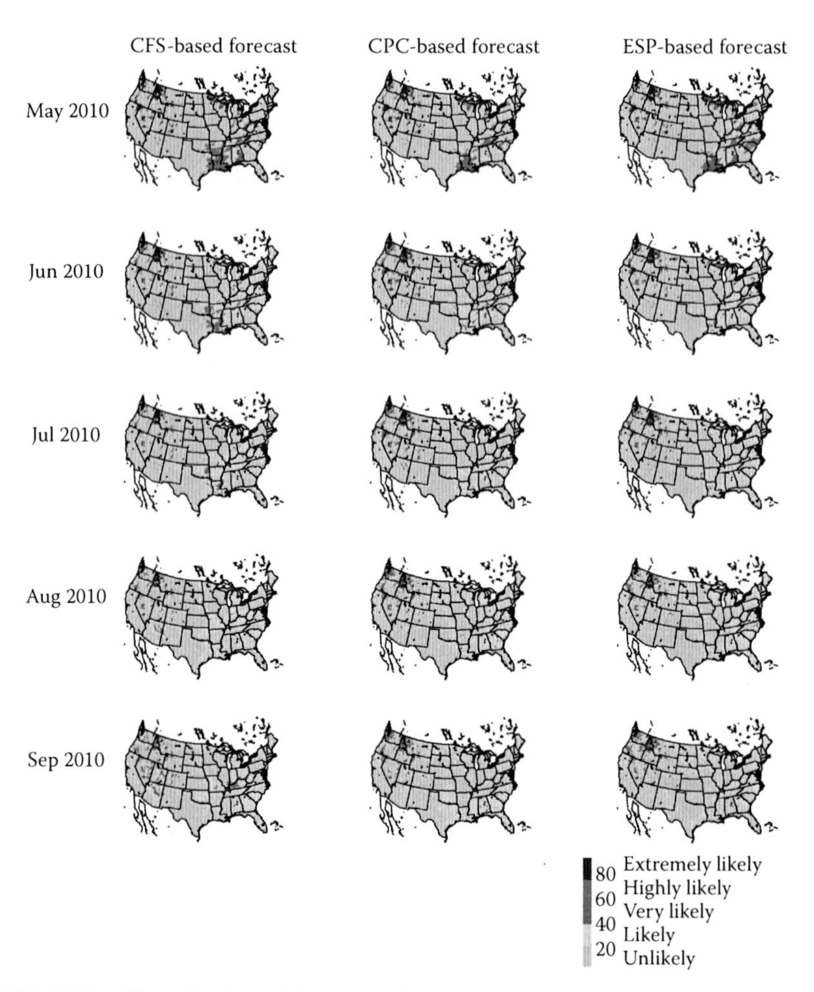

FIGURE 10.7 **(See color insert.)** Example of seasonal forecasts for May through September 2010, showing the likelihood of drought developing or persisting at lead times of 1–6 months. A drought is defined as soil moisture deficits below the 20th percentile, and the likelihood is based on ensemble forecast distributions. Forecasts are based on three methods: (1) CFS, (2) CPC official outlooks, and (3) ESP.

10.5 INTEGRATION OF NEW REMOTE SENSING DATA INTO NLDAS DROUGHT PRODUCTS

The NLDAS framework provides a mature platform for producing real-time fields of hydrologic variables in support of drought monitoring and as initial conditions for seasonal drought forecasting. These products have been evaluated through a series of studies that began with NLDAS-1 (e.g., Sheffield et al., 2003; Lohmann et al., 2004; Schaake et al., 2004) through more recent studies within NLDAS-2 (e.g., Xia et al., 2012) as presented in Section 10.3, which show that consistent depictions of large-scale historic drought events were characterized within this system. Nevertheless, a number

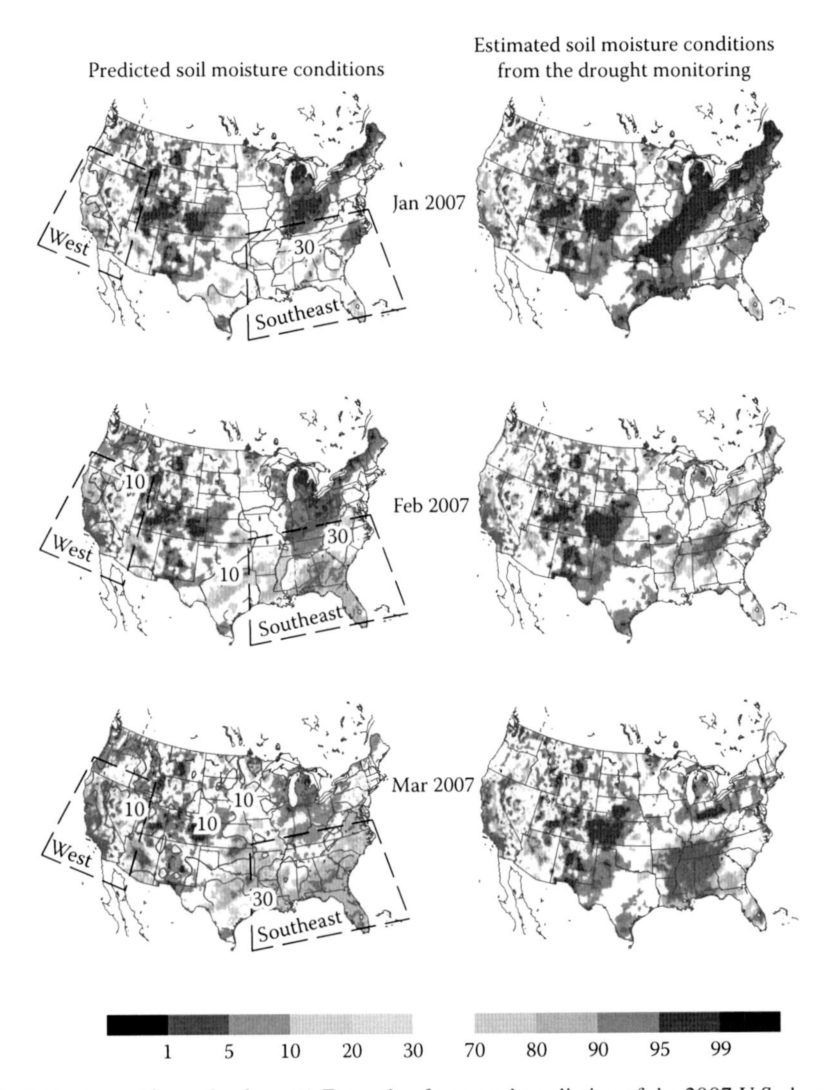

FIGURE 10.8 (See color insert.) Example of seasonal prediction of the 2007 U.S. drought (figure reproduced from Luo L. and E.F. Wood, *Geophys. Res. Lett.*, 34, L22702, 2007). Predictions of soil moisture percentiles (%) (left column) were made starting on January 1, 2007, using downscaled and bias-corrected CFS seasonal climate forecasts to drive the VIC model, and are compared to estimated soil moisture from the real-time drought monitoring (right column). Left column shows the mean of the most likely ensemble set (shaded) and their spread (contour). The boxes indicate regions where drought was most severe during early 2007.

of challenges remain, and improvements can be made, including understanding the differences in how each model represents the dynamics of drought development (as motivated by the large differences in drought statistics shown in Section 10.3) and improving the accuracy of the monitoring, especially in regions with few meteorological ground stations. The use of satellite-based remote sensing data can greatly benefit monitoring over areas with a sparse gauge network and at high elevations where there is high spatial variability in meteorological conditions. Remotely sensed observations provide exceptional spatial coverage at a relatively high temporal sampling interval over large areas where quality in situ data are limited. Currently, NLDAS-2 uses GOES-based downward solar radiation as a forcing and AVHRR data to parameterize the spatial and seasonal variation in vegetation. In this section, we explore the potential for integrating new sources of remote sensing data, specifically using remotely sensed data to estimate soil moisture, groundwater, and precipitation, which can be used to enhance NLDAS drought monitoring and prediction across the United States.

10.5.1 Microwave Soil Moisture Retrievals

Much progress has been made in recent years in retrieving terrestrial water cycle variables from space (Tang et al., 2009), and it is now possible to monitor all components, albeit with uncertainty and nonclosure of the water budget (Sheffield et al., 2009b). Nevertheless, it is possible to use these products to detect changes in moisture availability and the presence of drought. For soil moisture, long-term (decadal) products that merge information across satellites and sensors are being produced (e.g., Owe et al., 2008) and real-time products are available (e.g., Njoku et al., 2003). Remotely sensed soil moisture can be used in a number of different ways to improve drought monitoring: as a direct complement to in situ observations and modeled data, as well as through assimilation into LSMs. This can help in regions where gauges are sparse or where radar incorrectly identifies precipitation because of evaporation or advection before it hits the ground (McCabe et al., 2008). However, several challenges in using remotely sensed soil moisture data limit their use and dictate how they should be employed. For example, microwave soil moisture retrievals directly sample soil moisture conditions only in the top few cm of the soil profile and can be obtained only at relatively coarse spatial resolution (25–40 km) under relatively sparse vegetation cover.

Soil moisture can be retrieved at large scale (but coarse resolution) using satellite-borne passive and active microwave sensors. Various emission sources combine to provide the microwave brightness temperature (BT) that the satellite observes. Radiation is received from the atmosphere, vegetation, and the top layer of the soil, which is dependent on the moisture content via the sensitivity of the soil emissivity. Although BT is sensitive to soil moisture, these other sources of emissions must be taken into account. These sources can be modeled using a radiative transfer model, and a soil moisture value can be inferred with a 1 K change in BT roughly equivalent to a 2% change in soil moisture, depending on the microwave frequency. Several satellite-based microwave sensors have been used for retrieving soil moisture, including the Tropical Rainfall Measurement Mission (TRMM) Microwave Imager (TMI) on the NASA TRMM satellite, which was launched in 1997, and the Advanced Microwave Scanning Radiometer on the Earth Observing System (AMSR-E) aboard the NASA Aqua satellite that was launched in 2002. The biggest drawback of these current active radiometers is the emission depth of the microwave signal, which is dependent on the wavelength and is generally restricted to the top centimeter of soil. For drought applications, the primary interest is the integrated soil moisture over a greater depth (ideally the root zone) than the top centimeter. Soil moisture varies greatly vertically, and as a result, soil moisture conditions at a shallow soil depth may bear little resemblance to conditions in the total soil column. It is challenging to use soil moisture information estimated from these satellite-based systems to identify drought and to compare with data from (or when assimilating into) LSMs, which typically have top soil layers on the order of 5–10 cm or more.

In densely vegetated regions, the microwave signal from the underlying soil is attenuated, and the vegetation itself emits a signal, complicating the estimation of soil moisture from the satellite measurements. This generally results in retrievals being restricted to sparsely vegetated regions characterized by low vegetation biomass and water content. Current passive microwave sensors are generally in the C-band (4–8 GHz) (AMSR-E at 6.9 GHz) or X-band (8–12 GHz) range (TMI and AMSR-E at 10.7 GHz), although L-band (1–2 GHz) is much better in terms of lower attenuation through vegetation and a deeper effective soil emission depth. Figure 10.9 shows the regions of microwave-based soil moisture retrievability across the United States based on estimated vegetation water content for X- and L-band microwave radiometers. The upper limit for X-band

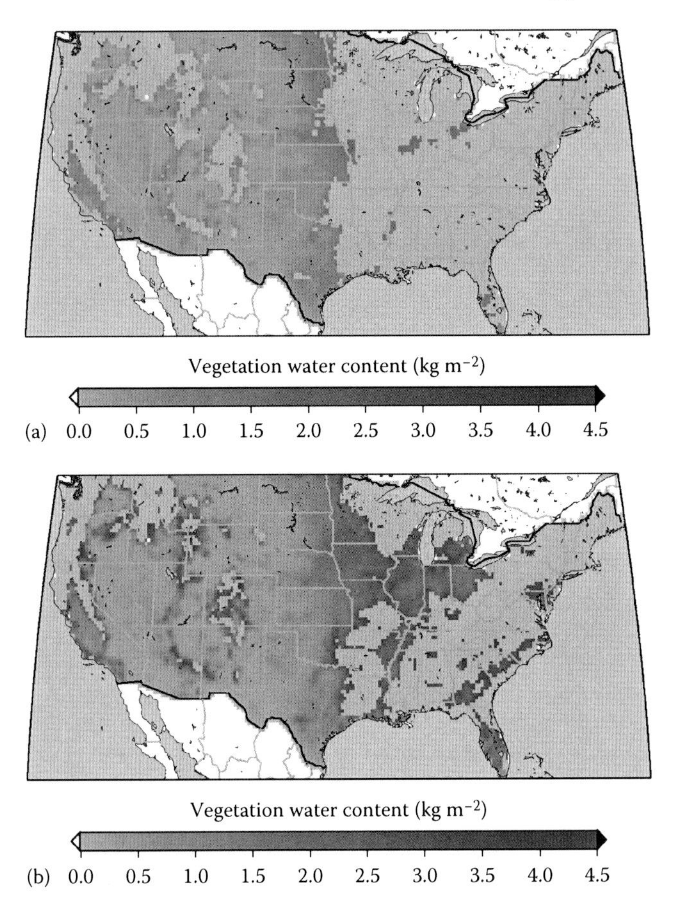

FIGURE 10.9 **(See color insert.)** Zones of applicability for microwave remote sensing retrievals of soil moisture based on penetration through vegetation for frequencies in (a) X-band (10.7 GHz) and (b) L-band (1.4 GHz). Vegetation is characterized by its vegetation water content (kg m^{-2}). Gray shading indicates areas where retrievals of soil moisture are not feasible.

(10.7 GHz) retrievals is 1.5–2 kg/m^2 (Njoku et al., 2003; Narayan et al., 2004) and for L-band (1.4 GHz) retrievals is 4–5 kg/m^2 (Kerr, 2007). As a reference, short grass is less than 1 kg/m^2; corn and soybeans can be up to 6 and 1–3 kg/m^2, respectively. There is a clear delineation in the middle of the United States in the zones of feasible soil moisture retrievals at 10.7 GHz (Figure 10.9a). For an L-band instrument, the extent of feasible zones expands as signals can penetrate higher vegetation water contents (Figure 10.9b). Despite this expanded geographic coverage, retrievals are still not possible in many areas, particularly in the densely vegetated eastern United States.

The spatial resolution of satellite-based microwave data, generally on the order of 40 km, is also a limiting factor even though data products are often provided at a higher 25 km spatial resolution because of oversampling. Because of the high spatial variability of soil moisture at a local scale, interpretation of 25–40 km data can be problematic. Landscapes can comprise many different land covers, often in complex spatial patterns, which results in the integration of emissions from multiple land cover surfaces at the coarse pixel level that may be contaminated by signals from water bodies and/or dense vegetation, resulting in unrepresentative soil moisture estimates. The presence of water bodies or dense vegetation will tend to give overestimates of soil moisture. This has obvious consequences when comparing to a point-based observation from a soil moisture probe or trying to infer soil moisture at subpixel scales. Other factors that hinder the retrievals include the presence of active precipitation, snow, and frozen soils. Current research is looking to combine products from multiple microwave satellite sensors to improve spatial and temporal coverage and resolution, including combining passive and active products (Das et al., 2011; Liu et al., 2011). However, these are generally still restricted to higher frequencies in the X-band because the C-band suffers from radio frequency interference (RFI) and cannot retrieve soil moisture over dense vegetation. The recently launched European Space Agency (ESA) Soil Moisture Ocean Salinity (SMOS) mission and the future planned NASA Soil Moisture Active and Passive (SMAP) mission will carry L-band instruments, which will increase temporal sampling and improve the effective emission depth to about five times deeper into the soil than the current TMI and AMSR-E X-band radiometers, which should provide better soil moisture estimates that are representative of deeper soil moisture conditions (Entekhabi et al., 2008).

Although there are challenges in determining where soil moisture can be retrieved from satellite-based microwave observations and how to interpret the values, the potential exists to use these data within a drought monitoring framework such as NLDAS-2. Figure 10.10 compares soil moisture retrievals from AMSR-E over the United States from 2002 to 2008 with NLDAS-2 model output. The retrieval is derived from AMSR-E BTs using the Princeton Land Surface Microwave Emission Model (LSMEM; Drusch et al., 2004). The model uses surface properties such as vegetation water content and soil texture and also the land surface states of temperature and soil moisture to estimate the top of atmosphere (TOA) BT that the satellite sensor would record. To estimate soil moisture from an actual satellite-observed BT, the model is run in "forward" mode by iterating over soil moisture values until the modeled BT matches the satellite observation. Figure 10.10a shows the range in soil moisture for all months over the full 7-year period and indicates that the largest sensitivity of the soil moisture retrievals is in the central United States. This sensitivity

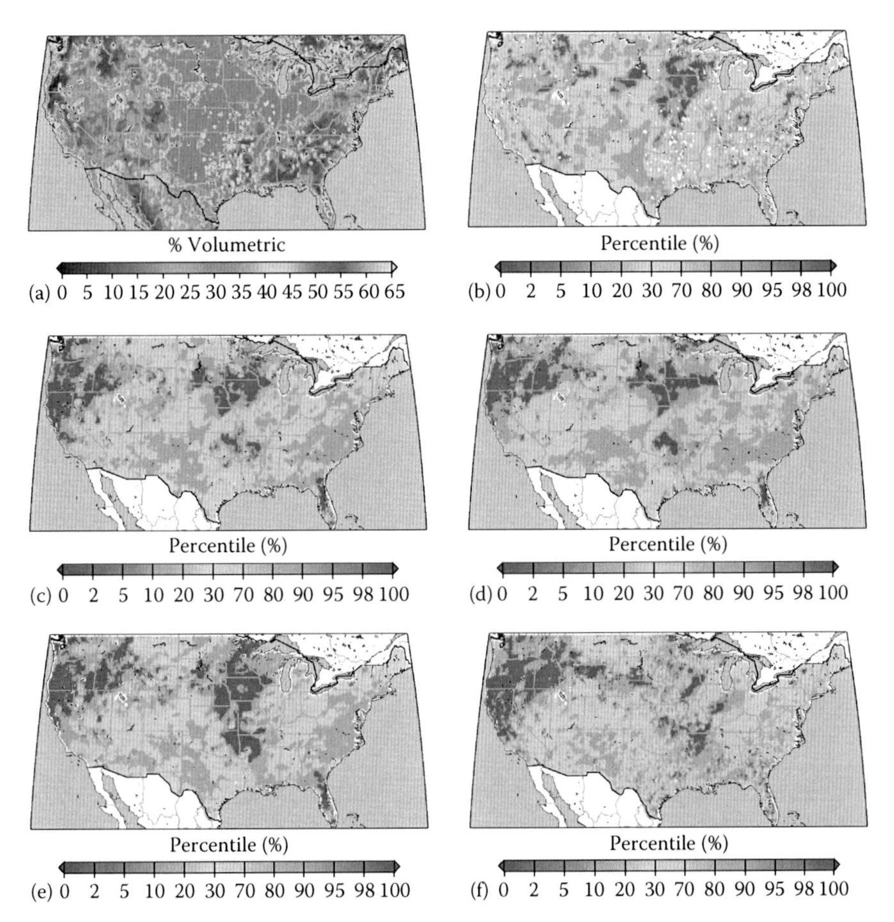

FIGURE 10.10 (See color insert.) (a) Dynamic range (% vol.) of AMSR-E daily soil moisture and (b–f) examples of monthly soil moisture percentiles for October 2007 for (b) AMSR-E, (c) Noah, (d) Mosaic, (e) SAC, and (f) VIC.

is derived from a combination of the sensitivity of the AMSR-E sensor, the LSMEM retrieval model, and the climate forcing variability. The rest of Figure 10.10 compares the AMSR-E-based soil moisture data for October 2007 with data from the four NLDAS-2 models taken from their top soil layer output. Since the AMSR-E retrievals and the modeled data could not be compared directly because they represent different soil layer thicknesses, they were normalized through the conversion to monthly percentiles based on the data from 2002 to 2008. To increase the sample size for calculating the percentiles, data from eight neighboring pixels are included, thus trading space for time. In this example, the AMSR-E data show remarkable similarity to the wet and dry regions depicted by the models. The largest differences are in regions of denser vegetation in the east, where the retrievals are expected to be less accurate, and the mountainous areas of the southwest, where terrain will affect the retrievals.

Time series (Figure 10.11) and correlation maps (Figure 10.12) between the retrievals and the models indicate consistency in the southwest United States and

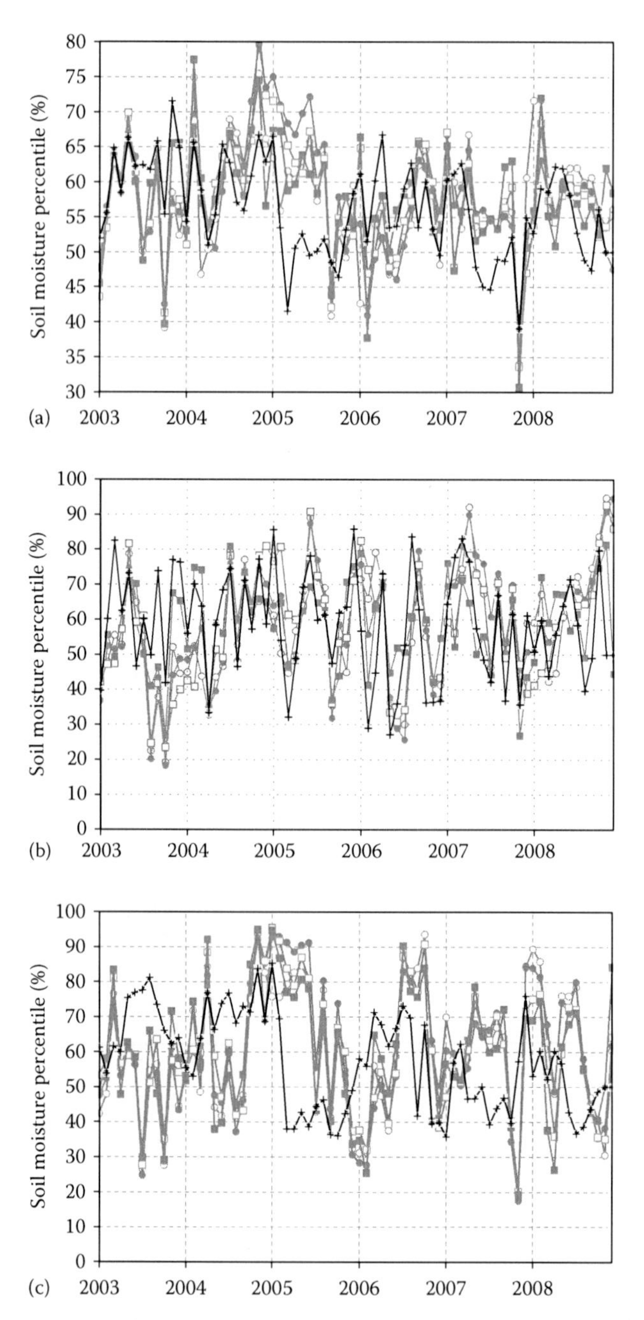

FIGURE 10.11 Time series of monthly soil moisture percentiles from AMSR-E and three NLDAS-2 LSMs (Noah, Mosaic, and SAC) for (a) the conterminous United States, (b) Northern Plains (40°–49°N, 95°–105°W), (c) Four Corners region (33°–40°N, 105°–115°W), and

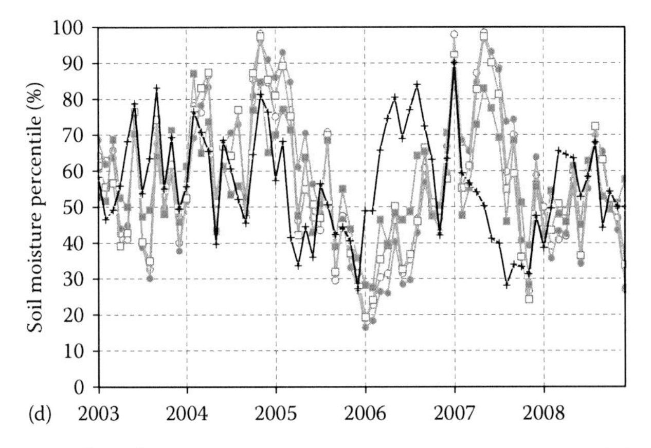

FIGURE 10.11 (continued) (d) Southern Plains/Texas (25°–37°N, 95°–105°W).

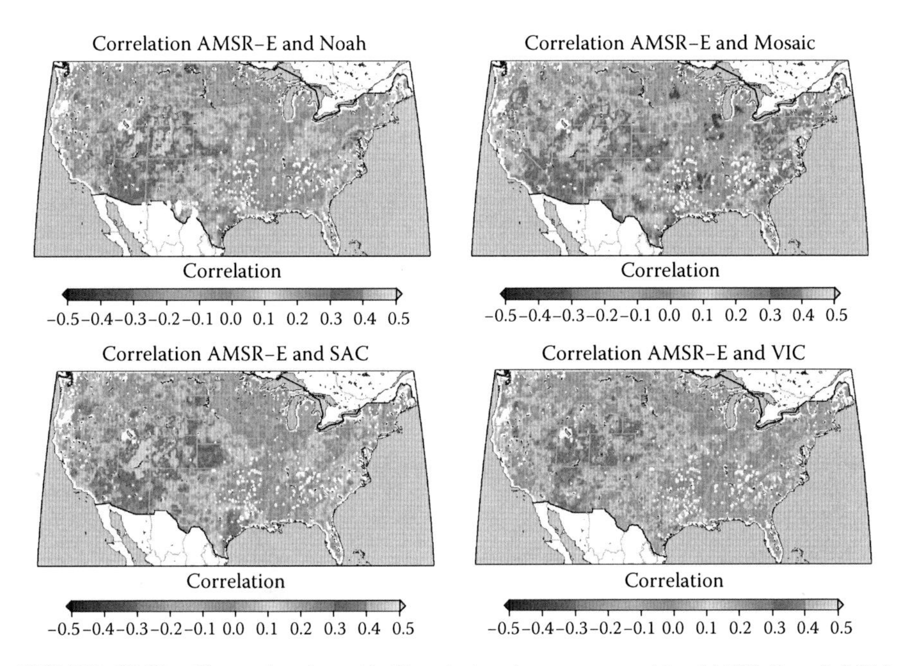

FIGURE 10.12 (See color insert.) Correlation between monthly AMSR-E and LSM monthly soil moisture percentiles for 2002–2008.

central Great Plains in line with the regions of retrievability presented in Figure 10.9. The correlation with Mosaic in the east is particularly strong despite the dense vegetation cover, which may be related to a decreasing trend in the AMSR-E and Mosaic data that may be an artifact of the short time period (2002–2008) of the comparison, rather than any physical connection. The Mosaic model also possesses a faster hydrologic cycle relative to the other models, instilling a more rapid connection between soil moisture and surface evaporation. Figure 10.11 shows reasonable agreement among the time series at regional and even national scales, although there are some inconsistencies. For example, there are considerable differences in 2005–2006 for the United States and the four corners regions and in 2006–2007 in the southern Plains–Texas region. For the latter, the differences may be related to the severe drought in 2006 and subsequent flooding conditions in 2007. In periods of extreme dry or wet conditions, the differences in the ASMR-E and model soil moisture estimates might be magnified. For example, the AMSR-E retrievals represent the top 1 cm of the soil, which will tend to become wet and dry more quickly than the underlying deeper (10 cm) layer represented by the models. In a wet period, standing water may contaminate the AMSR-E retrievals. Overall, however, there appears to be useful information in the retrieved values that reflects variation in wet and dry spells and may help improve model-based drought monitoring, especially in regions of sparse observational networks.

10.5.2 Synergy with Other Remote Sensing Signals of Drought

This book describes several newly developed remotely sensed drought products, many of which can be integrated into the NLDAS-2 either as improved inputs or assimilated signals of surface and subsurface moisture. In general, each of the drought products represents a different aspect of the hydrologic cycle or state of vegetation. In some cases (e.g., soil moisture and total water storage), they represent similar or overlapping quantities but provide complementary information that draws from the strengths of the individual sensor, retrieval algorithm, or characteristics of the retrieval. When combined, these mostly independent products can provide a more holistic view of drought and the hydrologic cycle in general, as well as allowing quantification of dependencies and feedbacks between components such as tracking the propagation of drought through the hydrologic and ecological systems.

A challenge for the scientific and user community is to determine the consistency among these different products and how they may be combined in useful ways to improve drought assessment. A flexible modeling system such as the NLDAS has the potential to provide the framework for merging these various products into a consistent, continuous in space and time, and robust picture of drought, by providing the hooks to tie the individual pieces together. From the perspective of the modeling, this can also be viewed as the use of remote sensing products to correct the errors in the models and their input data. Much work has already been carried out to merge remote sensing products with terrestrial modeling, some of which is discussed in this book (Chapters 7 and 11) in the context of drought. The rest of this section describes the potential to leverage from these activities to merge remote sensing and modeling within the NLDAS.

High-quality and high-resolution precipitation data are crucial for depicting the development and recovery of drought. Better estimates of precipitation in the NLDAS are likely to lead to better representation of land surface hydrology and drought, but this is dependent on the region and application. Over the United States, the density of ground observations of precipitation and other meteorological data is relatively high as compared to locations such as central Africa. However, even in the United States, gauge density is often not optimal for representing the spatial and temporal variability of precipitation and soil moisture. The problem of sparse gauge coverage is somewhat overcome in the NLDAS through the merging of gauge data with ground radar data. Nevertheless, radar coverage is not complete spatially and is prone to error for a number of reasons, including bright-band, elevation, and range effects that lead to a complex nonlinear relationship between radar reflectivity and rainfall rate at the surface (Krajewski and Smith, 2002). There is, therefore, potential for utilizing remote sensing–based estimates of precipitation to provide high-resolution complementary information to existing products used as input to the NLDAS (Chapters 12–15). In some regions, this is likely the best and sometimes the only source of precipitation information, and its potential has been demonstrated in several regional applications such as FEWS NET (Verdin et al., 2005) and the Princeton African Drought Monitor (Sheffield et al., 2008c).

In this chapter and Chapter 9, the potential of remotely sensed soil moisture as a drought assessment tool has been presented. There is further opportunity to exploit and extend this capability through assimilation into the NLDAS or similar land surface modeling system to provide a more complete view of drought and correct for model structural and input errors. Assimilation of remotely sensed soil moisture information from passive/active microwave into LSMs has been demonstrated previously (Houser et al., 1998; Crow and Wood, 2003; Reichle and Koster, 2005; Scipal et al., 2008) and can provide improvement in skill for assessment of both droughts and floods (Bolten et al., 2010; Brocca et al., 2010). Complementary information on soil moisture may be obtained from thermal infrared (TIR) remote sensing, which indirectly estimates soil moisture from the thermal response of the vegetation canopy to soil water stress. TIR soil moisture retrievals are described more fully in Chapter 7, which also elucidates their potential for drought monitoring. Microwave (passive and active) and TIR approaches have their strengths and limitations but together provide complementary information. Many of the issues described earlier regarding microwave-based soil moisture can be addressed with TIR approaches (including sampling of the root zone, skill in regions of denser vegetation, and higher spatial resolution) (Hain et al., 2011). Conversely, the limitations of TIR (such as lower temporal sampling due to cloud cover) can be partly addressed by the microwave approach. TIR retrievals have been demonstrated as useful for assimilation into LSMs (e.g., Crow et al., 2008). Further, the complementary information in both TIR and microwave retrievals has the potential to be mined in a joint data assimilation framework to provide improved estimates of soil moisture relative to assimilation of either in isolation (Hain, 2010; Li et al., 2010). Remotely sensed soil moisture can also provide complementary information to remotely sensed precipitation retrievals. For example, these retrievals represent the on-the-ground signature of actual rainfall as compared to remote sensing estimates that represent aboveground precipitation rates that are subject to advection before reaching the ground (McCabe et al., 2008).

The NLDAS LSMs do not explicitly model groundwater, although their parameterizations of baseflow represent the contribution of deeper soil layers to streamflow. The models are therefore subject to biases in their depiction of drought dynamics and especially in the potentially mediating effect of groundwater on drought propagation. Chapter 11 demonstrates how GRACE measurements of total water storage change (groundwater, soil moisture, surface water, snow, lakes, streams, and wetlands) provide useful information on total water storage dynamics and particularly groundwater. Chapter 11 also demonstrates how GRACE data, despite their coarse resolution, could be ingested into an LSM that possesses a groundwater component to help improve the depiction of longer-scale dynamics and may be crucial for improving drought assessment in groundwater-dominated regions. Similar to soil moisture retrievals, the blending of coarse-resolution GRACE data with higher-resolution modeling bridges the gap between observational sampling issues and the need for continuous and consistent drought information.

Further avenues for merging remote sensing information into the NLDAS to provide a consistent and more robust view of drought exist, such as with snow and vegetation products. For snow, this is particularly important in snow-dominated regions such as the western United States where water resources and agriculture are highly dependent on winter snow accumulation and timing of spring melt. Where local information on snowfall and accumulation is limited to gauges in valley bottoms or, at best, sparse high-elevation networks, remote sensing is an underexploited resource that can address some of these issues, as shown in Chapter 15. For vegetation, the current NLDAS models use a seasonal representation of vegetation phenology (in terms of LAI and other parameters) that is fixed from year to year. As well as being inconsistent with remote sensing–based estimates of vegetation stress, this also has implications for the simulation of soil moisture and hydrological drought in the models because of the vegetation controls on interception and transpiration. A simple approach to improving this is to incorporate remotely sensed vegetation information, such as NDVI, into the model inputs.

10.6 SUMMARY

The NLDAS-2 provides a temporally and spatially consistent, quantitative depiction of drought history, current conditions, and future seasonal changes. The use of observation-forced, physically based models enables all aspects of hydrological drought to be assessed and multiple models allow for the estimation of uncertainties. Comparison across models shows encouraging consistency in the depiction of large-scale drought events, although the development of drought at more localized scales appears to differ considerably across models despite the commonality of meteorological forcings and underlying landscape parameters. Improvements can be made, particularly through the increased use of remote sensing data. For example, remotely sensed soil moisture has the potential to augment the system, either directly as an additional monitoring variable or indirectly via assimilation. Despite the coarse spatial resolution and limited utility over areas with high vegetation biomass density, microwave-based remote sensing of soil moisture is responsive to precipitation and can discern between wet or dry periods at monthly to seasonal time scales, which is useful for drought monitoring. Microwave soil moisture retrievals may actually provide a better indication of wet areas than radar or in regions with sparse gauge networks, and can be used to augment NLDAS-based model estimates using an assimilation framework to merge the NLDAS and remote sensing soil moisture products. There is also potential to expand the system globally, particularly for regions such as Africa with sparse ground observations (Sheffield et al., 2008c), where there is heavy reliance on remote sensing to provide meteorological data for the forcings and hydrologic variables used for validation and assimilation. From a broader perspective, an assimilation approach within an NLDAS-type system is likely the most promising way forward for exploiting the breadth of complementary remote sensing products described in this book and providing a more consistent picture of drought.

ACKNOWLEDGMENTS

The authors acknowledge the support of the NOAA/OGP/CPPA, NOAA grant NA08OAR4310579, and the NLDAS team: Mike Ek, Youlong Xia, and Ken Mitchell (NCEP/EMC); Eric F. Wood and Justin Sheffield (Princeton University); Lifeng Luo (Michigan State University); Brian C. Cosgrove (NWS/OHD); D. Mocko and C. Alonge (NASA/Goddard Space Flight Center; GSFC); Dennis P. Lettenmaier and Ben Livneh (University of Washington); Pedro Restrepo, John Schaake, and Victor Koren (NWS/OHD); Kingste Mo and Huug Van den Dool (NWS/CPC); and Yun Fan (NWS/Office of Science and Technology; OST).

REFERENCES

Andreadis, K.M., E.A. Clark, A.W. Wood, A.F. Hamlet, and D.P. Lettenmaier. 2005. Twentieth-century drought in the conterminous United States. *Journal of Hydrometeorology* 6:985–1001.

Betancourt, J. 2003. The current drought (1999–2003) in historical perspective. *Southwest Drought Summit*, Northern Arizona University, Flagstaff, May 12–13. http://www.mpcer.nau.edu/megadrought/Betancourt%2520Abstract.pdf (accessed on May 2010).

Betts, A., F. Chen, K. Mitchell, and Z. Janjic. 1997. Assessment of the land surface and boundary layer models in two operational versions of the NCEP Eta model using FIFE data. *Monthly Weather Review* 125:2896–2916.

Bolten, J.D., W.T. Crow, X. Zhan, T.J. Jackson, and C.A. Reynolds. 2010. Evaluating the utility of remotely sensed soil moisture retrievals for operational agricultural drought monitoring. *IEEE Journal of Selected Topics in Applied Earth Observations and Remote Sensing* 3(1):57.

Brocca, L., F. Melone, T. Moramarco, W. Wagner, V. Naeimi, Z. Bartalis, and S. Hasenauer. 2010. Improving runoff prediction through the assimilation of the ASCAT soil moisture product. *Hydrology and Earth System Science* 14:1881–1893.

Burnash, R.J.C., R.L. Ferral, and R.A. McGuire. 1973. *A Generalized Stream Flow Simulation System: Conceptual Models for Digital Computers*. Joint Federal State River Forecast Center, Sacramento, California.

Chen, F., Z. Janjic, and K. Mitchell. 1997. Impact of atmospheric surface-layer parameterizations in the new land-surface scheme of the NCEP mesoscale Eta model. *Boundary-Layer Meteorology* 85:391–421.

Cosgrove, B.A., D. Lohmann, K.E. Mitchell, P.R. Houser, E.F. Wood, J.C. Schaake, A. Robock, C. Marshall, J. Sheffield, Q. Duan, L. Luo, R.W. Higgins, R.T. Pinker, J.D. Tarpley, and J. Meng. 2003. Real-time and retrospective forcing in the North American Land Data Assimilation System (NLDAS) project. *Journal of Geophysical Research* 108(D22):8842, doi:10.1029/2002JD003118.

Crow, W.T. and E.F. Wood. 2003. The assimilation of remotely sensed soil brightness temperature imagery into a land surface model using ensemble Kalman filtering: A case study based on ESTAR measurements during SGP97. *Advances in Water Resources* 26:137–149.

Crow, W.T., W.P. Kustas, and J.H. Prueger. 2008. Monitoring root-zone soil moisture through the assimilation of a thermal remote sensing-based soil moisture proxy into a water balance model. *Remote Sensing of Environment* 112(4):1268–1281.

Daly, C., R.P. Neilson, and D.L. Phillips. 1994. A statistical-topographic model for mapping climatological precipitation over mountainous terrain. *Journal of Applied Meteorology* 33:140–158.

Das, N., D. Entekhabi, and E. Njoku. 2011. An algorithm for merging SMAP radiometer and radar data for high resolution soil moisture retrieval. *IEEE Transactions on Geoscience and Remote Sensing* 49(5):1504–1512.

Day, G.N. 1985. Extended streamflow forecasting using NWSRFS. *Journal of Water Resources Planning and Management* 111:157–170.

Drusch, M., E.F. Wood, H. Gao, and A. Thiele. 2004. Soil moisture retrieval during the southern great plains hydrology experiment 1999: A comparison between experimental remote sensing data and operational products. *Water Resources Research* 40(2):21. doi:10.1029/2003WR002441.

Ek, M.B., K.E. Mitchell, Y. Lin, E. Rodgers, P. Grunman, V. Koren, G. Gayno, and J.D. Tarpley. 2003. Implementation of Noah land surface model advances in the National Centers for Environmental Prediction operational mesoscale Eta model. *Journal of Geophysical Research* 108:8851, doi:10.1029/2002JD003296.

Entekhabi, D., T.J. Jackson, E. Njoku, P. O'Neill, and J. Entin. 2008. Soil moisture active/passive (SMAP) mission concept. *Proceedings of SPIE* 7085:70850H, doi:10.1117/12.795910.

Gutman, G. and A. Ignatov. 1998. The derivation of the green vegetation fraction from NOAA/AVHRR data for use in numerical weather prediction models. *International Journal of Remote Sensing* 19:1533–1543.

Hain, C.R. 2010. Developing a dual assimilation approach for thermal infrared and passive microwave soil moisture retrievals. PhD. dissertation, Department of Atmospheric Science, University of Alabama.

Hain, C.R., W.T. Crow, J.R. Mecikalski, M.C. Anderson, and T. Holmes. 2011. An intercomparison of available soil moisture estimates from thermal-infrared and passive microwave remote sensing and land-surface modeling. *Journal of Geophysical Research – Atmospheres* (in press).

Hansen, M.C., R.S. DeFries, J.R.G. Townshend, and R. Sohlberg. 2000. Global land cover classification at 1 km spatial resolution using a classification tree approach. *International Journal of Remote* Sensing 21:1331–1364.

Houser, P.R., W.J. Shuttleworth, J.S. Famiglietti, H.V. Gupta, K.H. Syed, and D.C. Goodrich. 1998. Integration of soil moisture remote sensing and hydrologic modeling using data assimilation. *Water Resources Research* 34:3405–3420.

Joyce, R.J., J.E. Janowiak, P.A. Arkin, and P. Xie. 2004. CMORPH: A method that produces global precipitation estimates from passive microwave and infrared data at high spatial and temporal resolution. *Journal of Hydrometeorology* 5:487–503.

Kerr, Y.H. 2007. Soil moisture from space: Where are we? *Hydrogeology Journal* 15:117–120.

Koster, R. and M. Suarez. 1994. The components of the SVAT scheme and their effects on a GCM's hydrological cycle. *Advances in Water Resources* 17:61–78.

Koster, R. and M. Suarez. 1996. Energy and water balance calculations in the Mosaic LSM. *NASA Technical Memo* 104606 9:76.

Krajewski, W.F. and J.A. Smith. 2002. Radar hydrology: Rainfall estimation. *Advances in Water Resources* 25(8–12):1387–1394.

Kumar, S.V., C.D. Peters-Lidard, Y. Tian, J. Geiger, P.R. Houser, S. Olden, L. Lighty, J.L. Eastman, P. Dirmeyer, B. Doty, J. Adams, E. Wood and J. Sheffield. 2006. LIS—An interoperable framework for high resolution land surface modeling. *Environmental Modeling and Software* 21:1402–1415.

Lawford, R.G. 1992. Research implications of the 1988 Canadian Prairie Provinces drought. *Natural Hazards* 6:109–129.

Lawrimore, J., R.R. Heim, M. Svoboda, V. Swail, and P.J. Englehart. 2002. Beginning a new era of drought monitoring across North America. *Bulletin of the American Meteorological Society* 83:1191–1192.

Lawrimore, J. and S. Stephens. 2003. *Climate of 2002 Annual Review*. NOAA National Climatic Data Center, Asheville, North Carolina. http://lwf.ncdc.noaa.gov/oa/climate/research/2002/ann/drought-summary.html (accessed on June 2010).

Li, F., W.T. Crow, and W.P. Kustas. 2010. Towards the estimation of root-zone soil moisture via the simultaneous assimilation of thermal and microwave soil moisture retrievals. *Advances in Water Resources* 33(2):201–214.

Li, H., L. Luo, E.F. Wood, and J. Schaake. 2009. The role of initial conditions and forcing uncertainties in seasonal hydrologic forecasting. *Journal of Geophysical Research* 114:D04114, doi:10.1029/2008JD010969.

Liang, X., D.P. Lettenmaier, E.F. Wood, and S.J. Burges. 1994. A simple hydrologically based model of land surface water and energy fluxes for GCMs. *Journal of Geophysical Research* 99:14415–14428.

Liu, Y.Y., R.M. Parinussa, W.A. Dorigo, R.A.M. De Jeu, W. Wagner, A.I.J.M. van Dijk, M.F. McCabe, and J.P. Evans. 2011. Developing an improved soil moisture dataset by blending passive and active microwave satellite-based retrievals. *Hydrology and Earth System Sciences* 15:425–436.

Lohmann, D., K.E. Mitchell, P.R. Houser, E.F. Wood, J.C. Schaake, A. Robock, B.A. Cosgrove, J. Sheffield, Q. Duan, L. Luo, R.W. Higgins, R.T. Pinker, and J.D. Tarpley. 2004. Streamflow and water balance intercomparison of four land surface models in the North American Land Data Assimilation System (NLDAS). *Journal of Geophysical Research* 109:D07S91, doi:10.1029/2003JD003517.

Luo, L., A. Robock, K.E. Mitchell, P.R. Houser, E.F. Wood, J.C. Schaake, D. Lohmann, B. Cosgrove, F. Wen, J. Sheffield, Q. Duan, R.W. Higgins, R.T. Pinker, and J.D. Tarpley. 2003. Validation of the North American Land Data Assimilation System (NLDAS) retrospective forcing over the Southern Great Plains. *Journal of Geophysical Research* 108(D22):8843, doi:10.1029/2002JD003246.

Luo, L. and E.F. Wood. 2007. Monitoring and predicting the 2007 U.S. drought. *Geophysical Research Letters* 34:L22702, doi:10.1029/2007GL031673.

Luo, L. and E.F. Wood. 2008. Use of Bayesian merging techniques in a multi-model seasonal hydrologic ensemble prediction system for the Eastern U.S. *Journal of Hydrometeorology* 5:866–884.

McCabe, M.F., E.F. Wood, R. Wojcik, M. Pan, J. Sheffield, H. Gao, and H. Su. 2008. Hydrological consistency using multi-sensor remote sensing data for water and energy cycle studies. *Remote Sensing of Environment* 112(2):430–444.

Mesinger, F., G. DiMego, E. Kalnay, K. Mitchell, P. Shafran, W. Ebisuzaki, D. Jovic, J. Woollen, E. Rogers, E. Berbery, M. Ek, Y. Fan, R. Grumbine, W. Higgins, H. Li, Y. Lin, G. Manikin, D. Parrish, and W. Shi. 2006. North American regional reanalysis. *Bulletin of the American Meteorological Society* 87(3):343–360.

Mitchell, K.E., D. Lohmann, P.R. Houser, E.F. Wood, J.C. Schaake, A. Robock, B.A. Cosgrove, J. Sheffield, Q. Duan, L. Luo, R.W. Higgins, R.T. Pinker, J.D. Tarpley, D.P. Lettenmaier, C.H. Marshall, J.K. Entin, M. Pan, W. Shi, V. Koren, J. Meng, B.H. Ramsey, and A.A. Bailey. 2004. The multi-institution North American Land Data Assimilation System (NLDAS): Utilizing multiple GCIP products and partners in a continental distributed hydrological modeling system. *Journal of Geophysical Research* 109:D07S90, doi:10.1029/2003JD003832.

Myneni, R.B., R.R. Nemani, and S.W. Running. 1997. Estimation of global leaf area index and absorbed PAR using radiative transfer models. *IEEE Transactions on Geoscience and Remote Sensing* 35:1380–1393.

Narayan, U., V. Lakshmi, and E.G. Njoku. 2004. Retrieval of soil moisture from passive and active L/S band sensor during the soil moisture experiment in 2002 (SMEX02). *Remote Sensing of the Environment* 92(4):483–496.

NCDC (National Climate Data Center). 2009. Billion dollar climate and weather disasters, 1980–2009. NOAA National Climatic Data Center, Asheville, North Carolina. http://www.ncdc.noaa.gov/img/reports/billion/state2009.pdf (accessed on May 2010).

Njoku, E.G., T.J. Jackson, V. Lakshmi, T.K. Chan, and S.V. Nghiem. 2003. Soil moisture retrieval from AMSR-E. *IEEE Transactions on Geoscience Remote Sensing* 41(2):215–228.

Owe, M., R.A.M. de Jeu, and T.H.R. Holmes. 2008. Multi-sensor historical climatology of satellite derived global land surface soil moisture. *Journal of Geophysical Research* 113:F01002, doi:10.1029/2007JF000769.

Pan, M., J. Sheffield, E.F. Wood, K.E. Mitchell, P.R. Houser, J.C. Schaake, A. Robock, D. Lohmann, B. Cosgrove, Q. Duan, L. Luo, R.W. Higgins, R.T. Pinker, and J.D. Tarpley. 2003. Snow process modeling in the North American Land Data Assimilation System (NLDAS): 2. Evaluation of model simulated snow water equivalent. *Journal of Geophysical Research* 108(D22):8850, doi:10.1029/2003JD003994.

Pinker, R.T., J.D. Tarpley, I. Laszlo, K.E. Mitchell, P.R. Houser, E.F. Wood, J.C. Schaake, A. Robock, D. Lohmann, B.A. Cosgrove, J. Sheffield, Q. Duan, L. Luo, and R.W. Higgins. 2003. Surface radiation budgets in support of the GEWEX Continental-Scale International Project (GCIP) and the GEWEX Americas Prediction Project (GAPP), including the North American Land Data Assimilation System (NLDAS) project. *Journal of Geophysical Research* 108(D22):8844, doi:10.1029/2002JD003301.

Reichle, R.H. and R.D. Koster. 2005. Global assimilation of satellite surface soil moisture retrievals into the NASA Catchment land surface model. *Geophysical Research Letters* 32:L02404, doi:10.1029/2004GL021700.

Riebsame, W.E., S.A. Changnon, and T.R. Karl. 1991. *Drought and Natural Resources Management in the United States: Impacts and Implications of the 1987–89 Drought.* Boulder, Colorado: Westview Press.

Robock, A., L. Luo, E.F. Wood, F. Wen, K.E. Mitchell, P.R. Houser, J.C. Schaake, D. Lohmann, B. Cosgrove, J. Sheffield, Q. Duan, R.W. Higgins, R.T. Pinker, J.D. Tarpley, J.B. Basara, and K.C. Crawford. 2003. Evaluation of the North American Land Data Assimilation System over the Southern Great Plains during the warm season. *Journal of Geophysical Research* 108(D22):8846, doi:10.1029/2002JD003245.

Rodell, M., P.R. Houser, U. Jambor, J. Gottschalck, K. Mitchell, C.-J. Meng, K. Arsenault, B. Cosgrove, J. Radakovich, M. Bosilovich, J.K. Entin, J.P. Walker, D. Lohmann, and D. Toll. 2004. The global land data assimilation system. *Bulletin of the American Meteorological Society* 85:381–394.

Saha, S., S. Nadiga, C. Thiaw, J. Wang, W. Wang, Q. Zhang, H.M. van den Dool, H.-L. Pan, S. Moorthi, D. Behringer, D. Stokes, M. Pena, S. Lord, G. White, W. Ebisuzaki, P. Peng, and P. Xie. 2006. The NCEP climate forecast system. *Journal of Climate* 19:3483–3517.

Schaake, J.C., Q. Duan, V. Koren, K.E. Mitchell, P.R. Houser, E.F. Wood, A. Robock, D.P. Lettenmaier, D. Lohmann, B. Cosgrove, J. Sheffield, L. Luo, R.W. Higgins, R.T. Tinker, and J.D. Tarpley. 2004. An intercomparison of soil moisture fields in the North American Land Data Assimilation System (NLDAS). *Journal of Geophysical Research* 109:D01S90, doi:10.1029/2002JD003309.

Scipal, K., M. Drusch, and W. Wagner. 2008. Assimilation of a ERS scatterometer derived soil moisture index in the ECMWF numerical weather prediction system. *Advances in Water Resources* 31:1101–1112.

Sheffield, J., K.M. Andreadis, E.F. Wood, and D.P. Lettenmaier. 2009a. Global and continental drought in the second half of the 20th century: Severity-area-duration analysis and temporal variability of large-scale events. *Journal of Climate* 22(8):1962–1981.

Sheffield, J., C.R. Ferguson, T.J. Troy, E.F. Wood, and M.F. McCabe. 2009b. Closing the terrestrial water budget from satellite remote sensing. *Geophysical Research Letters* 36:L07403, doi:10.1029/2009GL037338.

Sheffield, J., G. Goteti, F. Wen, and E.F. Wood. 2004. A simulated soil moisture based drought analysis for the United States. *Journal of Geophysical Research* 109:D24108, doi:10.1029/2004JD005182.

Sheffield, J., M. Pan, E.F. Wood, K.E. Mitchell, P.R. Houser, J.C. Schaake, A. Robock, D. Lohmann, B. Cosgrove, Q. Duan, L. Luo, R.W. Higgins, R.T. Pinker, J.D. Tarpley, and B.H. Ramsey. 2003. Snow process modeling in the North American Land Data Assimilation System (NLDAS): 1. Evaluation of model-simulated snow cover extent. *Journal of Geophysical Research* 108(D22):8849, doi:10.1029/2002JD003274.

Sheffield, J. and E.F. Wood. 2007. Characteristics of global and regional drought, 1950–2000: Analysis of soil moisture data from off-line simulation of the terrestrial hydrologic cycle. *Journal of Geophysical Research* 112(D17115): 12. doi:10.1029/2006JD008288.

Sheffield, J. and E.F. Wood. 2008a. Global trends and variability in soil moisture and drought characteristics, 1950–2000, from observation-driven simulations of the terrestrial hydrologic cycle. *Journal of Climate* 21:432–458.

Sheffield J. and E.F. Wood. 2008b. Projected changes in drought occurrence under future global warming from multi-model, multi-scenario, IPCC AR4 simulations. *Climate Dynamics* 13(1):79–105.

Sheffield, J., E.F. Wood, D.P. Lettenmaier, and A. Lipponen. 2008c. Experimental drought monitoring for Africa. *GEWEX News* 18(3):4–6.

Svoboda, M., D. LeComte, M. Hayes, R. Heim, K. Gleason, J. Angel, B. Rippey, R. Tinker, M. Palecki, D. Stooksbury, D. Miskus, and S. Stephens. 2002. The drought monitor. *Bulletin of the American Meteorological Society* 83:1181–1190.

Tang, Q., H. Gao, H. Lu and D.P. Lettenmaier. 2009. Remote sensing: Hydrology. *Progress in Physical Geography* 33(4):490–509.

Trenberth, K.E. and G.W. Branstator. 1992. Issues in establishing causes of the 1988 drought over North America. *Journal of Climate* 5:159–172.

Verdin, J., C. Funk, G. Senay, and R. Choularton. 2005. Climate science and famine early warning. *Philosophical Transactions of the Royal Society B* 360:2155–2168.

Wang A., T.J. Bohn, S.P. Mahanama, R.D. Koster, and D.P. Lettenmaier. 2009. Multimodel ensemble reconstruction of drought over the continental United States. *Journal of Climate* 22:2694–2712.

Wang, A., D.P. Lettenmaier, and J. Sheffield. 2010. Soil moisture drought in China, 1950–2006. *Journal of Climate* (in press).

Wilhite, D.A. 2000. Drought as a natural hazard: Concepts and definitions. In *Drought: A Global Assessment*, ed. D.A. Wilhite, pp. 3–18. London, U.K.: Routledge Publishers.

Wood, A.W. and D.P. Lettenmaier. 2006. A testbed for new seasonal hydrologic forecasting approaches in the western U.S. *Bulletin of the American Meteorological Society* 87(12):1699–1712.

Wood, E.F., D.P. Lettenmaier, X. Liang, B. Nijssen, and S.W. Wetzel. 1997. Hydrological modeling of continental-scale basins. *Annual Review of Earth and Planetary Sciences* 25:279–300.

Xia, Y., K. Mitchell, M. Ek, J. Sheffield, E. Wood, L. Luo, B. Cosgrove, C. Alonge, H. Wei, J. Meng, B. Levenieh, D. Lettenmaier, V. Koren, Y. Duan, K. Mo, and Y. Fan. 2012. Continental-scale water and energy flux analysis and validation for the North-American Land Data Assimilation System Project Phase 2 (NLDAS-2), Part 1: Comparison analysis and application of model products. *Journal of Geophysical Research* (in review).

Yevjevich, V. 1972. *Stochastic Processes in Hydrology*. Fort Collins, Colorado: Water Resources Publications.

11 Satellite Gravimetry Applied to Drought Monitoring

Matthew Rodell

CONTENTS

11.1 INTRODUCTION

Near-surface wetness conditions change rapidly with the weather, which limits their usefulness as drought indicators. Deeper stores of water, including root-zone soil wetness and groundwater, portend longer-term weather trends and climate variations; thus, they are well suited for quantifying droughts. However, the existing in situ networks for monitoring these variables suffer from significant discontinuities (short records and spatial undersampling), as well as the inherent human and mechanical errors associated with the soil moisture and groundwater observation. Remote sensing is a promising alternative, but standard remote sensors, which measure various wavelengths of light emitted or reflected from Earth's surface and atmosphere, can only directly detect wetness conditions within the first few centimeters of the land's surface. Such sensors include the Advanced Microwave Scanning Radiometer–Earth Observing System (AMSR-E), C-band passive microwave measurement system on the National Aeronautic and Space Administration's (NASA) Aqua satellite, and the combined active and passive L-band microwave system currently under development for NASA's planned Soil Moisture Active Passive (SMAP) satellite mission. These instruments are sensitive to water as deep as the top 2 and 5 cm of the soil column, respectively, with the specific depth depending on vegetation cover. Thermal infrared (TIR) imaging has been used to infer water stored in the full root zone, with limitations: auxiliary information including soil texture is required, the TIR temperature versus soil water content curve becomes flat as wetness increases, and dense vegetation and cloud cover impede measurement. Numerical models of land surface hydrology are another potential solution, but the quality of output from such models is limited by errors in the input data and trade-offs between model realism and computational efficiency.

Water mass has a gravitational potential that, when integrated over a large enough area, can be great enough to alter the orbits of satellites. If those orbits are tracked precisely enough, the information can be used to infer redistributions of water both on and below the land surface regardless of depth. The capability to assess both surface and subsurface changes in water is one of the primary rationales for satellite missions dedicated to the measurement of earth's time variable gravity field. The Gravity Recovery and Climate Experiment (GRACE) is the first such mission. Launched in 2002, GRACE has already proved to be a game changer in the field of global hydrology. Pilot projects are now demonstrating how GRACE-derived terrestrial water storage (TWS) data can be downscaled and enhanced by integrating them with other data within sophisticated numerical land surface models (LSMs), and how the resulting fields of groundwater and soil moisture can be applied for drought monitoring.

This chapter is divided into eight sections, the next of which describes the theory behind satellite gravimetry. Following that is a summary of the GRACE mission and how hydrological information is gleaned from its gravity products. The fourth section provides examples of hydrological science enabled by GRACE. The fifth and sixth sections list the challenging aspects of GRACE-derived hydrology data and how they are being overcome, including the use of data assimilation. The seventh section describes recent progress in applying GRACE for drought monitoring, including the development of new soil moisture and drought indicator products, and that is followed by a discussion of future prospects in satellite gravimetry–based drought monitoring.

11.2 SATELLITE GRAVIMETRY

The gravitational force experienced at the Earth's surface varies in space and time, so that the gravity field of the whole Earth takes the form of a not-quite-smooth ellipsoid. It is said to have both static and time variable components. The static components are orders of magnitude stronger and include the total mass of the Earth and mass heterogeneities that only vary on geologic time scales, such as the distribution of the continents and the locations of mountain ranges and depressions in the crust (e.g., Dickey et al., 1997). Jeffreys (1952) was among the first to report the existence of the time variable component of the gravity field, noting that mass movements such as ocean tides changed gravitational potential. At that time, spatial variations in the gravity field were observed primarily with the aid of pendulums.

Mapping spatial heterogeneities in the Earth's gravity field was facilitated by the first artificial satellites, including Vanguard 1 and ANNA 1B, which began orbiting in the late 1950s and early 1960s. Satellite tracking via optical and Doppler techniques allowed scientists to compute departures from predicted orbits, and these departures were attributed to previously unobserved factors affecting the paths of the satellites, irregularities in the static gravity field in particular. In the 1980s and 1990s, orbit determination was accomplished by ground-to-satellite laser ranging (e.g., Yoder et al., 1983), which afforded more detailed assessments of the gravity field. Gutierrez and Wilson (1987) confirmed that perturbations in the orbits of the Lageos and Starlette satellites were caused by seasonal patterns in the distribution of atmospheric mass and water stored on the continents. Chao and O'Connor (1988) reached a similar conclusion in their study of the effects of seasonal changes in surface water on the Earth's rotation, length of day, and gravity field. Dickey et al. (1997) performed a comprehensive study of the potential benefits to hydrology and other disciplines of a dedicated satellite gravimetry mission designed to observe the time variable component of Earth's gravity field, and explored possible mission scenarios. Based in part on the recommendations of that study, NASA provided funding to develop a mission that would later be named the Gravity Recovery and Climate Experiment, or GRACE.

11.3 GRAVITY RECOVERY AND CLIMATE EXPERIMENT

GRACE, jointly operated by NASA and the German Aerospace Center, launched on March 17, 2002, and has continued to perform past its nominal mission lifetime of 5 years. It consists of two satellites in a tandem orbit about 200 km apart at 450–500 km altitude (Figure 11.1). A K-band microwave tracking system continuously measures changes in the distance between the satellites caused by heterogeneities in the gravity field, with a precision of better than 1 μm (Tapley et al., 2004). Nongravitational accelerations are monitored by onboard accelerometers, and the precise positions of the satellites are fixed via global positioning system (GPS). Each month, those measurements are used to produce a mathematical representation of the

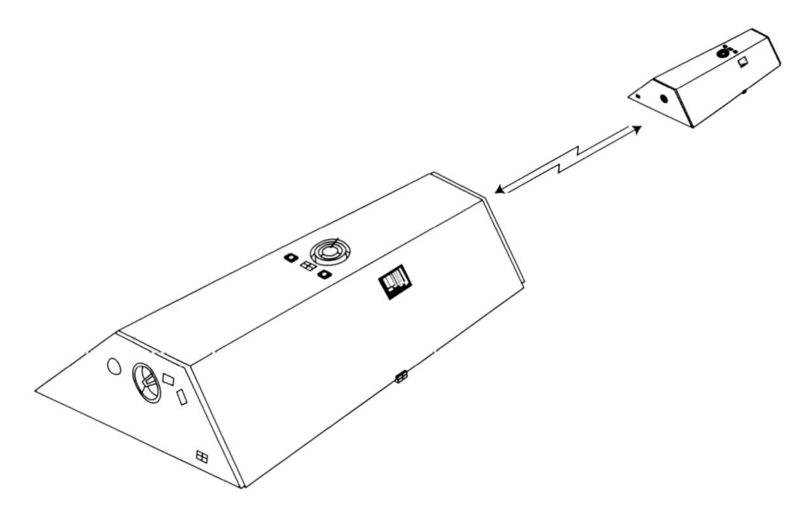

FIGURE 11.1 Artist's rendering of the GRACE satellite pair, which orbits the earth at about 500 km altitude. The second satellite follows about 200 km behind the first, and the precise separation is continuously measured by a K-band microwave tracking system. (From NASA, Washington, DC.)

global gravity field, as a set of coefficients (degree and order $\leq$ 120) to a spherical harmonic expansion that describes the shape of the geoid (the surface of constant gravitational potential best matching the mean sea surface). The expansion coefficients can be manipulated using numerical devices such as Gaussian averaging functions in order to isolate mass anomalies (deviations from the baseline temporal mean) over regions of interest (e.g., Wahr et al., 1998). Errors in such estimates are inversely related to the size of the region, being as small as 1–2 cm equivalent height of water over continental-scale river basins, and being large enough to overwhelm the hydrology signal as the area drops below ~150,000 km^2 (Wahr et al., 2006). As an alternative to the spherical harmonic technique, the inter-satellite ranging data can also be used directly to estimate regional "mass concentrations" without the need to first derive a global gravity field (Rowlands et al., 2005).

The main drivers of temporal variations in the gravity field are oceanic and atmospheric circulations and redistribution of terrestrial water via the hydrological cycle. By accounting for the first two using analysis models (another error source), GRACE scientists quantify anomalies in TWS, which is the sum of groundwater, soil moisture, surface water, snow, ice, and vegetation biomass. Glacial isostatic adjustment also must be considered in certain regions such as Hudson's Bay in Canada, and a major earthquake can produce a significant gravitational anomaly, but the timescales of most solid earth processes are too long to be an issue.

11.4 HYDROLOGICAL SCIENCE ENABLED BY GRACE

GRACE was designed as a geodesy mission, but it has supported many advances in water cycle science. Among the early results, Tapley et al. (2004) and Wahr et al. (2004) presented the first satellite-based estimates of column-integrated TWS variations at continental scales. Rodell et al. (2004) and Swenson and Wahr (2006a) took advantage of the fact that TWS change (ΔTWS) is a component of the terrestrial water budget equation, and demonstrated the estimation of evapotranspiration (ET) and atmospheric moisture convergence (MC), respectively, over large river basins as residuals of observation-based water budget analyses using the following equations:

$$ET = P - Q - \Delta TWS \tag{11.1}$$

$$MC = P - ET = Q + \Delta TWS \tag{11.2}$$

where P and Q are total precipitation and net streamflow over a specified time period.

Equation 11.2 ignores atmospheric water storage changes and horizontal transport of condensed water, which are normally much smaller than the other variables. As described in those papers, the actual calculations are somewhat more complicated because of the time-averaged nature of the GRACE observations. Syed et al. (2005) used a similar approach to estimate monthly discharge from the Amazon and Mississippi Rivers:

$$Q = MC - \Delta TWS \tag{11.3}$$

where MC is provided by an atmospheric analysis model.

One of the most valuable applications of GRACE has been accurate monitoring of mass loss from the Greenland and Antarctic ice sheets (e.g., Velicogna and Wahr, 2005, 2006; Luthcke et al., 2006; Ramillien et al., 2006). Likewise, Tamisiea et al. (2005) and Luthcke et al. (2008) showed that melt water from glaciers in southern Alaska contribute significantly to sea level rise.

GRACE continues to spawn innovative science. Crowley et al. (2006) used GRACE to confirm the previously hypothesized anticorrelation of interannual TWS variations on opposite sides of the southern Atlantic, such that droughts in the Amazon tend to be coincident with pluvials in the Congo River basin, and vice versa. Rodell and Famiglietti (2002), Yeh et al. (2006), and Rodell et al. (2007) described how GRACE could be used with ancillary information to generate time series of regionally averaged groundwater storage variations. Rodell et al. (2009) and Tiwari et al. (2009) later applied the technique to quantify massive groundwater depletion in northern India due to withdrawals for irrigation. Han et al. (2009) used a similar method to isolate surface water storage variations in the Amazon. Swenson (2010) evaluated satellite-derived precipitation products at high latitudes, where data for validation are scarce, using GRACE-derived time series of winter snow load. The potential for combining GRACE observations with data from other Earth observing missions is only just beginning to be realized. Current examples include detailed ice sheet monitoring through the combination of ICESAT and GRACE data, and terrestrial water budget studies that utilize satellite observations (Rodell et al., 2004; Sheffield et al., 2009). Future prospects include synthesis of GRACE TWS data with high resolution soil moisture observations from SMAP, which may enable more detailed drought assessments building on the data assimilation–based approach described in Section 11.7.

11.5 UNIQUE AND CHALLENGING ASPECTS OF GRACE DATA

The key to GRACE's success in delivering valuable hydrological data is its unique ability to measure changes in water stored at all levels on and beneath the Earth's surface. Downward-looking satellite remote sensors are limited by the depth of penetration of the measured light (visible, microwave, or otherwise), which is typically a few centimeters into the soil or perhaps a meter into the snowpack. Clouds, vegetation, and radio frequency interference can also be problematic. Because GRACE senses water through the orbital response of its two satellites to gravitational perturbations, it has no such limitations.

However, applying satellite gravimetry data for hydrology can be challenging. First, although remote sensing retrievals are always affected by mixed signals and indirect sensing of the variable of interest, most earth observing systems ultimately are able to distinguish individual variables, such as snow cover, soil moisture, or surface elevation, and deliver such specific products. Gravimetry-based TWS information, on the other hand, convolves changes in groundwater, soil water in all layers, surface water, snow, ice, and vegetation biomass into one integrated quantity. Second, an individual GRACE observation provides no indication of the absolute quantity of water stored in and on the land. It must be considered in relation to previous observations in order to discern relative changes or anomalies of water storage.

In contrast, SMAP is expected to deliver instantaneous estimates of the moisture content of the upper soil. Third, the spatial resolution of GRACE, which is about 150,000 km^2 at best (Rowlands et al., 2005; Swenson et al., 2006), is orders of magnitude coarser than most other satellite-based Earth observations. Future gravimetry missions may improve on this, but they are unlikely to approach the resolutions of optical (e.g., Landsat) and microwave (e.g., AMSR-E) remote sensors, which are on the order of meters and tens of kilometers, respectively. Fourth, the standard GRACE products are monthly means, with a lag of several weeks between the time of observation and the data product release. The hydrology community is accustomed to instantaneous, daily or better observations available in near-real time. Finally, at the time of this writing, the algorithms used to produce readily available global gridded hydrology maps were not sophisticated enough to address issues such as gravity signal leakage (blurring of time variable gravity features across the boundaries of adjacent regions). Extracting quantitatively reliable TWS time series from global GRACE gravity solutions has typically required the involvement of a trained geodesist. Recently, the GRACE science team has made it a priority to develop "off-the-shelf" research-quality hydrology products in order to accelerate the adoption of GRACE for hydrological research and applications including drought monitoring (Rodell et al., 2010).

11.6 DISAGGREGATING AND DOWNSCALING GRACE DATA

Satellite gravimetry data are quite different from and highly complementary to most other types of hydrology data. Many of the challenges to using GRACE-derived TWS data described earlier can be addressed by combining them with other information. Using the simple disaggregation method suggested by Rodell and Famiglietti (2002), Yeh et al. (2006) isolated groundwater storage variations averaged over Illinois by subtracting soil moisture and snow water storage (based on in situ network observations) from GRACE-based TWS. Rodell et al. (2007) accomplished the same for groundwater variations in the Mississippi River basin and its major subbasins using soil moisture and snow water storage estimated by numerical LSMs. Both studies assumed that surface water and biomass variations were insignificant in the central United States compared to groundwater, soil moisture, and TWS (Rodell and Famiglietti, 2001; Rodell et al., 2005).

A more sophisticated approach for separating GRACE-derived TWS into individual water storage components is to integrate these and other relevant data within an LSM via data assimilation. LSMs simulate the redistribution of water and energy incident on the land surface using physical equations of the processes involved. They incorporate soil, vegetation, and topographic parameters, and use meteorological fields such as precipitation and solar radiation as time-varying inputs (known as "forcing"). These forcing fields can be atmospheric analysis system outputs, observational data products, or a combination of the two. Advantages of LSMs include their spatial and temporal continuity, resolution, and low cost. However, their accuracy is limited by the quality of the input data, the model developers' understanding of the physics involved, and the simplifications necessary to represent complex Earth system processes in a computationally efficient manner. Further, most LSMs have a lower boundary at the bottom of the root zone, about 2–3 m below the land surface. Only models able to simulate groundwater storage are compatible with, and thus able to assimilate, GRACE-based TWS observations.

The objective of data assimilation is to combine multiple estimates of a state variable (TWS, in this case) to produce a single, more accurate estimate. Data assimilation combines the advantages of observations (ground- or remote sensing–based) and numerical modeling, in that the model fills spatial and temporal gaps in the observations, provides quality control, and enables data from disparate observing systems to be merged, while the observations anchor the results in reality. A commonly used data assimilation approach is the Kalman filter (Kalman, 1960), which computes a weighted average of a predicted value and a measured value, with the weights being inversely proportional to the uncertainty in each estimate. The ensemble Kalman filter (e.g., Reichle et al., 2002) uses a finite number of model trajectories (the ensemble) to estimate uncertainty in the model physics, parameters, and forcing data, with a value for each model spatial element (grid cell, catchment, or other discretization of the spatial model domain).

Whereas filters assimilate observations as they become available and update only the most recent model states, smoothers assimilate time series of observations to update model states over a specified period (Evensen and van Leeuwen, 2000; Dunne and Entekhabi, 2005). Smoothers are therefore well suited to GRACE observations, which are noninstantaneous. Zaitchik et al. (2008) developed an ensemble Kalman smoother approach for assimilating GRACE data into the Catchment LSM (Koster et al., 2000), in which the smoothing periods are identical to the epochs of the GRACE products (i.e., calendar months). The Catchment model was chosen because it simulates groundwater, which is uncommon among LSMs but essential for assimilation of TWS. The ensemble update is computed separately for each coarse scale GRACE observation. The basis for the distribution of information from the scale of the these observations (>150,000 km^2) to that of the model elements (roughly 2,500 km^2 for the Catchment LSM) and from TWS to its components (groundwater, soil moisture, and snow) is a matrix of cross-covariances between the TWS observations and model predicted component states at each location, which is computed a priori using all data in the period of record. This matrix has dimensions for TWS components and model elements, and it also varies in time.

Zaitchik et al. (2008) demonstrated that their data assimilation scheme successfully disaggregates and spatially and temporally downscales the GRACE-based TWS data into higher resolution fields of unconfined groundwater, surface and root-zone soil moisture, and snow water equivalent. Figure 11.2 shows that the assimilated results have much higher spatial resolution than the GRACE observations, and Figure 11.3 illustrates the disaggregation of water storage components and improvement in temporal resolution from monthly to nearly continuous. Zaitchik et al. (2008) compared the open loop (Catchment model run with no data assimilation) and assimilated output with observation-based groundwater and runoff data, and, significantly for the drought monitoring application described in this chapter, concluded that assimilating GRACE data improved the accuracy of the results (Figure 11.3). For time series of groundwater storage averaged over the Mississippi River basin, the coefficient of correlation increased from 0.59 to 0.70 and the root mean square error diminished from 23.5 to 18.5 mm equivalent height

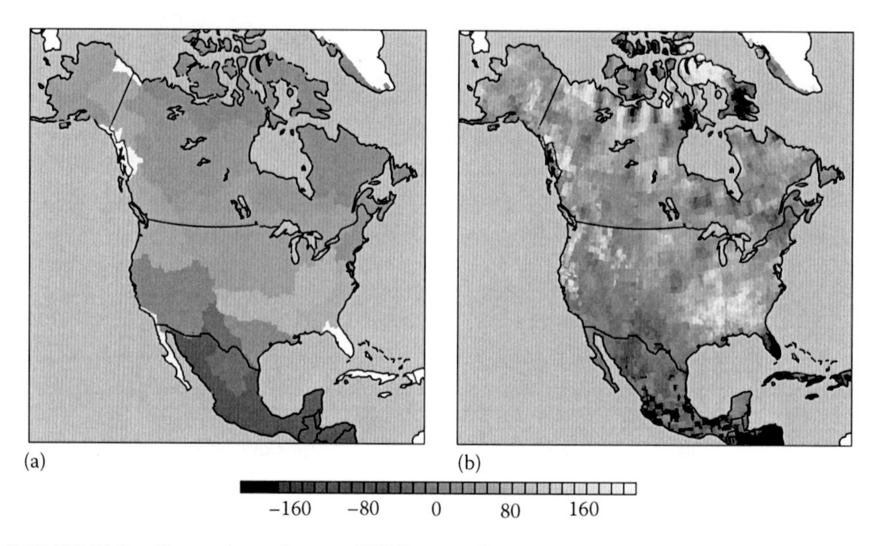

FIGURE 11.2 Comparison of maps of TWS anomalies (mm) over North America for May 2009 (a) at the resolution (major river basins) at which they are derived from GRACE data and (b) at the resolution (catchments intersected with an atmospheric forcing grid) of the data assimilation output. Note that white indicates no data, and thus Florida in panel B consists of model output with no GRACE data assimilated. (Bailing Li, University of Maryland, College Park, MD.)

of water. GRACE data assimilation also improved results in the four major subbasins of the Mississippi (not shown). Modeled runoff generally improved as well, but at a statistically insignificant level. The groundwater data used for evaluation were based on measurement records from 58 wells archived by the U.S. Geological Survey (USGS) (Rodell et al., 2007), while runoff data for the Mississippi River basin and three major subbasins were made available by the U.S. Army Corps of Engineers and USGS. Data assimilation has the additional benefit of extrapolating the GRACE data record to near-real time, using precipitation and other observation-based products that are available with minimal latency and taking advantage of the long "memory" (persistence or time scale of variability) of TWS.

11.7 DROUGHT MONITORING WITH GRACE

Standard drought quantification methods and products including the U.S. Drought Monitor (USDM; discussed further in a subsequent paragraph) rely heavily on in situ observations of precipitation, streamflow, and snowpack data. These largely constitute the point-based measures, gridded images generated via geostatistical interpolation, indices such as the Palmer Drought Severity Index, and soil moisture estimates from a simple water balance ("bucket") model that are used by the USDM authors to draw weekly maps, along with satellite-based vegetation greenness and any other reports (both objective and subjective) that the authors consider to be informative. Observations of groundwater and soil moisture are conspicuously lacking. Although drought is defined as an extended period with deficient precipitation, droughts are also influenced by precursor conditions, seasonality, the intensity of rain that

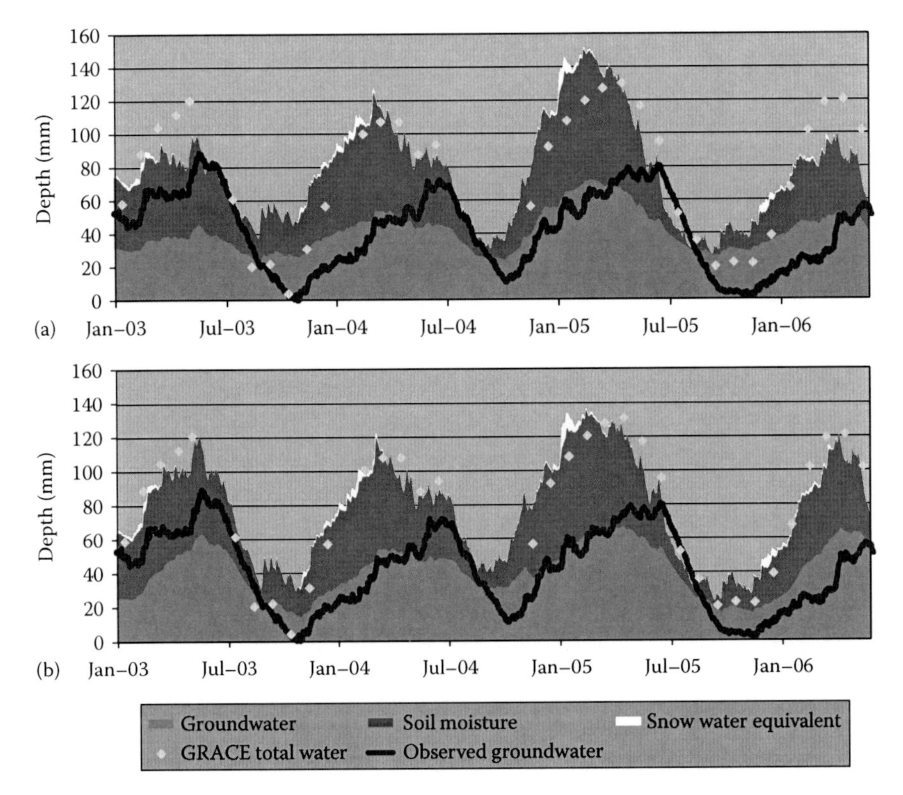

FIGURE 11.3 Groundwater, soil moisture, and snow water equivalent averaged over the Mississippi river basin from (a) open loop model output and (b) GRACE data assimilation. Also shown are daily, observation-based groundwater and monthly GRACE-derived TWS time series. GRACE-based and modeled TWS were adjusted to a common mean, as were observed and modeled groundwater. The coefficient of correlation between modeled and observed groundwater time series improved from 0.59 to 0.70 due to GRACE data assimilation. (Reprinted from Zaitchik, B.F. et al., *J. Hydrometeorol.*, 9, 3, 535, 2008. With permission.)

does fall, solar radiation, and other meteorological factors. LSMs are designed to account for all of these variables, and this has motivated the development of new LSM-based drought products. These products include the North American Land Data Assimilation System (NLDAS) Drought Monitor (presented in Chapter 10) and Princeton University's Drought Monitoring and Hydrologic Forecasting with the Variable Infiltration Capacity LSM (Luo and Wood, 2007). However, such products only consider the upper layers of the soil column (to about 2 m depth), and as with PDSI and USDM, they do not yet incorporate systematic observations of soil moisture, groundwater, or collective TWS.

Surface moisture conditions can fluctuate quickly with the weather, while the deeper components of TWS (e.g., groundwater) are well suited to drought quantification, particularly hydrologic droughts, because they integrate meteorological conditions over timescales of weeks to years. As mentioned before, GRACE is the only current satellite remote sensing mission able to monitor water below the first few centimeters of the land surface. Hence assimilating GRACE data into a groundwater inclusive LSM is appealing as a new drought monitoring method, combining observations of deep water storage with the resolution and timeliness of a numerical model. The resulting maps of relative groundwater storage and shallow and deep soil moisture address major gaps in current drought monitoring capabilities. In situ measurement records of these variables are often short or discontinuous, have poor spatial coverage, and may not be centralized and easily accessible. The GRACE data assimilation approach also yields maps of evaporation, transpiration, and runoff, which may be consulted as secondary indicators of drought.

The application of the GRACE data assimilation approach for drought monitoring has recently been explored by a NASA-funded investigation. Its objectives are to develop surface and root-zone soil moisture and groundwater drought indicators based on GRACE data assimilation results, and to assess the value of those indicators as inputs to the USDM and North American Drought Monitor (NADM) products. The USDM program was initiated in the late 1990s to centralize the drought monitoring activities conducted by federal, regional, and state entities in the United States (Svoboda et al., 2002). USDM drought maps are published on a weekly basis by a team of authors, and these are widely considered to be the premier drought products available for use by governments, farmers and other stakeholders, and the public, despite limited groundwater and soil moisture data as direct inputs. The NADM is based on the USDM concept and spans Canada, Mexico, and the United States.

Figure 11.4 summarizes the process of incorporating assimilated GRACE data into the Drought Monitor maps. Houborg and Rodell (2010) used groundwater well observations to corroborate the accuracy of the new drought indicator products, and at the time of this writing, we are evaluating the utility of these products as inputs to the short-term and long-term "objective blends" (amalgams of several drought indicators) that serve as baselines for the Drought Monitor maps. Their value will be determined through comparisons between the original and experimental objective blends and the final Drought Monitor products, and based on feedback from Drought Monitor authors and other end users.

The new indicators are generated as follows (Houborg and Rodell, 2010). First, an open loop simulation of the Catchment LSM is executed for the period of 1948 to near present using for input a meteorological forcing data set developed at Princeton University (Sheffield et al., 2006). This provides the climatological background for identifying droughts and quantifying their severity based on probability of occurrence. Monthly GRACE TWS anomaly fields (Swenson and Wahr, 2006b), which are defined to have a long-term zero mean at every location, are converted to absolute TWS fields by adding the time-mean total water storage field from the open loop LSM output. Thus the mean field of the assimilated TWS output is assured to be nearly identical to that of the open loop, which is to say that assimilating the GRACE data does not introduce a bias. We use a model restart file (set of initial conditions) from the open loop to initialize the GRACE data assimilating simulation (including the ensemble) when the first GRACE TWS data field is available in 2002.

Although bias in the assimilation is prevented by the method of converting TWS anomalies to model-appropriate water storage values, the GRACE data, and therefore

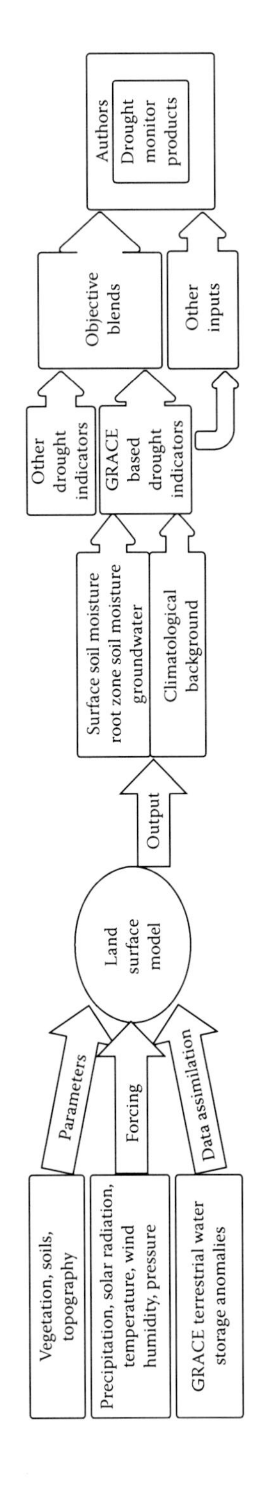

FIGURE 11.4 Flowchart illustrating the process for incorporating GRACE TWS data into the Drought Monitor products. "Other Inputs" include drought indicators and subjective reports that are used at the discretion of the Drought Monitor authors.

the assimilation results, could still have a larger or smaller range of variability than the open loop LSM results at any given location. This is an important consideration because drought monitoring involves the extremes. Therefore, a statistical adjustment must be applied to correct for differences in the range of variability between the assimilation-mode and open loop model output. This is accomplished by computing a cumulative distribution function (CDF) of wetness at each model pixel for the open loop and assimilation results during the overlapping period (2002 to near present) and mapping between the two CDFs. Finally, drought indicator fields for surface (top several centimeters) soil moisture, root-zone soil moisture, and groundwater are computed based on probability of occurrence in the 1948 present record. For ease of comparison with the Drought Monitor objective blends, droughts can be characterized from D0 (abnormally dry) to D4 (exceptional), corresponding to decreasing cumulative probability percentiles of 20%–30%, 10%–20%, 5%–10%, 2%–5%, and 0%–2%, and the results translated from the model grid onto coarser-resolution climate divisions. It should be noted that the original objective blends use a somewhat longer climatological background period, which begins in 1932.

Figure 11.5 shows examples of the new groundwater and soil moisture drought indicators in comparison with the original (non-GRACE) objective blends and USDM product for the epoch at the end of June 2005. The level of agreement and disagreement typifies that displayed by the various drought indicators and USDM product, although the patterns of disagreement are variable. In this example, all six maps generally show drought extending from Texas to Michigan and eastward to the mid-Atlantic coast, and another drought occurring in the northwest, with some differences in severity and extent (particularly in the case of the long-term objective blend). Note that the three GRACE-based maps (left panels) are not identical. The soil moisture percentile maps indicate drought conditions across all of Texas, while groundwater levels are still normal to high outside of easternmost Texas. Having maps of multiple TWS components will help the USDM authors to distinguish agricultural (shallow) droughts from hydrological (deep/water resources) droughts. The USDM currently depicts both types of drought on the same map. In some locations such as Idaho and the Four Corners area of the southwest, all three GRACE-based maps conflict with the USDM product. The objective blends are equivocal in these regions, suggesting that the new drought indicators would have been a valuable source of additional information for the USDM authors when they were drawing the map. A lack of independent, reliable data precludes determining which map is closest to the truth, but records indicate that in June 2005, both Idaho and the Four Corners area were near the end of multiyear droughts, the visible remnants of which may have colored reports from inhabitants of those regions.

11.8 FUTURE PROSPECTS

The GRACE data assimilation–based approach described earlier delivers spatially continuous fields of anomalies in groundwater storage—an ideal drought indicator—that were previously unavailable. If they prove to be skillful (early indications are positive) and there is demand, the approach could be expanded to the global scale. At present, most drought monitoring products are regional (few exist outside of North America)

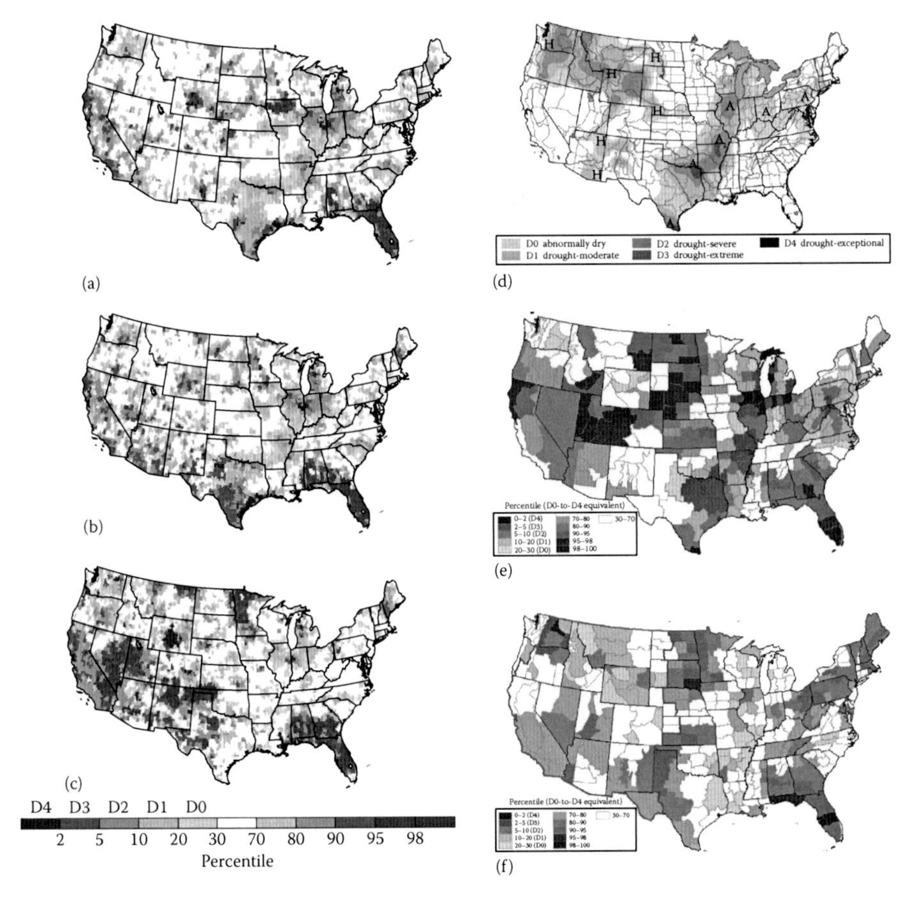

FIGURE 11.5 (See color insert.) Surface soil moisture (top several centimeters) (a), root-zone soil moisture (b), and groundwater storage (c) drought indicators, on a 0.25° grid, based on GRACE data assimilation results for June 25, 2005, compared with the U.S. Drought Monitor product for June 28, 2005 (d), and short-term (e) and long-term (f) objective blends (on U.S. climate divisions), which constitute the baselines for the Drought Monitor products, for June 25, 2005. (From Rasmus Houborg, University of Maryland, College Park, MD.)

and many drought stricken areas of the world lack adequate hydrological observing networks, so the potential benefits of a global product are even greater.

The approach could be refined in at least two ways. First, certain model parameters could be tuned to allow the LSM to better represent drought conditions. For example, early in the previously described project it was discovered that the range of TWS variability in the Catchment LSM was not large enough to represent severe multi-annual droughts. In certain locations, it would reach a limit of dryness. The solution was to increase the model's depth to bedrock parameter, which was determined to have minimal undesired impact on modeled fluxes. As evaluation of the results continues and as the approach is expanded beyond North America, it may be determined that other adjustments are necessary. Second, the GRACE data assimilation scheme could be implemented in the Community Land Model (CLM) and other groundwater inclusive LSMs as they are developed. That would facilitate assessment of uncertainty by revealing the degree to which the choice of LSM affects the simulation and detection of dry extremes. Kato et al. (2007) used in situ observations to demonstrate that the standard deviation of multiple LSM outputs provides a conservative estimate of uncertainty in those outputs.

GRACE has already persisted well beyond its designed lifetime of 5 years, and it is the only satellite mission able to measure temporal variations in the gravity field with the precision necessary to detect changes in TWS. Depending on battery and instrument health and fuel consumption for orbital adjustments, the mission might continue into the middle of this decade. NASA, the European Space Agency, and various other organizations have begun discussions of a next-generation satellite gravimetry mission, which would improve upon GRACE's horizontal resolution by perhaps a factor of 4 by replacing the microwave ranging system with a laser interferometer and flying at a lower altitude in atmospheric-drag-free spacecraft (NRC, 2007). Although the improved resolution will be valuable, it will not overcome the need for further downscaling via data assimilation. In the meantime, NASA has begun initial development of a follow-on to GRACE, with a nearly identical mission design, which would provide continuity in the data record while affording some small improvement in resolution due to lessons learned and basic technological advancements of the past 10 years. That mission could launch as soon as 2016 and enable gravimetry-based drought monitoring into the next decade.

ACKNOWLEDGMENTS

This work was funded by NASA's Decision Support Through Earth Science Research Results Program and GRACE Mission Science. The author thanks Rasmus Houborg for his work on the GRACE-based drought indicators and for contributing Figure 11.5; Ben Zaitchik, Rolf Reichle, and Bailing Li for development of the GRACE data assimilation capability; Jay Lawrimore, Richard Heim, Mark Svoboda, Brian Wardlow, Rich Tinker, and Matthew Rosencrans for assistance with drought indicator generation and evaluation; and Hiroko Kato Beaudoing and Jay Famiglietti for additional knowledge and assistance. GRACE land data were processed by Sean Swenson, supported by the NASA MEASURES Program, and are available at http://grace.jpl.nasa.gov.

REFERENCES

Chao, B.F. and W.P. O'Connor. 1988. Global surface-water-induced seasonal variations in the Earth's rotation and gravitational field. *Geophysical Journal of the Royal Astronomical Society* 94:263–270.

Crowley, J.W., J.X. Mitrovica, R.C. Bailey, M.E. Tamisiea, and J.L. Davis. 2006. Land water storage within the Congo Basin inferred from GRACE satellite gravity data. *Geophysical Research Letters* 33:L19402, doi:10.1029/2006GL027070.

Dickey, J.O., C.R. Bentley, R. Bilham, J.A. Carton, R.J. Eanes, T.A. Herring, W.M. Kaula, G.S.E. Lagerloef, S. Rojstaczer, W.H.F. Smith, H.M. van den Dool, J.M. Wahr, and M.T. Zuber. 1997. *Satellite Gravity and the Geosphere*. Washington, D.C: National Academy Press.

Dunne, S. and D. Entekhabi. 2005. An ensemble-based reanalysis approach to land data assimilation. *Water Resources Research* 41:W02013, doi:10.1029/2004WR003449.

Evensen, G. and P.J. van Leeuwen. 2000. An ensemble Kalman smoother for nonlinear dynamics. *Monthly Weather Review* 128:1852–1867.

Gutierrez, R. and C.R. Wilson. 1987. Seasonal air and water mass redistribution effects on Lageos and Starlette. *Geophysical Research Letters* 14(9):929–932.

Han, S.C., H. Kim, I.-Y. Yeo, P. Yeh, T. Oki, K.-W. Seo, D. Alsdorf, and S.B. Luthcke. 2009. Dynamics of surface water storage in the Amazon inferred from measurements of inter-satellite distance change. *Geophysical Research Letters* 36:L09403, doi:10.1029/2009GL037910.

Houborg, R. and M. Rodell. 2010. Integrating enhanced GRACE terrestrial water storage data into the U.S. and North American Drought Monitors. In *Proceedings of the American Society for Photogrammetry and Remote Sensing 2010 Annual Conference*. Bethesda, MD: ASPRS.

Jeffreys, H. 1952. *The Earth: Its Origin, History, and Physical Constitution*. London, U.K.: Cambridge University Press.

Kalman, R.E. 1960. A new approach to linear filtering and prediction problems. *Journal of Basic Engineering* 82(1):35–45.

Kato, H., M. Rodell, F. Beyrich, H. Cleugh, E. van Gorsel, H. Liu, and T.P. Meyers. 2007. Sensitivity of land surface simulations to model physics, land characteristics, and forcings, at four CEOP sites. *Journal of the Meteorological Society of Japan* 87A:187–204.

Koster, R.D., M.J. Suarez, A. Ducharne, M. Stieglitz, and P. Kumar. 2000. A catchment-based approach to modeling land surface processes in a general circulation model 1. Model structure. *Journal of Geophysical Research* 105:24809–24822.

Luo, L. and E.F. Wood. 2007. Monitoring and predicting the 2007 U.S. drought. *Geophysical Research Letters* 34:L22702, doi:10.1029/2007GL031673.

Luthcke, S.B., A.A. Arendt, D.D. Rowlands, J.J. Mccarthy, and C.F. Larsen. 2008. Recent glacier mass changes in the Gulf of Alaska region from GRACE mascon solutions. *Journal Glaciology* 54:767–777.

Luthcke, S.B., H.J. Zwally, W. Abdalati, D.D. Rowlands, R.D. Ray, R.S. Nerem, F.G. Lemoine, J.J. McCarthy, and D.S. Chinn. 2006. Recent Greenland ice mass loss by drainage system from satellite gravity observations. *Science* 314:1286–1289.

National Research Council. 2007. *Earth Science and Applications from Space: National Imperatives for the Next Decade and Beyond*, ed. R.A. Andres and B. Moore III, 456 pp, National Academic Press, Washington, D.C.

Ramillien, G., A. Lombard, A. Cazenave, E.R. Ivins, M. Llubes, F. Remy, and R. Biancale. 2006. Interannual variations of the mass balance of the Antarctica and Greenland ice sheets from GRACE. *Global and Planetary Change* 53:198–208.

Reichle, R., D. McLaughlin, and D. Entekhabi. 2002. Hydrologic data assimilation with the ensemble Kalman filter. *Monthly Weather Review* 130(1):103–114.

Rodell, M., B.F. Chao, A.Y. Au, J. Kimball, and K. McDonald. 2005. Global biomass variation and its geodynamic effects, 1982–1998. *Earth Interactions* 9(2):1–19.

Rodell, M., J. Chen, H. Kato, J.S. Famiglietti, J. Nigro, and C.R. Wilson. 2007. Estimating ground water storage changes in the Mississippi River basin (USA) using GRACE. *Hydrogeology Journal* 15:159–166, doi:10.1007/s10040-006-0103-7.

Rodell, M. and J.S. Famiglietti. 2001. An analysis of terrestrial water storage variations in Illinois with implications for the Gravity Recovery and Climate Experiment (GRACE). *Water Resources Research* 37:1327–1340.

Rodell, M. and J.S. Famiglietti. 2002. The potential for satellite-based monitoring of groundwater storage changes using GRACE: The High Plains aquifer, Central U.S. *Journal of Hydrology* 263:245–256.

Rodell, M., J.S. Famiglietti, J. Chen, S. Seneviratne, P. Viterbo, S. Holl, and C.R. Wilson. 2004. Basin scale estimates of evapotranspiration using GRACE and other observations. *Geophysical Research Letters* 31:L20504, doi:10.1029/2004GL020873.

Rodell, M., J.S. Famiglietti, and B.R. Scanlon. 2010. Realizing the potential of GRACE for hydrology. *EOS, Transactions of the American Geophysical Union* 91(10):96.

Rodell, M., I. Velicogna, and J.S. Famiglietti. 2009. Satellite-based estimates of groundwater depletion in India. *Nature* 460:999–1002, doi:10.1038/nature08238.

Rowlands D.D., S.B. Luthcke, S.M. Klosko, F.G.R. Lemoine, D.S. Chinn, J.J. McCarthy, C.M. Cox, and O.B. Anderson. 2005. Resolving mass flux at high spatial and temporal resolution using GRACE intersatellite measurements. *Geophysical Research Letters* 32:L04310, doi:10.1029/2004GL021908.

Sheffield, J., C.R. Ferguson, T.J. Troy, E.F. Wood, and M.F. McCabe. 2009. Closing the terrestrial water budget from satellite remote sensing. *Geophysical Research Letters* 36:L07403, doi:10.1029/2009GL037338.

Sheffield, J., G. Goteti, and E.F. Wood. 2006. Development of a 50-yr high-resolution global dataset of meteorological forcings for land surface modeling. *Journal of Climate* 19(13):3088–3111.

Svoboda, M., D. LeComte, M. Hayes, R. Heim, K. Gleason, J. Angel, B. Rippey, R. Tinker, M. Palecki, D. Stooksbury, D. Miskus, and S. Stephens. 2002. The Drought Monitor. *Bulletin of the American Meteorological Society* 83(8):1181–1190.

Swenson, S. and J. Wahr. 2006a. Estimating large-scale precipitation minus evapotranspiration from GRACE satellite gravity measurements. *Journal of Hydrometeorology* 7(2):252–270, doi:10.1175/JHM478.1.

Swenson, S. and J. Wahr. 2006b. Post-processing removal of correlated errors in GRACE data. *Geophysical Research Letters* 33:L08402, doi:10.1029/2005GL025285.

Swenson, S., P.J.F. Yeh, J. Wahr, and J. Famiglietti. 2006. A comparison of terrestrial water storage variations from GRACE with in situ measurements from Illinois. *Geophysical Research Letters* 33:L16401, doi:10.1029/2006GL026962.

Swenson, S. 2010. Assessing high-latitude winter precipitation from global precipitation analyses using GRACE. *Journal of Hydrometeorology* 11:405–420, doi:10.1175/2009JHM1194.1.

Syed, T.H., J.S. Famiglietti, J. Chen, M. Rodell, S.I. Seneviratne, P. Viterbo, and C.R. Wilson. 2005. Total basin discharge for the Amazon and Mississippi River basins from GRACE and a land-atmosphere water balance. *Geophysical Research Letters* 32:L24404, doi:10.1029/2005GL024851.

Tamisiea, M.E., E.W. Leuliette, J.L. Davis, and J.X. Mitrovica. 2005. Constraining hydrological and cryospheric mass flux in southeastern Alaska using space-based gravity measurements. *Geophysical Research Letters* 32:L20501, doi:10.1029/2005GL023961.

Tapley, B.D., S. Bettadpur, J.C. Ries, P.F. Thompson, and M.M. Watkins. 2004. GRACE measurements of mass variability in the Earth system. *Science* 305:503–505.

Tiwari, V.M., J. Wahr, and S. Swenson. 2009. Dwindling groundwater resources in northern India, from satellite gravity observations. *Geophysical Research Letters* 36:L18401, doi:10.1029/2009GL039401.

Velicogna, I. and J. Wahr. 2005. Greenland mass balance from GRACE. *Geophysical Research Letters* 32:L18505, doi:10.1029/2005GL023955.

Velicogna, I. and J. Wahr. 2006. Measurements of time-variable gravity show mass loss in Antarctica. *Science* 311:1754–1756.

Wahr, J., M. Molenaar, and F. Bryan. 1998. Time-variability of the Earth's gravity field: Hydrological and oceanic effects and their possible detection using GRACE. *Journal of Geophysical Research* 103(B12):30205–30230.

Wahr, J., S. Swenson, and I. Velicogna. 2006. The accuracy of GRACE mass estimates. *Geophysical Research Letters* 33:L06401, doi:10.1029/2005GL025305.

Wahr, J., S. Swenson, V. Zlotnicki, and I. Velicogna. 2004. Time-variable gravity from GRACE: First results. *Geophysical Research Letters* 31:L11501, doi:10.1029/2004GL019779.

Yeh, P.J.F., S.C. Swenson, J.S. Famiglietti, and M. Rodell. 2006. Remote sensing of groundwater storage changes in Illinois using the Gravity Recovery and Climate Experiment (GRACE). *Water Resources Research* 42:W12203, doi:10.1029/2006WR005374.

Yoder, C.F., J.G. Williams, J.O. Dickey, B.E. Schutz, R. Eanes, and B.D. Tapley. 1983. Secular variation of Earth's gravitational harmonic J_2 coefficient from Lageos and nontidal acceleration of Earth rotation. *Nature* 303:757–762.

Zaitchik, B.F., M. Rodell, and R.H. Reichle. 2008. Assimilation of GRACE terrestrial water storage data into a land surface model: Results for the Mississippi River Basin. *Journal of Hydrometeorology* 9(3):535–548, doi:10.1175/2007JHM951.1.

Part IV

Precipitation

12 Estimating Precipitation from WSR-88D Observations and Rain Gauge Data

Potential for Drought Monitoring

Gregory J. Story

CONTENTS

12.1 INTRODUCTION

Since its deployment, the precipitation estimates from the network of National Weather Service (NWS) Weather Surveillance Radars-1988 Doppler (WSR-88D) have become widely used. These precipitation estimates are used for the flash flood warning program at NWS Weather Forecast Offices (WFOs) and the hydrologic program at NWS River Forecast Centers (RFCs), and they also show potential as an input data set for drought monitoring. However, radar-based precipitation estimates can contain considerable error because of radar limitations such as range degradation and radar beam blockage or false precipitation estimates from anomalous propagation (AP) of the radar beam itself. Because of these errors, for operational applications, the RFCs adjust the WSR-88D precipitation estimates using a multisensor approach. The primary goal of this approach is to reduce both areal-mean and local bias errors in radar-derived precipitation by using rain gauge data so that the final estimate of rainfall is better than an estimate from a single sensor.

This chapter briefly discusses the past efforts for estimating mean areal precipitation (MAP). Although there are currently several radar and rain gauge estimation techniques, such as Process 3, Mountain Mapper, and Daily Quality Control (QC), this chapter will emphasize the Multisensor Precipitation Estimator (MPE) Precipitation Processing System (PPS). The challenges faced by the Hydrometeorological Analysis and Support (HAS) forecasters at RFCs to quality control all sources of precipitation data in the MPE program, including the WSR-88D estimates, will be discussed. The HAS forecaster must determine in real time if a particular radar is correctly estimating, overestimating, or underestimating precipitation and make adjustments within the MPE program so the proper amount of precipitation is determined. In this chapter, we discuss procedures used by the HAS forecasters to improve initial best estimates of precipitation using 24 h rain gauge data, achieving correlation coefficients greater than 0.85. Finally, since several organizations are now using the output of MPE for deriving short- and long-term Standardized Precipitation Indices (SPIs), this chapter will discuss how spatially distributed estimates of precipitation can be used for drought monitoring.

The U.S. Drought Monitor (USDM), which is considered the current state-of-the-art drought monitoring tool for the United States, is presently not designed for county-scale representations, yet its output is used by customers for critical decision making at this spatial scale. Thus, drought indicators are needed at the county and subcounty scale. The MPE estimates can be used as a "gold standard" precipitation product to compare with or validate other remote-sensing drought products, as long as the user understands the weaknesses of MPE. In the hands of a knowledgeable user, MPE provides information that no other existing drought tool can provide. With these products, we can look at detailed rainfall patterns and see how they correlate with evapotranspiration (ET) products across large areas, as well as identify localized areas of rainfall deficits over time. These data could also provide higher-resolution inputs for remote-sensing drought index formulations such as the Vegetation Drought Response Index (VegDRI) (Brown et al., 2008). VegDRI currently integrates SPI grids spatially interpolated from Applied Climate Information System (ACIS) gauge data, which characterize broadscale precipitation patterns but are often unrepresentative of county-scale level precipitation variations. Higher-spatial-resolution 4 km MPE observations are now available to enhance these types of tools and support local-scale drought monitoring and early warning activities that have been identified as a priority by the recently established National Integrated Drought Information System (NIDIS).

12.2 PAST EFFORTS IN DETERMINING MEAN AREAL PRECIPITATION

This chapter briefly discusses some of the reasons why the WSR-88D does not always estimate precipitation accurately and explain how HAS forecasters use the MPE PPS to determine the accuracy of radar precipitation estimates, as well as highlight some known issues with traditional rain gauge data. But before we look at the current state of ground-based radar rainfall estimation, an examination of past estimation techniques will be presented to gain an appreciation of the current algorithms.

12.2.1 Rain Gauge–Only Estimation

Before MPE, the RFCs only used rain gauge data to calculate basin-averaged MAP, which is the average depth of precipitation over a specific area for a given time period. This led to timing and location errors in the identification of heavy rainfall events, especially in a highly convective environment where intense rainfall often occurs over small core areas. Precipitation estimates were generated from discrete rain gauge observations using the Thiessen polygon method. This method attempted to calculate MAP, allowing for a nonuniform distribution of gauges by providing a weighting factor for each gauge. In basins where no rain gauges existed, this method was forced to use rain gauges that were outside the basin in question for its calculation. Although gauge-only analyses exist for drought monitoring in the United States at the climate division scale (e.g., the 1 month accumulated precipitation product at http://www.wrcc.dri.edu/spi/spi.html), these products are noisy, particularly in the western United States where gauge density is sparse with only a few observations per climate division. And since older radar systems described in the next section did not have the computer algorithms necessary to produce MAP, RFCs had no choice but to use a rain gauge–only methodology.

12.2.2 Radar Rainfall Estimation before the WSR-88D

Early radar systems (WSR-57, WSR-74S, and WSR-74C) came on line in 1973 and were used through 1993, but meteorologists at that time used a very crude technique for determining rainfall rates. These early radar networks would show rainfall and storm intensities using digital video integrator and processor (D/VIP) levels. The D/VIP levels were based on a predetermined value of returned power called the equivalent reflectivity, Z. A lookup table was used to establish rainfall rates for each D/VIP level. Radar operators would place a digital grid over the planned position indicator (PPI) radar scope and manually write in a value ranging from 0 to 6 that represented the maximum D/VIP level in each grid cell. The rectangular grid cells are known as manually digitized radar (MDR) boxes, which are based on a subgrid of the Limited Fine Mesh (LFM) model. The spatial resolution of the MDR grid cell was approximately 40 km. By contrast, the Hydrologic Rainfall Analysis Project (HRAP) grid now used by the WSR-88D has further improved the spatial resolution to ~4 km.

After the radar operators determined the maximum D/VIP level in each MDR box, they would transfer these values onto a paper overlay, which was usually a county boundary map. As an example, a D/VIP level of 5 meant the returned power from the echo had an equivalent reflectivity Z of between 50 and 57 decibels (dBZ). Next the operators would attempt to determine how much rain had accumulated. Using a reflectivity rainfall rate table, the hourly rainfall rate for this value would be found to be 4.5–7.1 in./h in a convective environment. They would then visually inspect the D/VIP levels over the past few hours and add the D/VIP levels together for longer-term rainfall estimates for specific counties. Using these early methods, considerable guesswork and manual analysis was involved in using radar to determine the amount of rainfall.

12.3 CURRENT ESTIMATION OF PRECIPITATION

12.3.1 Radar: The WSR-88D Precipitation Estimation Algorithm

Estimates from radar have become the base product for deriving mean areal, basin-averaged precipitation within the NWS. A photograph of a typical WSR-88D station is shown in Figure 12.1. The precipitation algorithm in the WSR-88D radar product generator (RPG) is complex, and given all the factors involved in radar sampling and performance, such as proper radar calibration and assumptions regarding radio wave propagation through the atmosphere, errors in radar precipitation estimates often occur. The precipitation algorithm contains dozens of adaptable parameters that control its performance (Fulton et al., 1998), improving accuracy over earlier radar estimation methods (Pereira Fo et al., 1988). The algorithm itself consists of five main scientific processing components (or subalgorithms) and an external independent support function called the precipitation detection function (NWS/ROC, 1999). The five scientific subalgorithms are (1) preprocessing, (2) determination of rainfall rate, (3) determination of rainfall accumulation, (4) rainfall adjustment, and (5) generation of precipitation products. The five subalgorithms are executed in sequence as long as the precipitation detection function determines that rain is occurring anywhere within a 230 km radius of the radar, which is referred to as the radar umbrella.

Once precipitation is detected, the first subalgorithm is executed: The base reflectivity data go through the preprocessing stage, which includes a quality control step that corrects for beam blockage using a terrain-based hybrid scan (O'Bannon, 1997) and checks for AP and biscan maximization (see Fulton et al., 1998 for more details).

FIGURE 12.1 A WSR-88D radar. (Photo courtesy of NOAA, Washington, DC.)

The reflected power returned to the radar (Z) is then assigned a rainfall rate (R) using a conversion known as a Z/R relationship. As the value Z increases, the R estimate in inches per hour increases exponentially based on the Z/R equation employed. Within this precipitation rate subalgorithm, more quality control is performed using a time continuity test, as well as corrections for hail and range degradation. Next, precipitation accumulations are determined through interpolation of scan-to-scan rain accumulation while simultaneously running clock-hour accumulations. Precipitation products are then generated and updated with each volume scan (NWS/ROC, 1999). An important end product is the hourly Digital Precipitation Array (DPA) product that provides 1 h estimates of rainfall on the 4 km HRAP grid discussed earlier. These DPAs are the one of four primary inputs to the MPE PPS program, a tool primarily used east of the Rocky Mountains, which will be discussed later in Section 12.4.

12.3.1.1 Problems with Radar-Based Precipitation Estimates

The WSR-88D precipitation algorithm is not without deficiencies and limitations, which all operational radars experience when attempting to estimate rainfall. Many factors that make accurate radar precipitation estimates difficult have been well documented (Wilson and Brandes, 1979; Hunter, 1996). The following text is a brief description of some of these factors and how they affect precipitation estimates.

12.3.1.1.1 Radar Reflectivity Calibration

Precipitation estimates can experience significant error if the reflectivity (i.e., value of returned power) from a rainfall target is too large or too small (Chrisman and Chrisman, 1999). The WSR-88D calibrates reflectivity before every volume scan using internally generated test signals. These calibration checks should maintain an accuracy of 1 dBZ, which translates to an accuracy of 17% in rainfall rates when the default Z/R relationship ($Z = 300R^{1.4}$) is employed. However, hardware problems (such as a change in actual transmitted power, or path loss of the returned power before reaching the receiver signal processor since the last off-line calibration) can cause significant changes in absolute calibration over time. Absolute calibration needs to be maintained because a change in Z of ±4 dBZ will result in doubling (or halving) the estimated R when the default Z/R relationship is used. Therefore, the WSR-88D Radar Operations Center (ROC) has developed absolute calibration procedures that are designed to ensure that reflectivity data are accurate to within ±1 dBZ.

12.3.1.1.2 Proper Use of Adaptable Parameters

As mentioned earlier, several adaptable parameters have a bearing on the precipitation algorithm, including parameters defining the Z/R relationship and the maximum precipitation rate (MXPRA). In the WSR-88D, the default Z/R relationship is the convective $Z = 300R^{1.4}$, and the default MXPRA is established at 53 dBZ, which equates to a maximum rainfall rate of ~104 mm/h (4 in./h) when the convective Z/R is employed. This value of MXPRA was established to eliminate the effects of hail contamination on rainfall estimates, as water-coated ice in clouds returns larger reflectivity values than liquid water alone would produce. However, extreme rainfall rates above the default MXPRA have been shown to occur when a deep warm cloud layer exists and warm rain processes prevail, which is most prevalent in tropical rainfall regimes where larger water drop size diameters exist (Baeck and Smith, 1998) and hail is absent. To compensate for this, radar operators have the option of using a different Z/R relationship called the Rosenfeld tropical Z/R ($Z = 250R^{1.2}$). When the tropical Z/R relationship is employed, significantly more rainfall is estimated for reflectivities higher than 35 dBZ (Vieux and Bedient, 1998). For example, the convective Z/R relationship yields a rainfall rate of 28 mm/h (1.10 in./h) when Z = 45 dBZ, while the tropical Z/R yields double the rainfall rate of 56 mm/h (2.22 in./h). Three additional Z/R relationships have been approved for use by the ROC: the Marshall–Palmer relationship ($Z = 200R^{1.6}$) for warm or arid climates where rainfall events are mostly stratiform in nature and two cool-season stratiform relationships (East $Z = 200R^{2.0}$ and West $Z = 75R^{2.0}$). Radar operators may also change the MXPRA parameter so that a higher rainfall rate will be used in the precipitation accumulation function to a maximum of 152 mm/h (6.00 in./h). In general, changes in the Z/R relationship have been shown to be extremely important in radar precipitation estimation (Fournier, 1999), while changes in MXPRA have far less impact.

Two other important adaptable parameters (RAINA and RAINZ) control when rainfall accumulations start and stop (Boettcher, 2006). Rainfall underestimation can occur if these parameters are set such that accumulations begin too late and/or end too early. RAINA is the minimum areal coverage of significant rain with a default setting of 80 km^2. RAINZ is the dBZ threshold that represents significant rain (i.e., the level of returned power for which you desire to begin radar rainfall accumulation) with a default setting of 20 dBZ. When the reflectivities of echoes are at or above RAINZ and the total areal coverage of returns meets or exceeds RAINA, the precipitation algorithm will accumulate rainfall. If these parameters are not adjusted for the rainfall type noted on any given day, this would have implications for drought monitoring. If a rain event is isolated (covering less than 80 km^2) or if the dBZ detected is less than the minimum defined level, then rainfall will not be accumulated. This could introduce a "dry bias" such that, if it is consistent over a period of time, it would indicate a signal drier than the rainfall that is actually received.

12.3.1.1.3 Hail Contamination, Bright Band, Snow, and Subcloud Evaporation

The presence of frozen or wet frozen precipitation can cause significantly enhanced reflectivity values (Wilson and Brandes, 1979). As hail stones grow in size, they become coated with water and reflect high amounts of power back to the radar, which can be significantly higher than the power returned from liquid precipitation present within the storm. The hail-contaminated higher power value results in an overestimation of the precipitation reaching the ground. Similarly, when ice crystals fall through the freezing level, their outer surfaces begin to melt. These water-coated ice crystals also produce abnormally high reflectivities, which lead to "bright band" enhancement (the layer of the atmosphere where snow melts to rain) and an overestimation of the precipitation.

Snowflakes are sampled fairly well by radar, but improper Z/R relationships can lead to an underestimation of the snowfall by the WSR-88D. A snow accumulation algorithm (SAA) has been added using a more representative relationship between reflectivity and frozen precipitation (Z/S relationship, identical to the East or West cool season stratiform Z/R relationship) to improve the water equivalent snowfall estimates. Vasiloff (2001) and Barker et al. (2000) provide more detailed review of the SAA.

Subcloud evaporation below the radar beam will also cause overestimation. This occurs when the rain falls into a dry subcloud layer and is most likely to occur in locations where clouds frequently have very high bases. In this situation, the rainfall estimate in the cloud may be relatively accurate, but the estimate will be too high if little or no rainfall reaches the ground. A prime example of this is virga (or dry microbursts).

12.3.1.1.4 Range Degradation

At far ranges, rainfall rates may be reduced because of signal degradation from partial beam filling that occurs when the radar beam widens with distance from the antenna and precipitation fills only part of the beam's field of view. Although the capability exists for range correction, it is currently not implemented on the WSR-88D pending scientific data to support accurate parameterization. Two other range degradation problems are more significant compared to partial beam filling. Certain rainfall types, such as stratiform rains (e.g., rainfall from clouds of extensive horizontal development as opposed to vertically developed convective clouds), show strong vertical reflectivity gradients. The stratiform gradient is positive until you get past the "bright band," and then it decreases sharply, leading to an overestimation of precipitation close to the radar and an underestimation with greater range. Orographic rain events also have sharp vertical reflectivity gradients as can certain rainfall events associated with distinct meteorological lifting surfaces such as a warm front. A rainfall event with a sharp vertical reflectivity gradient will show fairly strong range degradation. The reflectivity values decrease so rapidly with height within a cloud that the radar will have a higher degree of underestimation as the radar beam increases in altitude. In such rainfall events, the beam height becomes the largest single contributor to radar rainfall underestimations. Last, in stratiform rain events and with rains from thunderstorms that have small vertical height (usually 20,000 ft or less), a rainfall underestimation occurs due to the radar beam overshooting the precipitation at far ranges, which is a lack of detection problem. To compensate for this, the NWS set up the NEXRAD radar network with a spatial distribution of roughly 300 km apart. Figure 12.2 shows the WSR-88D radar coverage area for the United States. Notice that many sections of the western United States are without adequate radar coverage, which leads to unrepresentative precipitation estimates. Thus, radar- and range-dependent low precipitation biases can accumulate over time, leading to an underestimation of precipitation and a depiction of drier conditions. Users should understand this issue before using these estimates to evaluate drought conditions and other informational products.

12.3.1.1.5 Anomalous Propagation and Clutter Suppression

The WSR-88D displays reflectivity returns at locations assuming the beam is refracting normally in a standard atmosphere. At times, severe deviations from the standard atmosphere occur in layers with large vertical gradients of temperature and/or water vapor. When these deviations occur, super-refraction of the radar beam can result, and inaccurate calculations of actual beam height are made. These changes in refraction usually occur in the lower troposphere and can lead to persistent and quasi-stationary returns of high reflectivity either from ducting of the radar beam (where radio waves traveling through the lower atmosphere are curved to a value greater than the curvature of the earth) or from the beam coming in contact with the ground (Chrisman et al., 1995). This AP can lead to extreme precipitation accumulation estimates from false echoes. The WSR-88D does employ a clutter mitigation decision algorithm, which allows the radar operator to filter undesirable reflectivity returns, often from permanent targets near the radar (Maddox, 2010). However, this capability depends on the radar operator's ability to recognize the AP and invoke the algorithm. Improper or excessive use of clutter filtering may cause real meteorological echoes to be unnecessarily removed, leading to rainfall underestimation. This occurs most frequently when real rainfall targets are embedded in or near areas of AP, which is common behind a line of strong thunderstorms. Also, precipitation estimates from nonmeteorological targets (such as wind farms) are still observed on precipitation products, as certain targets that exhibit motion are not removed using current clutter filtering techniques. Figure 12.3 shows an example of AP across the south-central United States caused by superrefraction of the beams of several radars. Note the widespread light rainfall indicated over Oklahoma and central and deep south Texas and heavy rain over the Gulf of Mexico. No rainfall was actually occurring at this time. For hydrologic applications, this false rainfall is eliminated by conducting further data quality control external to the WSR-88D and is performed within the MPE PPS at RFCs.

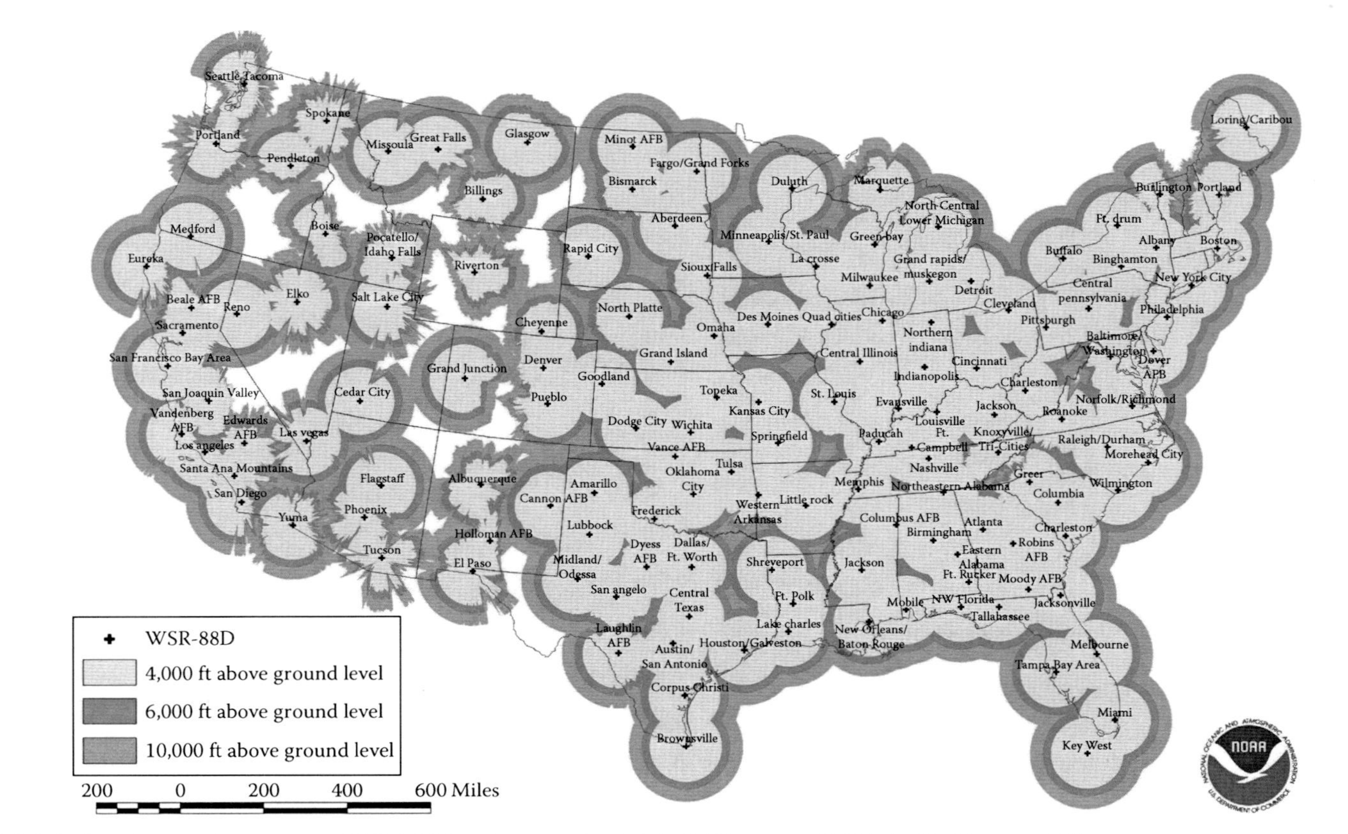

FIGURE 12.2 Geographic coverage of WSR-88D radars over the CONUS. (Courtesy of NOAA, Washington, DC.)

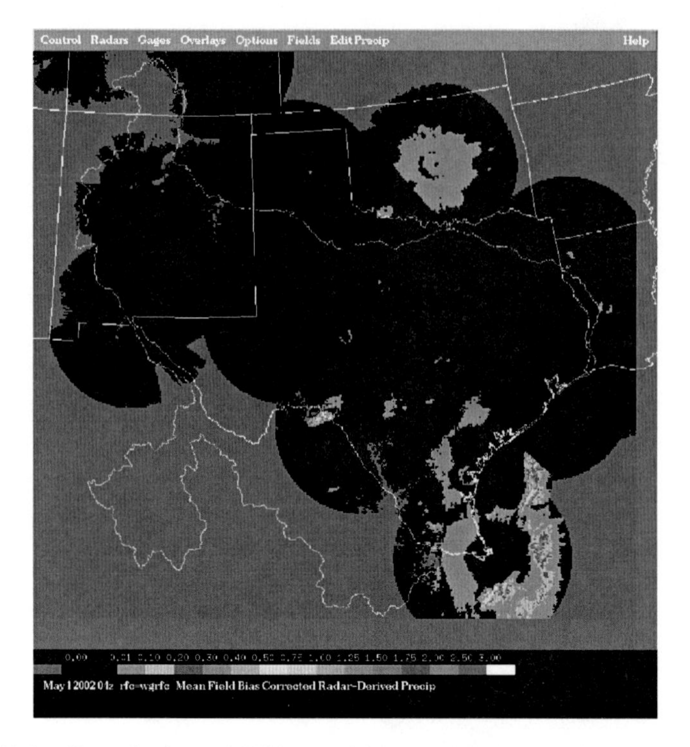

FIGURE 12.3 **(See color insert.)** Widespread false precipitation, or AP, shown on the MPE radar mosaic. (Photos courtesy NOAA/NWS, Silver Spring, MD.)

12.3.1.1.6 Beam Blockage

Beam blockage is a major problem where radars are situated near mountains and is unavoidable in many western U.S. locations. For radials (portions of the circular scan of the radar at a set elevation angle) with a blockage of no more than 60% in the vertical and 2° or less in azimuth, corrections are made to the reflectivities and are increased by 1–4 dBZ in the range bins beyond the obstacle, depending on the percentage of the blockage. Many sites have beam blockages of more than 60% and greater than 2° in azimuth, and this correction cannot be applied. Instead, the WSR-88D employs a terrain-based hybrid scan (O'Bannon, 1997), so radials that experience this high degree of beam blockage use the next higher elevation slice (complete scan of the radar at a set elevation angle) for the PPS for that radial (up to a maximum elevation angle of 3.4°, which is the fourth elevation slice aboveground). However, if a higher elevation slice is employed, range degradation is more likely, leading to underestimation of the precipitation. As a result, precipitation underestimation is common from radars located near mountains. The problem has been mitigated at some sites by installing radars on a peak. However, in this situation, the lowest elevation slices are so high above valleys that near-surface precipitation is not detected, which leads to the underestimation of rainfall from clouds of low vertical extent. Figure 12.2 also illustrates the gaps in radar coverage over the western United States due to the mountainous terrain.

12.3.1.1.7 Attenuation

The radar corrects for gaseous attenuation of the microwave radar signal, leaving a wet radar dome covering the antenna and intervening precipitation as the principal attenuators of energy to and from the target. Although this attenuation for S-band radars (10 cm wavelength) is considered to have minimal impacts on rainfall estimation, Ryzhkov and Zrnic (1995) show results indicating that attenuation may have a greater impact on rainfall estimates than previously thought. Signal attenuation could be one reason why rainfall is often underestimated during extremely heavy rain events due to reduced reflectivity returns, but it is difficult to quantify exactly how much the rainfall rates are reduced.

12.3.1.1.8 Polarization

The current WSR-88D is a single horizontal linear polarized radar. Dual polarization radar measurements of a specific differential phase at two orthogonal polarizations (horizontal and vertical) have shown improved skill in rainfall estimation compared to single polarization radars using Z/R relationships (Zrnic and Ryzhkov, 1999). Additional hydrometeor microphysical information can be inferred from the addition of vertical polarization measurements to obtain differential reflectivity, which aids in determining the size and type of liquid or frozen water particles (e.g., precipitation such as rain, sleet, hail, or snow), which would lead to improved precipitation estimation. A retrofit for the WSR-88D to implement dual polarization on a national scale is slated for 2011–2013. It has been determined that adding dual polarization capability to the WSR-88D will provide improved rainfall estimation for floods and drought and additional benefits that include improved hail detection for discriminating between liquid and frozen hydrometeors, rain/snow discrimination for winter weather, data retrieval from areas of partial beam blockage to improve services in mountainous terrain, and removal of nonweather artifacts such as birds and ground clutter to improve overall data quality for the precipitation algorithm.

12.3.1.2 Benefits of Radar-Based Precipitation Estimates

In spite of the limitations and some of the issues related to radar-based precipitation estimates, there are valid reasons for using them. A recent study by Krajewski et al. (2010) summarized the operational capability of radar to provide quantitative rainfall estimates with potential applications not only in hydrology but also in drought monitoring by improving gridded standard precipitation indices. Radar has the ability to show the spatial and temporal distribution of rainfall more accurately than other traditional sensors such as rain gauges. The timing and intensity of the rainfall is more easily determined because of the availability of hourly and subhourly estimates. Radar also provides a more accurate determination of rainfall location, which is critical for providing more local-scale information to the drought community about spatial variations in rainfall patterns and the identification of more localized areas experiencing precipitation deficits. This is far superior to waiting for 24 h rain gauge data to be reported and performing only a single calculation of MAP over a predefined geographic area (e.g., a river basin), as was the standard operating procedure in the past.

12.3.2 Rain Gauge Networks

Rain gauge networks form a supplemental source of precipitation data concurrent to gridded precipitation estimates. Two basic types of rain gauge networks support NWS hydrologic operations. One network has the ability to transmit rainfall data in near real time, while the other stations report 24 h data once a day. These two types of networks will be discussed separately as follows.

12.3.2.1 Near-Real-Time Gauges

Several near-real-time rain gauge networks with the ability to report precipitation hourly or even at 15 min intervals exist. These include the Automated Surface Observing System (ASOS) rain gauges at airports, data collection platforms operated by the U.S. Geological Survey, and mesonet alert systems maintained by various cities, states, and river authorities. Although these gauges are part of different networks, they all use tipping bucket gauges (Figure 12.4a) to automate the quantification of precipitation amounts.

Unfortunately, although these data are important, they are not without error, which can be introduced by wind, tipping bucket losses, poor siting (e.g., blockage from buildings, trees, and other tall vegetation), frozen precipitation, electronic signal malfunctions, mechanical problems, and timing/coding issues related to the transmission of rainfall data. Linsley et al. (1982) showed that strong winds will cause all rain gauges, regardless of type, to undercatch the precipitation. For example, approximately a 10% loss is estimated at a 10 mph wind speed, with losses often exceeding 50% at wind speeds over 39 mph. To help compensate for losses, ASOS tipping bucket gauges have a shield around them to disrupt the air flow over the top of the gauge (see Figure 12.4b). Tipping bucket gauges also tend to underreport intense rainfall when the rainfall rate exceeds the bucket's rate to discard the captured rain (~1.5 s). Thus, they cannot be calibrated for 0.01 of an inch precision or well calibrated for high rainfall rates. Maintenance is also an issue because many gauges are located in remote locations and frequent site visits by technicians may not be possible. In general, automated gauges provide good quality rainfall data if the gauges have good exposure, are well maintained, are recording when the air temperature is above freezing, when wind conditions are relatively light (15 mph or less), and the rainfall rate is not in excess of 4 in/h.

12.3.2.2 Daily Reporting Gauges

Gauge networks that report daily, 24 h rainfall totals are usually submitted by human observers who typically use a nontipping bucket type of gauge. Data received from these networks are considered to be of higher quality than the data received from the hourly automated networks partially because of the standard 4 in rain gauge or a weighing gauge used by the observers, which are typically free from some of the errors commonly encountered with tipping bucket gauges. The two best known daily gauge networks are the NWS Cooperative Observer (COOP) network and the Community Collaborative Rain, Hail and Snow (CoCoRaHS) network. We will discuss how these data are used to improve precipitation estimates produced by the RFC later in Section 12.4.5.

(a)

(b)

FIGURE 12.4 Tipping bucket rain gauge (a) and ASOS tipping bucket rain gauge with wind shield (b). (Photos courtesy of NOAA/NWS, Silver Spring, MD.)

12.4 RADAR-BASED MULTISENSOR PRECIPITATION ESTIMATOR PRECIPITATION PROCESSING SYSTEM

The main purpose of the MPE PPS is to take the raw hourly DPAs from the WSR-88Ds and perform additional quality control to achieve the best radar-based precipitation estimates possible for inclusion into the NWS River Forecast System (NWSRFS) for the primary purpose of river streamflow prediction. These estimates also hold considerable potential for providing both spatially and temporally explicit information about precipitation patterns and deficits over an extended period of time, which would greatly enhance the drought community's monitoring capabilities beyond the spatially interpolated precipitation grids generated from station observations that are currently used in operational monitoring systems. The following sections are a brief overview of the three PPS stages.

12.4.1 Three Stages of MPE Precipitation Processing

12.4.1.1 Stage I of the MPE PPS

The first PPS stage ingests the hourly 4 km DPA data that are generated by the WSR-88D, selecting the DPA that is timed closest to the top of each hour. The only quality control applied to the DPA data is features associated with the WSR-88D precipitation algorithm itself. Some of these features were discussed in Section 12.3.1, but for a more detailed discussion, see Story (1996).

12.4.1.2 Stage II of the MPE PPS

The second PPS stage calculates and applies a bias adjustment factor based on a comparison of rain gauge readings and radar precipitation estimates (Seo et al., 1999). Two biasing techniques are derived in the PPS: a mean-field bias and a local bias. The mean-field bias represents the ratio of the sum of all positive (nonzero) rain gauge data over the radar umbrella from the previous x number of hours to the sum of all nonzero DPA rainfall estimates at the corresponding gauge locations over the same temporal sampling window. The size of the temporal window x is specified by the adaptable parameter "mem-span" (memory span in hours, determined as a function of how widespread the rainfall is, how many gauges are available for sampling, and how long ago since it last rained). The MPE program calculates a mean-field bias for 10 memory spans, ranging from the current hour (instantaneous bias) to 10,000,000 h (climatological bias). The program also has an adaptable parameter that tells MPE which bias calculated from the 10 memory spans to apply to the DPA file. The default for this adaptable parameter is a minimum of 10 radar-rain pairs (called N-Pairs) for a mean-field bias to be applied to the "raw" radar rainfall estimate. If there are 10 or more N-Pairs for mem-span 1, the program uses the bias calculated from the radar-gauge pairs from the current hour. If there are no 10 N-Pairs for the current hour, the program goes back in time until a mem-span is found where 10 radar-gauge pairs are achieved. A time-weighting factor is applied to older N-Pairs so that the most recent data carry the most weight in these calculations. For example, if the bias calculated from mem-span 720 is used, the program had to go back between 168 (the maximum number of hours from the previous mem-span) and 720 h to find enough rain events that had at least 10 N-Pairs, which would include all nonzero radar-gauge pairs from the past 30 days. In general, the denser the rain gauge network is, the shorter the mem-span, unless a drought is in progress or the radar samples an area in a dry climate. In times of drought, the mem-span continues to increase over time as few N-Pairs are achieved, leading to the possibility that when it does rain again, the bias calculation will be inappropriate. The goal of MPE is to capture the temporal variability of the bias for different rainfall regimes to allow for the variability of radar precipitation estimates. A detailed description of all MPE functionality can be found in the MPE Editor User's Guide (NWS/OHD/HL, 2010).

In short, the larger the number of rain gauges located under a radar umbrella, the better chance the program has of obtaining nonzero radar/rain gauge pairs and calculating a mean-field bias. Under radar umbrellas that have a large number of hourly rain gauges available, the calculated MPE mean-field bias adjustment factor is a good indicator of whether a radar is over- or underestimating rainfall. A bias of 1.00 means that the MPE program has accepted the radar estimates as correct. If the mean-field bias adjustment is greater than 1.00, the radar is underestimating compared to its associated gauges, and if a bias is less than 1.00, the radar is overestimating rainfall. This factor is used to either increase or decrease the precipitation estimates in the MPE mean-field bias adjusted analysis.

In addition to the mean-field bias (one bias for each radar), a local bias technique is also calculated in the MPE program, assigning a bias correction factor for each HRAP grid box (or cell) in the MPE area. Like the mean-field bias, local bias values are computed by comparing gauge values to raw radar estimates. They are also processed over 10 memory spans, selecting the memory span whose bias value has at least 10 contributing gauge/radar pairs falling within a 40 km radius circle around each HRAP grid box for which a bias factor is being computed. The resulting grid of local bias values is then applied to the raw radar mosaic (similar to how the mean-field bias is applied) to produce the local bias–corrected radar mosaic. By computing the bias for each HRAP grid box, local geographical and microclimatological effects on rainfall can be accounted for (Seo and Breidenbach, 2002). Because of this accounting, the chosen default MPE field at many RFCs is the *local bias multisensor field* (i.e., the combination of the local bias radar mosaic and a gauge-only analysis).

In addition to the biased radar mosaics, a gauge-only gridded field is derived using hourly rain gauge observations, which must be quality controlled at this stage (Fulton et al., 1998). Tools exist within MPE (such as a gauge table) that allow HAS forecasters to detect rain gauge readings that subjectively appear to be inaccurate. Although rain gauge data are often referred to as "ground truth," these data also have known deficiencies, as mentioned in the previous section. However, the West Gulf RFC (WGRFC) HAS forecasters have found that most rain gauge data received are of acceptable quality and can be used (with some caution) to make accurate bias adjustments during most events. If any gauge reading appears incorrect (e.g., when radar fields are nonzero and a gauge reads zero), it is removed by the HAS forecaster, and all the MPE fields are regenerated. This may cause a change in the bias adjustment factors for one or more radars and in the gauge-only fields. The end result of this second stage is an adjusted radar precipitation estimate for each WSR-88D defined in the MPE program.

12.4.1.3 Stage III of the MPE PPS

In stage three of the PPS, the adjusted radar fields (those derived in Stage II, which were discussed in the previous section) are merged with the derived gauge-only field to calculate the final multisensor fields. The multisensor field of the specific radar site is then mosaicked with the multisensor fields of other radar sites to obtain the final multiradar precipitation map. Two primary multisensor fields are created in MPE, one for each biasing technique described in the previous section. The HAS forecaster makes a determination of which multisensor field is estimating correctly each hour (to use as our best estimate field, discussed further in the next two sections).

These multisensor fields are created for the areal extent covered by each RFC and are used daily by the National Centers for Environmental Prediction (NCEP) to generate a national Stage IV quantitative precipitation estimation (QPE) product. The HAS forecaster has other quality control options within the MPE program, such as the removal of AP. For a more detailed discussion of precipitation processing, see Story (2000).

12.4.2 Q2, the Next-Generation QPE

The WGRFC has been experimenting with a new precipitation estimation technique called Q2, which is the second technique derived by research meteorologists at the National Severe Storms Laboratory (NSSL). The National Mosaic and Multisensor QPE (NMQ) project is a joint initiative between the NSSL and other entities (such as the Federal Aviation Administration [FAA] and the University of Oklahoma). The National Mosaic and Q2 system is an experimental system designed to improve QPE and eventually very short-term Quantitative Precipitation Forecasts (QPF). For detailed information on the system, readers are referred to the NMQ web site at http://nmq.ou.edu. The NMQ ingests data from 128 WSR-88D stations every 5 min, quality controls the radar data, and derives a vertical profile of reflectivity from each radar. Analyses are done on eight tiles of radar data that are stitched together to form a continental U.S. (CONUS) three-dimensional (3-D) grid. Hybrid scan reflectivity and other products (such as a composite reflectivity map and precipitation flag product) are then derived to produce the experimental Q2 products. The products (such as QPE accumulations for the current hour or several hours of up to 72 h) are then translated over to the 4 km HRAP grid. The Q2 products hold several advantages over traditional radar-based estimates, with two primary advantages including an AP removal technique and rainfall estimates beyond the nominal 230 km range of the DPA files that are used in regions where radar umbrellas do not overlap. Because of these advantages, WGRFC HAS forecasters have the option of implementing Q2 as our final best estimate field.

12.4.3 Satellite Precipitation Estimates

The MPE also ingests satellite-derived precipitation estimates from the National Environmental Satellite, Data, and Information Service (NESDIS). The Hydroestimator is an automated technique, initially designed for large, moist thunderstorm systems, which uses Geostationary Operational Environmental Satellite (GOES) infrared (IR) imagery cloud top brightness temperatures (Scofield and Kuligowski, 2003). Pixels with the coldest IR temperatures are assigned the heaviest rainfall rates at the surface. Numerous other factors, including the cloud-top geometry, the available atmospheric moisture (precipitation efficiency), stability parameters from weather models, radar, and local topography, are used to further adjust the rain rates. Although caution should be used in drawing conclusions about radar performance based on satellite-derived precipitation estimates, HAS forecasters can confirm radar performance if the precipitation estimates from both sources are in close agreement. However, correlation coefficients comparing 24 h satellite precipitation estimates (SPEs) to 24 h rain gauges show the lowest correlation of any of the remote-sensing fields (biased radar estimates or Q2) used by HAS forecasters. Therefore, satellite-based estimates have the most benefit over land areas where no or limited observations of precipitation (e.g., radar, Q2, or rain gauges) are available (e.g., border area of Mexico). SPEs can be used without bias correction, or can be corrected for local biases using the techniques described earlier for radar. And like the other fields previously discussed, the option exists to integrate SPEs into our final best estimate precipitation field by performing polygon edits. One example of when SPEs would be integrated is when lightning data indicate thunderstorm activity and SPEs are the only field estimating rainfall in this location. As a result, the final best estimate field is based on a combination of radar-based multisensor fields from DPA files, Q2, and SPEs.

12.4.4 Final Postanalysis Quality Control Technique

Hundreds of 24 h COOP rainfall reports and CoCoRaHS observations are available for postanalysis of the MPE results. Direct comparisons of the MPE and observer rainfall totals shortly after 12 Coordinated Universal Time (UTC) each morning allow HAS forecasters to determine areas where the MPE estimates may be too low or too high. Forecasters can raise or lower estimates in specific hours in order to produce a 24 h estimate that is more consistent with 24 h gauge reports. The goal is to achieve a "general" level of acceptable error in the estimates. Programs are run that show the correlation coefficient and percent bias of MPE estimates, which vary by time and location. The goal is to modify the estimates to achieve correlation coefficients of greater than 0.85. Most initial estimates are low (meaning the 24 h gauge reports are higher than MPE) and have correlation coefficients of less than 0.85. When initial MPE estimates are raised or lowered, the inherent error of most estimates is improved to the desired correlation. Since these data are to be used for improved drought monitoring, removal of the traditional underestimation is crucial. If these biases are not mitigated, a false identification of the onset of drought might occur over time.

12.5 DROUGHT MONITORING: HOW THESE ESTIMATES CAN BE USED TO DETERMINE CURRENT LOCATIONS OF DROUGHT

12.5.1 NWS Southern Region Precipitation Analysis Project

In the early and mid-2000s, NWS Southern Region offices began to display the gridded MPE output maps on the Internet, and the data became available for download a short time later. Initially, these pages graphically showed the short-term observed and climatic trends of precipitation across the southern region (from New Mexico eastward to Tennessee, Georgia, and Florida). In 2009, this project was expanded to include the entire CONUS and Puerto Rico. The national-level products can be found on the Advanced Hydrologic Prediction Service (AHPS) web site (http://water.weather.gov). Tools are also available to compare MPE estimates to normal rainfall over different timescales (http://water.weather.gov/precip/), which can provide valuable insight into detailed spatiotemporal patterns of precipitation deficits to characterize both short- and long-term drought conditions.

"Departure from Normal" and "Percentage of Normal" products are generated by using simple grid mathematics, where the "Normal" data set is respectively subtracted from or divided into the "Observed" data set. "Observed" data are derived from output (e.g., from MPE or similar PPSs) from 12 NWS RFCs. "Normal" precipitation is derived from Parameter-elevation Regressions on Independent Slopes Model (PRISM) climate data (Gibson et al., 2002), which represent a 30 year period of record (1971–2000). The data sets were created as a unique knowledge-based system that uses point measurements of precipitation, temperature, and other climatic factors to produce continuous, digital grid estimates of monthly, yearly, and event-based climatic parameters. This unique analytical tool incorporates point data, a digital elevation model, and expert knowledge of complex climatic extremes, including rain shadows, coastal effects, and temperature inversions. In order to fill in areas that have radar-coverage gaps in the mountainous western United States, gauge reports are plotted against long-term climatic PRISM precipitation data, and amounts between gauge locations are spatially interpolated (more information about this method is available at http://www.cnrfc.noaa.gov/products/rfcprismuse.pdf). The derived precipitation products (specifically, "Departure from Normal" and "Percentage of Normal" products) can provide useful contextual information to identify the amount and magnitude of precipitation deficits that can be used for drought monitoring.

Figure 12.5 shows an example of a percent of normal rainfall graphic from December 2010 across the southern United States. This month was exceptionally dry, and this graphic depicts few areas where percent of normal precipitation

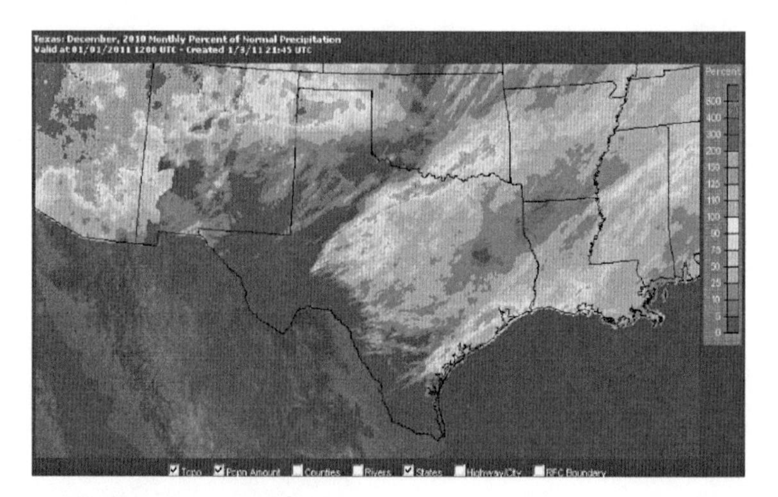

FIGURE 12.5 **(See color insert.)** Percent of normal rainfall for the southern United States from the AHPS precipitation analysis page for December 2010. (Image courtesy of NOAA/NWS, Silver Spring, MD.)

exceeded 100% (the upper Texas Gulf coast near Houston was one area). Of note are the large regions where the percent of normal precipitation was less than 50% of normal, specifically from northern Louisiana into east-central Texas, and across the Texas/Mexican border to western Texas. This indicates a strong dry signal, collocated with an extreme drought category designated on the USDM (not shown) for these locations.

12.5.2 Advanced Hydrologic Prediction Service

Before 2009, all radar-based product data displayed by the Southern Region Precipitation Analysis Project were considered to be "experimental." To make these data "operational," the data pages were packaged into a nationwide program known as the AHPS, a new and essential component of the NWS Climate, Water, and Weather Services. AHPS is a web-based suite of products that display drought magnitude and uncertainty of occurrence, based on the range of potential outcomes computed from historical hydrometeorological data and current conditions using an ensemble streamflow prediction model. These new products are enabling the USDM, National Drought Mitigation Center (NDMC), government agencies, private institutions, and individuals to make more informed decisions about risk-based policies and actions to mitigate the dangers posed by droughts. Although these products were not designed specifically for drought monitoring, the high-spatial-resolution precipitation information they provide has substantial potential to support this application. For example, the office of the Texas State Climatologist creates a gridded 4 km resolution and a county-scale resolution SPI from the AHPS precipitation analyses data (http://atmo.tamu.edu/osc/drought/). A more detailed description of the SPI grid generation using the AHPS is provided by Nielsen-Gammon and McRoberts (2009).

Traditionally, coarse resolution SPI maps derived from spatial interpolations of point-based gauge data have been used for drought monitoring, as shown in Figure 12.6a. In Figure 12.6b, the 4 km SPI maps generated from radar-based precipitation data depict considerably more spatially detailed precipitation variations, which provide considerably more local-scale information about precipitation deficits that is more appropriate for county to subcounty decision making related to drought. In brief, the SPI map generated from AHPS precipitation analyses is created using the following process. Initially, a cluster analysis is performed to determine Texas precipitation normals by location and season. A frequency distribution is then calculated for each location and season, from which high-resolution gridded frequency distributions are produced (using PRISM data over higher terrain of west Texas and roughly 1500 COOP stations in Texas and surrounding states). Finally, accumulations of precipitation are computed, creating 4 km and county-aggregated SPI for various time periods from 2 to 24 months, and related products such as an SPI blend, an SPI blend 1 week change map, and a percent of normal precipitation map.

The primary motivation for using AHPS precipitation data in this project was to facilitate local-scale drought monitoring for Texas. Climate division-scale drought monitoring tools are wholly inadequate for the state, and even ACIS gauge data are too coarse and unrepresentative in many areas. For example, the USDA applies the

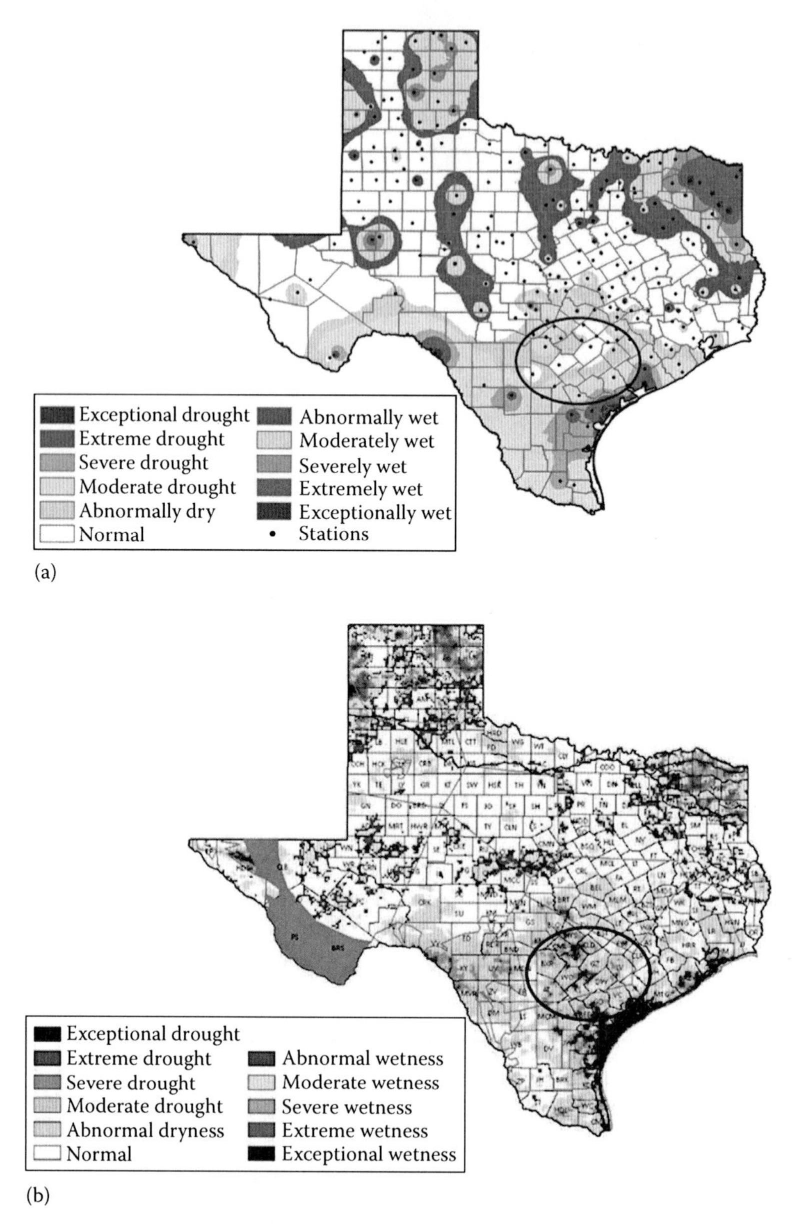

FIGURE 12.6 **(See color insert.)** An 8 week SPI map interpolated from station-based precipitation data (a) and an 8 week SPI map derived from 4km precipitation from MPE (b) (Image courtesy of Dr. John Nielsen-Gammon) for early September 2009 during the severe drought in southern Texas, as shown by the USDM map on September 7, 2009 (c). The circle highlights an area of exceptional drought in the USDM that is shown to have near-normal conditions in the interpolated SPI map (a) but clearly had localized areas of severe drought conditions that were detected in the SPI map based on higher-resolution, radar-based precipitation observations (b).

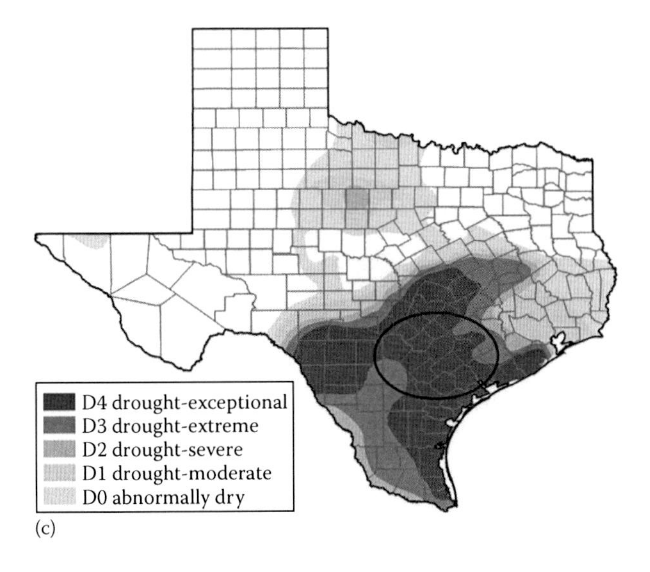

FIGURE 12.6 (continued) **(See color insert.)**

USDM map on a county scale for its drought relief decisions, yet the USDM and other existing drought index tools do not have the sufficient spatial resolution to enable estimation of drought at this spatial scale within Texas. An example of the SPI blend for Texas during the 2009 drought can be seen in Figure 12.6b. During this drought, the MPE-based SPI blend was able to accurately highlight the locations of most severe drought in Texas. Gauges within these hardest-hit areas, as indicated by our MPE products, were indeed experiencing historic drought severity based on an analysis of the period-of-record data, while stations adjacent to these areas were not. Nine counties (Nueces, San Patricio, Aransas, Refugio, Jackson, Calhoun, Bee, Brazoria, and Goliad) experiencing unprecedented drought severity were identified in southern Texas along the Gulf coast using MPE data, even though most of those counties did not have long-term precipitation records because of the sparse number of COOP stations that had a long history in that region (Nielsen-Gammon, August 2010, personal communication). Without the long-term precipitation records, SPI blends based on MPE data provided information that improved the assessment of the severity of the local drought situation.

In contrast, Figure 12.6a shows a station-based SPI map with data taken from the NDMC's Drought Atlas for the same time period as Figure 12.6b. Eight-week SPI data from 475 weather stations (226 stations in Texas and 249 stations from the surrounding states to minimize edge effects during the spatial interpolation of the SPI point data) were used to generate the map in Figure 12.6a. The station-based SPI shows the overall location of exceptionally wet conditions (northeast Texas and the eastern Texas panhandle) and exceptional drought (middle Texas Gulf Coastal region). But it is clear that the station-based SPI missed how widespread the extreme and exceptional drought conditions were across south-central Texas. For example, in several counties in south-central Texas where the station-based SPI showed normal conditions to abnormally dry conditions (Comal County eastward to Gonzales County), the MPE-based blended SPI showed severe to extreme drought conditions. Since hourly gauge data are incorporated into the final multisensor MPE, the drought features that appear over this area in Figure 12.6b should be representative of relative precipitation patterns (and deficits) at a local subcounty scale because ground-based precipitation observations are considered in the adjusted, radar-based precipitation fields. Across this area of Texas, notable rainfall discrepancies among stations during the defined SPI interval were likely due to the convective nature of the rainfall in this region, with the interstation variations being relatively consistent with the drought/nondrought patterns depicted in Figure 12.6b. The USDM map for September 8, 2009 (Figure 12.6c), reaffirms the severe drought conditions over this area, classifying these counties in the most severe drought class (D4, an exceptional drought that is defined as a one in 50 year event). Further visual analysis of the MPE-derived SPI map of the area reveals many subtle subcounty variations in dryness that are not detected in the station-based SPI map. Many counties in southern Texas have pockets of both drought and nondrought conditions in the radar-generated SPI map that cannot be spatially resolved using traditional interpolated maps from station-based observations.

The use of 4 km precipitation data provides a more accurate depiction of the breadth and scope of the Texas drought conditions in 2009. This result suggests that the improved spatial resolution of this information will be a tremendous benefit for local-scale drought monitoring activities by characterizing detailed subcounty spatial variations in precipitation deficits. The 4 km precipitation and other derivative products such as the SPI will also be extremely valuable in areas with sparse weather station networks and for counties with large areas that commonly experience considerable within-county climate variations.

12.6 CONCLUSIONS

Over the past several years, advancements have been made in both radar-based precipitation sensing and multisensor estimation processing techniques. Further improvements will be made in radar precipitation estimation with the implementation of dual polarization in the next few years. New rainfall rate algorithms such as Q2 have also been implemented within the MPE PPS. This chapter has discussed the benefit that improved, quality-controlled, and finer-scale precipitation data can have in drought monitoring by detailing deficits in rainfall with greater spatial resolution that is not available using gauge-based SPI data alone.

East of the Continental Divide, RFCs derive estimates of precipitation using a multisensor approach. Hourly precipitation estimates from WSR-88D radars are compared to ground rainfall gauge reports, and a bias (correction factor) is calculated and applied to the radar field. The radar and gauge fields are combined into a “multisensor field,” which is quality controlled on an hourly basis. In areas with limited or no radar coverage, SPE can be incorporated into this multisensor field, and the SPE can also be biased against rain gauge reports. In mountainous areas west of the Continental Divide, a different method is used to derive the estimates of precipitation. Gauge reports are plotted against long-term climatological precipitation (PRISM data), and derived amounts are interpolated between gauge locations.

Studies have shown (Seo, 1999; Seo and Breidenbach, 2002) that algorithms that combine sensor inputs—radar, gauge, and satellite—yield more accurate precipitation estimates than those that rely on a single sensor (i.e., radar only, gauge only, and satellite only). Although it is not perfect, the MPE data set is one of the best sources of timely, high-resolution precipitation information available. Still, users should understand the inherent weaknesses of this data set before using it in drought monitoring applications, especially those that require a high degree of accuracy.

Many quantitative measures of drought have been developed in the United States, depending on the sector impacted, the region being considered, and the particular application. Although different definitions and measures of drought exist, they all originate from a deficiency of precipitation resulting from an unusual weather pattern. Therefore, using an improved source of precipitation data such as MPE 4 km products would lead to a better determination of the onset, intensity, and geographic and temporal evolution of drought.

Several of the Palmer indices and the SPI are useful for describing drought on varying temporal scales (i.e., weeks, months, or years). On a climate-division scale, a standard suite of products including the NCDC's SPI, the CPC's soil moisture–related drought severity index, and the Western Region Climate Center's SPI exist. On a station scale, the U.S. Geological Survey provides gauge-based streamflow data, and the High Plains Regional Climate Center produces a 30 day SPI using daily data from ACIS that incorporates COOP observer and automated weather data. Satellite-based tools such as VegDRI (Brown et al., 2008) that assist in agricultural-related drought monitoring also rely on precipitation data as a primary input. Collectively, these drought indices have relied on gauge-based data and have not provided indices representative of county- to subcounty-scale drought information because of the coarse spatial resolution inputs. The higher-resolution 4 km precipitation data produced by MPE can be used to replace the traditional point or interpolated precipitation products in the development of these indices to provide a more detailed characterization of drought patterns. This holds the potential to advance local-scale drought monitoring activities as prioritized by NIDIS, as well as improve current state-of-the-art monitoring tools such as the USDM, which was initially designed to classify broadscale, national drought patterns but is increased being relied upon for county and subcounty drought information. With the goal of improved drought monitoring, Texas A&M University, North Carolina State University, and Purdue University received a USDA award to improve the long-term calibration of the AHPS MPE analyses, and take the SPI products beyond Texas to include at least the eastern parts of the United States (i.e., south-central and eastern sections). The project began in January 2011, with tangible results expected a few months after that.

ACKNOWLEDGMENT

The views expressed are those of the author of this chapter and do not necessarily represent those of NOAA or the NWS.

REFERENCES

Baeck, M. and J. Smith. 1998. Rainfall estimation by the WSR-88D for heavy rainfall events. *Weather Forecasting* 13:416–436.

Barker, T., P. Felsch, T. Mathewson, C. Sullivan, and M. Zenner. 2000. Test of the WSR-88D snow accumulation algorithm at Weather Field Office Missoula. *Technical Attachment, NWS Western Region* 00–13:1–6.

Boettcher, J. 2006. PPS parameters & their impact on rainfall estimates. *NEXRAD Now Newsletter* 16:9–11.

Brown, J.F., B.D. Wardlow, T. Tadesse, M.J. Hayes, and B.C. Reed. 2008. The vegetation drought response index (VegDRI): A new integrated approach for monitoring drought stress in vegetation. *GIScience and Remote Sensing* 45(1):16–46.

Chrisman, J. and C. Chrisman. 1999. An operational guide to WSR-88D reflectivity data quality assurance. WSR-88D Radar Operations Center paper. http://www.roc.noaa.gov/WSR88D/PublicDocs/Publications/Reflectivity_Quality_Assurance.pdf (accessed December 13, 2011).

Chrisman, J.N., D. Rinderknecht, and R. Hamilton. 1995. WSR-88D clutter suppression and its impact on meteorological data interpretation. *Preprints, First WSR-88D User's Conference*. WSR-88D Radar Operations Center, Norman, OK. http://www.roc.noaa.gov/WSR88D/PublicDocs/Publications/Legacy_Clutter_paper.pdf (accessed July 20, 2011).

Fournier, J. 1999. Reflectivity-rainfall rate relationships in operational meteorology. National Weather Service Technical Memo, National Weather Service, Tallahassee, FL.

Fulton, R., J. Breidenbach, D.-J. Seo, D. Miller, and T. O'Bannon. 1998. The WSR-88D rainfall algorithm. *Weather Forecasting* 13:377–395.

Gibson, W.P., C. Daly, T. Kittel, D. Nychka, C. Johns, N. Rosenbloom, A. McNab, and G. Taylor. 2002. Development of a 103-year high-resolution climate data set for the conterminous United States. *Reprints, 13th AMS Conference on Applied Climatology*, American Meteorological Society, Portland, OR, pp. 181–183.

Hunter, S. 1996. WSR-88D radar rainfall estimation: Capabilities, limitations and potential improvements. *National Weather Association Digest* 20(4):26–36.

Krajewski, W., G. Villarini, and J. Smith. 2010. Radar-rainfall uncertainties: Where are we after thirty years of effort? *Bulletin of the American Meteorological Society* 91:87–94.

Linsley, R., M. Kohler, and J. Paulhus. 1982. *Hydrology for Engineers*. 3rd edn, pp. 55–61. New York, McGraw-Hill.

Maddox, A. 2010. The importance of proper clutter filtering. *NEXRAD Now Newsletter* 19:8–12.

Nielsen-Gammon, J. and B. McRoberts. 2009. Tracking drought in Texas: Kicking it down a notch. Sixth US Drought Monitor Forum, Austin, TX. http://atmo.tamu.edu/osc/library/osc-pubs/drought_products.pdf (accessed December 13, 2011).

NWS/OHD/HL. 2010. *MPE Editor User's Guide, Build 9.2*. NWS/OHD Hydrologic Laboratory, Silver Spring, MD.

NWS/ROC. 1999. *WSR-88D Interactive Training Modules: Volume 5; Build 10 WSR-88D Products*. NWS/ROC Operations Training Branch, Norman, OK, CD-ROM.

O'Bannon, T. 1997. Using a 'terrain-based' hybrid scan to improve WSR-88D precipitation estimates. *Preprints, 28th Conference on Radar Meteorology*, Austin, TX, pp. 506–507. American Meteorological Society, Boston, MA.

Pereira Fo, A., K. Crawford, and C. Hartzell. 1988. Improving WSR-88D hourly rainfall estimates. *Weather Forecasting* 13:1016–1028.

Ryzhkov, A. and D. Zrnic. 1995. Precipitation and attenuation measurements at a 10cm wavelength. *Journal of Applied Meteorology* 35:2121–2134.

Scofield, R.A. and R.J. Kuligowski. 2003. Status and outlook of operational satellite precipitation algorithms for extreme-precipitation events. *Weather Forecasting* 18:1037–1051.

Seo, D.-J. 1999. Real-time estimation of rainfall fields using radar rainfall and rain gauge data. *Journal of Hydrology* 208:37–52.

Seo, D.-J. and J. Breidenbach. 2002. Real-time correction of spatially nonuniform bias in radar rainfall data using rain gauge measurements. *Journal of Hydrometeorology* 3:93–111.

Seo, D.-J., J. Breidenbach, and E. Johnson. 1999. Real-time estimation of mean field bias in radar rainfall data. *Journal of Hydrology* 223:131–147.

Story, G. 1996. The use of the hourly digital precipitation array at the West Gulf River Forecast Center. NWS/WGRFC, Fort Worth, TX, p. 15.

Story, G. 2000. Determining WSR-88D precipitation algorithm performance using the Stage III precipitation processing system. *NWA Electronic Journal of Operational Meteorology*, FTT2. http://www.nwas.org/ej/2000/2000.php (accessed July 20, 2011).

Vasiloff, S. 2001. WSR-88D performance in northern Utah during the winter of 1998–1999. Part 1: Adjustments to precipitation estimates. *National Weather Service Technical Attachment, National Weather Service Western Region* 01–02:1–7.

Vieux, B. and P. Bedient. 1998. Estimation of rainfall for flood prediction from WSR-88D reflectivity: A case study, 17–18 October 1994. *Weather Forecasting* 13:407–415.

Wilson, J. and E. Brandes. 1979. Radar measurement of rainfall: A summary. *Bulletin of the American Meteorological Society* 60:1048–1058.

Zrnic, D. and A. Ryzhkov. 1999. Polarimetry for weather surveillance radars. *Bulletin of the American Meteorological Society* 80:389–406.

13 Precipitation Estimation from Remotely Sensed Information Using Artificial Neural Networks

Application to Drought Monitoring and Analysis

Amir AghaKouchak, Kuolin Hsu, Soroosh Sorooshian, Bisher Imam, and Xiaogang Gao

CONTENTS

13.1 INTRODUCTION

Improving our understanding of weather and climate, along with the development of reliable and uninterrupted precipitation measurement techniques, is essential for the proper assessment of droughts. Precipitation plays a dominant role in the global hydrologic cycle and is one of the key variables in drought monitoring and analysis. In fact, drought is often defined as a prolonged period of deficient precipitation with respect to the average expected values. The drought phenomenon is usually described using drought detection and monitoring indices. Typically, droughts are categorized into three major classes: (a) agricultural, (b) meteorological, and (c) hydrological. Agricultural drought is related to the total soil moisture deficit, while meteorological drought is identified by lack of precipitation as the main indicator. Hydrological drought, on the other hand, is characterized by a shortage of streamflow, as well as groundwater supplies. Droughts have significant socioeconomic impacts that may vary for different sectors and at different spatiotemporal scales. For example, lack of precipitation over a 3 month period may be considered as a significant agricultural drought, although it may not be considered as significant from a hydrological viewpoint. Furthermore, the definition of drought may differ between various climate regions with different climatic properties.

Obviously, all three categories of drought are related to sustained lack of precipitation; hence, having accurate, long-term, and timely precipitation data is fundamental to drought prediction and analysis. The precipitation deficit can be described using the Standardized Precipitation Index (SPI, McKee et al., 1993; Hayes et al., 1999), which is commonly used in drought analysis. Many other indices based on other variables (e.g., soil moisture, evapotranspiration, and streamflow) are defined to describe other aspects of drought duration and severity (Karl, 1983; Mo, 2008; Shukla and Wood, 2008). The main obstacle in utilizing soil moisture and runoff from models (e.g., Land Data Assimilation Systems, LDAS) for drought classification is that the runoff and soil moisture–based indices depend on the type of model and forcing. For a given forcing, different models may estimate relatively similar drought indices. However, models driven with different forcing may lead to significant differences in the estimated drought indices (Robock et al., 2004; Dirmeyer et al., 2006).

Currently, the precipitation-based SPI is one of the most commonly used drought measures because of its simplicity and applicability (WMO, 2009; WCRP, 2010). One attractive feature of the SPI is that, if the precipitation data are available in near real time, the drought information can be provided in a timely manner for operational drought-monitoring applications. The true strength of the SPI is that precipitation anomalies can be calculated over flexible timescales in a consistent fashion across multiple locations. Traditionally, the SPI and other precipitation-based drought indices have been estimated based on long-term rain-gauge measurements. However, rain gauges are sparsely distributed in most parts of the world, and gauge measurements often are not representative of true areal precipitation distribution. This can be quite limiting for various drought-related analyses, including deriving the spatial extent of drought, probability of occurrence of multiple extreme droughts, spatial dependence, and tail behavior of drought indices. When precipitation data are not spatially representative because of sparse coverage of gauges, the precipitation analysis becomes method dependent and, hence, may lead to a great deal of uncertainty (WCRP, 2010). These issues become even more significant when gauge-only precipitation analyses are used as a forcing of models designed to monitor and analyze droughts. The value of satellite-derived precipitation data is the improved ability to capture spatial variations of precipitation beyond the traditional surfaces interpolated from gauge data.

In recent years, the issue of climate change and its impact on extreme climate events (e.g., droughts and floods) has drawn considerable attention worldwide. Cubasch et al. (2001) reported that the frequency and intensity of extreme climate events, particularly precipitation, would likely increase mainly because of the increase of greenhouse gases and aerosols. A warmer climate has a greater water-holding capacity of the atmosphere, which in turn leads to an increased chance of intense precipitation and flooding (Barnett et al., 2006). Using numerical simulations, Gong and Wang (2000) showed that an increase in the greenhouse gases may result in more extreme rainfall events, but it may not necessarily reduce the duration of extreme droughts. Similar findings, indicating increased climate extremes (e.g., flooding and drought), have been reported in other studies (Groisman et al., 1999; Frich et al., 2002; Wilby and Wigley, 2002; Schmidli and Frei 2005; Zolina et al., 2005; Alexander et al., 2006; Boo et al., 2006; Gao et al., 2006; Solomon et al., 2007; Zhang et al., 2008; Caprio et al., 2009). Furthermore, most global climate models predict increased winter wetness and summer dryness and droughts (Barnett et al., 2006).

In order to account for the potential impact of climate change, new modeling techniques and data resources with higher spatial and temporal resolutions may be required. In addition to rain gauge stations, weather radar systems provide high-spatiotemporal-resolution precipitation data. However, the spatial density and coverage of current gauge and radar networks are inadequate for monitoring global precipitation and for evaluating weather and climate models at global and continental scales. In fact, in most parts of the world, radar installations for precipitation measurements are not available. With the currently available radar systems, regional-scale studies can be performed only across the continental United States, western Europe, and a few other locations across the globe. Even in the United States, many regions (e.g., the western United States) suffer from poor radar coverage because of significant mountain blockage (Maddox et al., 2002). The limitations of rain gauges and weather radar systems highlight the importance of satellite-based global precipitation data for drought-related studies.

13.2 SATELLITE-BASED PRECIPITATION RETRIEVAL

The first meteorological satellite, the Television and Infrared Observation Satellite (TIROS-1), was launched in the 1960s. Since then, many satellite missions sponsored by the National Aeronautics and Space Administration (NASA), the National Oceanic and Atmospheric Administration (NOAA), and several other international space agencies have led to an increase in the availability of remotely sensed global observations and have provided valuable weather information to the hydrometeorological community. Satellites are either geostationary (GEO), hovering over the same location on earth, or polar-orbiting, revisiting the same swath of the earth at regular intervals (Kidd, 2001). GEO satellites (first launched in 1974) orbit over the equator at altitudes of 35,880 km while moving at the speed of the earth's rotation and are used to sense meteorological variables continuously in the visible and infrared (VIS/IR) wavebands (Levizzani, 2008). Polar-orbiting (low earth orbit, LEO) weather satellite series circle the earth at a typical altitude of 850 km in a north–south (or south–north) path. These satellites are positioned in sun-synchronous orbits, and thus are able to visit any location on earth twice each day (depending on the swath width and resolution of sensors). LEO satellite sensors have been conceived to provide passive microwave (PMW) data for rainfall estimation. These satellites offer more accurate precipitation estimates than the GEO satellites. On the other hand, GEO satellites transmit weather data continuously over the hemispherical disk, which is crucial for short-term monitoring and forecasting of weather systems.

GEO satellites provide IR and VIS scans of cloud-top temperature and albedo with high temporal resolutions of less than 30 min. Although they do not offer direct estimates of precipitation, the cloud-top temperatures, along with other information, can be used to retrieve precipitation estimates. The indirect nature of precipitation retrieval based on GEO satellites gives rise to the uncertainty of precipitation estimates. The PMW sensors on LEO platforms, on the other hand, offer more direct and, therefore, more accurate instantaneous precipitation estimates. However, PMW sensors scan each location only once or twice per day, which may not be frequent enough to capture precipitation variability. The low temporal sampling frequency of polar-orbiting satellite sensors can be compensated for by effective integration of information from multiple satellites, including geosynchronous sensors, in order to obtain better temporal and spatial coverage of diurnal rainfall patterns. In the past three decades, different satellite rainfall-retrieval algorithms, for example, Tropical Rainfall Measuring Mission (TRMM) 3B42, Huffman et al., 2007; CPC (Climate Prediction Center) MORPHing (CMORPH) technique, Joyce et al., 2004; Precipitation Estimation from Remotely Sensed Information using Artificial Neural Networks (PERSIANN) (Hsu et al., 1997, Sorooshian et al., 2000, Hong et al., 2004), have been developed to retrieve estimates of precipitation. For comprehensive reviews of satellite rainfall estimation algorithms and product intercomparisons, the interested reader is referred to Tian et al. (2009), Adler et al. (2001), Levizzani and Amorati (2002), Tian et al. (2007), and Kidd (2001). Validation and verification studies show that merging multiple channels of satellite information not only compensates for the weaknesses of both PMW and IR sensors, but also exploits the advantages of both types of sensors in precipitation retrieval (Vincent and Soille, 1991; Xie and Arkin, 1995; Huffman et al., 1997; Xie and Arkin, 1997; Vicente et al., 1998; Xu et al., 1999; Sorooshian et al., 2000; Turk et al., 2000; Adler et al., 2001; Ba and Gruber, 2001; Huffman et al., 2001; Tapiador et al., 2004; Turk et al., 2006; Levizzani et al., 2007).

Currently, satellite-based precipitation data are actively being used for drought monitoring and analysis. The experimental African Drought Monitor (Sheffield et al., 2008), developed by the Land Surface Hydrology Group at Princeton University, provides near-real-time monitoring of land-surface hydrological conditions. The input precipitation to the model is from the PERSIANN and TRMM data sets; model output includes quantitative assessment of water-budget components, along with near-real-time drought conditions. Using rainfall data provided by the TRMM, Anderson et al. (2008) investigated droughts based on vegetation response with the precipitation. Their results showed that higher water stress coincided with critical months in the Normalized Difference Water Index (NDWI) and Enhanced Vegetation Index (EVI). Anderson et al. (2008) further concluded that remotely sensed data can be utilized in the analysis of drought effects on vegetation. Paridal et al. (2008) also utilized TRMM data to detect and assess drought in the Philippines, and the results highlighted the potential significance of satellite-derived data for drought-related analysis. Zhang et al. (2008) employed CMORPH precipitation data to estimate drought severity in northern China. In this work, satellite-based precipitation data, along with other remotely sensed information, were successfully used to monitor drought conditions based on the Plant Water Stress Index (PWSI; Idso et al., 1981).

The potential significance of satellite-derived data in current and future drought research highlights a critical need to improve the currently available precipitation estimation algorithms. Clearly, any major advances in reliable and accurate satellite-based precipitation-retrieval algorithms may lead to significant improvements in drought monitoring, analysis, and decision making. This chapter describes a precipitation-retrieval algorithm that integrates multichannel observations from several satellites and data sources. The PERSIANN algorithm, introduced here, utilizes precipitation estimates from geosynchronous IR images and PMW data to generate near-real-time global precipitation data sets. Furthermore, the PERSIANN Cloud-patch Classification System (PERSIANN-CCS), which integrates a cloud-patch-based classification method with the original PERSIANN algorithm, is discussed. It is well known that remotely sensed data have several advantages over the traditional ground-based measurement systems, including higher spatial resolution and uninterrupted temporal coverage. However, satellite-based precipitation data have not yet been fully integrated into drought monitoring, analysis, and decision-making systems mainly because of (a) the relatively short length of the satellite data record, (b) undetermined uncertainties associated with satellite-based products, and (c) a general lack of experience in the application of satellite precipitation data. Therefore, this document envisions future research activities toward providing long-term satellite-based precipitation data with higher spatial and temporal resolutions. Furthermore, potential applications of satellite-based data to drought monitoring and analysis are addressed. It is hoped that this introduction motivates greater utilization of satellite data in drought monitoring, particularly in describing their spatial extent and temporal occurrences.

13.3 PERSIANN ALGORITHM

The current operational PERSIANN algorithm utilizes an Artificial Neural Network (ANN) classification and approximation approach to derive precipitation estimates. ANN models are capable of modeling highly nonlinear and complex systems and have been applied in modeling environmental systems dating back to the early 1990s (Govindaraju and Rao, 2000). Since then, many precipitation-retrieval algorithms based on the ANN have been developed (Hsu et al., 1997; Bellerby et al., 2000; Hong et al., 2004; Tapiador et al., 2004; Coppola et al., 2006). The PERSIANN algorithm was initially based on thermal IR (10.2–11.2 μm) imagery (cloud-top temperature) from GEO satellites, while the current operational version is extended to include microwave (MW) data from a wide range of spectral bands (approximately 10–85 GHz) aboard LEO platforms for calibration and parameter adjustment.

The algorithm transforms long-wave IR images to precipitation estimates in three steps by (1) extracting the mean and variance of the cloud-top brightness temperatures in the vicinity of the target pixel, (2) classifying the extracted features (e.g., cloud-top temperature patterns), and (3) translating the classified features to precipitation rates. The main motivation to integrate IR and PMW data is to benefit from the strengths and compensate for the weaknesses of both types of sensors. The PMW-based retrieval algorithms provide more accurate instantaneous rainfall; however, the sampling frequency is limited to few times per day for any specific location. On the other hand, IR-derived estimates provide high-temporal-resolution (~3–5 km) sampling of precipitation, but with less accuracy regarding total rainfall amount. The main reason for lower accuracy in the IR-based estimates is the fact that precipitation rate is derived through a statistical relationship between IR estimates of cloud-top temperature and mean precipitation rate (indirect measurement). Merging the two data sets (IR and PMW) enables us to use the strengths of both sensors for monitoring precipitation events.

The operational version of the PERSIANN algorithm runs in two parallel modes: (a) “simulation mode,” which generates precipitation estimates at the 0.25° spatial resolution and every 30 min based on the geosynchronous satellite IR data and (b) the “continuous update mode,” which continuously updates the algorithm parameters using PMW instantaneous precipitation estimates (Hsu et al., 1997). To perform PMW-based adjustment, the PERSIANN coverage is divided into overlapping subregions. The IR estimates are then calibrated based on the instantaneous PMW data that fall within the same subregion. For more detailed information, the interested reader is referred to Hsu et al. (1997). The continuous update mode is designed to improve the quality of the IR-derived precipitation estimates. In the simulation mode, the algorithm uses gridded IR images from multiple satellites such as the East and West Geostationary Operational Environmental Satellites (GOES) and the Geostationary Meteorological Satellite (GMS-5), provided by the NOAA CPC (see Janowiak et al., 2001, for details). The PMW-based precipitation estimates, required for the continuous update mode, are derived from the available LEO satellites. It is noted that, if necessary, GEO data can be obtained at up to 15 min temporal intervals; however, PMW data are not available at higher temporal resolutions for parameter calibration. Validation studies indicate that precipitation estimates are significantly improved by adjusting IR-based data with PMW information (Hsu et al., 1999).

The operational PERSIANN precipitation estimates cover 60°S–60°N globally, with a temporal resolution of 3 h and a spatial resolution of 0.25° since 2000. This product is further accumulated to various product scales (e.g., 6 h, daily) for hydrological applications. Figure 13.1 shows the components of the PERSIANN data-retrieval algorithm. Figure 13.2 displays diurnal distribution of summer 2002 precipitation pre- and postadjustment using PMW data (here, TRMM Microwave Imager [TMI]) across southwestern Texas. In Figure 13.2, the solid black line shows the diurnal distribution derived from radar data. The dashed black and solid gray

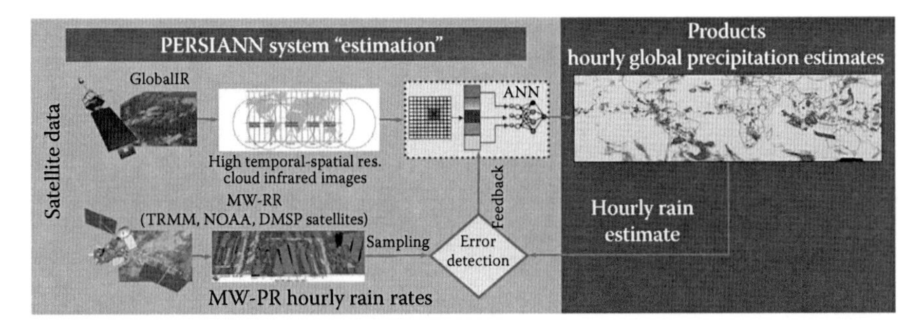

FIGURE 13.1 **(See color insert.)** Components of the PERSIANN data-retrieval algorithm. (Modified after Hsu, K. et al., *Water Resour. Res.*, 45, 2038, 2009.) [MW, microwave; PR, precipitation radar].

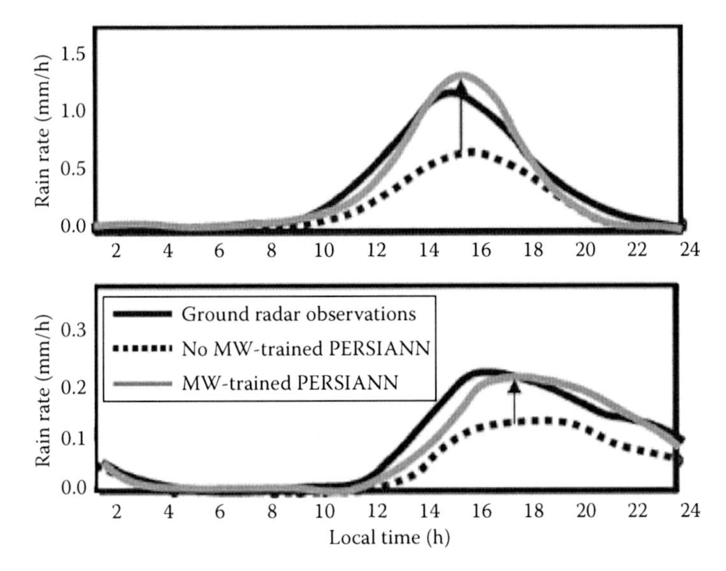

FIGURE 13.2 Diurnal distribution of summer 2002 precipitation pre- and post-MW (TMI) adjustment (location: southern Florida [upper panel], southwestern Texas [lower panel], and United States—Modified after Hsu, K. et al., *Water Resour. Res.*, 45, 2038, 2009.)

lines demonstrate the diurnal precipitation pattern pre- and post-MW adjustment, respectively. As shown, the MW adjustment scheme corrects the underestimated diurnal precipitation pattern toward the reference data (i.e., radar estimates).

Generally, the approach for estimating precipitation from IR-satellite data relies on statistical relationships between IR estimates of cloud-top brightness temperature and mean precipitation rate. However, such statistical relationships are associated with large uncertainties because of the variability in factors such as atmospheric conditions and cloud type, height, and thickness. Precipitation estimation can be improved by building a multidimensional input–output data space using multiple precipitation-related sources of information from LEO and GEO satellites. The PERSIANN algorithm relates satellite imagery from multiple platforms and wavebands to precipitation estimates through intelligently "learning" about their relationships using an automatic "training process." The learning process is accomplished by providing the network with the desired outcome (derived from radar networks and gauge measurements) for a given set of input variables.

Other major satellite precipitation products exist, generated using data sets similar to the PERSIANN algorithm but applied in different ways. The NASA TRMM Multi-satellite Precipitation Analysis (TMPA), for example, collects PMW images of rainfall from polar-orbiting satellites and fills the remaining gaps in a bracket of 3h by IR-retrieved rain estimates (Huffman et al., 2007). This product has several versions, including the TMPA-RT (real time) and an adjusted product that combines precipitation estimates from multiple satellites and gauge analyses.

Another mature precipitation product is the NOAA CMORPH technique (Joyce et al., 2004), which derives precipitation features from available PMW observations that are then advected in space and time using IR data. In other words, this algorithm employs IR scans to transport the observed PMW-derived precipitation estimates when PMW data are not available. Based on a time-weighting interpolation between PMW scans, the intensity and patterns of the precipitation estimates are derived in the intervening half-hour periods. The CMORPH algorithm propagates precipitation features forward and backward in time from the previous and following PMW scans.

Unlike the aforementioned two products, the PERSIANN algorithm uses grid IR scans of the global geosynchronous satellite network provided by NOAA CPC (Janowiak et al., 2001) as the main source of information. The IR-based estimates are then calibrated and adjusted based on PMW data from LEO satellites (e.g., TMI aboard TRMM, Special Sensor Microwave/Imager [SSM/I] on Defense Meteorological Satellite Program [DMSP], Advanced Microwave Scanning Radiometer-Earth Observing System [AMSR-E] on Aqua spacecraft, and Advanced Microwave Sounding Unit-B [AMSU-B] aboard the NOAA satellite series).

13.4 PERSIANN CLOUD CLASSIFICATION SYSTEM

To further improve the precipitation estimates from the GEO satellites, a cloud-patch-based classification system has been integrated with the PERSIANN algorithm described earlier, allowing precipitation retrieval with a higher resolution of 0.04°. In this algorithm, the retrieval is performed at the cloud-patch scale instead of the pixel scale. In other words, clouds are categorized to patches based on their top temperature levels. The precipitation rate is then derived for each patch based on the training algorithm described in the previous section. The algorithm, known as the PERSIANN-CCS, includes four computational steps: (1) IR-based cloud pattern segmentation; (2) extraction of coldness, geometry, and texture features from the IR cloud patches; (3) clustering of patch features; and (4) precipitation retrieval. A detailed description on the processes and computational steps can be found in Hong et al. (2004) and Hsu et al. (2007). A brief overview is provided here.

13.4.1 Cloud Pattern Segmentation

Based on a watershed segmentation approach developed by Vincent and Soille (1991), the PERSIANN-CCS cloud-segmentation method divides the half-hour IR image into separate patches. The classification algorithm detects the local minima of the cloud-top temperature. For every local minimum, the algorithm connects the surrounding neighboring pixels until all of the local minima are separated into distinct cloud patches. The approach is similar to delineating a watershed based on the topographical information. In this method, however, the segmentation is based on cloud-top temperature.

13.4.2 Cloud Feature Extraction

The classified cloud patches are assigned classes in three different feature categories: (1) coldness, (2) geometry, and (3) texture. Based on these class assignments, representative features including cloud size, cloud shape, cloud height (based on the coldest temperature within the patch), surface textures, and surface gradients are extracted.

13.4.3 Clustering of Patch Features

Based on the similarities of cloud patches derived in the previous two steps, a clustering approach, known as the Self-Organizing Feature Map (SOFM), is used to classify patch features (e.g., size, shape, height, surface textures, and gradients) into a number of cloud-patch categories (Kohonen, 1995; Hsu et al., 1997, 1999). In other words, cloud properties are determined in the previous steps, and in this step, similar patch features are linked together for precipitation estimation. After training of the network, cloud patches with similar features and characteristics are assigned to the same category (i.e., a classification).

13.4.4 Precipitation Retrieval

In the fourth step, precipitation rate for combined rainfall and snow is retrieved by assigning a rainfall distribution to each classified cloud-patch category based on available historical records of IR-based precipitation estimates and surface precipitation measurements. For example, historical records of NEXRAD radar data over the continental United States and PMW rainfall estimates (Ferraro and Marks, 1995; Kummerow et al., 1998) are used to derive the precipitation distributions of each classified IR patch group. In the current version of PERSIANN-CCS, the Probability Matching Method (PMM, Atlas et al., 1990; Rosenfeld et al., 1994) is employed to match the relationship between IR temperature data and the hourly precipitation estimates in each classified patch group. Finally, an exponential function with five parameters is fitted to the temperature-precipitation relationship to retrieve precipitation based on IR temperature (Hong et al., 2004; Hsu et al., 2007). The parameters of the exponential function are estimated using the Shuffled Complex Evolution (SCE-UA) algorithm (Duan et al., 1992). The algorithm produces half-hour precipitation rates that can be accumulated over a longer temporal interval (e.g., 1 h, 3 h). Figure 13.3 displays example PERSIANN-CCS 0.04° (top panel) and PERSIANN 0.25° (bottom panel) rainfall rates (mm/h). As shown, PERSIANN-CCS provides more detailed information on precipitation pattern.

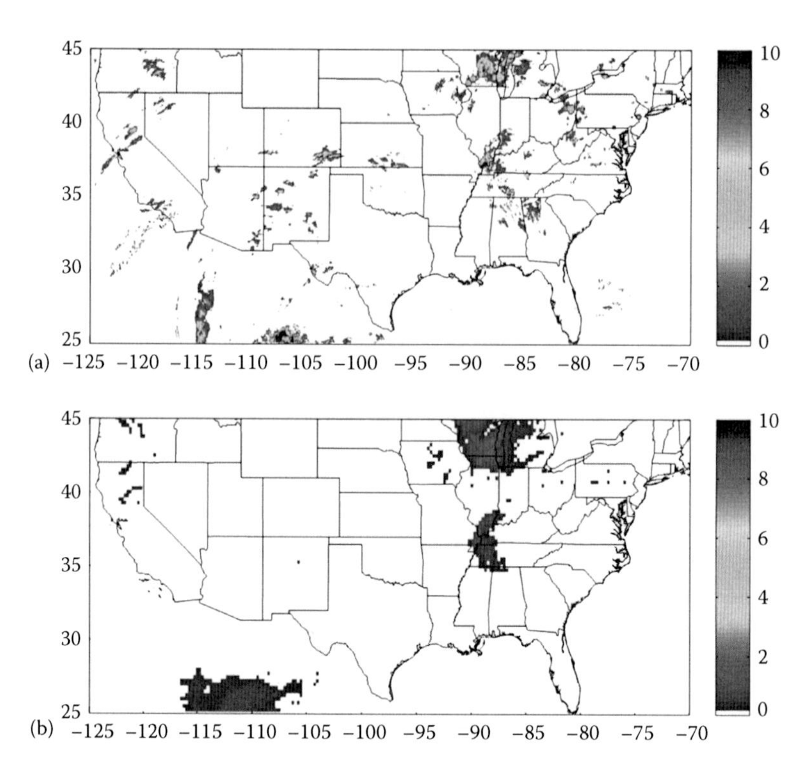

FIGURE 13.3 **(See color insert.)** Example: PERSIANN-CCS 0.04° (a) and PERSIANN 0.25° (b) rainfall rates (mm/h)—date: July 25, 2005.

13.5 MULTISENSOR PRECIPITATION ESTIMATES: MERGING AND ADJUSTMENT

As mentioned earlier, satellite precipitation estimates offer excellent spatial coverage; however, their applications are often limited because of unknown levels of uncertainty associated with interpreting IR and PMW data with respect to ground-based rainfall measurements. On the other hand, rain-gauge data provide fairly accurate measures of precipitation over a much smaller spatial scale. Ground radar networks also provide valuable precipitation information. Because each measurement technique has its own advantages and disadvantages, one may choose to integrate different sources (multisensor) to obtain the optimal estimates of precipitation. Currently, a grid-based adjusted version of the PERSIANN product, based on the Global Precipitation Climatology Project (GPCP) version 2 (Adler et al., 2003), is available for practical applications (hereafter, referred to as PERSIANN-MBA [Monthly Bias Adjusted]). As mentioned previously, PERSIANN precipitation estimates are available at a 0.25° spatial resolution. To remove biases in the PERSIANN data, the satellite-based precipitation estimates are accumulated into monthly data with a spatial resolution of 2.5° commensurate with the GPCP data set. The overall monthly bias is removed in the PERSIANN-MBA product by multiplying the adjustment factors to PERSIANN precipitation estimates. The adjustment factor is the sum of PERSIANN estimates

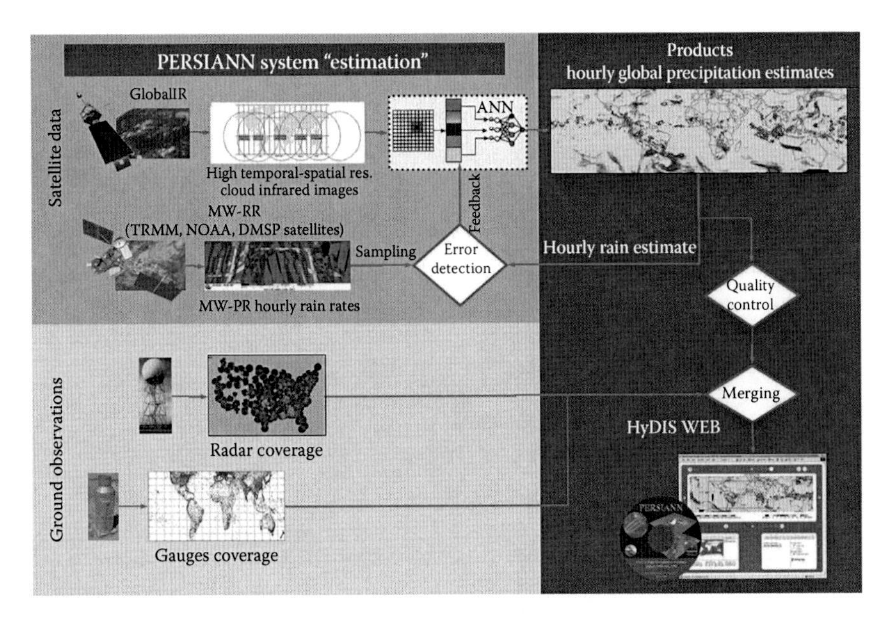

FIGURE 13.4 Components of the PERSIANN-MBA version.

over the sum of GPCP data over a given month. The PERSIANN-MBA fields are then downscaled to the original 0.25° grid resolution. The spatial downscaling is based on weighting factors derived from rain rates in the original PERSIANN data. In other words, the volume of precipitation is adjusted over a certain period of time, while the patterns of precipitation are dominated by the observed PERSIANN fields. Figure 13.4 shows the components of the PERSIANN-MBA version.

Verification analyses show that the adjusted version is significantly improved with respect to the unadjusted rainfall estimates. Figure 13.5 presents time

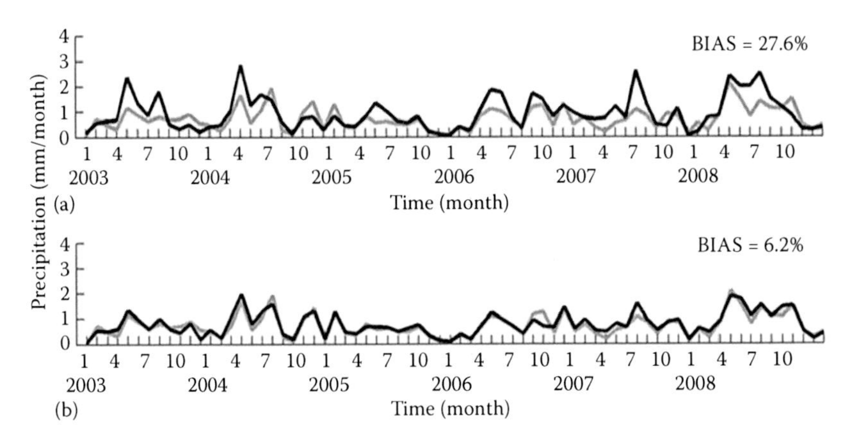

FIGURE 13.5 Time series of monthly precipitation estimates over the Illinois River basin based on (a) PERSIANN algorithm and (b) PERSIANN-MBA algorithm introduced earlier. The gray lines show the monthly averaged multisensor radar reference data. (See Behrangi, A. et al., *J. Hydrol.,* 397, 225, 2011, for more details.)

series of monthly averaged precipitation rates over the Illinois River basin for the PERSIANN unadjusted and PERSIANN-MBA. The gray lines show the monthly averaged multisensor radar/gauge-adjusted baseline data, which are an independent data set used for verification. A visual comparison confirms that, although the PERSIANN algorithm (Figure 13.5a) can capture the general monthly precipitation patterns, the bias-adjusted version (Figure 13.5b) exhibits a better agreement with gauge-adjusted radar multisensor data. The unadjusted PERSIANN overestimates the amount of precipitation, particularly during the spring and summer months (Figure 13.5a). The PERSIANN-MBA shows substantial improvement in capturing both the monthly variation and the total amount of precipitation. The improvement in the overall bias (ratio of total satellite over radar precipitation accumulations) is reduced from 27.6% to 6.2% through this adjustment process.

13.6 HyDIS WEB-BASED DATA INTERFACE

PERSIANN data are available through the NASA-funded Hydrological Data and Information Systems (HyDIS, http://hydis8.eng.uci.edu/hydis-unesco/; see Figure 13.6). PERSIANN-CCS is available through the Global Water and Development Information (GWADI) web server (http://hydis.eng.uci.edu/gwadi/). The GWADI online data visualization and acquisition web-mapping server has been developed with functionalities that allow user interaction with the available data, including visualization and extracting pixel, watershed, and country queries, and downloading data for a specific location. Figure 13.7 displays a sample precipitation report from the GWADI web server that includes precipitation accumulations and pattern development over a 72 h period (ending January 14, 2011, 16:30 PST) across the northwestern Amazon (top panels), as well the monthly long-term climatology of the region (bottom panel).

13.7 REMOTE SENSING PRECIPITATION FOR DROUGHT ANALYSIS

As a case study, we present an example of applying PERSIANN rainfall to drought analysis. The SPI is derived from PERSIANN data for each 0.25° pixel across the contiguous United States using data from 2000 to 2009. The SPI is designed to identify precipitation deficits at various timescales (e.g., 3, 6, 12, 24, and 48 months). Although a long-term precipitation record is needed for deriving SPI reliably, the following example shows that when using 10 years of data, one can obtain reasonable estimates of drought condition. To obtain SPI, the precipitation record is fitted to a gamma distribution, and the accumulated gamma probability is then transferred to a normal-distributed cumulative density function (CDF). As a result, SPI is standardized to a normal distribution having zero mean and a standard deviation of one (McKee et al., 1993; Tsakiris and Vangelis, 2004). The SPI classification of wet-dry intensity is shown in Table 13.1. The SPI represents the amount of precipitation in a normalized scale, over a given timescale. High SPI values (>2.0) represent high precipitation accumulation over a certain period of time (e.g., 3, 6, 12 month period), whereas SPI near zero refers to normal precipitation (near climatologic mean value). On the other hand, low SPI values (<–2.0) indicate less precipitation than the climatologic mean over the period of interest. In other words, a sequence of negative SPI values in time implies that climate is in the dry status (drought), while a consecutive number of positive SPIs show a wet climate scenario (McKee et al., 1993).

Figure 13.8 displays the SPI estimates computed using PERSIANN satellite data (left panels) and the National Climatic Data Center (NCDC) Global Historical Climatology Network (GHCN) gauge data set (right panels). In Figure 13.8, the

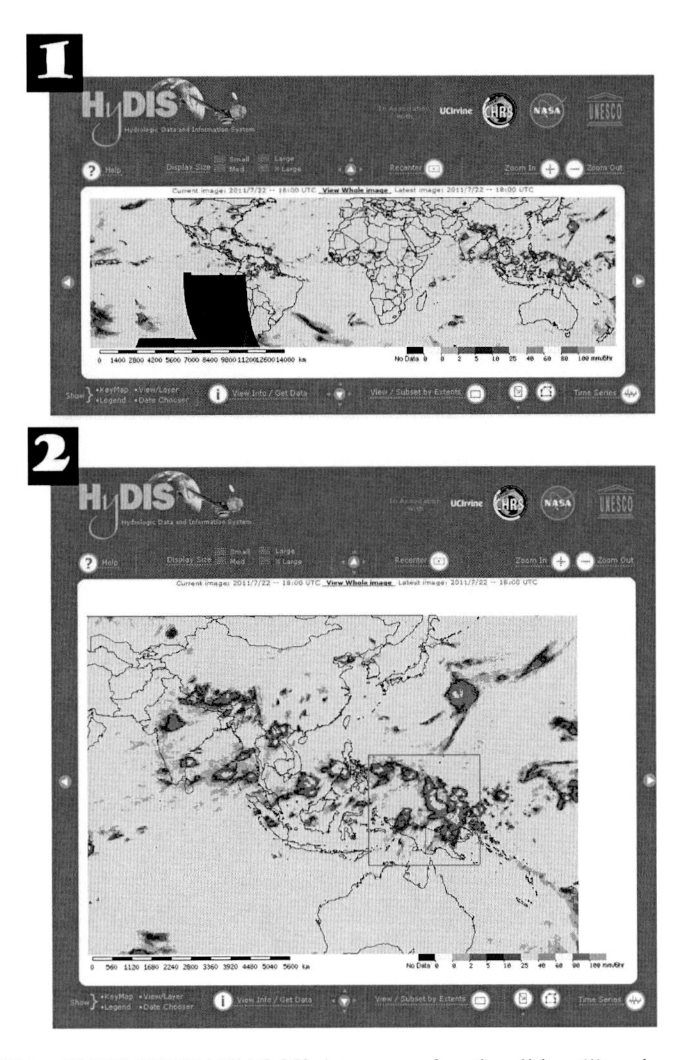

FIGURE 13.6 HyDIS PERSIANN 0.25° data server functionalities: (1) main access page (http://hydis8.eng.uci.edu/hydis-unesco/), (2) navigation includes interactive zooming,

(continued)

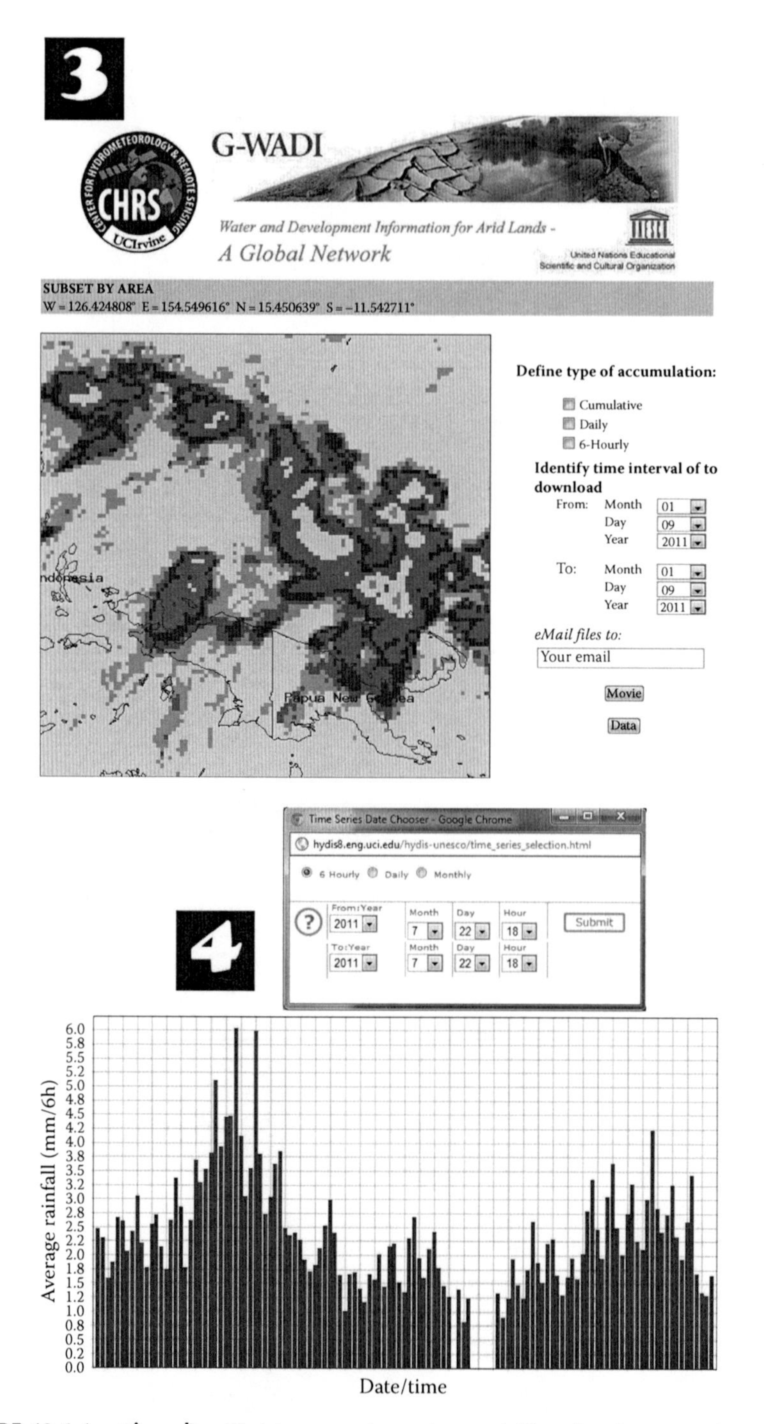

FIGURE 13.6 (continued) (3) data access by region, and (4) regional average time series plot and data access.

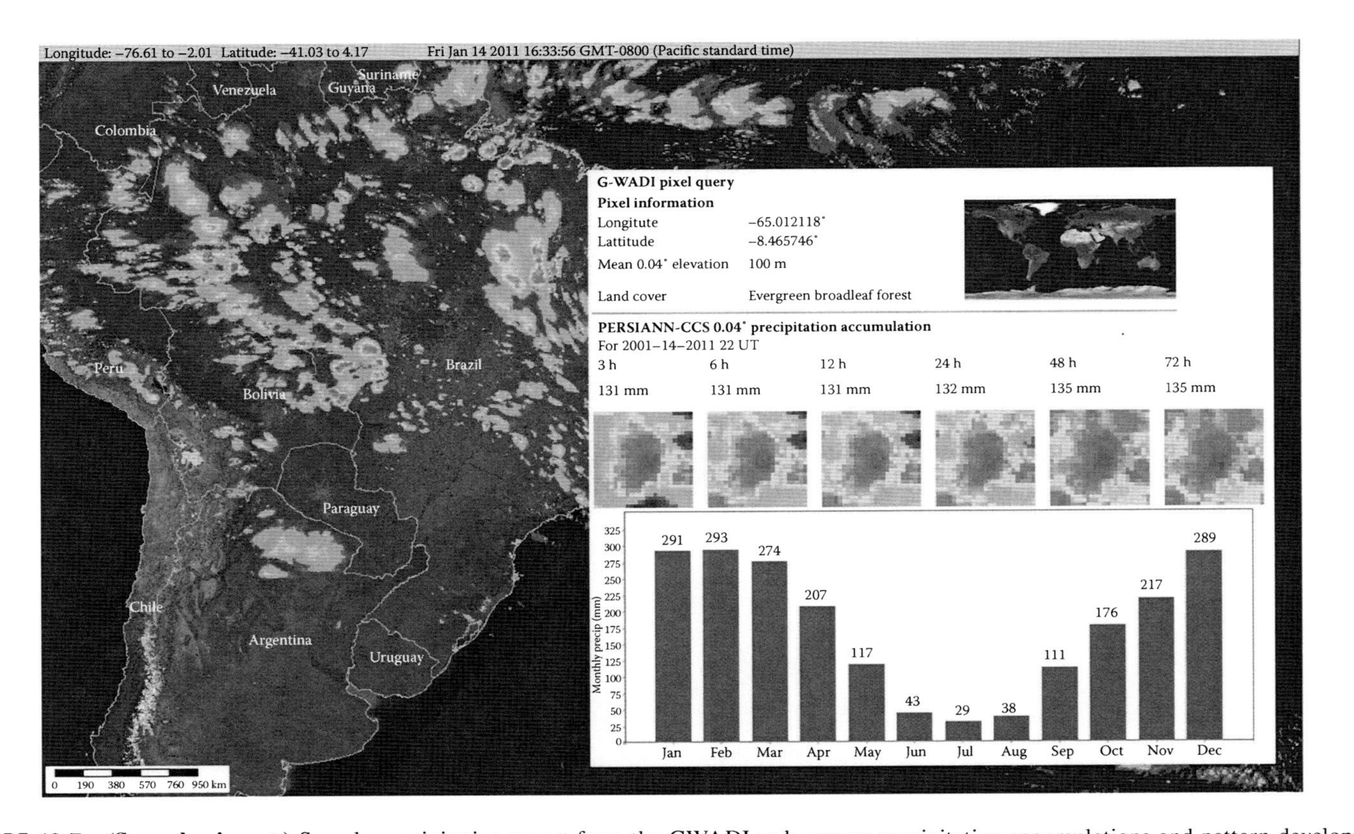

FIGURE 13.7 **(See color insert.)** Sample precipitation report from the GWADI web server: precipitation accumulations and pattern development over a 72 h period (ending January 14, 2011, 16:30 PST) across northwestern Amazon (top panels) and monthly long-term climatology of the region (bottom panel).

TABLE 13.1
SPI Classification of Wet and Dry Climate Status

SPI Index Value	Climate Condition
>2.0	Extremely wet
1.5–2.0	Very wet
1.0–1.5	Moderately wet
−1.0 to 1.0	Near normal
−1.0 to −1.5	Moderate drought
−1.5 to −2.0	Severe drought
<−2.0	Extreme drought

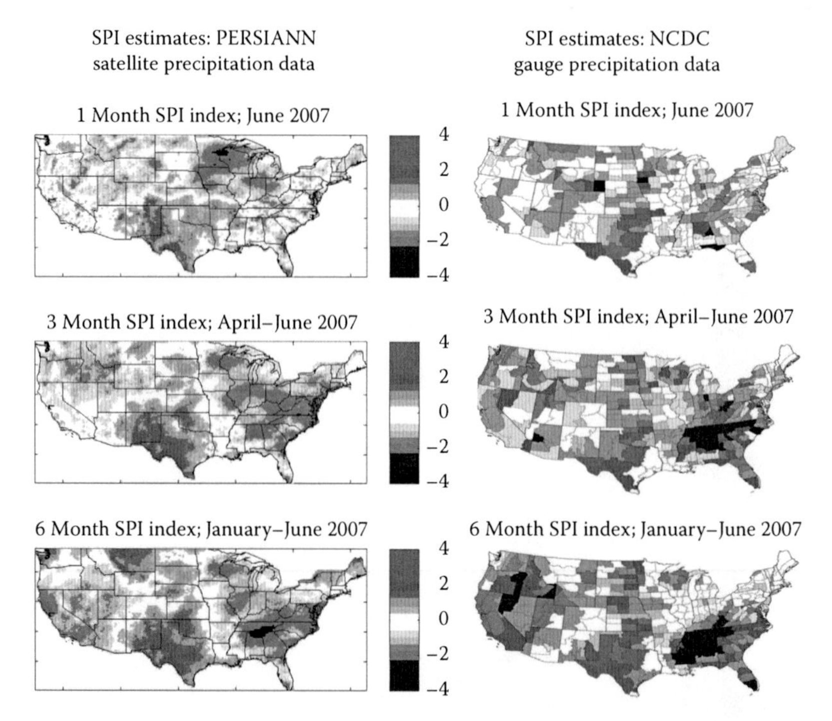

FIGURE 13.8 **(See color insert.)** SPI estimates using PERSIANN satellite data (left panels) and the NCDC GHCN data set (right panels).

upper, middle, and lower panels show 1 month (June 2007), 3 month (April–June 2007), and 6 month (January–June 2007) SPI, respectively. Because they use gauge data, the NCDC analyses are generated at the climate division scale, while the PERSIANN satellite-based SPI analyses are performed at the pixel scale (0.25°). General patterns of wet and dry conditions are similar in both types of analysis, showing extreme dry conditions in the southeast and wetter conditions in the central

United States during 2007. Some differences may be attributable to the difference in normalization interval. The NCDC-based SPI is normalized using long-term gauge data, while only 10 years of data (2000–2009) are used in the PERSIANN SPI computations. Overall, Figure 13.8 indicates that satellite precipitation estimates have the potential to be used in drought monitoring and analysis, particularly over remote areas where no other means of reliable information is available.

For reliable climate research, including drought analysis, long-term global precipitation observations are of primary importance. As mentioned earlier, one of the limitations of the satellite-derived precipitation estimates is their relatively short length of the data record. The narrative of satellite precipitation estimation is one of continuous progress toward increasingly better quality and resolution. NOAA's CPC Merged Analysis of Precipitation (CMAP, Xie and Arkin, 1997) and NASA's GPCP (Huffman et al., 1997, 2001; Adler et al., 2003) data sets, with their quasi-global coverage and long-term time periods (1979–present), have contributed greatly to weather and climate change studies. However, the coarse resolutions (2.5° spatial and monthly temporal resolutions) limit their ability to capture the spatial and dynamic details of precipitation. The emergence of high-resolution satellite precipitation algorithms and the subsequent publication of their products are due largely to advances in sensor technologies, beginning in 1997 with the launch of the PMW sensor carried by the TRMM satellite. Because of their reliance on satellite data, many high-resolution precipitation data sets do not have the duration necessary for weather and climate studies at a global scale.

Several high-resolution satellite precipitation algorithms rely on the synergy between the superior quality but infrequent low orbit observations and the high sampling rate of IR cloud-top temperature data. By design, these high-resolution precipitation estimation algorithms (e.g., TMPA, CMORPH) require the availability of both IR and PMW observations, which restricts their applicability for reconstruction of precipitation records during the pre-1997 era. By way of contrast, the PERSIANN system can be viewed as a two-phase system: (1) a training phase in which an ANN algorithm is trained using available observations including high-quality PMW data, and (2) an application phase during which the ANN can be applied to IR-only data without the need for PMW observations. As such, the PERSIANN algorithm is uniquely positioned to utilize historical archives of IR observations to produce precipitation estimates for the pre-PMW era. Currently, efforts are under way to reconstruct a long-term PERSIANN data set for weather and climate research and applications using this approach to complete a longer, retrospective time series of estimates. It is hoped that reconstruction of the PERSIANN algorithm will bridge the gap between high-resolution precipitation data that have been estimated over the past 10+ years and the need for long-term data required for global drought analysis.

13.8 FUTURE DEVELOPMENTS

Thus far, a great deal of effort has been made to validate and improve PERSIANN precipitation products. Although the use of satellite data in practical applications seems to be very promising, extensive research is required to further improve the available data sets. Previous studies show that most unadjusted precipitation products, including PERSIANN, tend to overestimate precipitation, particularly over warm months (AghaKouchak et al., 2011). The overestimations of all satellite products are even more significant for extreme precipitation rates. However, the false alarm ratios (an index of false precipitation) indicate that PERSIANN, along with most other products, has less skill identifying precipitation correctly over cold months (AghaKouchak et al., 2011). Therefore, efforts are under way to develop more sophisticated adjustment techniques to further improve the reliability of the data for practical applications. Additionally, new sources of information, including VIS channels, IR data from the NASA Moderate Resolution Imaging Spectroradiometer (MODIS) images, and satellite-based cloud classification fields from NASA CloudSat, are being incorporated to derive precipitation more accurately.

13.9 SUMMARY

Reasonable management and decision making in the discipline of water resources require reliable prediction of drought duration and severity in order to manage drought-sensitive systems. Several climate predictions suggest an increase in drought duration and severity in the future. Additionally, there is evidence that drought has been increasing on a global basis (WCRP, 2010). With increasing population and water demand, as well as potential impacts of climate change, droughts and their consequences are of major concern across the globe. One of the main obstacles in developing algorithms for characterization and prediction of droughts is the lack of information on drought mechanisms. There is general agreement that among current drought models, no single methodology is accepted as the best model to predict/characterize droughts and their severity (Kao and Govindaraju, 2010). Furthermore, some drought-assessment approaches may not be consistent with others because drought measures are based on one or more indicator(s) that may differ from model to model (e.g., precipitation, evapotranspiration, or baseflow). Currently, the SPI method is frequently being used because of its simplicity and flexibility. An attractive feature of SPI is that, if the precipitation is available in real time, the drought information can be provided in a timely manner. Satellite sensing is capable of providing better overall spatial and temporal coverage of rainfall distribution than that of rain gauges. Furthermore, it is expected that satellite-precipitation data will provide additional information for drought monitoring and analysis.

In this chapter, the development of the PERSIANN and PERSIANN-CCS systems and their continuing evolution during the past decade were presented. The main advantage of PERSIANN precipitation estimates is the possibility of reconstructing long-term records relevant to drought-related research and practice, with quasi-global coverage. Furthermore, the adaptive training method built into PERSIANN allows for optimal use of the effective sampling capability of GEO satellites, as well as high-quality PMW sensors in LEO. The evolution of PERSIANN has continued with the development of the cloud classification system (PERSIANN-CCS), which now allows precipitation estimation on regional cloud-patch systems instead of on local pixel-based precipitation retrieval. The PERSIANN-CCS algorithm separates the thermal IR cloud-top temperature data into disjointed cloud patches for which patch features are categorized into different classes. Using independent sources of data, a specific rainfall distribution is assigned to each classified category.

This study also demonstrated the use of PERSIANN rainfall data for drought monitoring and analysis. Using 0.25° PERSIANN precipitation estimates, SPIs were derived for the contiguous United States. The results showed that the general patterns of SPIs derived from PERSIANN data were in reasonable agreement with those obtained from the NCDC GHCN gauge data. It is well known that drought analyses require high-quality, long-term precipitation records. Currently, interdecadal drought analysis using high-resolution satellite-based precipitation data is not possible because of limitations in length of available records. Efforts are under way by the authors to provide PERSIANN data continuously for the past three decades for which satellite IR images are available. We therefore anticipate that the improved long-term (multidecadal), high-resolution PERSIANN data will provide a valuable quasi-global data set for drought monitoring and applications.

REFERENCES

Adler, B., G. Huffman, A. Chang, R. Ferraro, P. Xie, J. Janowiak, B. Rudolf, U. Schneider, S. Curtis, D. Bolvin, A. Gruber, J. Sussking, P. Arkin, and E. Nelkin. 2003. The version-2 Global Precipitation Climatology Project (GPCP) monthly precipitation analysis (1979– present). *Journal of Hydrometeorology* 4:1147–1167.

Adler, R., C. Kidd, G. Petty, M. Morrisey, and H. Goodman. 2001. Intercomparison of global precipitation products: The Third Precipitation Intercomparison Project (PIP-3). *Bulletin of the American Meteoological Society* 82:1377–1396.

AghaKouchak, A., A. Behrangi, S. Sorooshian, K. Hsu, and E. Amitai. 2011. Evaluation of satellite-retrieved extreme precipitation across the central United States. *Journal of Geophysical Research* 116:D02115, doi:10.1029/2010JD014741.

Alexander, L., X. Zhang, T. Peterson, J. Caesar, B. Gleason, A.M.G.K. Tank, M. Haylock, D. Collins, B. Trewin, F. Rahimzadeh, A. Tagipour, K.R. Kumar, J. Revadekar, G. Griffiths, L. Vincent, D.B. Stephenson, J. Bum, E. Aguilar, M. Brunet, M. Taylor, M. New, P. Zhai, M. Rusticucci, and J.L. Vazquez-Aguirre. 2006. Global observed changes in daily climate extremes of temperature. *Journal of Geophysical Research* 111:D05109.

Anderson, L., Y. Malhi, L. Aragao, and S. Saatchi. 2008. Spatial patterns of the canopy stress during 2005 drought in Amazonia. *Proceedings of IEEE International Geoscience and Remote Sensing Symposium* 1–12:2294–2297.

Atlas, D., D. Rosenfeld, and D. Short. 1990. The estimation of convective rainfall by area integrals. 1. The theoretical and empirical basis. *Journal Geophysical Research* 95:1153–2160.

Ba, M. and A. Gruber. 2001. GOES Multispectral Rainfall Algorithm (GMSRA). *Journal of Appied Meteorology* 29:1120–1135.

Barnett, D., S. Brown, J. Murphy, D. Sexton, and M. Webb. 2006. Quantifying uncertainty in changes in extreme event frequency in response to doubled CO2 using a large ensemble of GCM simulations. *Climate Dynamics* 26:489–511.

Behrangi, A., B. Khakbaz, T.C. Jaw, A. AghaKouchak, K. Hsu, and S. Sorooshian. 2011. Hydrologic evaluation of satellite precipitation products at basin scale. *Journal of Hydrology* 397:225–237.

Bellerby, T., M. Todd, D. Kniveton, and C. Kidd. 2000. Rainfall estimation from a combination of TRMM precipitation radar and GOES multispectral satellite imagery through the use of an artificial neural network. *Journal of Applied Meteorology* 39:2115–2128.

Boo, K., W. Kwon, and H. Baek. 2006. Change of extreme events of temperature and precipitation over Korea using regional projection of future climate change. *Geophysical Research Letters* 33:281–290, L01701.

Caprio, J., H. Quamme, and T. Kelly. 2009. A statistical procedure to determine recent climate change of extreme daily meteorological data as applied at two locations in Northwestern North America. *Climatic Change* 92:65–81.

Coppola, E., D. Grimes, M. Verdecchia, and G. Visconti. 2006. Validation of improved TAMANN neural network for operational satellite-derived rainfall estimation in Africa. *Journal of Applied Meteorology* 45:1557–1572.

Cubasch, U., G. Meehl, G. Boer, R. Stouffer, M. Dix, A. Noda, C.A. Senior, S. Raper, and K.S. Yap. 2001. Projections of future climate change. In *Climate Change 2001: The Scientific Basis. Contribution of Working Group I to the Third Assessment Report of the Intergovernmental Panel on Climate Change*, eds. J.T. Houghton, Y. Ding, D.J. Griggs et al., pp. 526–582, Cambridge, MA: Cambridge University Press.

Dirmeyer, P., X. Gao, Z. Gao, T. Oki, and M. Hanasaki. 2006. The Global Soil Wetness Project (GSWP-2). *Bulletin of the American Meteorological Society* 87:1381–1397.

Duan, Q., S. Sorooshian, and V. Gupta. 1992. Effective and efficient global optimization for conceptual rainfall-runoff model. *Water Resources Research* 28:1015–1031.

Ferraro, R. and G. Marks. 1995. The development of SSM/I rain-rate retrieval algorithms using ground-based radar measurements. *Journal of Atmospheric Oceanic Technology* 12:755–770.

Frich, P., L. Alexander, P. Della-Marta, B. Gleason, M. Haylock, A.M.G.K. Tank, and T. Peterson. 2002. Observed coherent changes in climatic extremes during the second half of the twentieth century. *Climate Research* 19:193–212.

Gao, X., J. Pal, and F. Giorgi. 2006. Projected changes in mean and extreme precipitation over the Mediterranean region from a high resolution double nested RCM simulation. *Geophysical Research Letters* 33:L03706.

Gong, D. and S. Wang. 2000. Severe summer rainfall in China associated with the enhanced global warming. *Climate Research* 16:51–59.

Govindaraju, R. and A. Rao. 2000. *Measurement of Precipitation from Space: EURAINSAT and Future*. Dordrecht, the Netherlands: Kluwer Academic Publishers.

Groisman, P., T. Karl, D. Easterling, R. Knight, P. Jameson, K.J. Hennessy, R. Suppiah, C.M. Page, J. Wibig, K. Fortuniak, V.N. Razuvaev, A. Douglas, E. Forland, and P.M. Zhai. 1999. Changes in the probability of heavy precipitation: Important indicators of climatic change. *Climatic Change* 42:243–283.

Hayes, M., M. Svoboda, D. Wilhite, and O. Vanyarkho. 1999. Monitoring the 1996 drought using the Standardized Precipitation Index. *Bulletin of the American Meteorological Society* 80:429–438.

Hong, Y., K. Hsu, X. Gao, and S. Sorooshian. 2004. Precipitation estimation from remotely sensed imagery using Artificial Neural Network-Cloud Classification System. *Journal of Applied Meteorology* 43:1834–1853.

Hsu, K., X. Gao, S. Sorooshian, and H. Gupta. 1997. Precipitation estimation from remotely sensed information using artificial neural networks. *Journal of Applied Meteorology* 36:1176–1190.

Hsu, K., H. Gupta, X. Gao, and S. Sorooshian. 1999. A neural network for estimating physical variables from multi-channel remotely sensed imagery: Application to rainfall estimation. *Water Resources Research* 35:1605–1618.

Hsu, K., Y. Hong, and S. Sorooshian. 2007. Rainfall estimation using a cloud patch classification map. In *Measurement of Precipitation from Space: EURAINSAT and Future,* eds. V. Levizzani, P. Bauer, and F. Turk, pp. 329–242. Dordrecht, the Netherlands: Springer.

Hsu, K., H. Moradkhani, and S. Sorooshian. 2009. A sequential Bayesian approach for hydrologic model selection and prediction. *Water Resources Research* 45:2038–2059.

Huffman, G., R. Adler, P. Arkin, A. Chang, R. Ferraro, A. Gruber, J. Jamowiak, A. McNab, B. Rudolf, and U. Schneider. 1997. The Global Precipitation Climatology Project (GPCP) combined precipitation dataset. *Bulletin of the American Meteorological Society* 78:5–20.

Huffman, G., R. Adler, D. Bolvin, G. Gu, E. Nelkin, K. Bowman, Y. Hong, E.F. Stocker, and D.B. Wolff. 2007. The TRMM Multi-satellite Precipitation Analysis: Quasi-global, multiyear, combined-sensor precipitation estimates at fine scale. *Journal of Hydrometeorology* 8:38–55.

Huffman, G., R. Adler, M. Morrissey, D. Bolvin, S. Curtis, R. Joyce, B. McGavock, and J. Susskind. 2001. Global precipitation at one-degree daily resolution from multisatellite observations. *Journal of Hydrometeorology* 2:36–50.

Idso, S.B., R.D. Jackson, P.J. Pinter Jr., R.J. Reginato, and J.L. Hatfield. 1981. Normalizing the stress degree day parameter for environmental variability. *Agricultural Meteorology* 24:45–55.

Janowiak, J., R. Joyce, and Y. Yarosh. 2001. A real-time global half-hourly pixel-resolution infrared dataset and its applications. *Bulletin of the American Meteorological Society* 82:205–217.

Joyce, R., J. Janowiak, P. Arkin, and P. Xie. 2004. CMORPH: A method that produces global precipitation estimates from passive microwave and infrared data at high spatial and temporal resolution. *Journal of Hydrometeorology* 5:487–503.

Kao, S. and R. Govindaraju. 2010. A copula-based joint deficit index for droughts. *Journal of Hydrology* 380(1–2):121–134.

Karl, T. 1983. Some spatial characteristics of drought duration in the United States. *Journal of Climate and Applied Meteorology* 22:1356–1366.

Kidd, C. 2001. Satellite rainfall climatology: A review. *International Journal of Climatology* 21:1041–1066.

Kohonen, T. 1995. *Self-Organizing Map*. New York: Springer-Verlag.

Kummerow, C., W. Barnes, T. Toshiaki, J. Shiue, and J. Simpson. 1998. The Tropical Rainfall Measuring Mission (TRMM) sensor package. *Journal of Atmospheric and Oceanic Technology* 15:809–817.

Levizzani, V. 2008. Satellite clouds and precipitation observations for meteorology and climate. In *Hydrological Modelling and the Water Cycle*, eds. S. Sorooshian, K.-L. Hsu, E. Coppola et al., pp. 49–68. Berlin, Germany: Springer.

Levizzani, V. and R. Amorati. 2002. A review of satellite-based rainfall estimation methods. A look back and a perspective. *Multiple Sensor Precipitation Measurements, Integration, Calibration, and Flood Forecasting(MUSIC)*. http://www.geomin.unibo.it/hydro/music/reports/d6.1_satellite%20rainfall%20overview.pdf (accessed on March 15, 2011).

Levizzani, V., P. Bauer, and F. Turk. 2007. *Measurement of Precipitation from Space: EURAINSAT and the Future*. Dordrecht, the Netherlands: Springer.

Maddox, R., J. Zhang, J. Gourley, and K. Howard. 2002. Weather radar coverage over the contiguous United States. *Weather Forecasting* 17:927–934.

McKee, T.B., N.J. Doesken, and J. Kleist. 1993. The relationship of drought frequency and duration to time scales. *Proceedings of the Eighth Conference of Applied Climatology*, Anaheim, CA, pp. 179–184. Boston, MA: American Meteorological Society.

Mo, K. 2008. Model based drought indices over the United States. *Journal of Hydrometeorology* 9:1212–1230.

Paridal, B.R., W.B. Collado, R. Borah, M.K. Hazarika, and L. Sarnarakoon. 2008. Detecting drought-prone areas of rice agriculture using a MODIS-derived soil moisture index. *GIScience and Remote Sensing* 4(1):109–129.

Robock, A., L. Luo, E. Wood, F. Wen, K. Mitchell, P. Houser, J.C. Schaake, D. Lohmann, B. Cosgrove, J. Sheffield, Q. Duan, R.W. Higgins, R.T. Pinker, J.D. Tarpley, J.B. Basara, and K.C. Crawford. 2004. Evaluation of the North American Land Data Assimilation System over the southern Great Plains during the warm season. *Journal of Geophysical Research* 108:doi10.1029/2002JD003245.

Rosenfeld, D., D. Wolff, and E. Amitai. 1994. The window probability matching method for rainfall measurements with radar. *Journal of Apply Meteorology* 33:683–693.

Schmidli, J. and C. Frei. 2005. Trends of heavy precipitation and wet and dry spells in Switzerland during the 20th century. *International Journal of Climatology* 25:753–771.

Sheffield, J., E.F. Wood, D.P. Lettenmaier, and A. Lipponen. 2008. Experimental drought monitoring for Africa. *GEWEX News* 18(3):4–6.

Shukla, S. and A. Wood. 2008. Use of a standardized runoff index for characterizing hydrologic drought. *Geophysical Research Letters* 35:L02405, doi: 10.1029/2007GL032487.

Solomon, S., D. Qin, M. Manning, M. Marquis, K. Averyt, M. Tignor, and H.L. Miller. 2007. Projections of future climate change. In *Climate Change 2007: The Physical Science Basis. Contribution of Working Group I to the Fourth Assessment Report of the Intergovermental Panel on Climate Change*. New York: Cambridge University Press.

Sorooshian, S., K. Hsu, X. Gao, H. Gupta, B. Imam, and D. Braithwaite. 2000. Evolution of the PERSIANN system satellite-based estimates of tropical rainfall. *Bulletin of the American Meteorological Society* 81(9):2035–2046.

Tapiador, K.C., K. Hsu, and F. Narzano. 2004. Neural networks in satellite rainfall estimation. *Meteorological Applications* 11:83–91.

Tian, Y., C. Peters-Lidard, B. Choudhury, and M. Garcia. 2007. Multitemporal analysis of TRMM-based satellite precipitation products for land data assimilation applications. *Journal of Hydrometeorology* 8(6):1165–1183.

Tian, Y., C. Peters-Lidard, J. Eylander, R. Joyce, G. Huffman, R. Adler, K. Hsu, F.J. Turk, M. Garcia, and J. Zeng. 2009. Component analysis of errors in satellite-based precipitation estimates. *Journal of Geophysical Research* 114:D24101, doi:10.1029/2009JD011949.

Tsakiris, G. and H. Vangelis. 2004. Towards a drought watch system based on spatial SPI. *Water Resources Management* 18:1–12.

Turk, F., P. Bauer, E. Ebert, and P. Arkin. 2006. Satellite-derived precipitation verification activities within the International Precipitation Working Group (IPWG). *Preprints, 14th Conference on Satellite Meteorology and Oceanography*, Atlanta, GA. Boston, MA: American Meteorological Society.

Turk, F., G. Rohaly, J. Hawkins, E. Smith, A. Grose, F. Marzano, A. Mugnai, and V. Levizzani. 2000. Analysis and assimilation of rainfall from blended SSM/I, TRMM and geostationary satellite data. *Proceedings of the 10th Conference on Satellite Meteorology and Oceanography*, Long Beach, CA, pp. 66–69. Boston, MA: American Meteorological Society.

Vicente, G., R. Scofield, and W. Menzel. 1998. The operational GOES infrared rainfall estimation technique. *Bulletin of the American Meteorological Society* 79:1883–1898.

Vincent, L. and P. Soille. 1991. Watersheds in digital spaces: An efficient algorithm based on immersion simulations. *IEEE Transactions on Pattern Analysis and Machine Intelligence* 13:583–598.

WCRP. 2010. A WCRP white paper on drought predictability and prediction in a changing climate: Assessing current predictive knowledge and capabilities, user requirements and research priorities. *World Climate Research Programme*, Barcelona, Spain, March 2–4, 2011. http://drought.wcrp-climate.org/workshop/Paper.shtml

Wilby, R. and T. Wigley. 2002. Future changes in the distribution of daily precipitation totals across North America. *Geophysical Research Letters* 29(1135):doi:10.1029/2001GL013048.

WMO. 2009. *Inter-Regional Workshop on Indices and Early Warning Systems for Drought*. Lincoln, Nebraska, December 8–11, 2009. Geneva, Switzerland: World Meteorological Organization. http://www.wmo.int/pages/prog/wcp/agm/meetings/wies09/wies09_present.php

Xie, P. and P. Arkin. 1995. An intercomparison of gauge observations and satellite estimates of monthly precipitation. *Journal of Applied Meteorology* 34:1143–1160.

Xie, P. and P. Arkin. 1997. Global precipitation: A 17-year monthly analysis based on gauge observations, satellite estimates, and numerical model outputs. *Bulletin of the American Meteorological Society* 78:2539–2558.

Xu, L., X. Guo, S. Sorooshian, P. Arkin, and B. Imam. 1999. A microwave infrared threshold technique to improve the GOES precipitation index. *Journal of Applied Meteorology* 38:569–579.

Zhang, Q., C.-Y. Xu, Z. Zhang, Y. Chen, C.-L. Liu, and H. Lin. H. 2008. Spatial and temporal variability of precipitation maxima during 1960–2005 in the Yangtze River basin and possible association with large-scale circulation. *Journal of Hydrology* 353:215–227.

Zolina, O., S. Clemens, A. Kapala, and S. Gulev. 2005. On the robustness of the estimates of centennial-scale variability in heavy precipitation from station data over Europe. *Geophysical Research Letters* 32:L14707, doi:10.1029/2005GL023231.

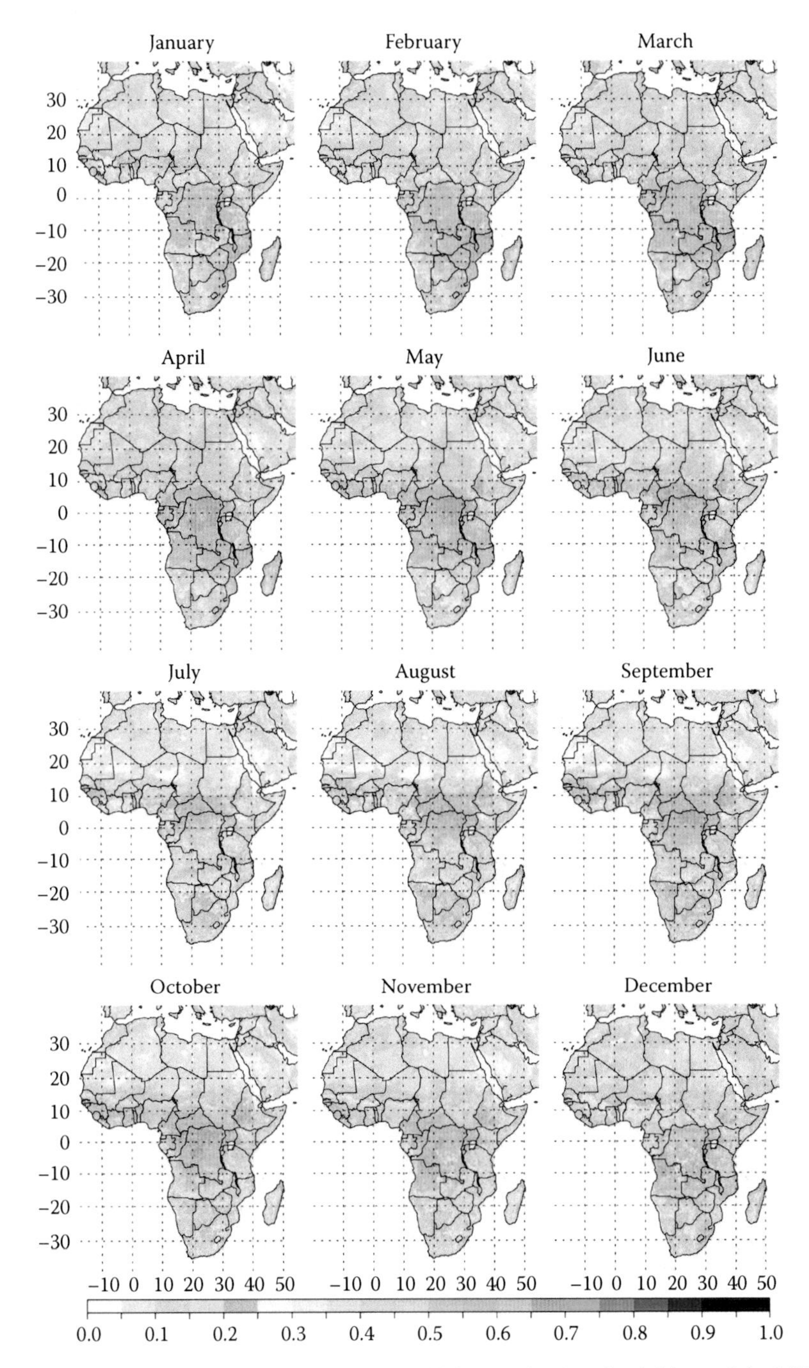

FIGURE 2.2 Example of typical monthly NDVI time-series data for Africa and the Middle East. In general, areas of high NDVI or high vegetation density are represented in shades of green while areas of low NDVI/low vegetation density such as semiarid lands and Sahara and Arabian deserts are shown in shades of yellow to brown. The patterns change seasonally from January through December. (Data produced by GIMMS Group at NASA/GSFC, Greenbelt, MD.)

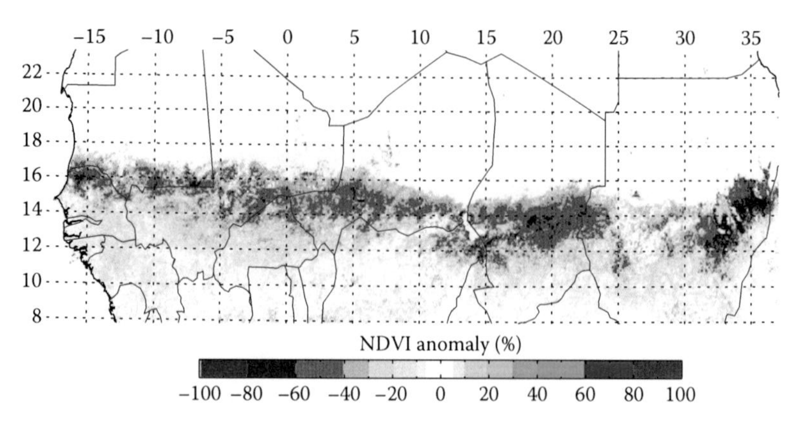

FIGURE 2.5 Growing season (July–October) NDVI anomaly for the Sahel region showing the large areal extent of the Sahelian drought in 1984. Before AVHRR NDVI data became available, such regional-to-continental mapping of drought extent and patterns was not possible. (Adapted from Anyamba, A. and Tucker, C.J., *J. Arid Environ.*, 63, 596, 2005.)

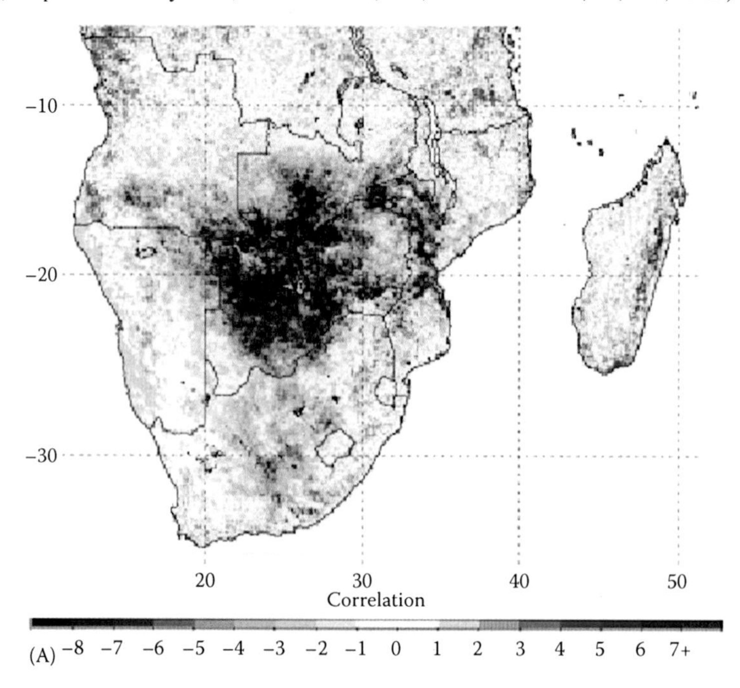

FIGURE 2.6 Principal component analysis results of monthly NDVI anomaly time series for southern Africa for the period 1986–1990, showing the drought spatial pattern in (A) and the associated temporal loadings for the first principal component in (B). This component accounts for 9.73% of the total variance of the anomaly time series. The temporal loadings (B) represent the correlation between each image in the time series with the component spatial pattern in the map (A). The component loadings show a positive correlation with the drought (negative) spatial component pattern in the map (A) between late 1986 and late 1987 and negative correlation (wetter or greener than normal conditions) between 1988 and 1990 with the spatial component pattern (A). This component pattern is related to interannual variability rainfall associated with the ENSO phenomenon. The temporal loadings are highly correlated ($r = 0.80$) with ENSO that is represented by the Oceanic Niño Index (ONI). (Reconstructed after Anyamba, A. and Eastman, J.R., *Int. J. Rem. Sens.*, 17, 2533, 1996.)

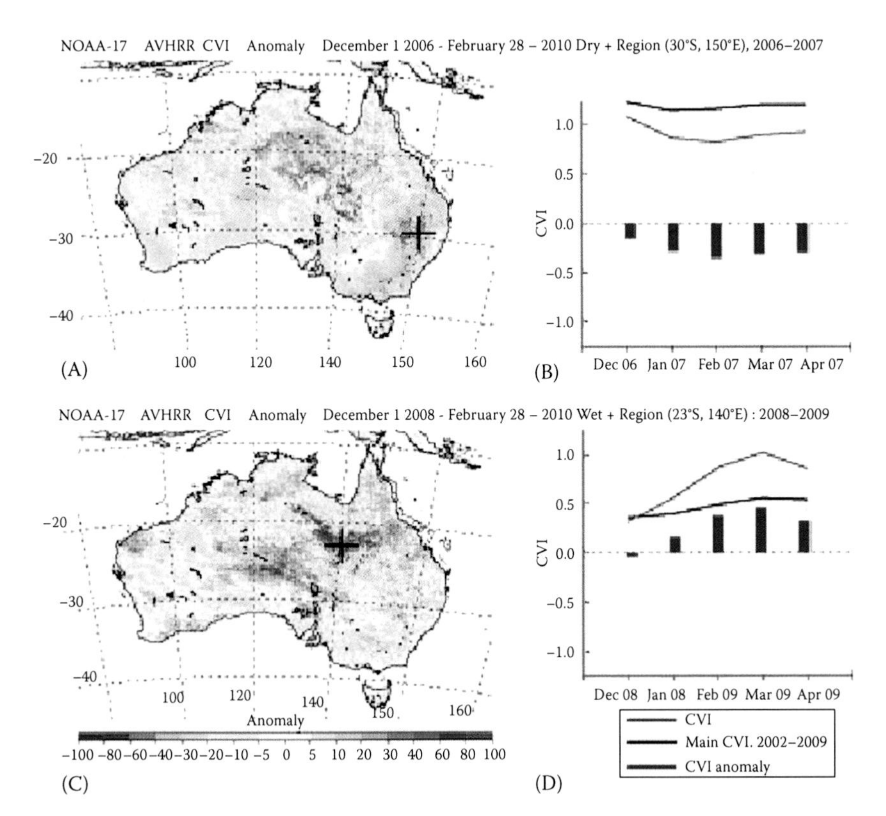

FIGURE 2.7 Cumulative NDVI anomalies (CVI) for Australia, showing the cumulative nature of drought from December 2006 to 2007 (A) and the wetter/greener-than-normal conditions from December 2008 to February 2009 (C). Cumulative time-series profiles are shown for a drought location (B) and a wet location (D).

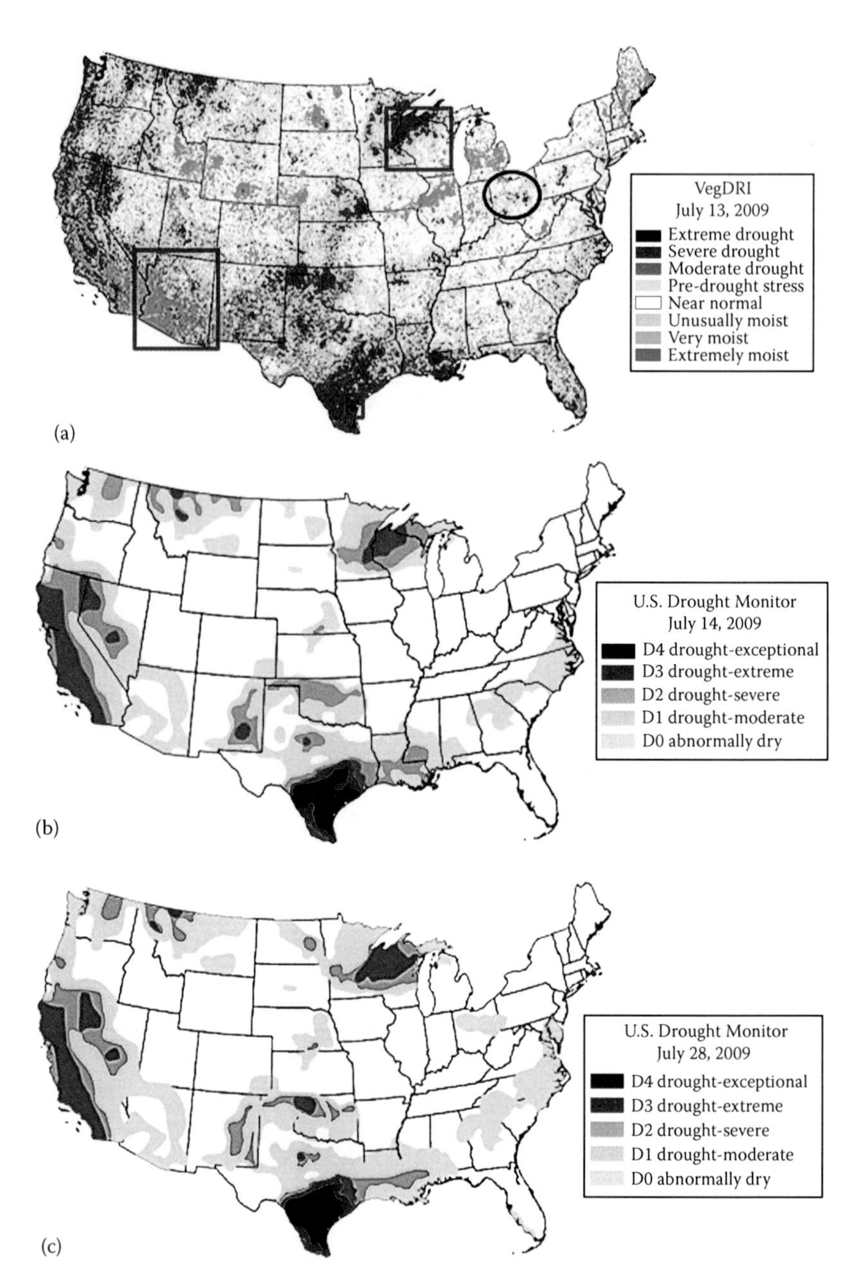

FIGURE 3.4 VegDRI map (a) for July 13, 2009, and USDM maps for July 14 (b) and July 28 (c), 2009, over the continental United States. The black circle highlights an area of central Ohio that was classified as predrought stress in the VegDRI map but lagged by 2 weeks in the USDM maps, which did not show abnormally dry conditions until late July. The red boxes on the VegDRI map delineate the geographic extent of the local case study areas presented later in Section 3.4.

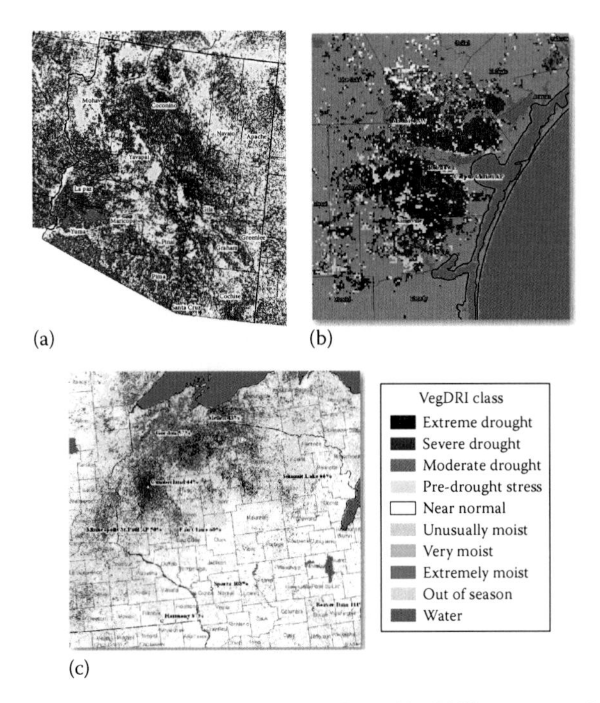

FIGURE 3.5 Local-scale VegDRI results on June 29, 2009, over south Texas (a), on November 2, 2009, over the state of Arizona (b), and on August 10, 2009, over eastern Minnesota and northern Wisconsin (percentages for highlighted locations represent the percent of historical average precipitation received at those locations in 2009) (c).

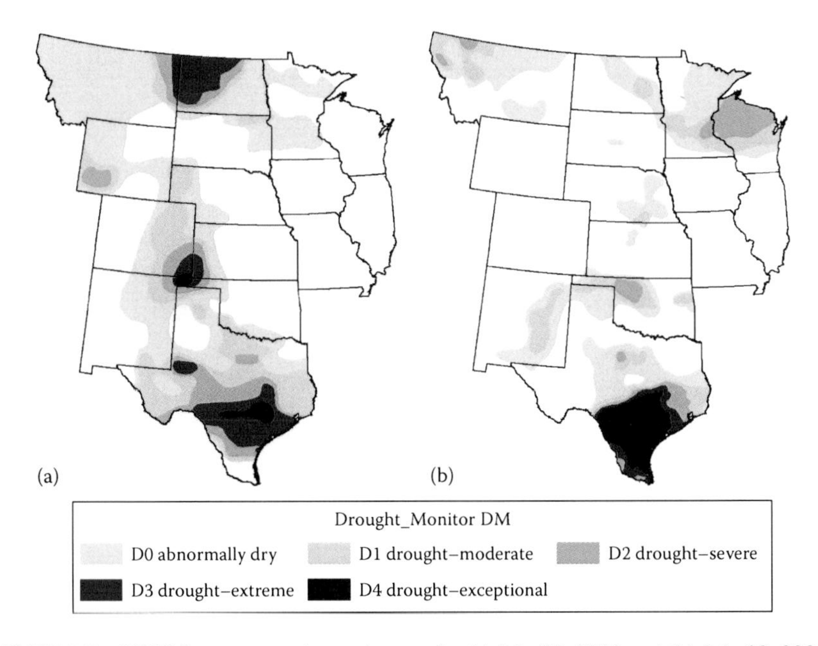

FIGURE 4.5 USDM maps over the study area for (a) July 29, 2008 and (b) July 28, 2009.

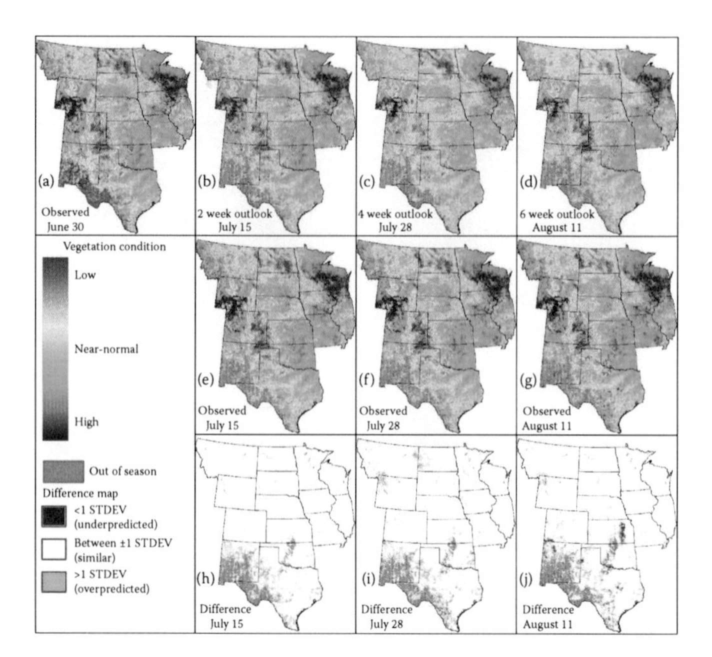

FIGURE 4.6 (a) Observed SSG for June 30, 2008; (b), (c), and (d) are 2, 4, and 6 week outlooks; (e), (f), and (g) are observed SSG for July 15, July 28, and August 11 that correspond to the 2, 4, 6 week outlooks, respectively; and (h), (i), and (j) show the difference between the predicted and observed greenness for the corresponding 2, 4, and 6 week outlooks, respectively.

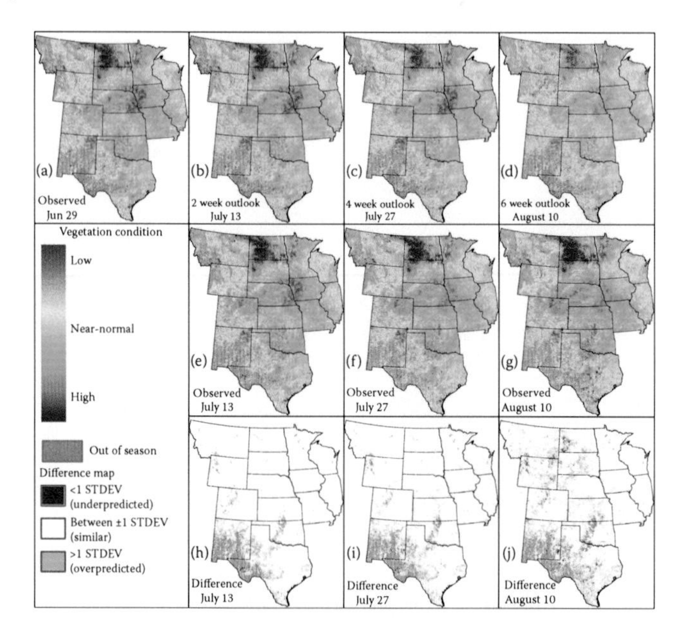

FIGURE 4.7 (a) Observed SSG for June 29, 2009; (b), (c), and (d) are 2, 4, and 6 week outlooks; (e), (f), and (g) are observed SSG for July 13, July 27, and August 10 that correspond to the 2, 4, 6 week outlooks, respectively; and (h), (i), and (j) show the difference between the predicted and observed greenness for the corresponding 2, 4, and 6 week outlooks, respectively.

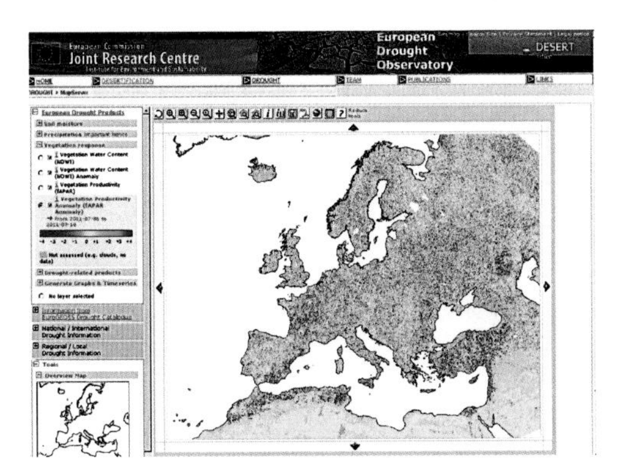

FIGURE 5.1 The EDO prototype map server (http://edo.jrc.ec.europa.eu), showing the fAPAR anomaly map over Europe for the July 1–10, 2011, period. Red indicates areas of anomalously low fAPAR.

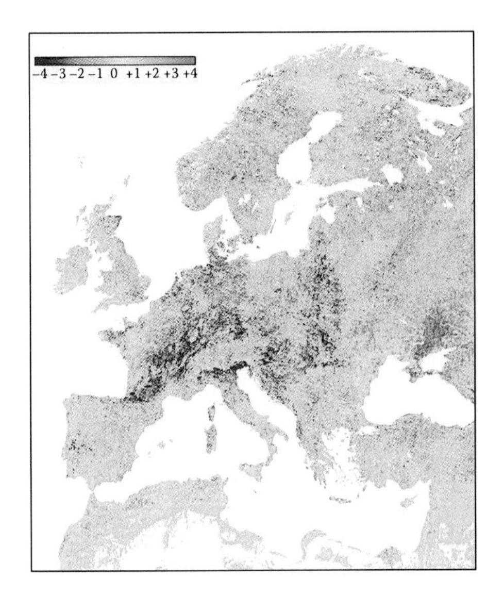

FIGURE 5.2 fAPAR anomaly image for the August 11–20, 2003, period, showing considerably lower than average fAPAR values across much of Europe (red areas) due to severe vegetation water stress.

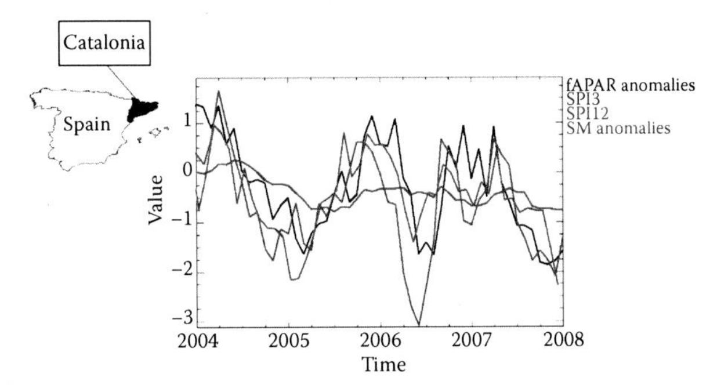

FIGURE 5.4 Detection of a prolonged drought event in Catalonia (Spain) using fAPAR and soil moisture anomalies and SPI (3 and 12 month) products.

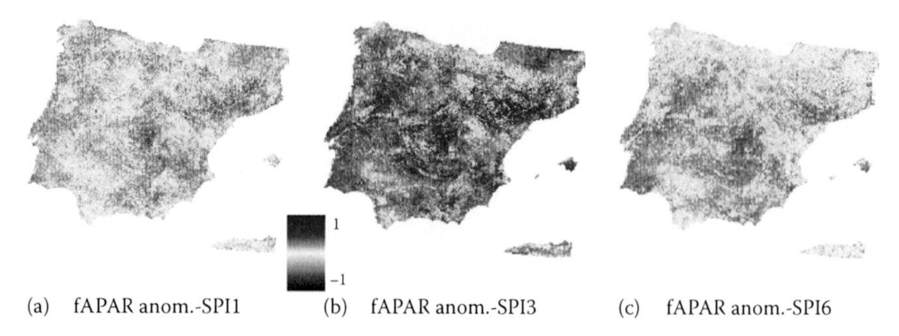

FIGURE 5.5 Correlation of fAPAR anomalies with SPI1 (a), SPI3 (b), and SPI6 (c) over the Iberian Peninsula for the period April 1998 to October 2008, using 5 km SPI data from the EDO.

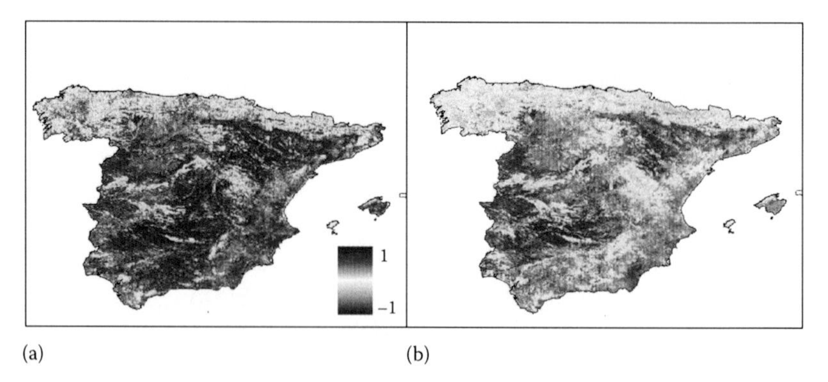

FIGURE 5.6 Correlation between the fAPAR anomalies-SPI3 (a) and NDVI anomalies-SPI3 (b) over Spain, using 2 km SPI data from the University of Valencia.

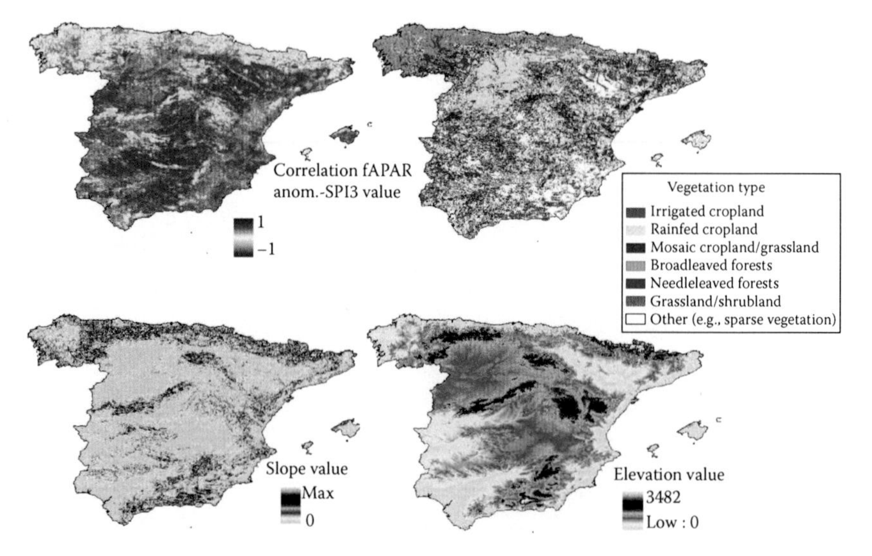

FIGURE 5.7 From top left, clockwise: maps of MERIS fAPAR-SPI3 correlation, land cover, slope, and elevation.

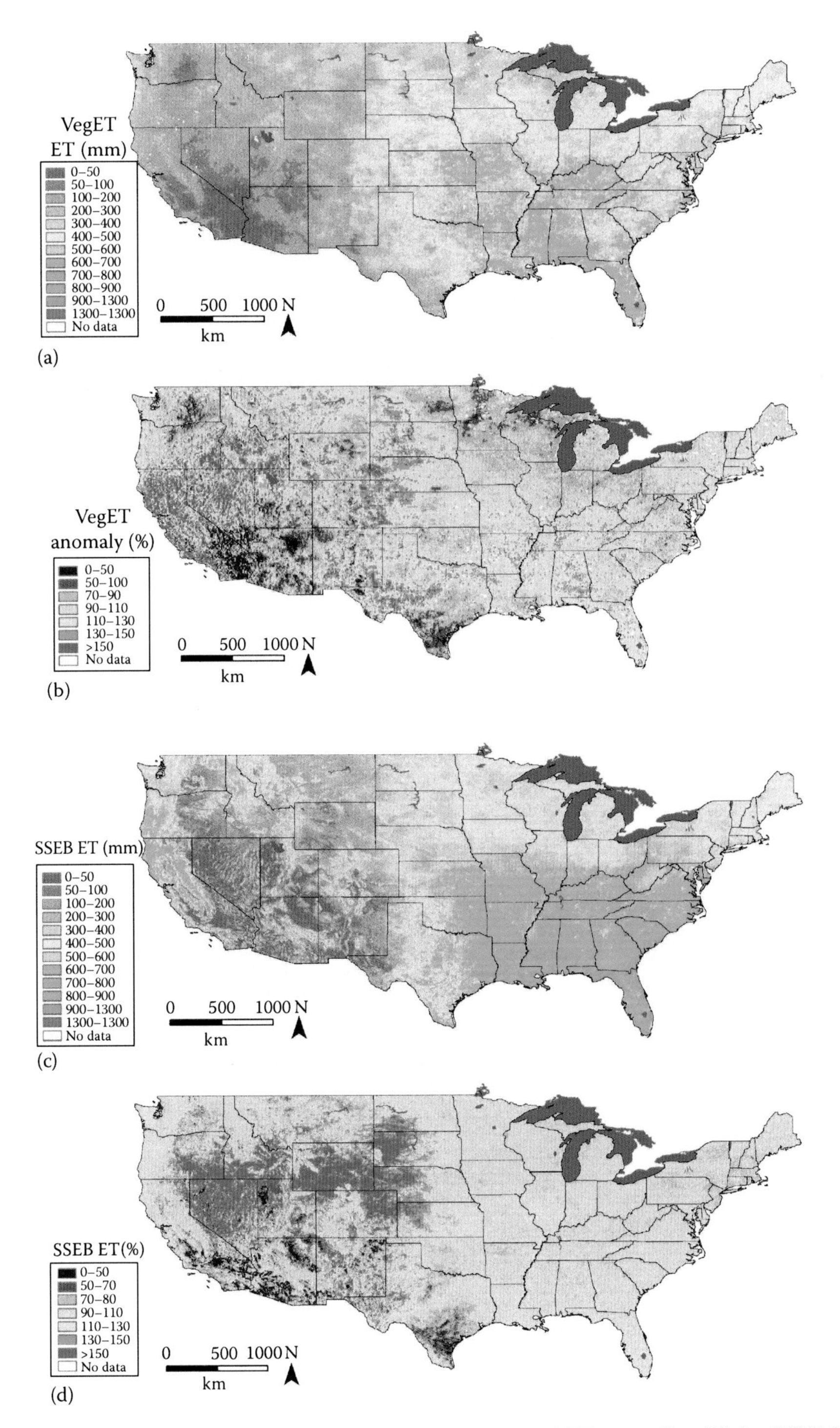

FIGURE 6.3 Seasonal (April–October) total ET_a (mm) and ET anomalies (%) for CONUS in 2009: (a) seasonal VegET ET_a, (b) seasonal SSEB ET_a, (c) seasonal VegET ET_a anomaly, and (d) seasonal SSEB ET_a anomaly.

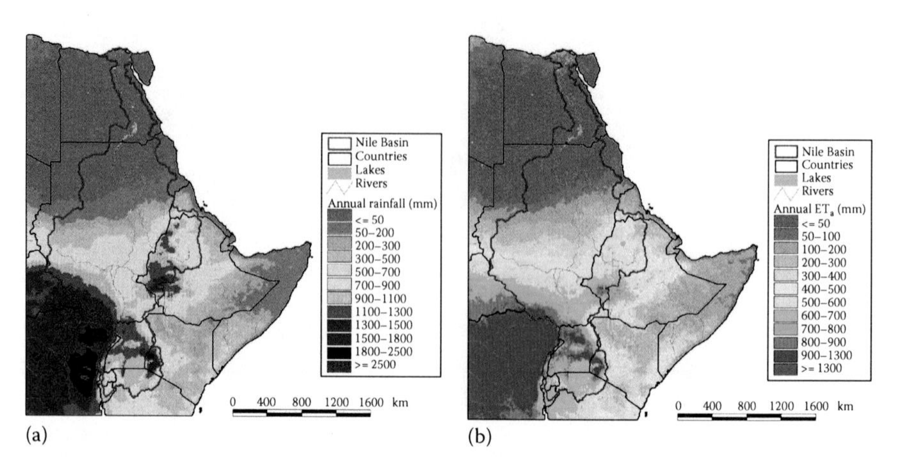

FIGURE 6.5 Spatial distribution of satellite-derived annual rainfall in northeastern Africa (median of 2001–2007) (a) and annual ET_a from the VegET model (median from the same period as the rainfall) (b).

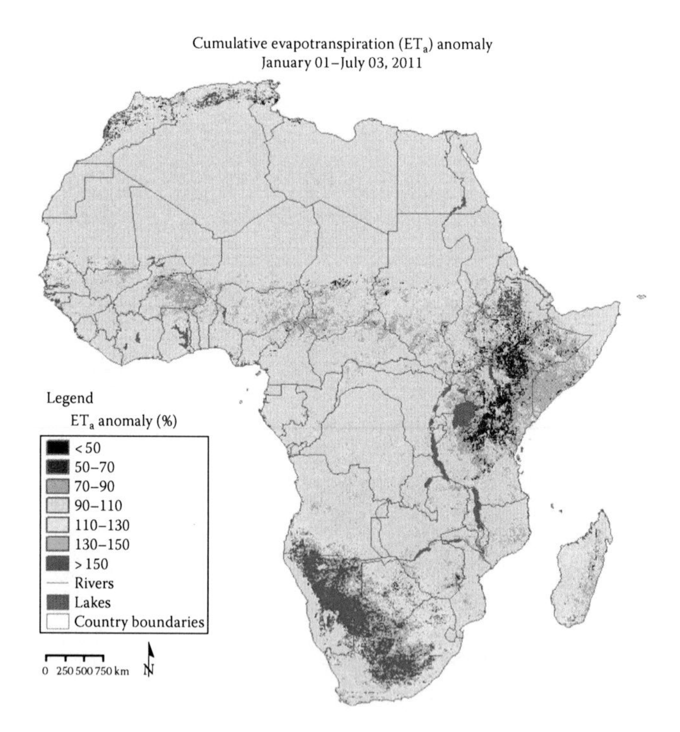

FIGURE 6.6 Africa-wide seasonal anomaly of ET_a from the SSEB model output for 2011 as of July 3, 2011 (January 1–July 3). SSEB ET anomaly is operationally processed and posted on a FEWS NET website regularly on an 8 day time step.

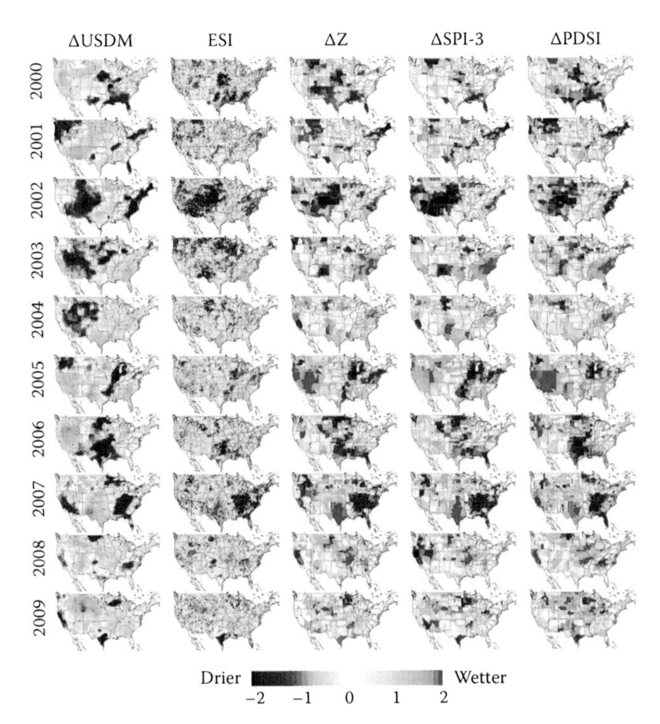

FIGURE 7.2 Seasonal (26 week) anomalies in USDM, ESI, Z, SPI-3, and PDSI for 2000–2009. All indices are presented as z-scores or standard deviations from mean values determined over the 2000–2009 period.

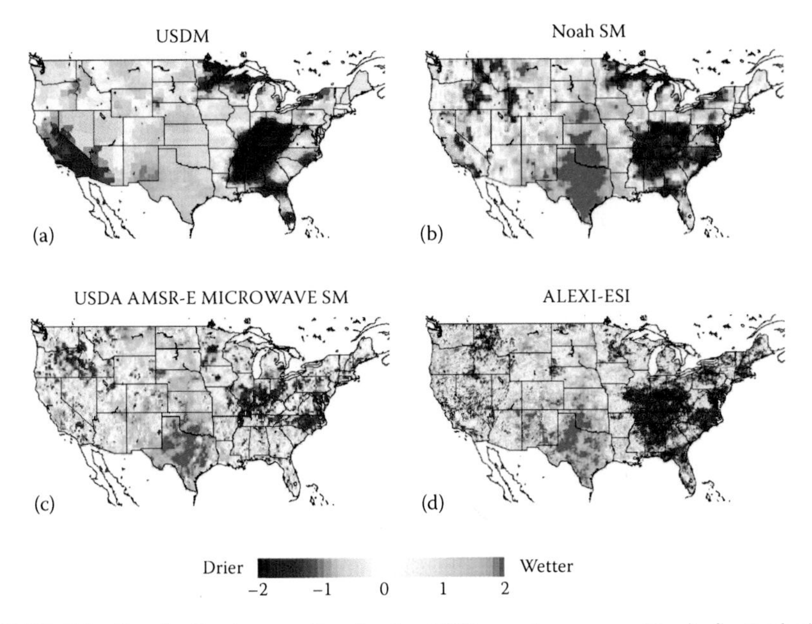

FIGURE 7.3 Standardized anomalies for the 2007 growing season (April–September) in (a) the USDM drought classes, (b) soil moisture predicted by the Noah LSM, (c) USDA AMSR-E passive MW soil moisture retrieval, and (d) ALEXI-ESI.

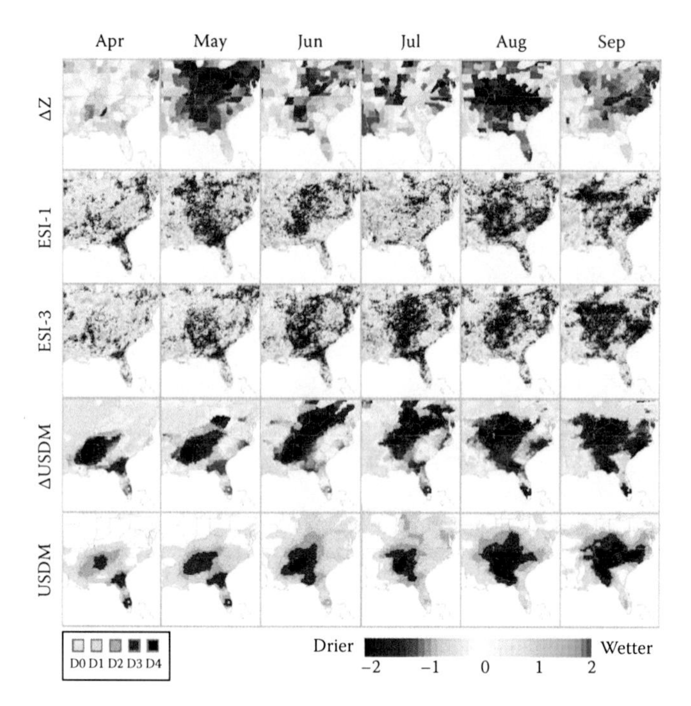

FIGURE 7.4 Maps of monthly drought indicators during April–September 2007, focusing on the severe drought event that occurred in the southeastern United States.

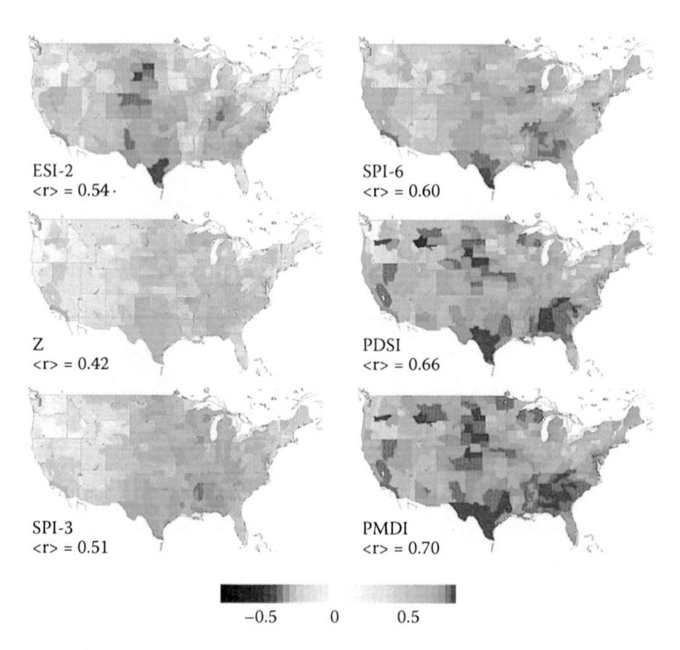

FIGURE 7.5 Coefficient of temporal correlation between monthly maps of USDM anomalies and other drought indices included in the intercomparison for 2000–2009. Domain-averaged values of correlation coefficient are indicated as <r>. To compute correlations, the ESI has been aggregated up to the climate division scale—the native resolution of the precipitation-based indices.

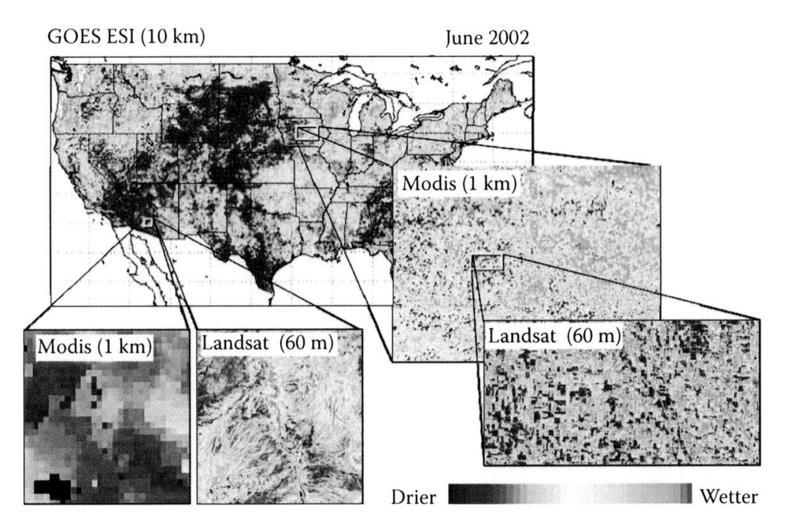

FIGURE 7.6 Comparison of spatial information content provided by ESI fields generated from GOES, MODIS, and Landsat TIR imagery. In this figure, red indicates drier conditions and green indicates wetter conditions.

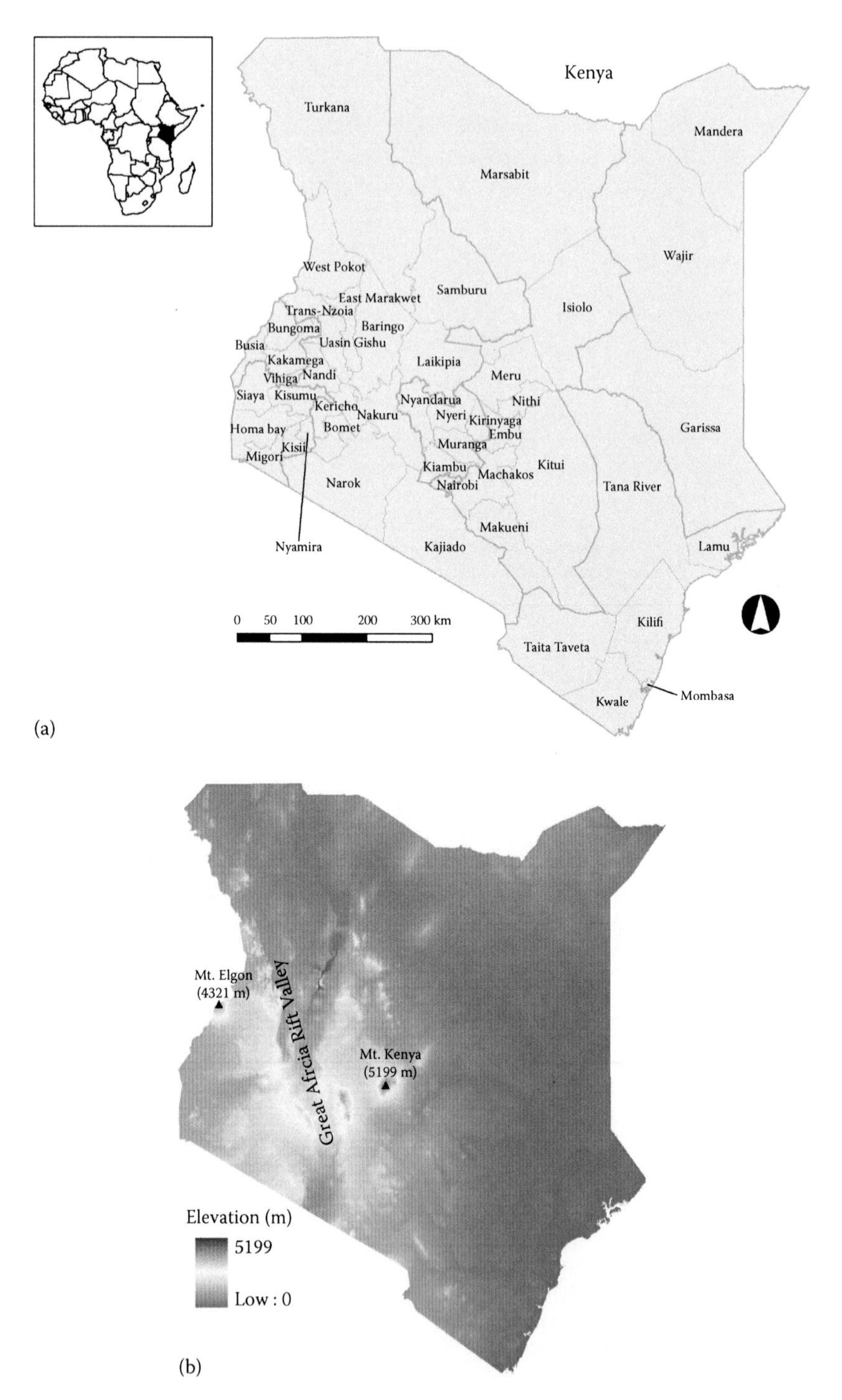

FIGURE 8.1 Administrative boundaries of Kenya (a) and topographic map of Kenya (b). Forty-seven districts span eight provinces outlined in dark brown on the administrative map. The western and eastern highlands of Kenya are divided by the Great African Rift Valley shown on the topographic map.

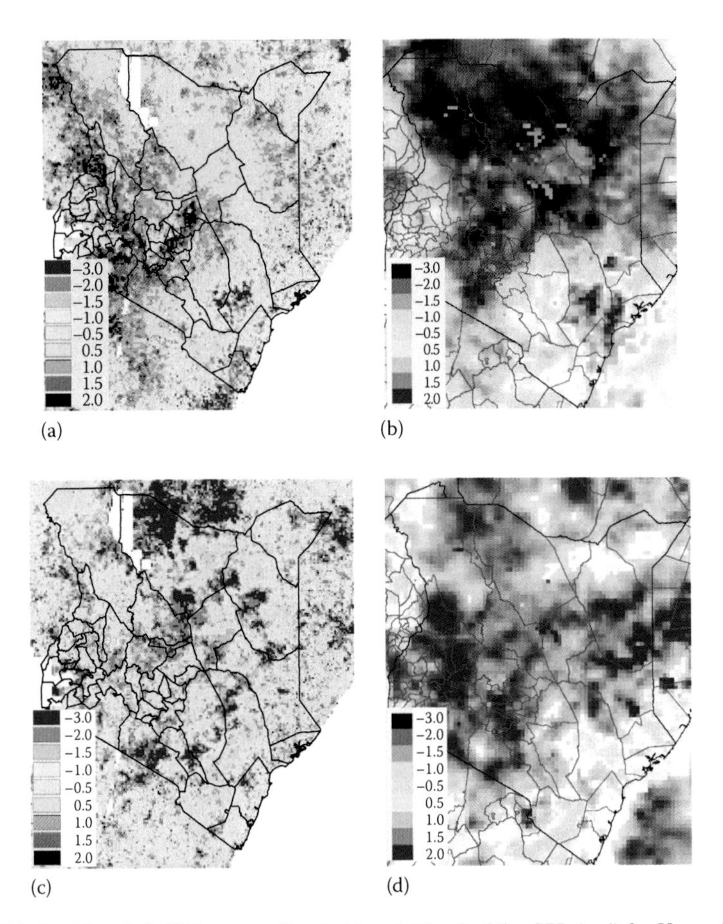

FIGURE 8.5 May–July ESIc anomalies (a, b) and March–May SPI, (c, d) for Kenya in 2000 and 2003, respectively. Values are expressed as Z-scores of ESIc and the gamma probability of rainfall. Areas in white indicate missing data/bad values.

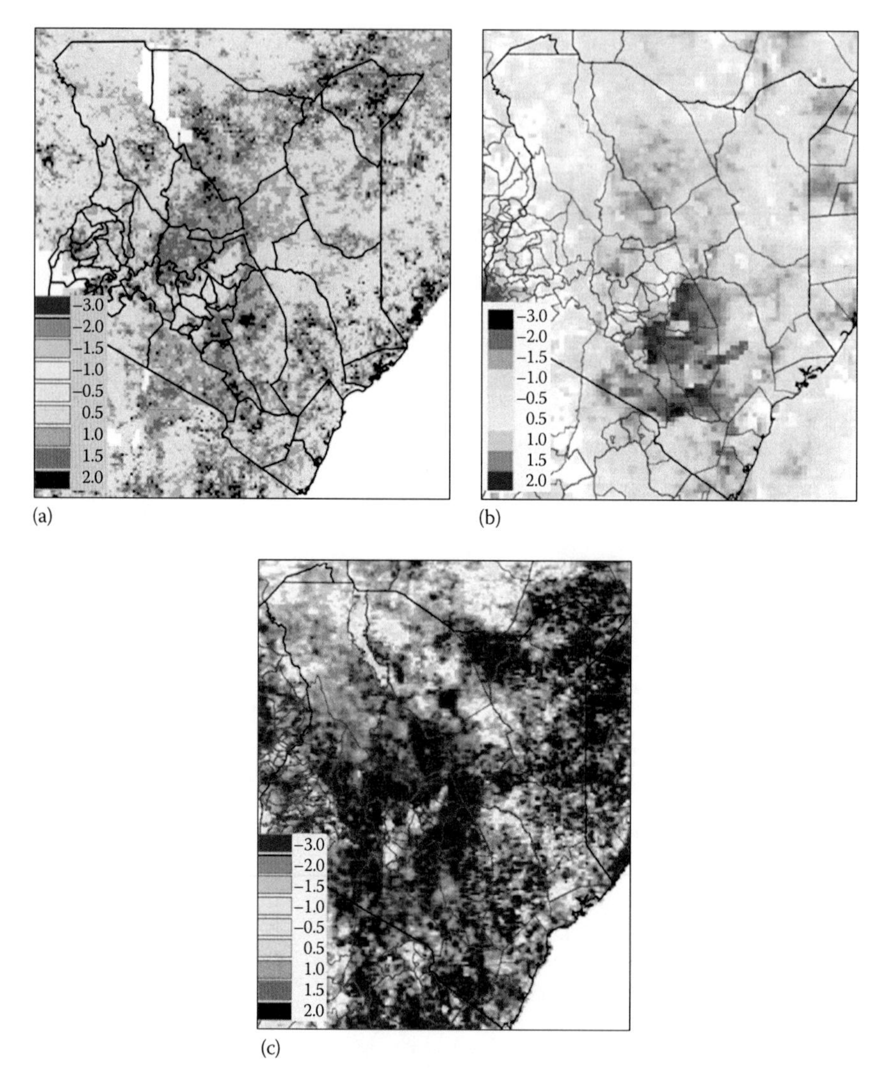

FIGURE 8.6 Map of Kenya in 2009 showing May–July ESI_c anomalies (a), March–May SPI (b), and May–July MODIS LST anomalies (c).

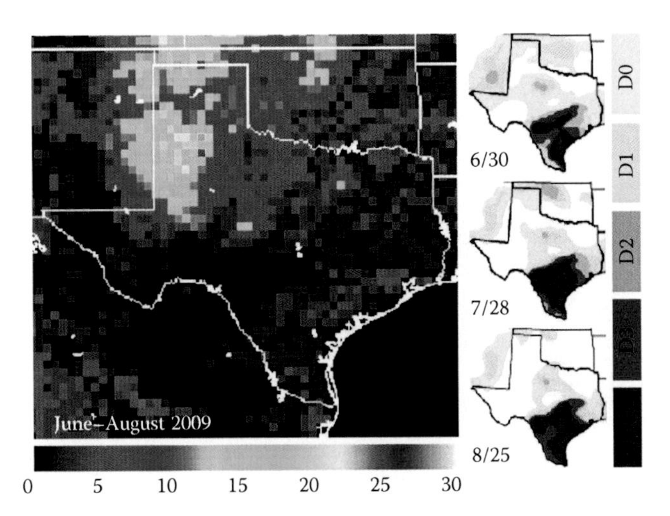

FIGURE 9.4 Effective precipitation frequency (%) measured by QSCAT for the period June–August 2009 (left panel) and drought levels from D0–D4 from the USDM for weeks ending on the marked dates in 2009 (right panels). The USDM drought levels include D0 for abnormally dry, D1 for moderate drought, D2 for severe drought, D3 for extreme drought, and D4 for exceptional drought. (Ref. Svoboda et al. 2002.)

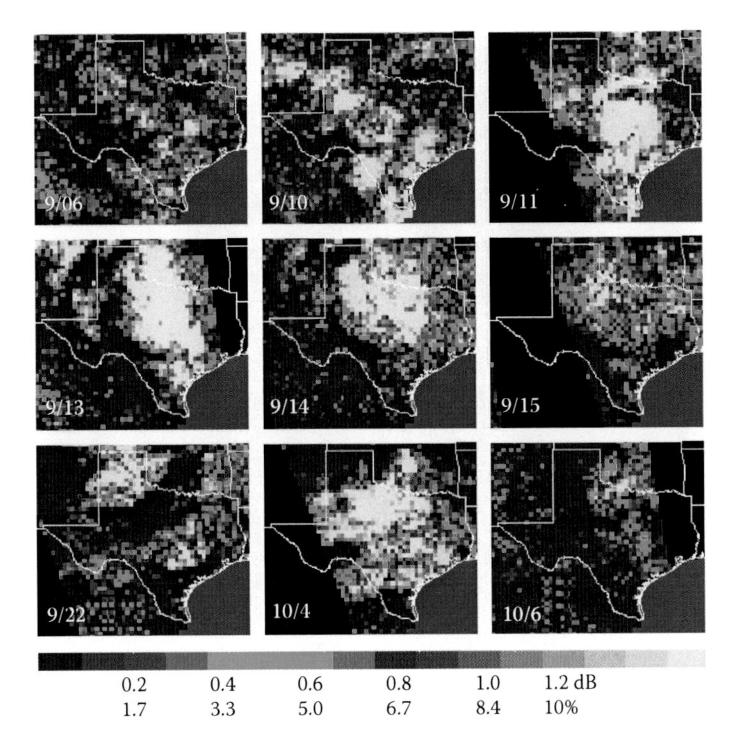

FIGURE 9.5 SMC measured by QSCAT with the vertical polarization along ascending orbits in September to early October 2009. The color scale represents backscatter change in dB and volumetric SMC in % with the Lonoke rating.

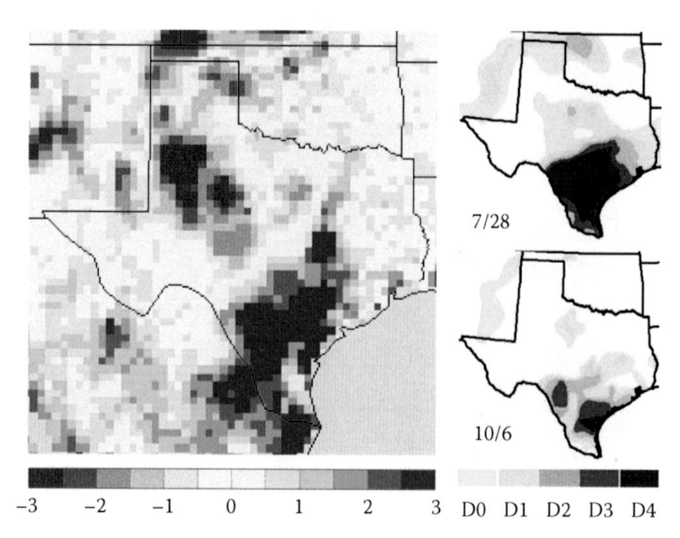

FIGURE 9.6 Difference of AMSR-E monthly averaged soil moisture in % of m_v (September 7 to October 6, 2009) and m_v (June 29 to July 28, 2009) showing seasonal SMC (left panel), and drought condition change between USDM drought maps in July and in September 2009 (right panels).

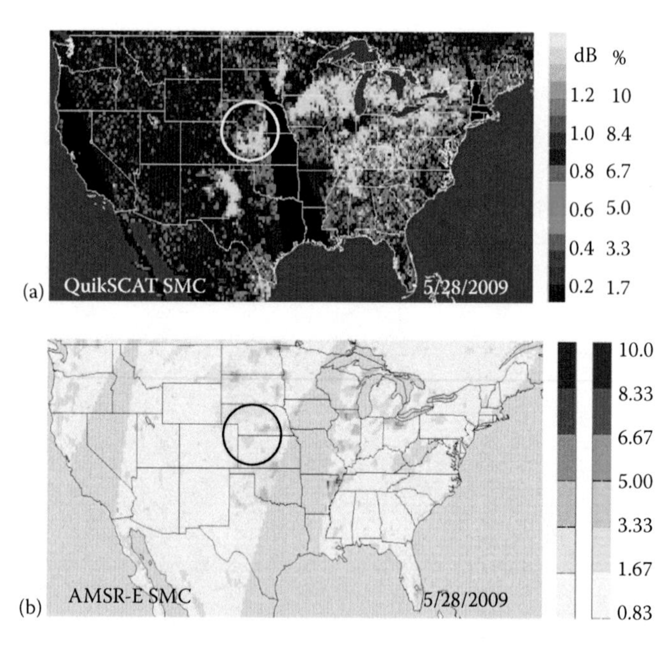

FIGURE 9.7 SMC on May 28, 2009, compared to the 2-week average between May 14 and 28, 2009, observed by (a) QSCAT SMC represented by backscatter change in dB and by volumetric moisture change in % from the Lonoke rating, and (b) AMSR-E by volumetric moisture change in % with yellow brown for drier and cyan blue for wetter conditions. The SMC maps are compared with Stage-4 24 h precipitation measurements (NMQ, 2009) at 12:00 UTC in inches for (c) May 28, 2009, and (d) and May 27, 2009.

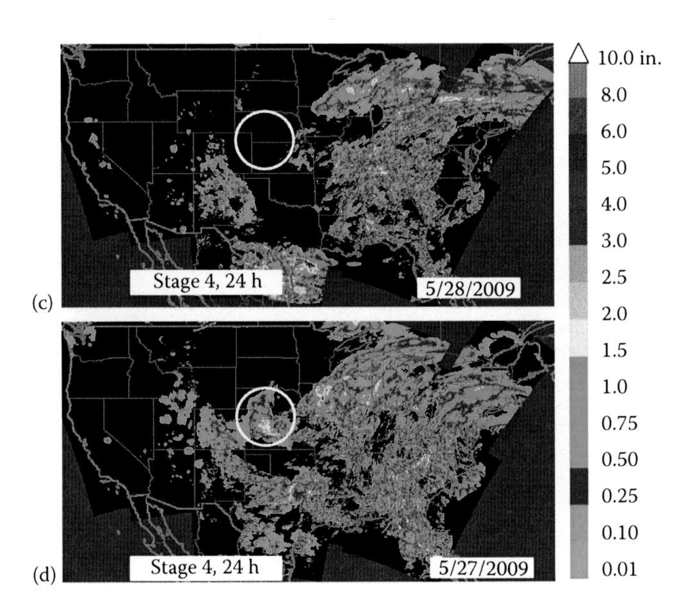

FIGURE 9.7 (continued)

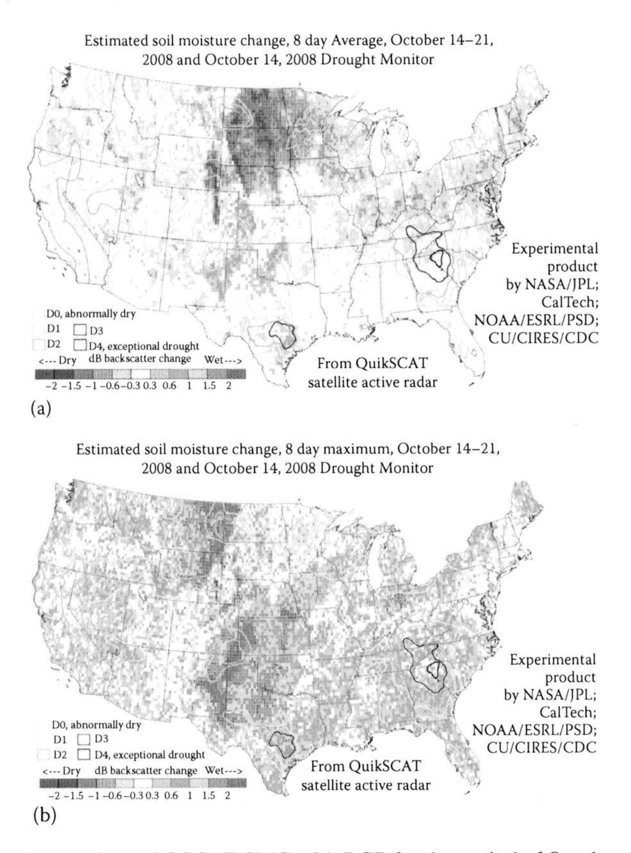

FIGURE 9.8 Comparison of QSCAT SMC with RGP for the period of October 14, 2008, and the ensuing 7 days: (a) mean SMC, (b) max SMC, and

(continued)

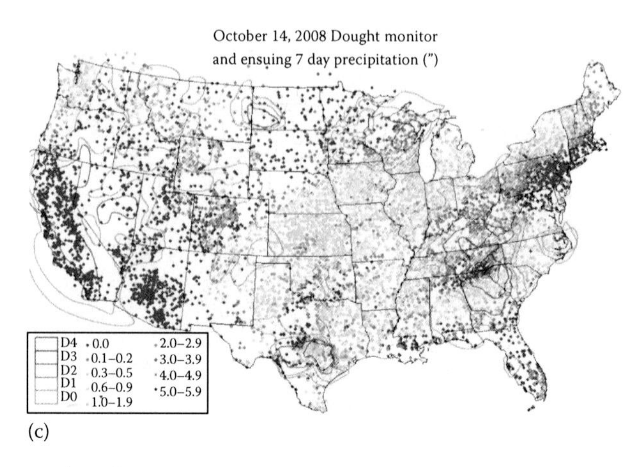

FIGURE 9.8 (continued) (c) RGP used in making USDM maps.

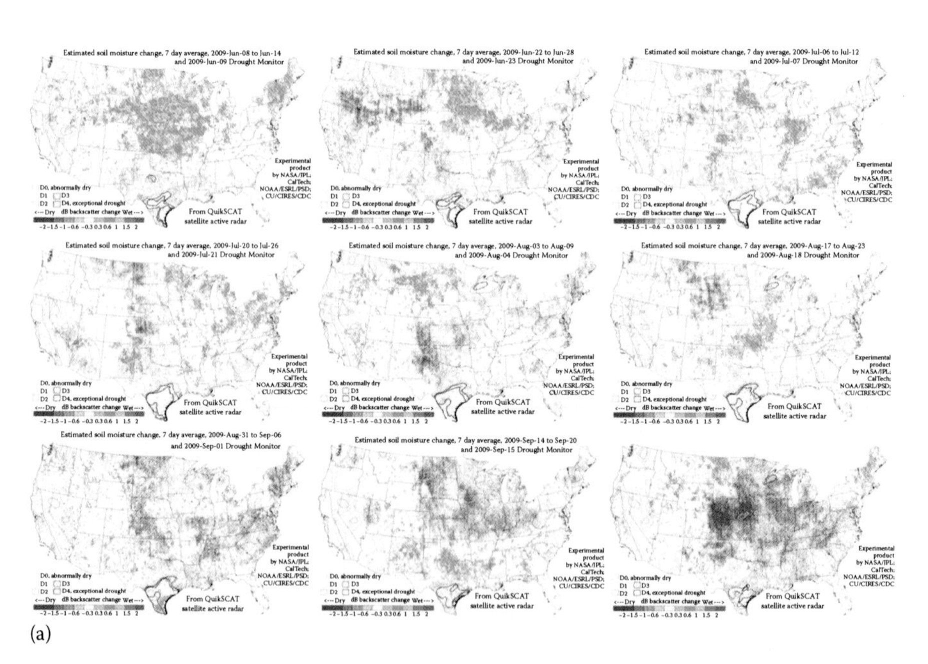

FIGURE 9.9 Weekly QSCAT mean SMC maps (a) and USDM maps (b) for the growing season in June–October 2009 (skipping a map once every other week).

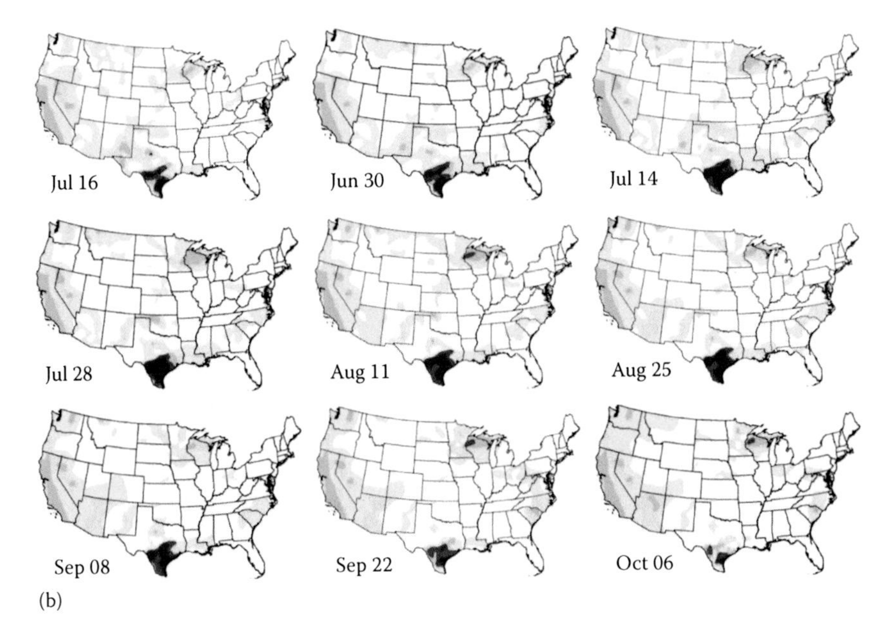

FIGURE 9.9 (continued)

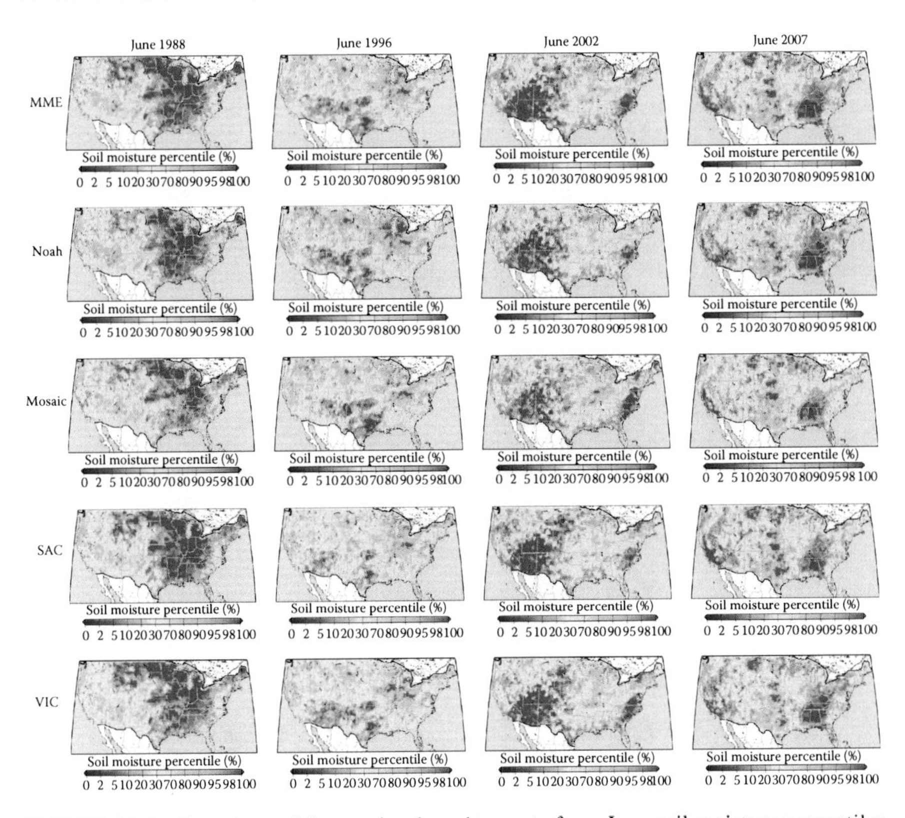

FIGURE 10.4 Snapshots of four major drought events from June soil moisture percentiles from the MME and the four models. Columns are (1) 1988, (2) 1996, (3) 2002, and (4) 2007.

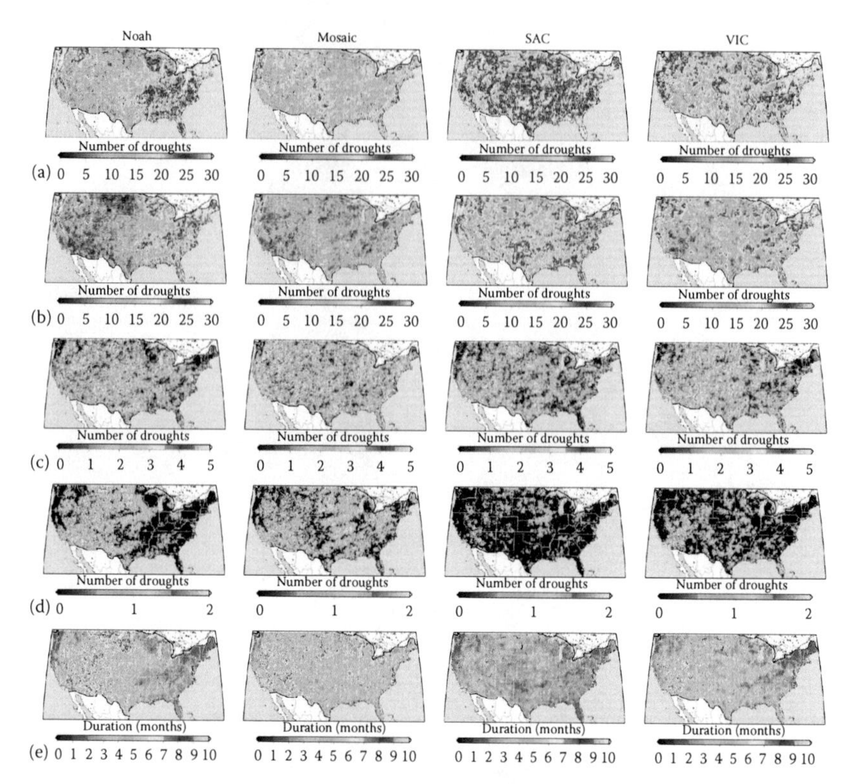

FIGURE 10.5 Statistics of drought duration and frequency for the four LSMs for 1979–2008 calculated from monthly soil moisture percentiles. A drought is defined at each grid cell when the soil moisture percentile drops below 20%. (a) Total number of droughts, (b) number of short-term (1–3 month duration) droughts, (c) number of medium-term (7–12 month duration) droughts, (d) number of long-term (>12 month duration) droughts, and (e) the mean drought duration.

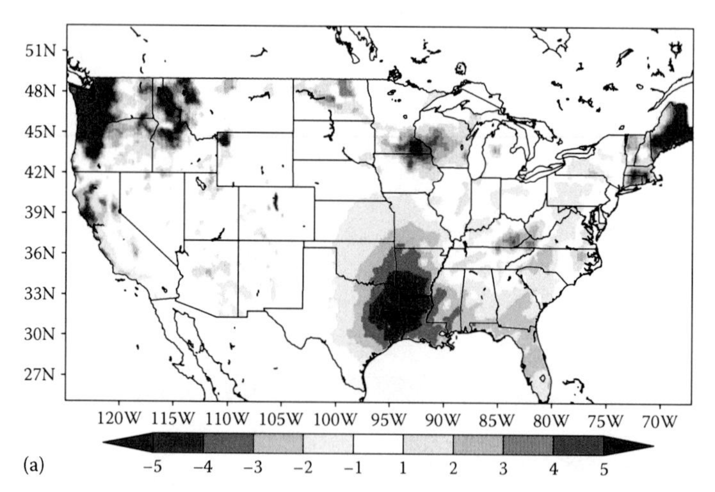

FIGURE 10.6 Example of output fields from the NLDAS-2 drought monitor (http://www.emc.ncep.noaa.gov/mmb/nldas/drought/), showing anomaly data for the week ending on December 16, 2010, for (a) precipitation and multi-model averages of (b) evapotranspiration, (c) runoff, (d) streamflow, (e) soil moisture, and (f) snow water equivalent.

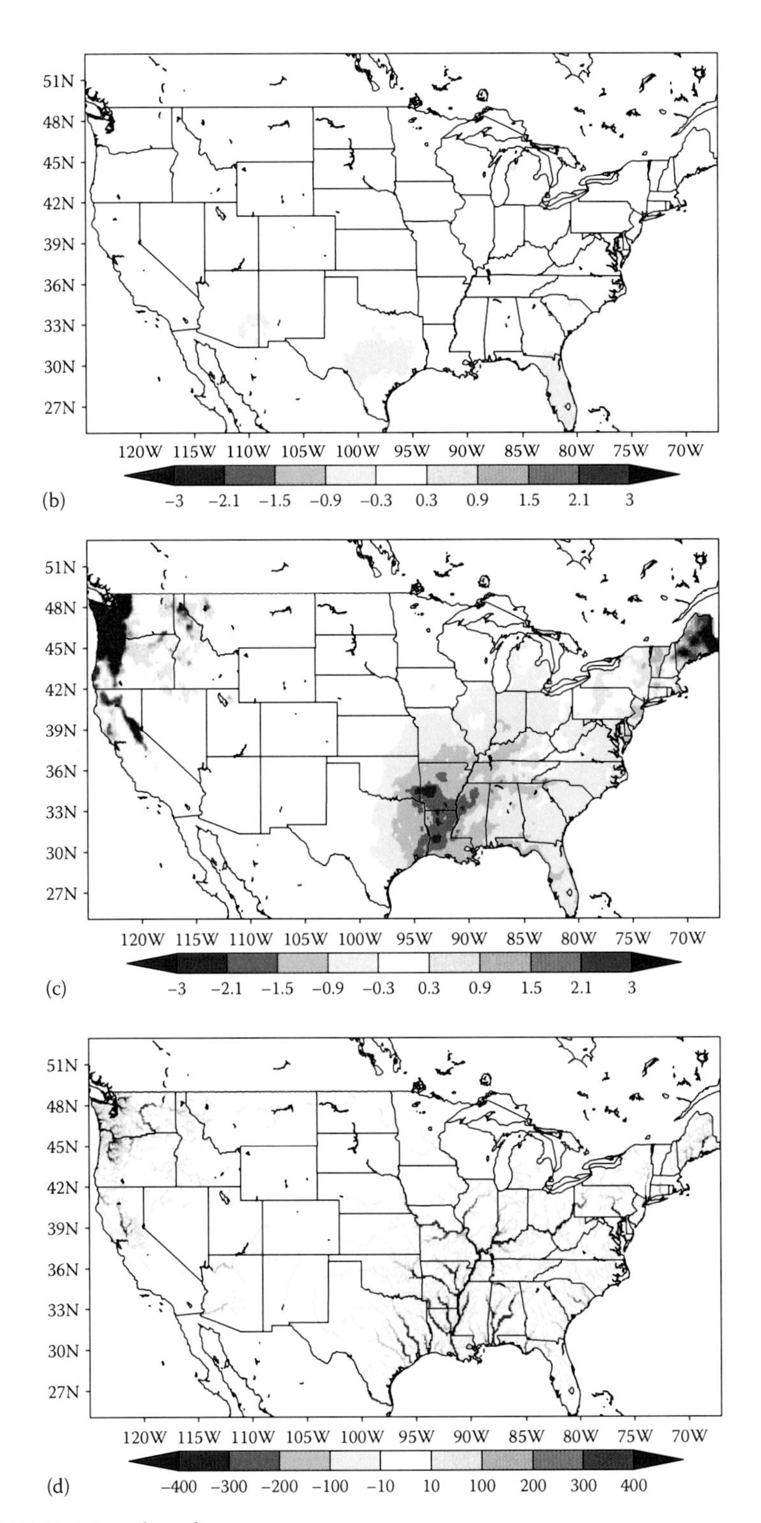

FIGURE 10.6 (continued)

(continued)

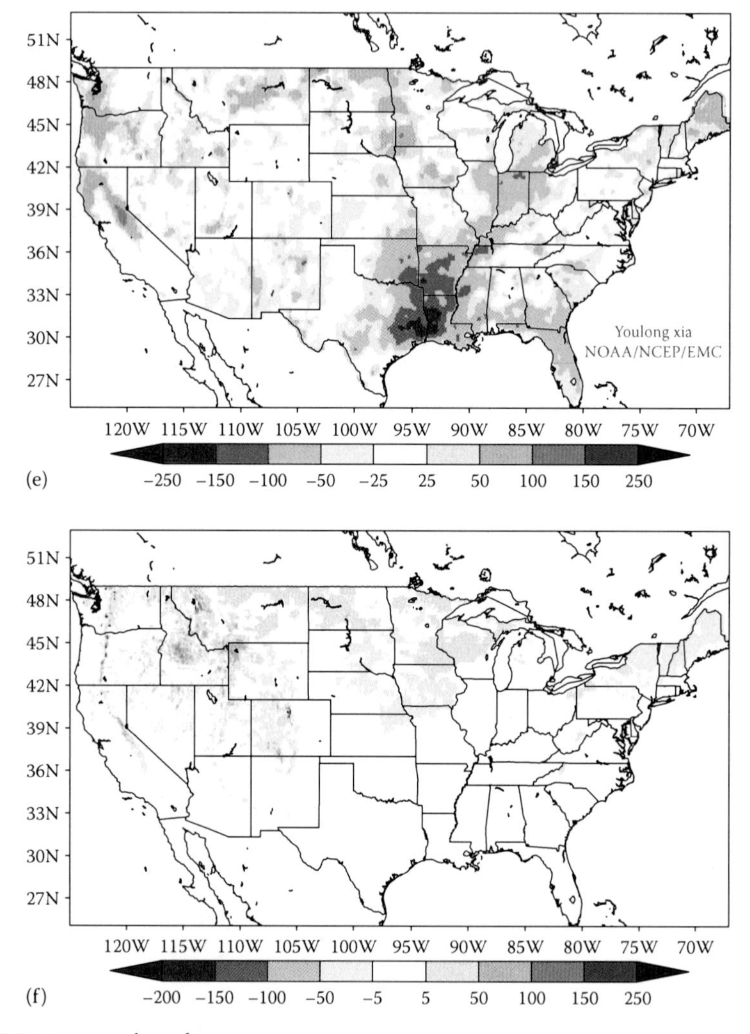

FIGURE 10.6 (continued)

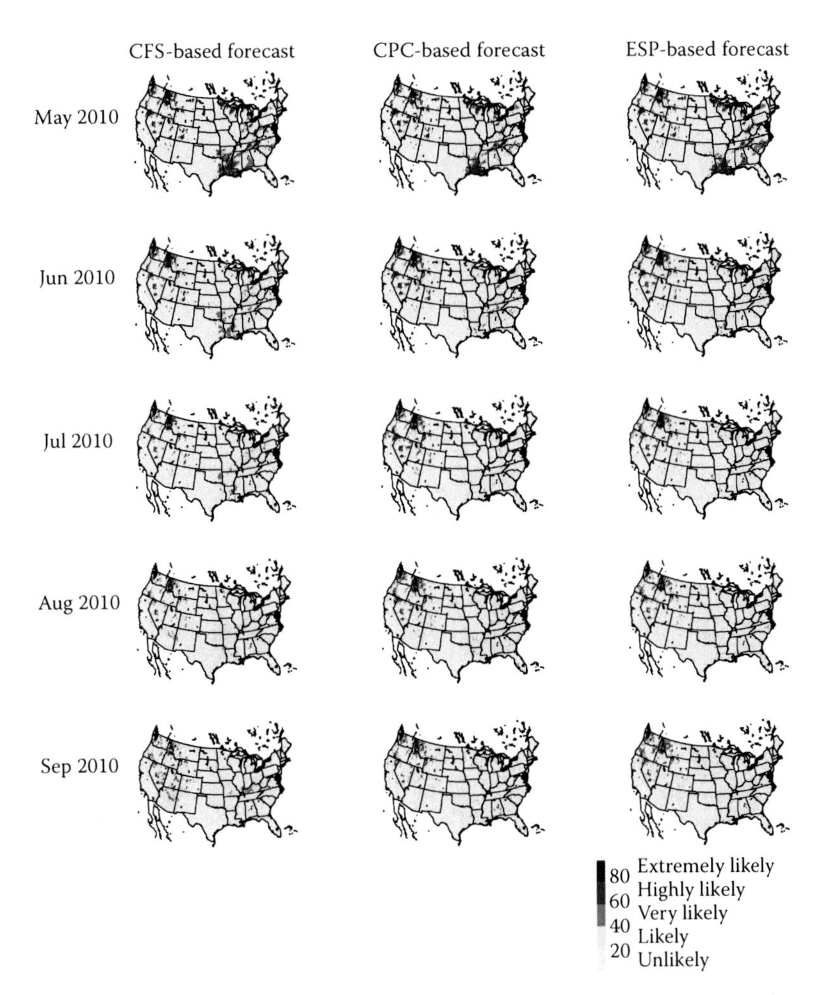

FIGURE 10.7 Example of seasonal forecasts for May through September 2010, showing the likelihood of drought developing or persisting at lead times of 1–6 months. A drought is defined as soil moisture deficits below the 20th percentile, and the likelihood is based on ensemble forecast distributions. Forecasts are based on three methods: (1) CFS, (2) CPC official outlooks, and (3) ESP.

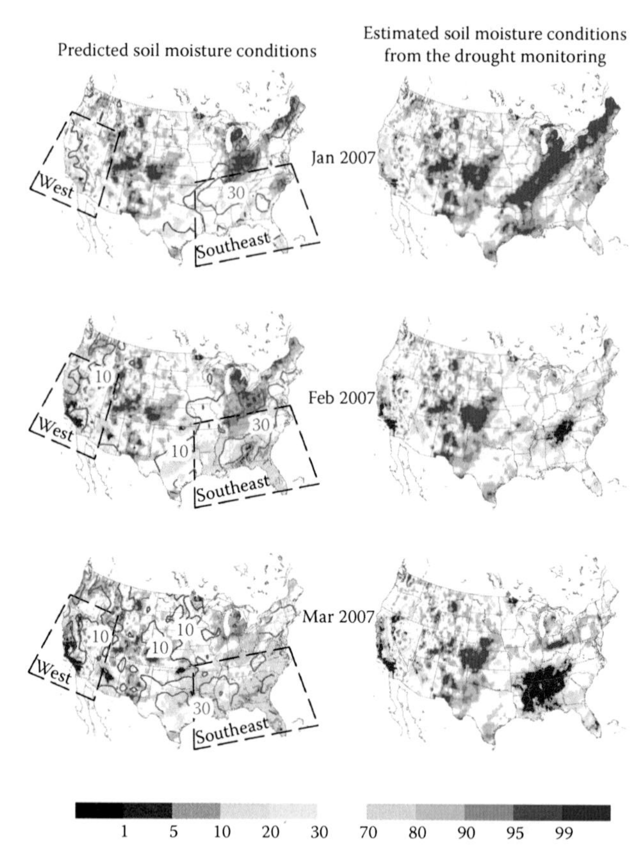

FIGURE 10.8 Example of seasonal prediction of the 2007 U.S. drought (figure reproduced from Luo L. and E.F. Wood, *Geophys. Res. Lett.*, 34, L22702, 2007). Predictions of soil moisture percentiles (%) (left column) were made starting on January 1, 2007, using downscaled and bias-corrected CFS seasonal climate forecasts to drive the VIC model, and are compared to estimated soil moisture from the real-time drought monitoring (right column). Left column shows the mean of the most likely ensemble set (shaded) and their spread (contour). The boxes indicate regions where drought was most severe during early 2007.

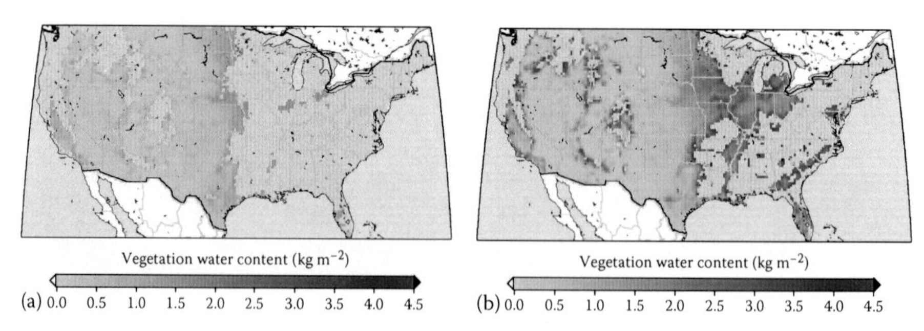

FIGURE 10.9 Zones of applicability for microwave remote sensing retrievals of soil moisture based on penetration through vegetation for frequencies in (a) X-band (10.7 GHz) and (b) L-band (1.4 GHz). Vegetation is characterized by its vegetation water content (kg m^{-2}). Gray shading indicates areas where retrievals of soil moisture are not feasible.

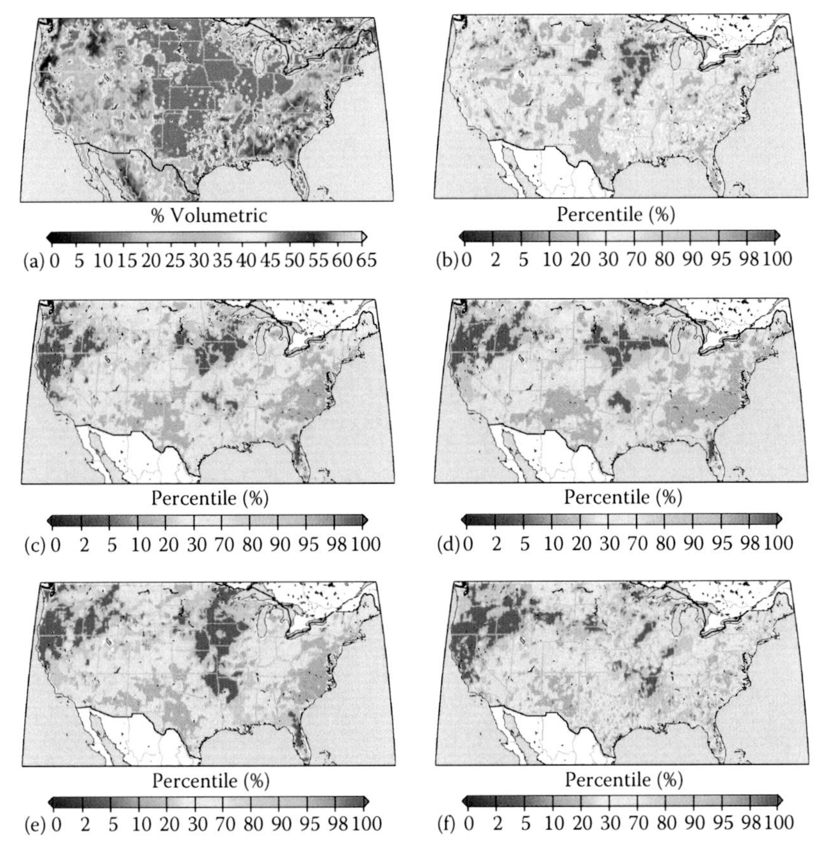

FIGURE 10.10 (a) Dynamic range (% vol.) of AMSR-E daily soil moisture and (b–f) examples of monthly soil moisture percentiles for October 2007 for (b) AMSR-E, (c) Noah, (d) Mosaic, (e) SAC, and (f) VIC.

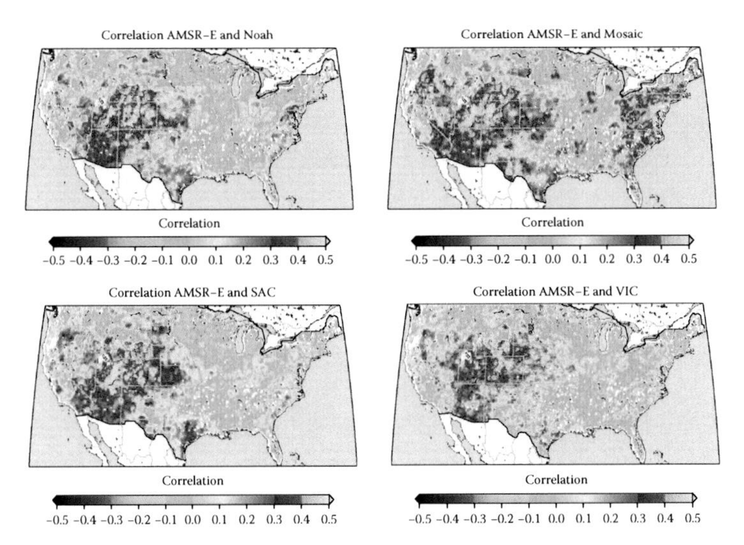

FIGURE 10.12 Correlation between monthly AMSR-E and LSM monthly soil moisture percentiles for 2002–2008.

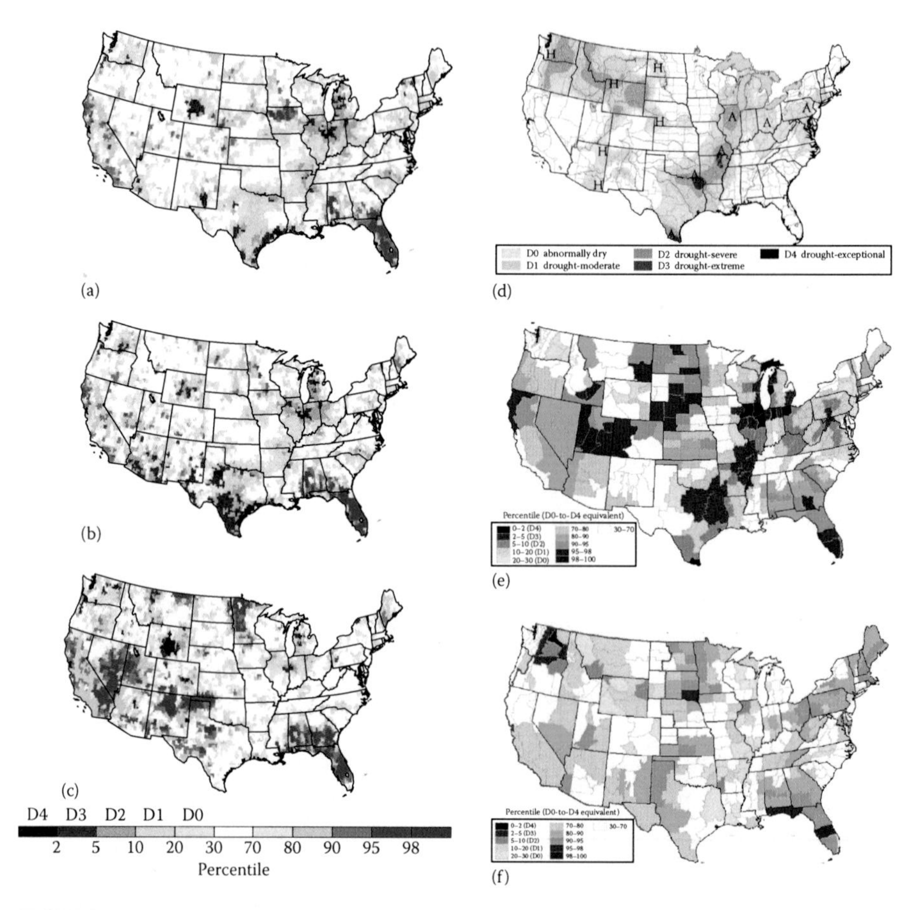

FIGURE 11.5 Surface soil moisture (top several centimeters) (a), root-zone soil moisture (b), and groundwater storage (c) drought indicators, on a 0.25° grid, based on GRACE data assimilation results for June 25, 2005, compared with the U.S. Drought Monitor product for June 28, 2005 (d), and short-term (e) and long-term (f) objective blends (on U.S. climate divisions), which constitute the baselines for the Drought Monitor products, for June 25, 2005. (From Rasmus Houborg, University of Maryland, College Park, MD.)

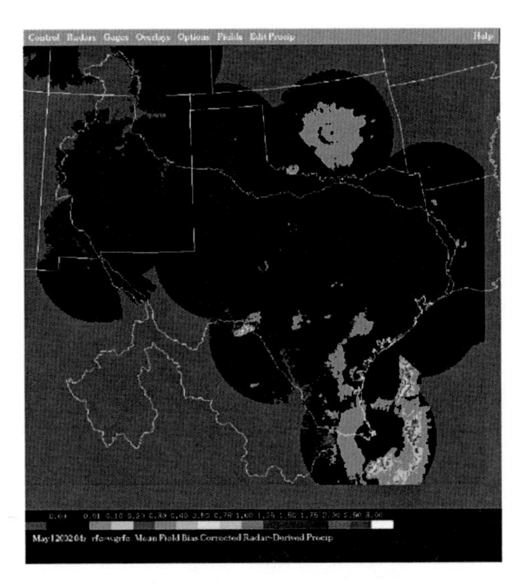

FIGURE 12.3 Widespread false precipitation, or AP, shown on the MPE radar mosaic. (Photos courtesy NOAA/NWS, Silver Spring, MD.)

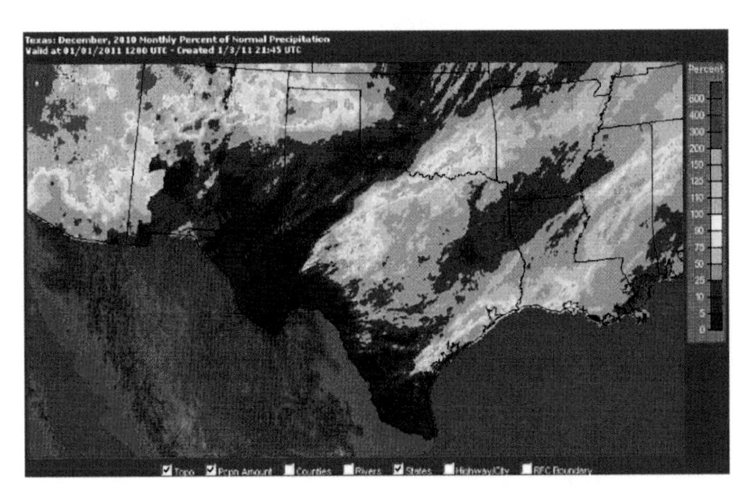

FIGURE 12.5 Percent of normal rainfall for the southern United States from the AHPS precipitation analysis page for December 2010. (Image courtesy of NOAA/NWS, Silver Spring, MD.)

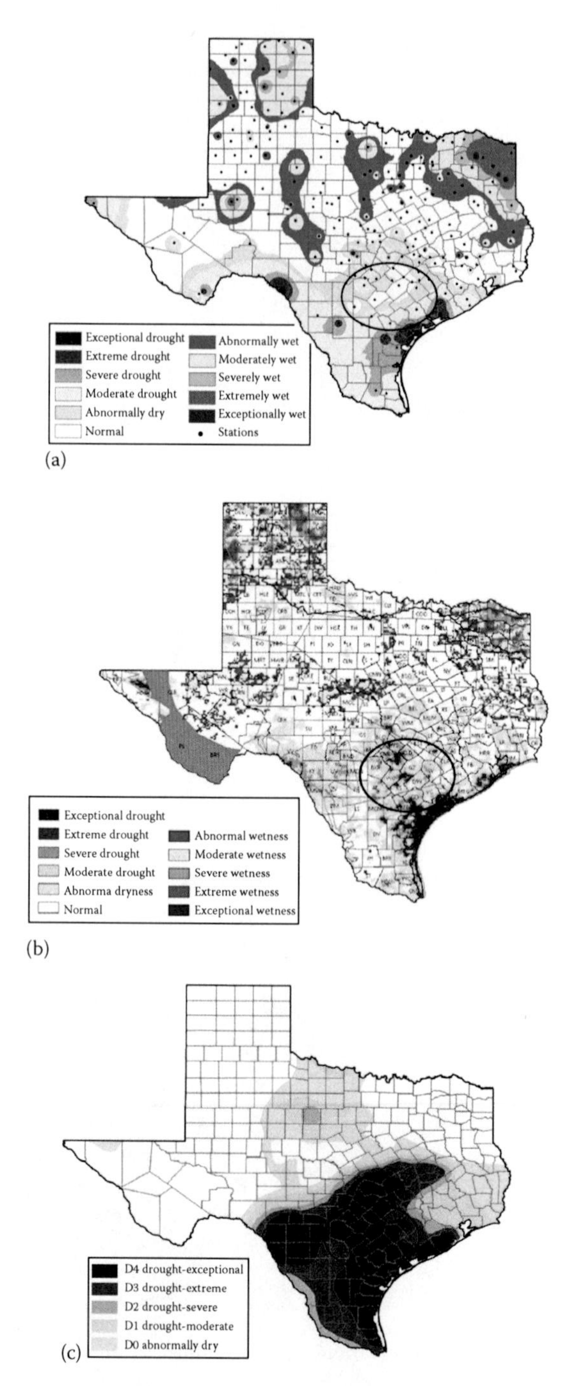

FIGURE 12.6 An 8 week SPI map interpolated from station-based precipitation data (a) and an 8 week SPI map derived from 4 km precipitation from MPE (b) (Image courtesy of Dr. John Nielsen-Gammon) for early September 2009 during the severe drought in southern Texas, as shown by the USDM map on September 7, 2009 (c). The circle highlights an area of exceptional drought in the USDM that is shown to have near-normal conditions in the interpolated SPI map (a) but clearly had localized areas of severe drought conditions that were detected in the SPI map based on higher-resolution, radar-based precipitation observations (b).

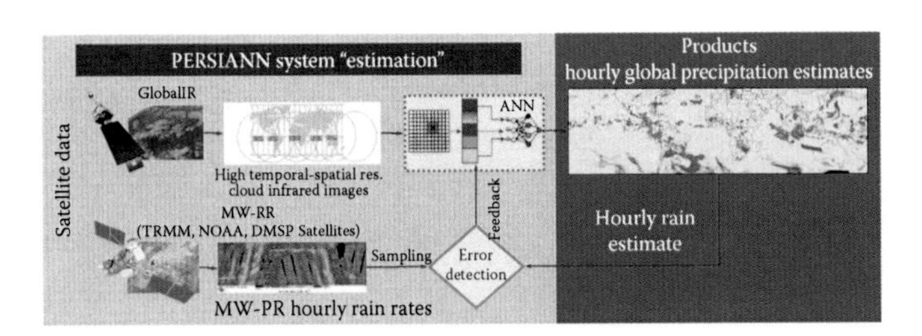

FIGURE 13.1 Components of the PERSIANN data-retrieval algorithm. (Modified after Hsu, K. et al., *Water Resour. Res.*, 45, 2038, 2009.) [MW, microwave; PR, precipitation radar].

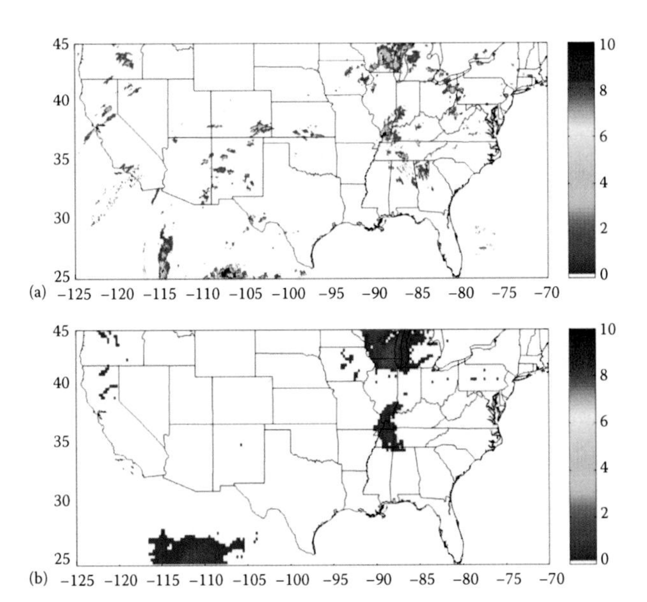

FIGURE 13.3 Example: PERSIANN-CCS 0.04° (a) and PERSIANN 0.25° (b) rainfall rates (mm/h)—date: July 25, 2005.

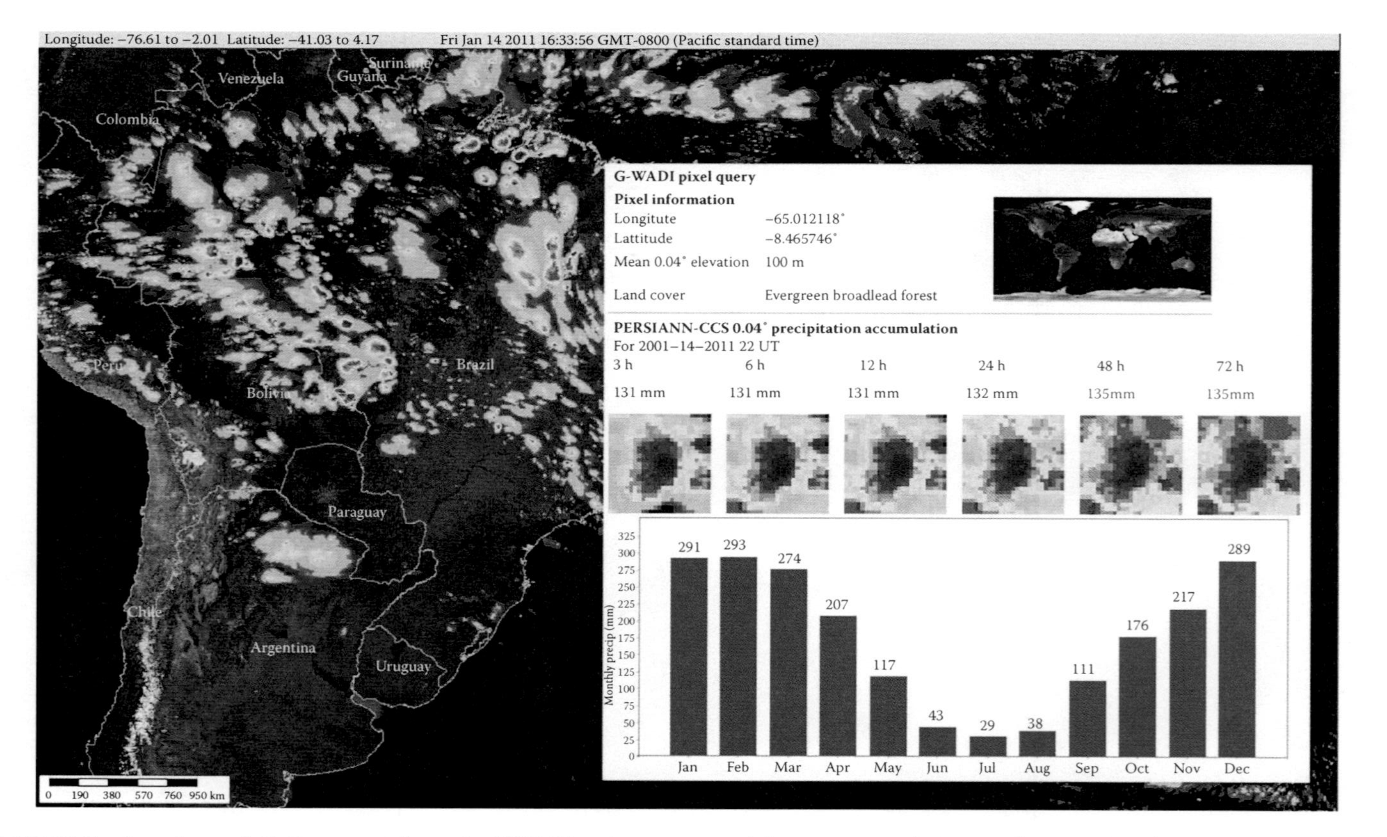

FIGURE 13.7 Sample precipitation report from the GWADI web server: precipitation accumulations and pattern development over a 72 h period (ending January 14, 2011, 16:30 PST) across northwestern Amazon (top panels) and monthly long-term climatology of the region (bottom panel).

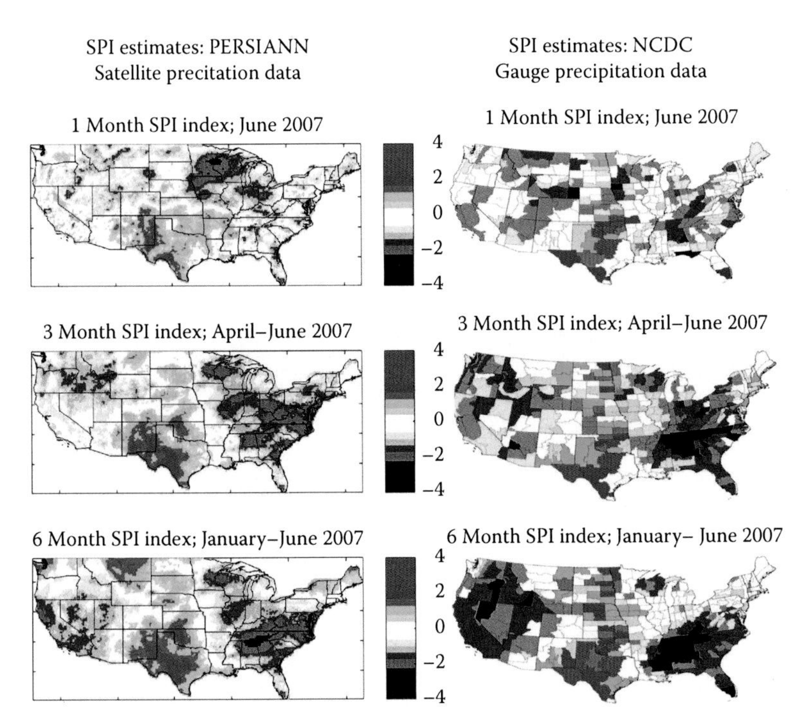

FIGURE 13.8 SPI estimates using PERSIANN satellite data (left panels) and the NCDC GHCN data set (right panels).

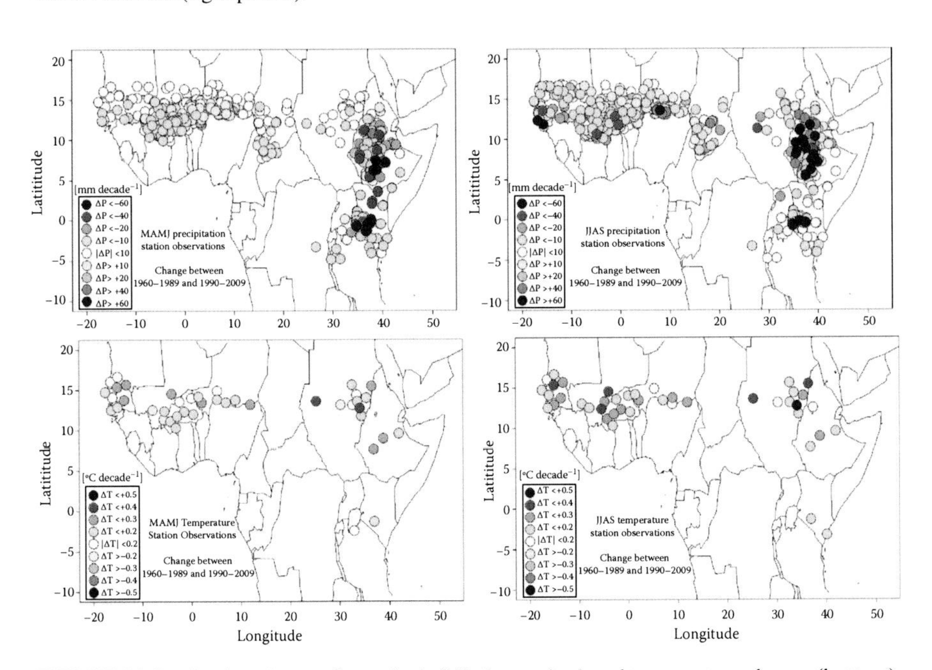

FIGURE 14.2 Station observations of rainfall change (top) and temperature change (bottom) between the 1960–1989 average and the 2000–2009 average for MAMJ (left) and JJAS (right).

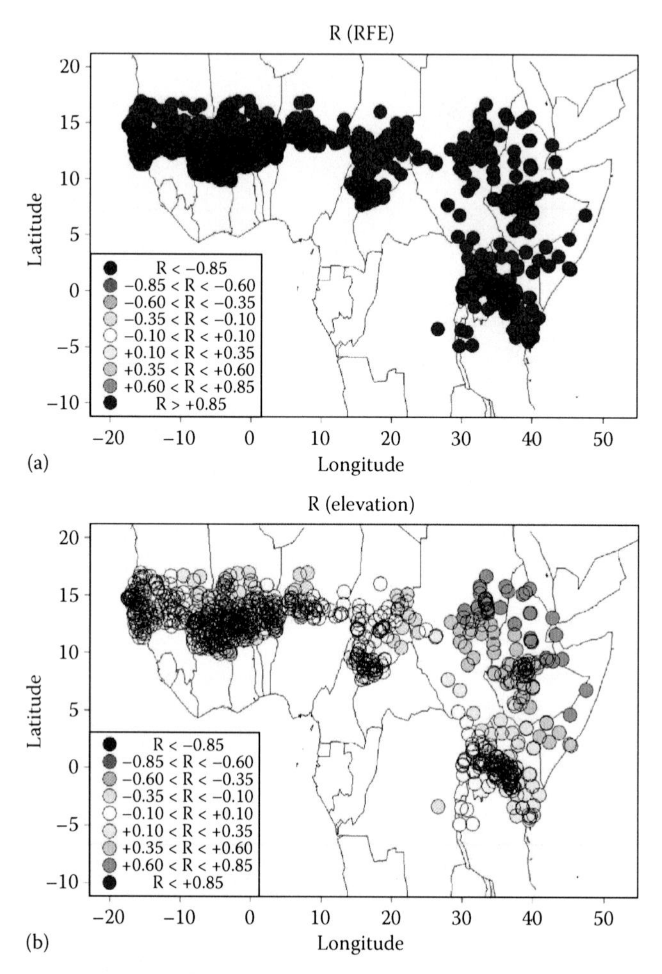

FIGURE 14.4 Comparison of RFE2 and elevation correlations with JJAS precipitation normals showing the local correlation between station observations of JJAS rainfall and RFE2 means (a) and elevation (b). The median local correlation between the station means and RFE2 means was 0.91. Local correlation computations are based on a d_{max} value of 1500 km^2.

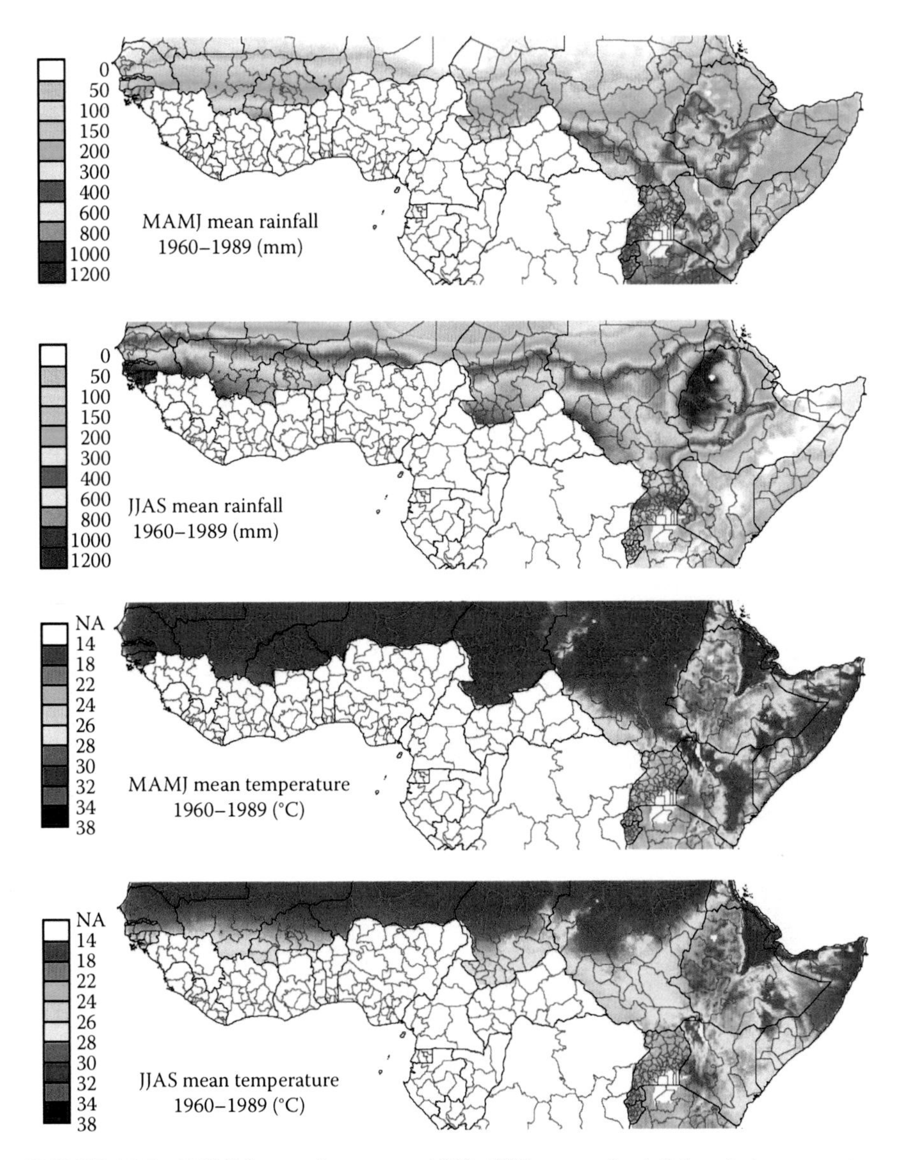

FIGURE 14.5 FCLIM maps for average 1960–1989 seasonal rainfall and air temperature for the MAMJ and JJAS seasons.

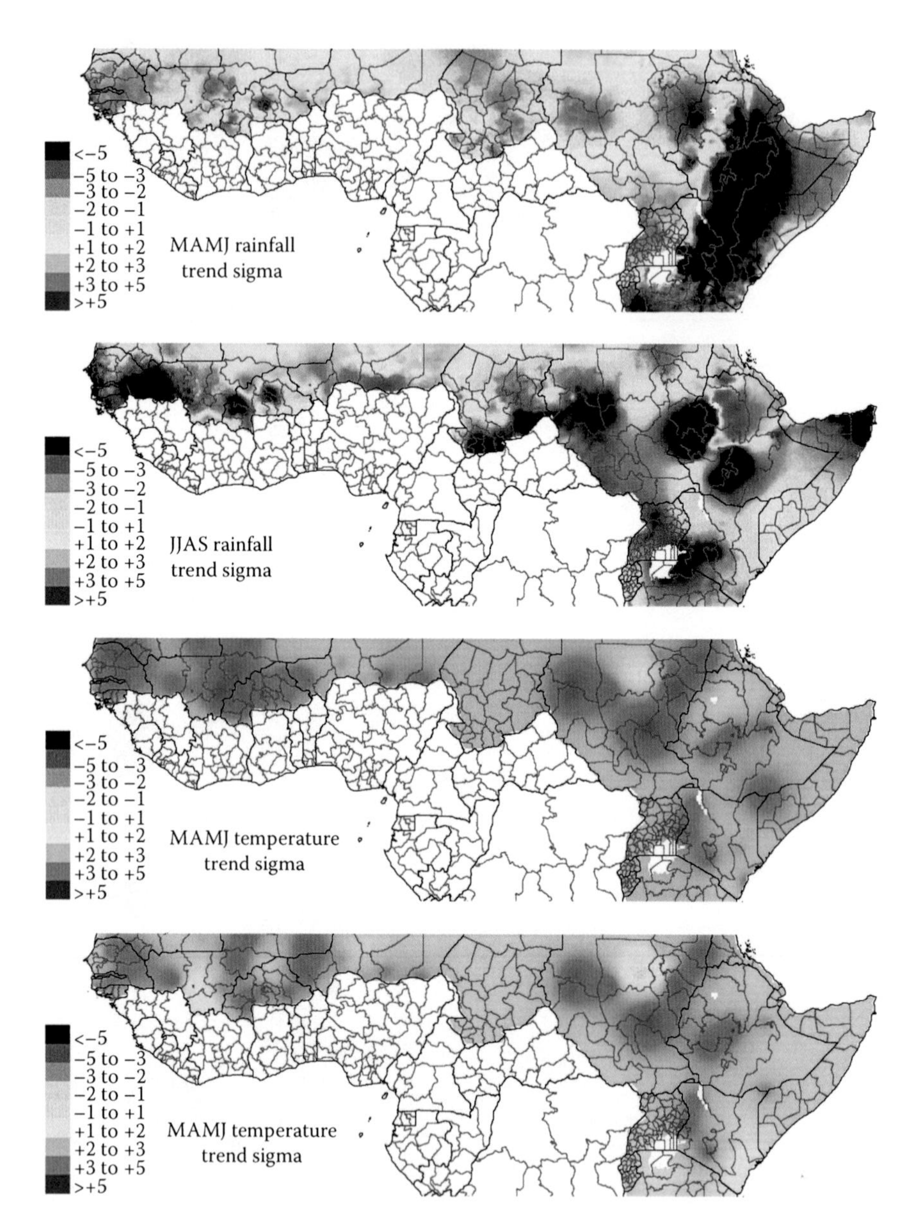

FIGURE 14.6 Rainfall and temperature sigma fields. Sigma values are the estimated trend fields divided by the interpolation standard error.

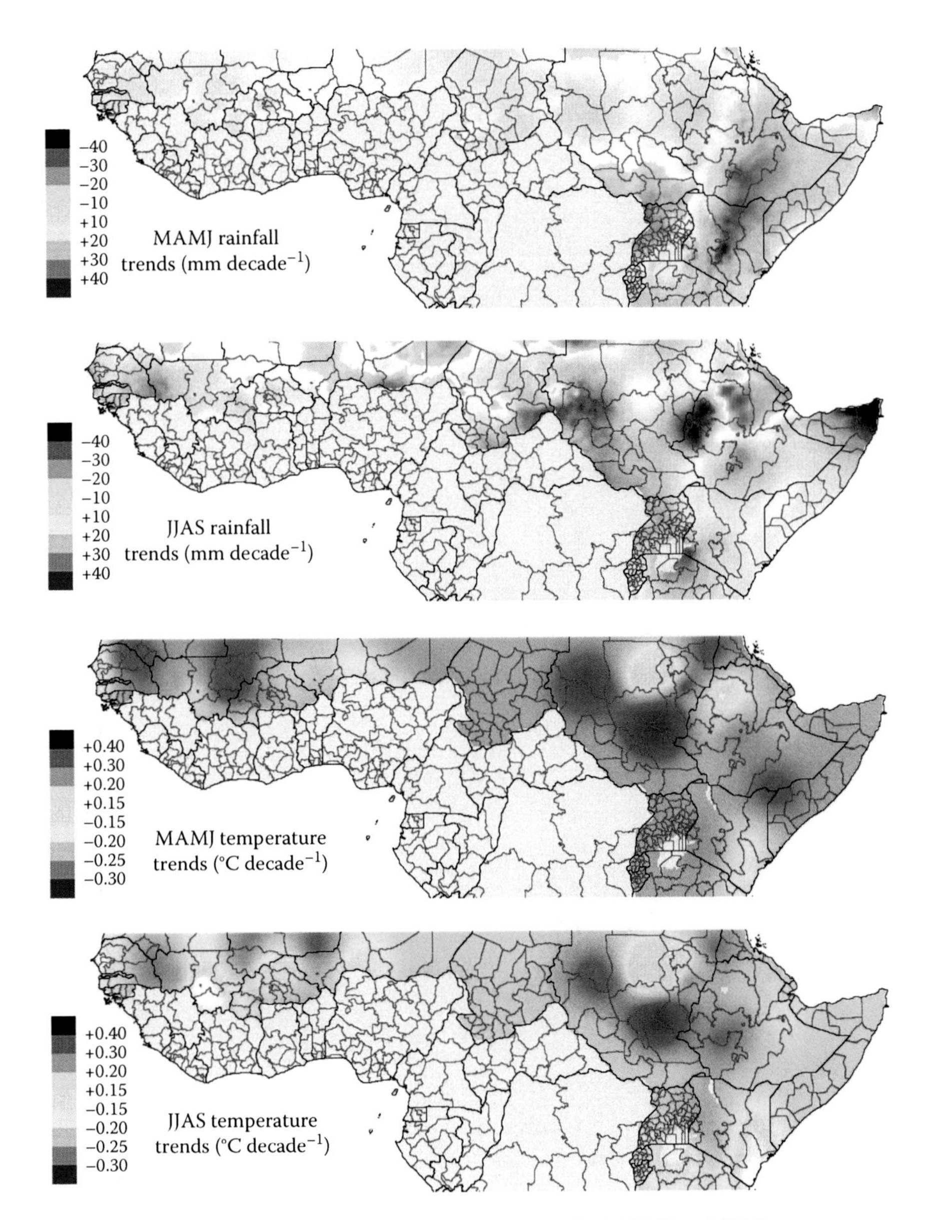

FIGURE 14.7 The 1960–2009 rainfall and trend maps for MAMJ and JJAS.

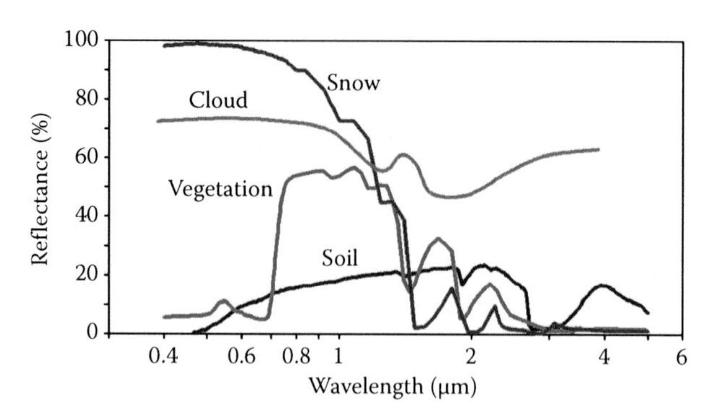

FIGURE 15.2 Spectral reflectance for different land surface cover types including snow.

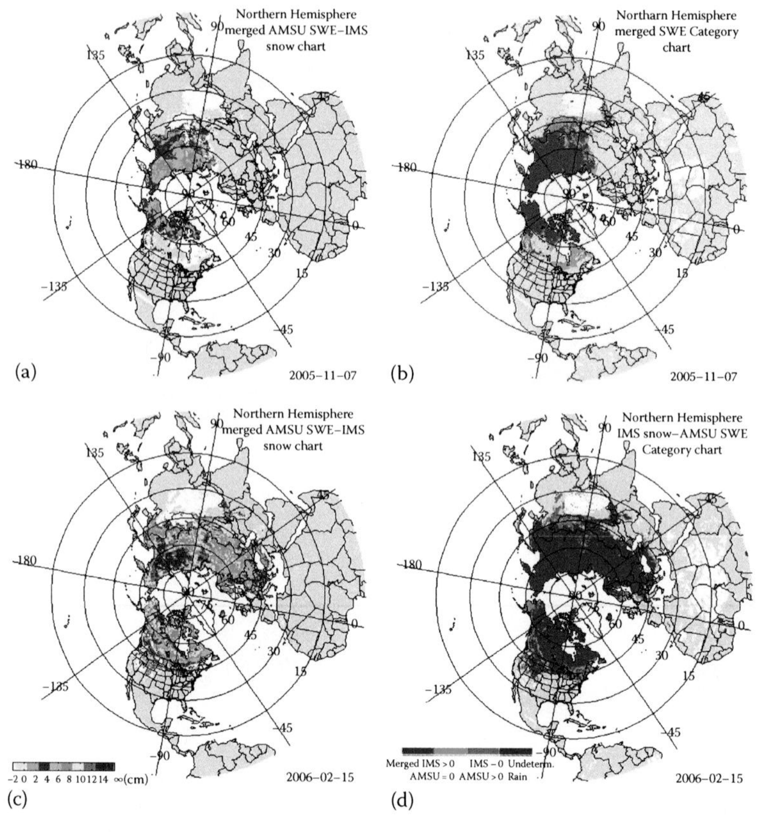

FIGURE 15.5 Intercomparisons between MW-derived and IMS snow charts. (From Kongoli, C. et al., *Hydrol. Process.*, 21, 1597, 2007.) Maps a and c depict the MW-derived SWE blended with IMS-derived SCA for early and midwinter, respectively, and maps b and d depict SCA overlays of IMS and MW-derived products for early and midwinter, respectively. The SCA overlay maps (b and d) show coincident MW- and IMS-derived SCA in blue, MW-underestimated areas in green, and MW-overestimated SCA in brown (as compared to IMS taken as ground truth reference).

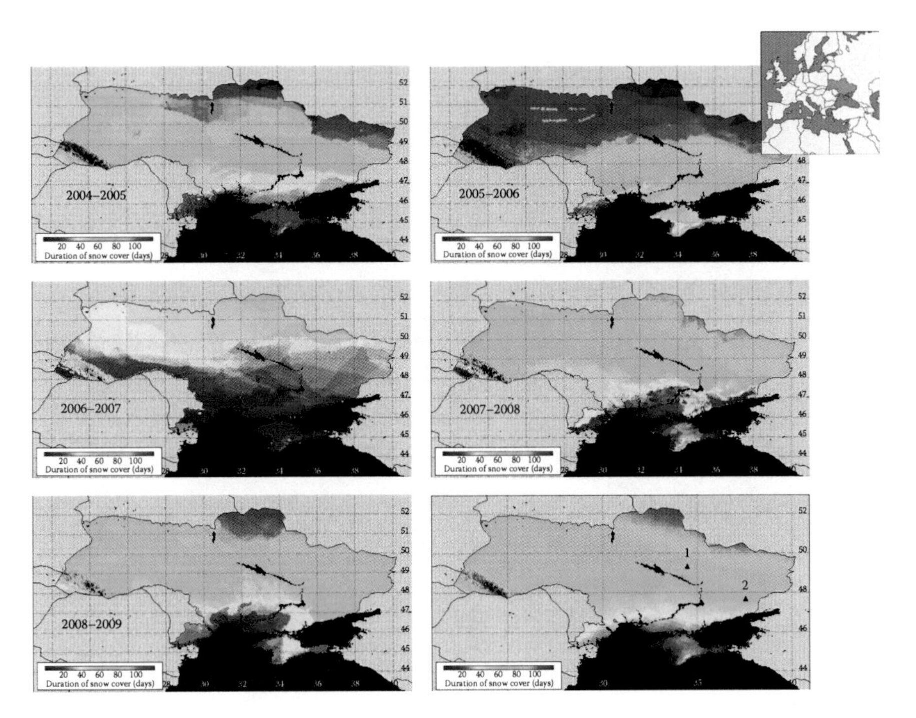

FIGURE 15.7 Maps of seasonal snow cover duration over Ukraine for selected years and the mean seasonal snow cover duration for the years 2000–2009. The upper right image shows the location of Ukraine in Europe. Triangles in the map of the mean duration of snow cover (in the lower right) indicate the location of Poltava (1) and Donetsk (2).

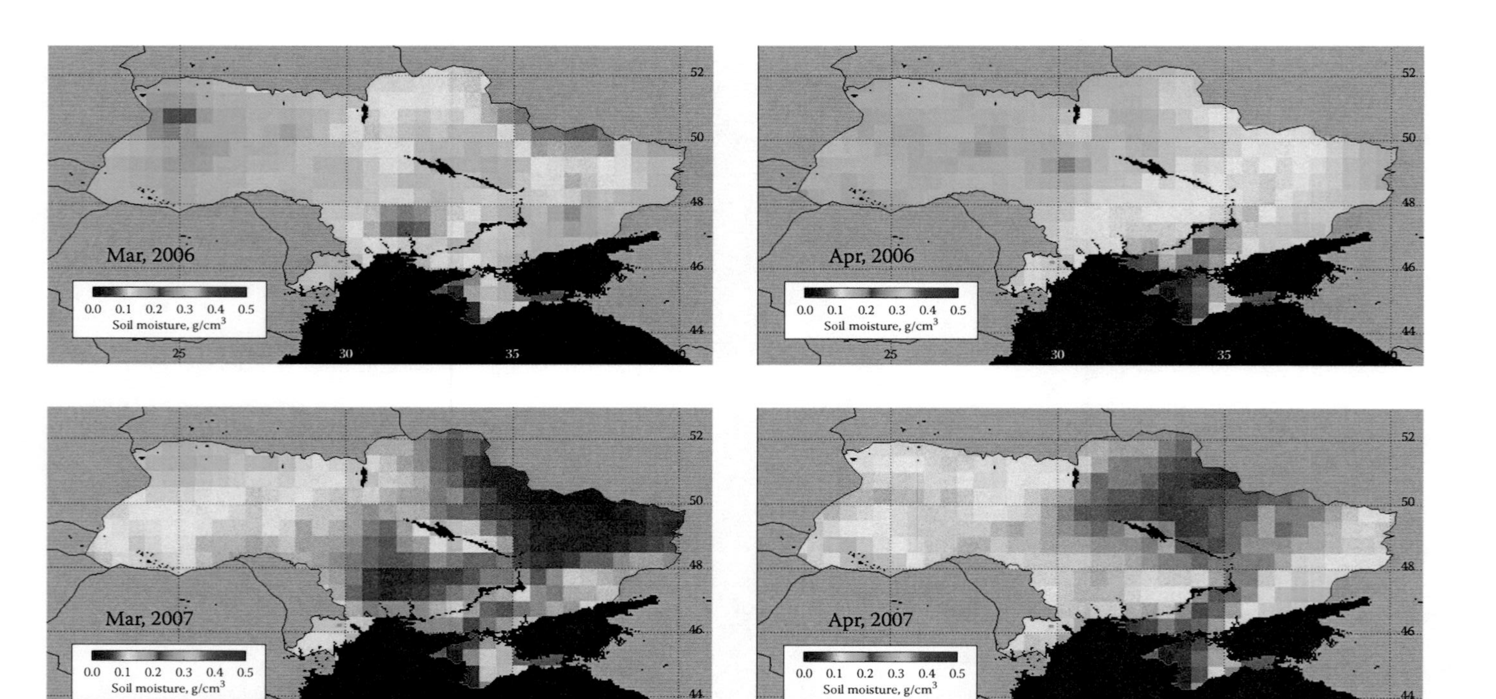

FIGURE 15.9 Monthly average soil moisture over Ukraine for the months of March and April of 2006 and 2007. Soil moisture was retrieved from AMSR-E Aqua data.

14 Mapping Recent Decadal Climate Variations in Precipitation and Temperature across Eastern Africa and the Sahel

Christopher Funk, Joel Michaelsen, and Michael T. Marshall

CONTENTS

14.1 INTRODUCTION

This chapter presents a novel interpolation approach that combines long-term mean satellite observations, station data, and topographic fields to produce grids of climate normals and trends. The approach was developed by the Climate Hazard Group (CHG) at the University of California, Santa Barbara (UCSB), to support food security analyses for the U.S. Agency for International Development's (USAID) Famine Early Warning Systems Network (FEWS NET). The resulting FEWS NET Climatology (FCLIM) combines moving window regressions (MWRs) with geostatistical interpolation (kriging). Satellite and topographic fields often exhibit strong local correlations with in situ measurements of air temperature and rainfall. The FCLIM method uses these relationships to develop accurate and unbiased temperature and rainfall maps. The geostatistical estimation process provides standard error fields that take into account the density and spatial distribution of the point observations. These error fields are especially important when evaluating climate trends. Numerous climate change analyses present trend evaluations without assessing spatial uncertainty. In many of these studies, the number of recent observations can be very low, potentially invalidating the results. This study presents analyses for the Sahelian and eastern African rainfall and air temperatures. The results indicate significant rainfall declines in Sudan, Ethiopia, and Kenya. Every country exhibits significant increases in average air temperatures, with Sudan warming the most. This chapter concludes with a short discussion of how these results are being used to guide climate change adaptation, with a case study focused on Ethiopia.

14.1.1 Mapping Decadal Variations Supports Adaptive Management

Although our capabilities to monitor and mitigate seasonal fluctuations in climate are fairly well developed, our capacity to monitor and respond to decadal climate variations is much more limited. Decadal fluctuations in temperature and rainfall can be associated with recurrent drought events, sapping the resilience of rural communities and helping to reinforce a spiral of increasing poverty. Although such decadal fluctuations are the epitome of a "slow-onset disaster," they are difficult to detect for three reasons: limited observation networks, low signal-to-noise ratios, and confounding societal factors.

At present, good station observations are the ultimate foundation for detecting decadal climate anomalies. Although satellite fields and climate model output can help guide analyses, they may contain trends and shifts driven by changes in earth observing systems or model assumptions. Unfortunately, the number of weather stations is declining in Africa. Typically, there is a disjunction between "weather monitoring" and "climate monitoring" systems, with weather station data only slowly being integrated into climatological databases, making the monitoring of recent multiyear droughts difficult.

The fact that the magnitude of a multiyear drought signal is typically relatively small compared to the size of interannual variations makes decadal monitoring even harder. The shifting composition of observing networks further complicates the problem, especially in areas of mountainous terrain. A lowland station may be characteristically 10° cooler and four times drier than a nearby highland station. Intermittent station reporting combined with interpolation of "raw" station values can create a large and completely spurious source of climate variation as low-hot-dry and high-cool-wet stations report alone or together. This problem can be addressed by interpolating station anomalies, as opposed to "raw" station values. By splitting the interpolation into two components—a long-term mean field and a seasonally varying anomaly field—the results are more accurate and less prone to contamination by changes in the spatial distribution of weather stations.

A third "detection" issue associated with multiyear droughts is the role played by societal and ecological trends. A drought arises when demand for water exceeds supply. Increasing water requirements, often associated with rapidly expanding populations and the use of water for power generation and irrigation, can lead to increasingly frequent water shortages. Similarly, land degradation and poor soil coupled with water management practices can lead to low water-use efficiencies, effectively increasing the frequency of droughts.

In order to respond effectively to climate change—whether natural or produced by greenhouse gases and aerosols—we need to develop our capacities to monitor, understand, and manage the factors related to multiyear drought. The work presented here focuses on using satellite data to map decadal climate variations. These results can support adaptive management practices and help identify emergent food security hot spots.

Although it has long been recognized that satellite data greatly improve our ability to observe drought on seasonal time scales, this chapter demonstrates the important role that remote sensing data can play in identifying trends and the emergence of new drought-prone areas exposed to a greater risk of multiyear drought events. This identification can play an important role in combating chronic poverty, malnutrition, and hunger in developing countries. Historically, most food- and water-insecure populations tend to live along the boundaries of semiarid regimes. Shifts in these boundaries can place new population groups at risks. Satellites directly observe geophysical data related to climate gradients and thus provide a valuable guide to mapping trends in climate gradients. At present, the standard climate products used to evaluate climate trends such as the Climatic Research Unit grids (New et al., 1999, 2000) rely solely on station data and physiographic predictors such as elevation. This chapter shows that satellite observations provide a valuable addition to the climatologists' toolbox, improving our ability to monitor and respond to drought on decadal time scales.

14.1.2 Project Objectives

The FCLIM methodology described here provides accurate, unbiased gridded estimates of recent trends in precipitation and temperature for two critical growing seasons in the east African and Sahel regions of Africa (Figure 14.1): the March–April–May–June (MAMJ) and the June–July–August–September (JJAS) seasons. A formal geostatistical framework, incorporating estimates of how information propagates in space (the spatial variogram, which expresses spatial covariance as a measure of distance), allows us to determine standard error maps and

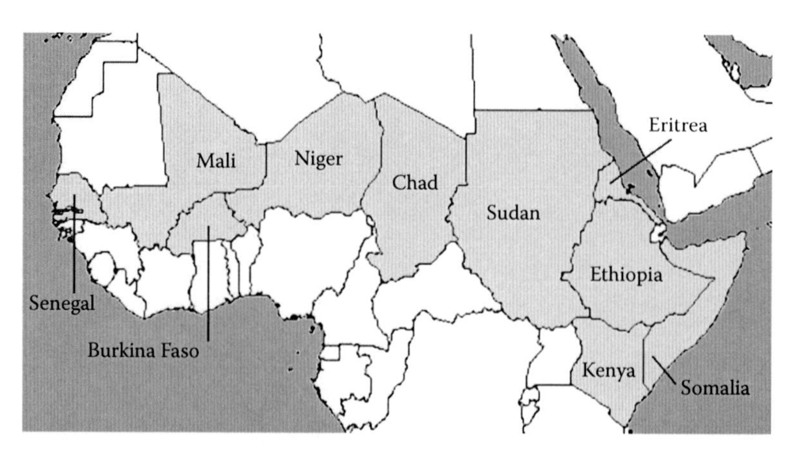

FIGURE 14.1 Map of the African countries examined in this study.

optimal estimates based on an optimal spatial interpolation technique (kriging). For the relatively dense Sahelian precipitation gauge network, the data were also mapped using an MWR technique developed by the CHG at UCSB (Funk et al., 2007). The MWR has previously been used to produce high-resolution climatologies supporting the USAID FEWS NET. The work here extends this approach to enhance precipitation trend analyses, which is important in enhancing drought monitoring capabilities. The use of long-term averages of satellite estimates of rainfall and temperature observations to guide the interpolation process is also explored. A further enhancement involves an explicit interpolation of the "at-station" trends, which facilitates estimation of uncertainties associated with the derived trend maps. Most previous analyses (including our own) have first produced a time series of monthly or seasonal rainfall and temperature grids and then analyzed trends in the resulting data layers. This can obscure uncertainties associated with the gridding process. The FEWS NET trend analysis (FTA) circumvents this problem through an explicit analysis of four season/climate variable combinations.

The MAMJ period corresponds to the critical "Long" rainy season in Kenya and the less-critical "Belg" season in Ethiopia. The JJAS period corresponds with Ethiopia's main "Meher" growing season, as well as the primary growing season across the Sahelian countries in western Africa. The Sahelian countries studied here stretch from Senegal in the west, across Mauritania, Mali, Burkina Faso, Niger, Chad, and Sudan to the nations of the Greater Horn: Eritrea, Djibouti, Ethiopia, Somalia, Kenya, Rwanda, and Burundi. Previous FEWS NET analyses have focused on downward rainfall trends in parts of the Greater Horn (primarily southern Ethiopia, central-eastern Kenya, and Somalia) (Funk et al., 2003a, 2005, 2008; Verdin et al., 2005; Williams and Funk, 2011) that suggest that the warming Indian Ocean causes hot dry air to descend across these regions, reducing MAMJ and JJAS rainfall. The work presented here updates this analysis through 2009, extends the geographic scope across the Sahel, and incorporates an additional set of observations for 100 stations provided by the Ethiopian National Meteorological Agency (NMA).

The temperature analysis used in this study is a new component of FEWS NET climate change research. The temperature data were based solely on the Global Historical Climate Network (GHCN) station archive, quality controlled via automatic screening procedures. This temperature data set is substantially less dense than the rainfall archive, and the associated results presented here should be seen as a first attempt by FEWS NET to quantify warming trends. This work will be augmented in the future with a denser set of in situ observations.

The FTA data set consists of a set of 24 grids at 0.1° resolution covering eastern Africa and the Sahel. The FTA provides sets of accurate high-resolution climatologies (FCLIM-A) together with estimates of recent climate trends (FCLIM-TR). The combination of these fields can enable us to identify emerging risk areas. While we have examined time series of seasonal rainfall totals (FCLIM-TS) in several previous papers (see Section 14.3.1), this study focuses on an explicit mapping of the rainfall and temperature trends themselves (in mm or °C per decade), along with techniques for explicitly quantifying the standard error estimates for the trend values. These standard error estimates (in mm or °C per decade), representing the geostatistical station support for a given location's trend estimate, are a unique contribution of this work to the climate change literature. Most climate change analyses evaluate trends without due regard to the impact of sparse and changing station support. We can quantify a trend's degree of certainty expressing the values in terms of standard error "sigma" values. These unitless sigma values are derived by dividing the trend estimate by the associated kriging standard errors. Kriging is a particular form of spatial interpolation that provides optimal estimates and spatial estimates of standard errors, based on the station distribution and distances. Projections of mean climate fields for the year 2025 are produced by multiplying the decadal trend estimates by five and adding the results to the 1960–1989 mean fields. Such predictions, which assume persistence of recent climate trends, certainly encompass a large, and essentially unknown, uncertainty. Nonetheless, if our analyses of the link between Indian Ocean warming and east African rainfall (Funk et al., 2008; Williams and Funk, 2011) are correct, plausible quasi-linear links exist between greenhouse gas emissions, warming in the central Indian Ocean, and reduced rainfall in southern Sudan, southern Ethiopia, Kenya, and Somalia. In any case, the decadal trend mapping approach described here should provide a useful contribution, helping us to routinely monitor and anticipate decadal climate variations.

14.1.3 Relevant Previous FEWS NET Analyses

One focus of FEWS NET research has been the evaluation of climate change and vulnerability trends in food-insecure eastern and southern Africa. This work began with the creation of historical rainfall time series for Africa (Funk et al., 2003b; Funk and Michaelsen, 2004). In 2003, the predictive potential of early growing season rainfall in Ethiopia was evaluated and provided USAID with food balance projections (Funk et al., 2003a). This analysis revealed two disturbing tendencies. First, agriculturally critical regions of Ethiopia had experienced substantial declines in seasonal precipitation. Second, population growth/food balance analyses suggested that Ethiopia would face chronic and increasing food deficits. This study was followed up with a careful analysis of more than a thousand eastern African rainfall gauge observations, suggesting that a warming Indian Ocean was likely to produce increasing dryness in extremely vulnerable areas of eastern and southern Africa. Satellite observations of vegetation greenness also exhibited these declines (Funk and Brown, 2005).

More recently, Funk et al. (2008) suggested that the warming in the Indian Ocean is likely to be at least partially caused by anthropogenic greenhouse gas emissions. Thus, further rainfall declines across parts of eastern and southern Africa appear likely. For eastern Africa, these drought projections run counter to the recent fourth Intergovernmental Panel for Climate Change (IPCC) assessment. Brown and Funk (2008) argue that climate change assessments, based on inaccurate global climate precipitation fields, probably understate the global agricultural risks of the warming Pacific and Indian Oceans. The interaction of growing populations and limited potential water and cultivated areas increases food and water insecurity, amplifying the impacts of drought. Funk and Brown (2009) focused on the global risks implied by these tendencies, with the overarching view that "early warning" must embrace both the short-term opportunities provided by the timely detection of food shocks as well as an effective tracking of the slow impacts of our changing climate. Funk and Verdin (2009) used this integrated approach to document both the recent rainfall deficits and the long-term declines across eastern Kenya. More detailed climate analyses link Ethiopian and Kenyan drying to warming in the Indian Ocean and overturning circulations bringing dry hot stable air masses down across parts of the Horn of Africa (Williams and Funk, 2011).

14.2 DATA

The FCLIM method incorporates climate, satellite, and physiographic data using a total of 10 specific input variables (listed in Table 14.1). Following is a brief description of each variable.

14.2.1 Climate Data

Two dense rainfall station data sets were provided for Ethiopia and the Sahel by the Ethiopian NMA (~100 stations) and the Centre Régional Agrhymet (~700 stations). These stations were augmented by rainfall records from the GHCN archive and United Nations' Food and Agriculture Organization's FAOCLIM database. For average air temperature, only data from the GHCN were used. Overall, records of 1339 rainfall stations and 178 temperature stations were examined. Observations were quality controlled both via visual comparison with neighboring stations and automated screening for extreme values. This rainfall database has a station density that is an order of magnitude greater than the density found in standard station archives such as the GHCN. Note, however, that the region being examined is vast (6.9 million km^2, or two-thirds the size of the United States including Alaska) and the station density is still extremely low (~1 rainfall station for every 5,000 km^2 and 1 temperature observation for every 40,000 km^2). With data this sparse, satellite observations play a critical role in the accurate mapping of climate and climate trend gradients, literally helping to connect the dots.

TABLE 14.1
Summary of Station Observations, Satellite Fields, and Topographic Data Sets Used in This Analysis

Data Products	Acronym	Dates	Sources
Station observations			
1. Seasonal rainfall [mm]		1960–2009	Ethiopian Nat. Met. Agency, Agrhymet, GHCN, FAO, GTS
2. Seasonal air temperature [°C]		1960–2009	GHCN
Satellite observations			
3. MODIS land surface temperatures [°C]	LST	2003–2009	NASA
4. Meteosat infrared brightness temperatures—10th percentile [°C]	IR10	2001–2009	NOAA/CPC
5. Meteosat infrared brightness temperatures—90th percentile [°C]	IR90	2001–2009	NOAA/CPC
6. Merged rainfall estimates v. 2 [mm]	RFE2	2001–2009	NOAA/CPC
Physiographic predictors			
7. Latitude [°]	Lat		
8. Longitude [°]	Lon		
9. Elevation [m]	Elev		USGS HYDRO1K
10. Slope [m per m]	Slp		USGS HYDRO1K

14.2.2 Satellite Data

Four satellite fields were used to improve the spatial resolution and precision of the gridded climate data. The high correlations between our in situ data and these fields supported regression models to "connect the dots" of the very sparse station observations in our study site, guiding the rainfall and temperature FCLIM and the rainfall FTA modeling. The temperature gauge density was not sufficient to support the use of satellite and topographic data in the derivation of the FTA. One objective of this study was to evaluate the relative merits of these remote sensing data sets in guiding spatial interpolation and mapping drought trends. Land Surface Temperature (LST) maps at 1 km resolution were produced by the LST group at UCSB using thermal infrared (TIR) data collected by the Moderate Resolution Imaging Spectroradiometer (MODIS). The MODIS instruments are in polar orbit around the earth, providing day and night imagery of the earth's surface. Because the energy observed by a satellite depends on the temperature and emissivity of the emitting object, as well as atmospheric effects, the MODIS LST algorithm uses nighttime/daytime image pairs, observations at multiple wavelength bands (11 and 12 μm), atmospheric corrections, and radiative transfer models to estimate surface skin temperatures (Wan and Dozier, 1989). In addition to LST, TIR (11 μm) brightness temperatures from geostationary Meteosat weather satellites were also used in our regression modeling. Ten years (2001–2009) of half-hourly Meteosat ~0.04° TIR data (Janowiak et al., 2001) were processed into seasonal images representing the warm (90th percentile) and cold (10th percentile) TIR brightness temperatures at each location. The cold 10th percentile IR maps (referred to as IR10) represent the spatial pattern of cold upper level clouds. These fields were used to guide our estimates of rainfall. The warm 90th percentile IR values (referred to as IR90) tend to isolate emissions from the earth's surface, providing gradient information related to the spatial pattern of LST values. These fields were used to guide our estimates of air temperature.

Multisatellite rainfall estimates (RFE2) from NOAA CPC (Xie and Arkin, 1997) were also used as potential guides to the FCLIM and FTA estimates. The RFE2 data set blends data from two passive microwave sensors (the Special Sensor Microwave/Imager and the Advance Microwave Sounding Unit), cold cloud duration rainfall estimates based on Meteosat TIR data (Janowiak et al., 2001), and Global Telecommunication System (GTS) rainfall values to produce daily estimates of rainfall.

14.2.3 Physiographic Data

Four physiographic indicators were also used as potential predictor variables for precipitation and temperature: latitude, longitude, elevation, and slope. Mean elevation and slope fields were derived on a 0.05° grid by aggregating HYDRO1K elevation derivatives (Verdin and Greenlee, 1996). The four satellite fields (LST, IR10, IR90, and RFE2) were resampled to the same grid. This set of predictors allowed us to compare the performance of the traditional physiographic predictors with the newly available long-term mean fields from satellites. Traditional climatologies only use physiographic data and station observations (New et al., 1999). This study's hypothesis was that the use of satellite fields would enhance the spatial accuracy of our estimates. Although physiographic data are commonly used to guide interpolations of mean air temperature and precipitation, the link between these variables and precipitation is indirect and variable in space. High elevations and steep slopes can, on average, experience more precipitation, but this does not always hold. Presumably, satellite observations, which are much more closely related to the physics of the associated processes, would provide a better basis for spatial prediction.

14.3 FEWS NET CLIMATOLOGY METHOD

Several strategies have been evaluated for mapping climate fields in Africa, Asia, and central America. Early efforts focused on blending interpolated station data with output from an internal-gravity waved-based model of orographic rainfall enhancement

(Funk et al., 2003b). Initial studies suggested that this orographic model could be successfully blended with satellite rainfall estimates (Funk et al., 2004). Extensive analysis of systematic bias within the CPC's RFE2 led to the realization that although the satellite record often struggles to correctly estimate the absolute magnitude of in situ observations, it is often very accurate in terms of representing the geographically "local" slopes of precipitation and temperature. Satellites are effective in determining areas that are relatively wet or warm from areas that are relatively dry or cool. The FCLIM uses satellite mean fields to guide the spatial interpolation of station data for point estimates of long-term means and decadal trends. This procedure has been used to guide trend analyses of Kenyan and Ethiopian rainfall (Funk et al., 2007, 2008; Funk and Verdin 2009).

In this chapter, we describe the FCLIM approach and extend its application to include long-term mean air temperature fields and precipitation trend fields. The FCLIM estimates have two major components. The first component uses MWRs (described in Section 14.3.1) to create a "first cut" estimate of the gridded field. The second component (described in Section 14.3.2) uses either kriging or a modified inverse distance weighting interpolation to create grids of regression model residuals. Cross-validation is then used to quantify the "at-station" estimation errors, while kriging standard error fields quantify the spatial uncertainty associated with the gridded FCLIM and FTA spatial predictions. In general, the FCLIM modeling process generally follows six phases:

1. *Data collection.* Collect and quality control all available station data.
2. *Parameter estimation and visualization.* Estimate the statistic of interest for the station data. Here we have used 1960–1989 historical means and 1960–2009 historical trends, but other parameters could also be used, such as medians, percentiles, or frequency. Point maps of the station means and trend values help guide the geospatial modeling of these parameters.
3. *Selection of optimal satellite and physiographic predictors.* Cross-validation is used with the modified inverse distance weighting (IDW) procedure to examine the fit of various parameter combinations. Visual examination of localized correlation plots (described in Section 14.3.1) can help the parameter selection process. The full output for various combinations of predictors should be examined. Prediction fields that change dramatically because of the selection of predictors are likely to indicate overfitting. Similar to typical regression applications, the FCLIM modeling process is iterative and "hands-on," guided by the modeler's expertise. As this chapter suggests, because satellite fields physically correspond directly to climate variables, they typically emerge as the best sources of predictors, and this allows the satellite-enhanced FCLIM fields to perform substantially better than traditional climate surfaces.
4. *Error analysis of trend surfaces.* The kriging procedure provides a measure of interpolation uncertainty based on the spatial pattern and spatial covariation of the station data. The magnitude of these standard errors can be meaningfully compared to the magnitude of the predicted trends. This is especially important in evaluating the trend fields. Given sparse data, can

we truly make claims about trends at locations without stations? This work shows that satellite data can substantially reduce spatial prediction errors.

5. *Climatic interpretation of the trend surfaces.* The next phase evaluates the resulting FCLIM and FTA maps based on our understanding of the physical process and previous research. The question is, do the results appear to be plausible, given our knowledge of the climate system (mean climate, recent changes in circulation and atmospheric chemistry) and independent, corroborating data sets?
6. *Interpretation and analysis of the trend surfaces.* Given the mapped trends, FEWS NET scientists interpret the likely food security impacts. Such analysis depends, in part, upon the climatological means. For example, semiarid crop growing areas are more sensitive to rainfall reductions than wetter regions, because a small reduction in rainfall may substantially increase the frequency of crop failures. Areas with steep topography, and mean temperatures that change rapidly across space, may be less influenced by a 1° temperature increase than are flat extremely warm regions. Underlying vulnerability, water security, and malnutrition can also exacerbate the impact of climate changes.

Section 14.2 (Data) summarized the data collection procedure. Phases two and three will be discussed in the following section and followed by a summary of phases four through six in Section 14.4 (Results).

14.3.1 Examining the At-Station Trends of Temperature and Precipitation

Before modeling and interpolating the data, it is worthwhile to examine trends at station locations to build confidence in the final results. Without such analysis, there is always a risk of introducing spurious "structure" into maps that does not, in fact, exist in the training data. Although sparse, the set of air temperature observations used in this study exhibits consistent increases over the period 1960–2009. Histograms of the station-based temperature trends (not shown) show that almost every station exhibited substantial temperature increases, typically ranging from 0.05°C to 0.5°C per decade. Simple averages of station data across the countries suggest that the Sudan–Niger–Mali area has experienced an increase of more than 1.0°C from 1969 to 2009, while the Kenya–Ethiopia area has experienced an increase of about 0.7°C in the same 30 year period. The magnitude of the temperature increases is equal to or greater than the interannual temporal standard deviations of the station data, averaged across the same groups of countries (0.5°C for Kenya–Ethiopia, 0.65°C for Sudan–Niger–Mali). Such warming can disrupt the seasonal cycle of crops, draw more water from the soil and plants—exacerbating drought conditions and reducing the amount of grain produced.

The precipitation trends tended to vary within and between countries (Figure 14.2). Dark red circles denote large recent decreases in rainfall or increases in temperature. There is fairly high level of congruence between the two seasons. In both MAMJ and JJAS, rainfall appears to be diminishing across Kenya and Ethiopia, with some smaller declines in rainfall appearing across the Sahel in JJAS.

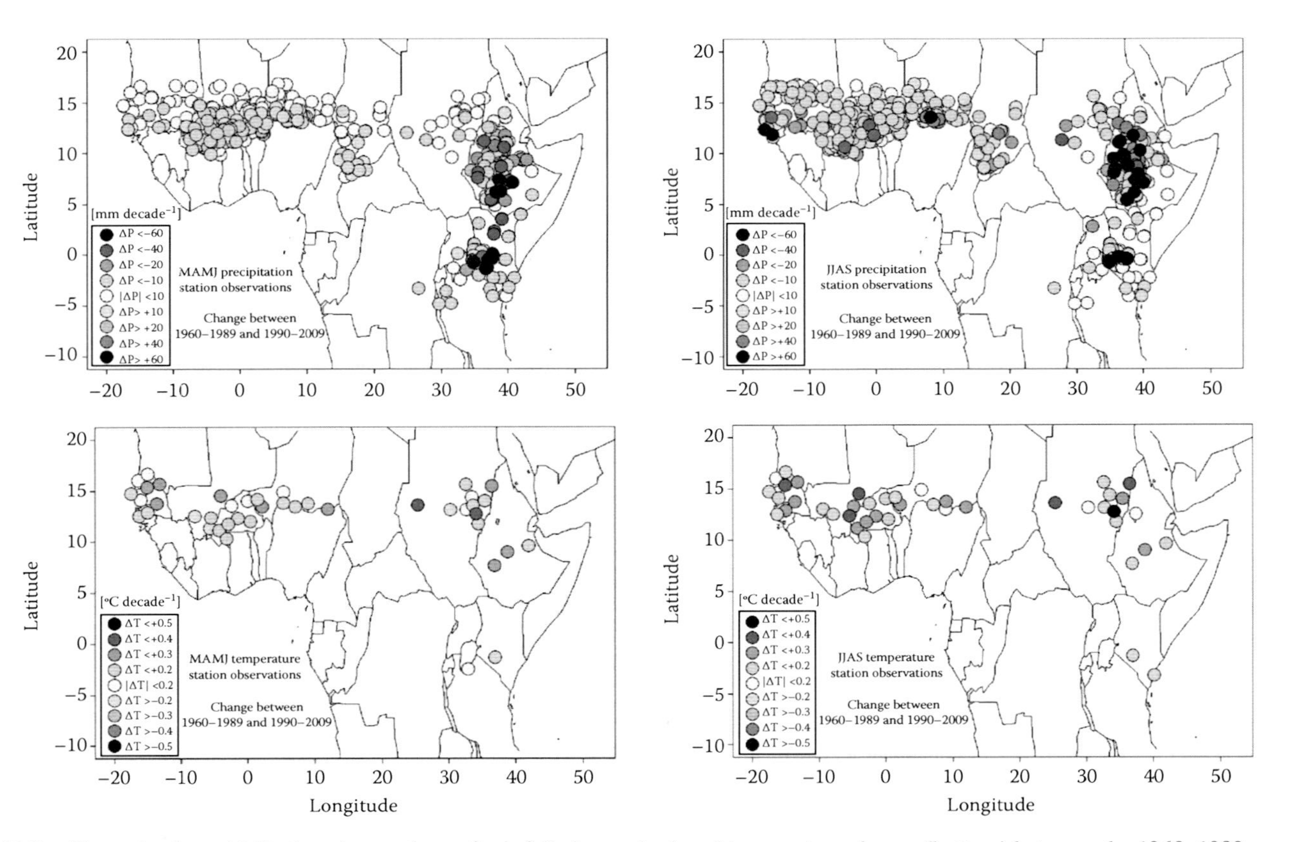

FIGURE 14.2 **(See color insert.)** Station observations of rainfall change (top) and temperature change (bottom) between the 1960–1989 average and the 2000–2009 average for MAMJ (left) and JJAS (right).

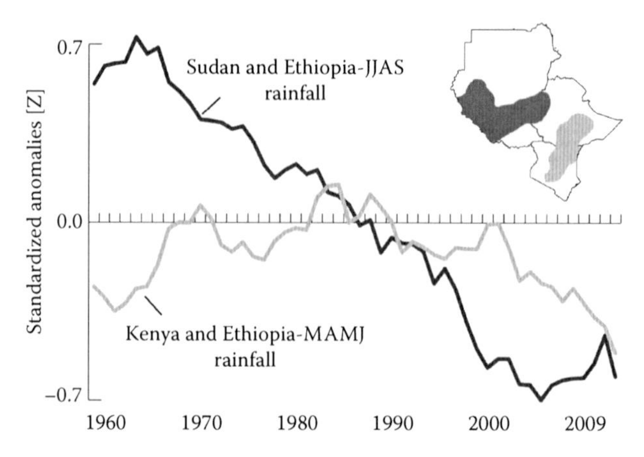

FIGURE 14.3 Time series of rainfall z-scores for two areas/seasons with declining rainfall: southern Sudan and southwestern Ethiopia during JJAS and central Kenya and southern Ethiopia during MAMJ. Time series have been smoothed with a 20 year running mean.

Since each climate observing station is essentially an independent "vote," the combined evidence suggests that significant drying has occurred in Ethiopia and Kenya. However, some stations do not exhibit downward trends in regions that appear, on average, to be drying. This could perhaps be the result of inaccurate station data or the local impact of terrain features. However, almost every temperature record shows increases in temperature over the recent era, with temperature trends ranging between 0.1°C and 0.4°C per decade. Time series of interpolated rainfall data for the most affected regions of Kenya–Ethiopia (MAMJ) and Sudan–Ethiopia (JJAS) are shown in Figure 14.3, with the rainfall expressed in terms of standardized anomalies (i.e., z-scores) based on the 1960–1989 time period. Over the past 20 years, both regions have experienced large (greater than 0.5 Z) standardized decreases in main growing season rainfall, implying much more frequent occurrence of drought. The magnitude of the observed trends can be quite large, with the JJAS rainfall declining by more than 100 mm since the mid-1970s. By the year 2025, a temperature trend of 0.2°C per decade would be associated with a warming of 1°C since 1975. By 2020, a 50 mm per decade decreasing trend would be associated with a 250 mm decline in rainfall.

The objective in this chapter is to show how satellite data can be used to provide spatially explicit maps of these trends to complement the point-based observations, accompanied by estimates of the interpolation accuracy. To this end, a blend of MWRs and geostatistical kriging was developed.

14.3.2 Selection of Optimal Satellite and Physiographic Predictors

The core of the FCLIM model fitting is based on local spatial correlations between station data and satellite and physiographic predictors. At a local scale, satellite fields typically exhibit a strong spatial covariance with in situ observations, and this can be used to make accurate long-term mean and trend maps. Satellite fields also typically

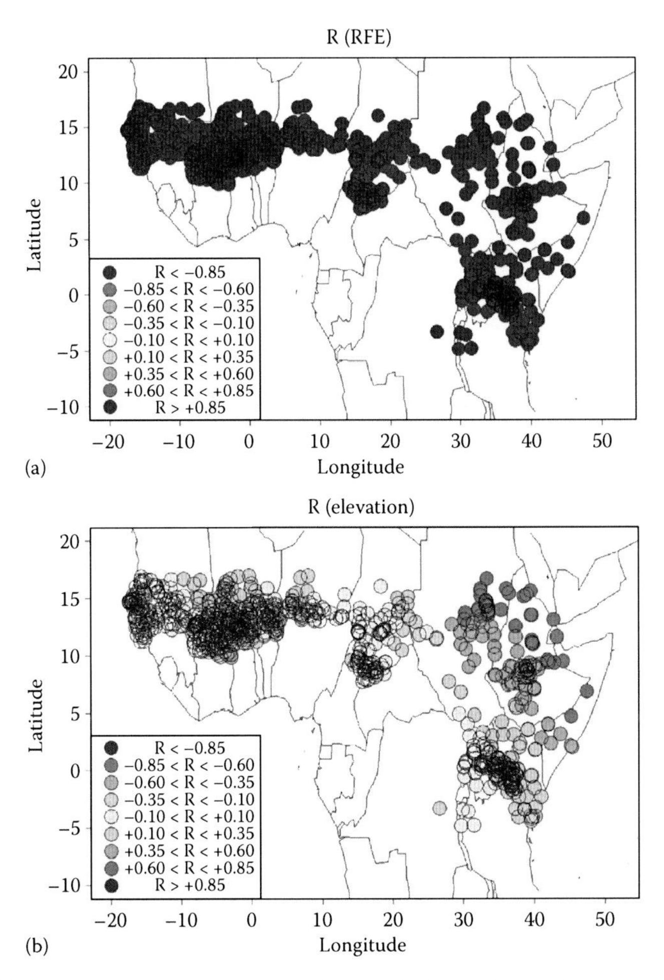

FIGURE 14.4 (See color insert.) Comparison of RFE2 and elevation correlations with JJAS precipitation normals showing the local correlation between station observations of JJAS rainfall and RFE2 means (a) and elevation (b). The median local correlation between the station means and RFE2 means was 0.91. Local correlation computations are based on a d_{max} value of 1500 km².

exhibit local spatial correlations that are much stronger and more consistent than physiographic fields. An example of this is shown in Figure 14.4. The Figure 14.4a shows the local correlation between 1960–1989 average JJAS precipitation and long-term (2001–2009) average JJAS RFE2 data. Despite a very complex precipitation landscape, involving steep rainfall gradients and complex orographic features, the local correlation exceeds 0.85 at every station location. In situ observations of mean rainfall are highly correlated with satellite estimates, and, thus, satellite fields can be used to guide interpolations in between station locations. This can be strongly contrasted with elevation, which forms the basis of most standard global climatologies. Station rainfall data over most of the Sahel are poorly correlated with elevation

(Figure 14.4b). This chapter examines how spatial satellite information (Figure 14.4a) may be integrated with traditional geostatistical estimation procedures.

The idea of local correlation is central to this work. The local correlation concept is defined here and then expanded to include multivariate regression. Local correlations and regressions are defined based on a weighted function of distance, for which we use a cubic function of distance (d) relative to some maximum distance (d_{max}):

$$w = 0 \quad if\ d > d_{max}$$

$$w = \left(1 - \left(\frac{d}{d_{max}}\right)^3\right)^3 \quad if\ d < d_{max} \tag{14.1}$$

For a given location, a set of observations (y) falling within a radius d_{max} is selected. Typically, y is a temporal statistic such as a long-term mean. A corresponding vector of predictors (x) is derived by extracting pixel values at these locations; x is typically either a long-term average of a satellite field or a static physiographic feature (e.g., elevation). Although local spatial correlation could be calculated solely based on x and y, anomalies weighted by distance were used instead. This weighting spatially focused the correlation on the target location (Equation 14.1). Anomalies at distances approaching d_{max} are forced to 0, while those located at our target center (i.e., near a station location) receive weights of 1. Local anomalies are calculated by first estimating the local mean of y (μ_y) and x (μ_x), and then creating a centered set of spatial anomalies for the observations ($y' = y - \mu_y$) and independent data ($x' = x - \mu_x$) for all station locations within a radius d_{max}. Given a set of weights (w) based on Equation 14.1, and defining "*" as the element-by-element multiplication operator, a set of weighted anomalies for y $\left(y'_w = w*y'\right)$ and x $\left(x'_w = w*x'\right)$ are generated and the local spatial correlation estimated between x′ and y′, where y′ represents a set of nearby station observations weighted by distance from the target center and x′ represents the associated predictor pixels also weighted by distance from the target center. The weighting procedure allows us to capture local variations in correlation structures, such as the positive relationships between [X = elevation] and [Y = rainfall] observed in the Ethiopian highlands (Figure 14.4b).

Figure 14.4 shows the local correlation values between RFE2 and elevation (Figure 14.4b) and JJAS average 1960–1989 mean rainfall (Figure 14.4a). Similar sets of local correlations may be produced for all the candidate predictors (Table 14.1). The median correlation values between satellite/geotopographic fields and station temperature and rainfall data are shown Table 14.2. RFE2 and P10 IR data are the best predictors of mean rainfall (median $r \sim 0.9$), and the LST and P90 fields are best predictors of average air temperatures (median $r \sim 0.8$).

Although the results shown in Figure 14.4 and Table 14.2 were estimated at the station locations, the same procedure may be carried out on a regular grid of locations. In the full FCLIM gridding procedure, moving window correlations or regressions are estimated at each target grid cell (i.e., each 0.1° cell across the maps). For clarity, this section first describes the FCLIM method using a single predictor and then expands our discussion to include a multivariate estimation procedure.

TABLE 14.2
Median Cross-Validated Absolute Correlation Values between Satellite/Geotopographical Fields (Vertical List) and Station Data (Horizontal) for the JJAS Season

	Temperature JJAS	Rainfall JJAS
RFE2	0.74	0.92
LST	0.86	0.87
P10	0.74	0.90
P90	0.90	0.84
Lon	0.15	0.25
Lat	0.82	0.79
Elev	0.23	0.13

Building on the localized correlation coefficients, we can create a local estimate of **y** based on x (y_{est}):

$$y_{est} = \mu_y + r_l\left(x'_w, y'_w\right)\sigma_y \sigma_x^{-1} \tag{14.2}$$

where
μ_y is the local mean of y
σ_y is the local standard deviation of the stations
σ_x is the local standard deviation of the predictors
$r_l\left(x'_w, y'_w\right)$ is the local correlation as described earlier

In practice, a more complicated relationship between x and y can be used, based on transforms of the data and/or a nonlinear (typically spline-based) estimator. These transforms were not used here because of the large spatial domain analyzed and the spatial nonstationarity of distributions and relationships across the region. MWR estimates are typically produced for every grid cell. Each station value is then paired with the closest grid cell, and residuals are estimated. A model semivariogram may then be fit to the residuals and a geostatistical interpolation technique (kriging) used to produce a grid of residual values. Because the model semivariogram explicitly quantifies the spatial decorrelation with increasing distance, kriging produces spatial maps of standard error, contingent on the spatial distribution of the observation network. The MWR and residual fields are summed, creating an estimate that combines correlated independent predictors and the spatial covariance of the in situ observations themselves. The estimate is calculated as

$$\text{FCLIM} = y_{est} + k \tag{14.3}$$

where k represents the interpolated residuals from our local regression. The estimates that are produced are referred to as FCLIM values.

The FCLIM methodology, which comprises Equations 14.1 through 14.3, uses the information in the station data in three ways. First, a local mean (μ_y), centered at location l, approximates the general magnitude of our estimated value. The local relationship between y and x is then used to further refine our estimate (in Equation 14.2), taking advantage of any local correlation between y and x. Finally, we include an interpolation step (Equation 14.3), which incorporates the values at the stations and adds fine resolution information to our results. Areas near observing sites will be adjusted toward the in situ values.

14.3.3 Novel Use of Satellite Data

In practice, the FCLIM methodology is typically invoked using a multivariate set of predictors, potentially including both physiographic variables (latitude, longitude, elevation, and slope) and satellite observations of rainfall, infrared brightness temperatures, and LST (Table 14.3). Instead of focusing on ability of these data sets to represent *temporal* variations in weather, the FCLIM approach focuses on the ability of these variables to represent *spatial* gradients of temperature and precipitation.

The FCLIM approach involves exploratory analysis and selection of a relatively small number (typically 4–6) of predictors and the identification of a characteristic scale (d_{max}) determined primarily by station density. For each location l, a matrix of

TABLE 14.3
Local Regression Models and Cross-Validated Skill Estimates

	Width [km]	Number of Stations	Regression R^2	Regression Standard Error	Predictors
MAMJ 1960–1989 temperature	2000	111	0.95	1.0°C	Lon, lat, elev, LST, P90
Temperature trends		57	0.00	0.13°C	
JJAS 1960–1989 temperature	2000	113	0.92	1.2°C	Lon, lat, elev, LST, P90
Temperature trends		56	0.07	0.13°C	
MAMJ 1960–1989 rainfall	1500	894	0.88	68 mm	Lon, lat, elev, RFE2, LST, P10, P90
Rainfall trends		536	0.34	13 mm	Lon, lat, elev, LST, P10, P90
JJAS 1960–1989 rainfall	1500	960	0.93	72 mm	Lon, lat, elev, RFE2, LST, P10, P90
Rainfall trends		587	0.22	21 mm	Lon, lat, elev, LST, P10, P90

centered predictors (x'_w) and a vector of observed values (y'_w) can be used to identify a local multivariate regression using

$$y_{est} = b_o + b^T x'_w \tag{14.4}$$

Again, either IDW or geostatistical kriging can be used to interpolate the residuals. At each grid location, the final FCLIM estimate combines the local regression intercept (b_o), a vector of local regression slope values (b), a vector of local gridded satellite and physiographic predictors (x), and a local estimate of the kriged residuals (k):

$$\text{FCLIM} = b_o + b^T x + k \tag{14.5}$$

Comparisons of kriging and IDW suggest similar levels of accuracies. In this work, IDW is used with automated cross-validation procedures to assess accuracy, and geostatistical kriging is applied within the final analyses. The kriging procedure produces maps of expected standard error.

14.3.4 Cross-Validation and Model Fitting

As with any regression procedure, model fitting is an important part of the modeling procedure. This can be especially true when working with spatially correlated data, which is the typical case when working with geographic information. To address this need, the FCLIM process uses interactive visualization of correlations (see Figure 14.4), cross-validation, and both visual and statistical evaluation of the resulting output fields. As with other regression techniques, FCLIM can be automated, but works best when expert analysts guide the procedure. A 10-fold cross-validation technique is used where 10% of the station data are withheld for validation and the remaining 90% of the data used to fit the full FCLIM model. This process is repeated 10 times to produce a robust estimate of the model accuracy. Cross-validation and examination of the gridded output fields is used to identify successful model combinations.

14.3.5 Final Outputs: FCLIM Maps of Means and Trends

The full FCLIM procedure, which includes the MWR and kriging, was used to model the MAMJ and JJAS temperature and precipitation 1960–1989 station averages. The resulting FCLIM-average fields are unique in that they use satellite data to help guide the interpolation of long-term mean station data. Another novel aspect of this work is the explicit FCLIM mapping of the station-based precipitation trend estimates (FCLIM-TR). Most trend analyses tend to interpolate station observations and then evaluate low frequency variations in the gridded data. This approach makes it difficult to accurately assess the implications of sparse and changing observation networks. This new approach applies the multivariate FCLIM methodology (Equations 14.3 and 14.4) directly to station-based estimates of trends for MAMJ and JJAS rainfall. The air temperature network was not sufficiently dense to support this level of detailed analysis, and the at-station trends were interpolated using kriging. The kriging procedure explicitly defines the spatial variation of the interpolated information; this produces maps of the associated standard errors. This allows us to say that at location l, the trend has most likely been y_{est}, with a 95% confidence interval of $\pm 1.96\sigma_{y_{est}}$. It is important to note, however, that there are many types of uncertainty not encapsulated here related to measurement errors and the fundamental nonstationarity of the climate itself, which is likely the most important. Despite these remaining uncertainties, the maps presented in this study, when used together, can assist in identifying emerging climate patterns that necessitate rapid response and adaptation strategies.

In addition to the core data products, eight other products are provided through the FCLIM methodology: four "sigma" fields and four "future climate" fields. The four sigma fields represent the ratio of the estimated trends and standard errors, which allow the relative magnitude of the trends to be compared to the spatial interpolation uncertainty. Areas with sigma values having absolute values of greater than 2 can be considered as quite reliable. Another four fields are derived by multiplying the decadal trend maps by 5 and adding the result to the 1960–1979 mean fields. These maps depict 2025 climate surfaces, assuming recent trends persist. Such maps are extremely useful for analyzing the drought implications of rainfall trends. For example, many crops require a certain minimum seasonal rainfall total, and the 2025 projections can help identify areas where specific types of agriculture may no longer be viable in the future. This information can then help guide adaptation efforts. A specific example of this type of analysis is given in Section 14.4.5, where the spatial drought implications of the observed drying trends (shown in Figures 14.2 and 14.3) are evaluated.

14.4 RESULTS

14.4.1 Local Regression Model Results

Cross-validation and model exploration led to the selection of localized regression models for all four long-term mean fields and the two rainfall trend surfaces (Table 14.3). The dense rainfall gauge network supported complex regression models that performed very well for long-term mean fields ($R^2 \approx 0.9$). While variance explained for the rainfall trends was substantially lower ($R^2 \approx 0.3$), the regression models still reduced the overall interpolation errors. These models also provided some guidance in areas devoid of station observations. The long-term mean temperature values were fit very well ($R^2 \geq 0.9$). For the sparse set of temperature trend observations, cross-validation analysis suggested that the combination of regression and interpolation models performed similarly to either regression or interpolation models alone. The very low density of observing sites did a poor job of restraining the regression model results, and the selection of different predictor variables created substantially different temperature trend maps. Parsimony led us, therefore, to simply adopt the interpolation model for this component. Future research, with a more complete climate record, should lead to a more sophisticated analysis and spatially complex picture of recent temperature trends.

14.4.2 FCLIM Mean Fields

Figure 14.5 shows the four FCLIM mean fields for the 1960–1989 period. This time period was chosen because these decades had the highest station density of rainfall and temperature observations. It is worth pausing to consider the spatial structure of the climate data displayed, as this will inform our discussion of the observed rainfall and temperature trends. In MAMJ, the region is generally dry, except for a "butterfly" pattern of higher rainfall in Kenya, spanning the shore of Lake Victoria and the central highlands, and along the southern escarpment of Ethiopian highlands.

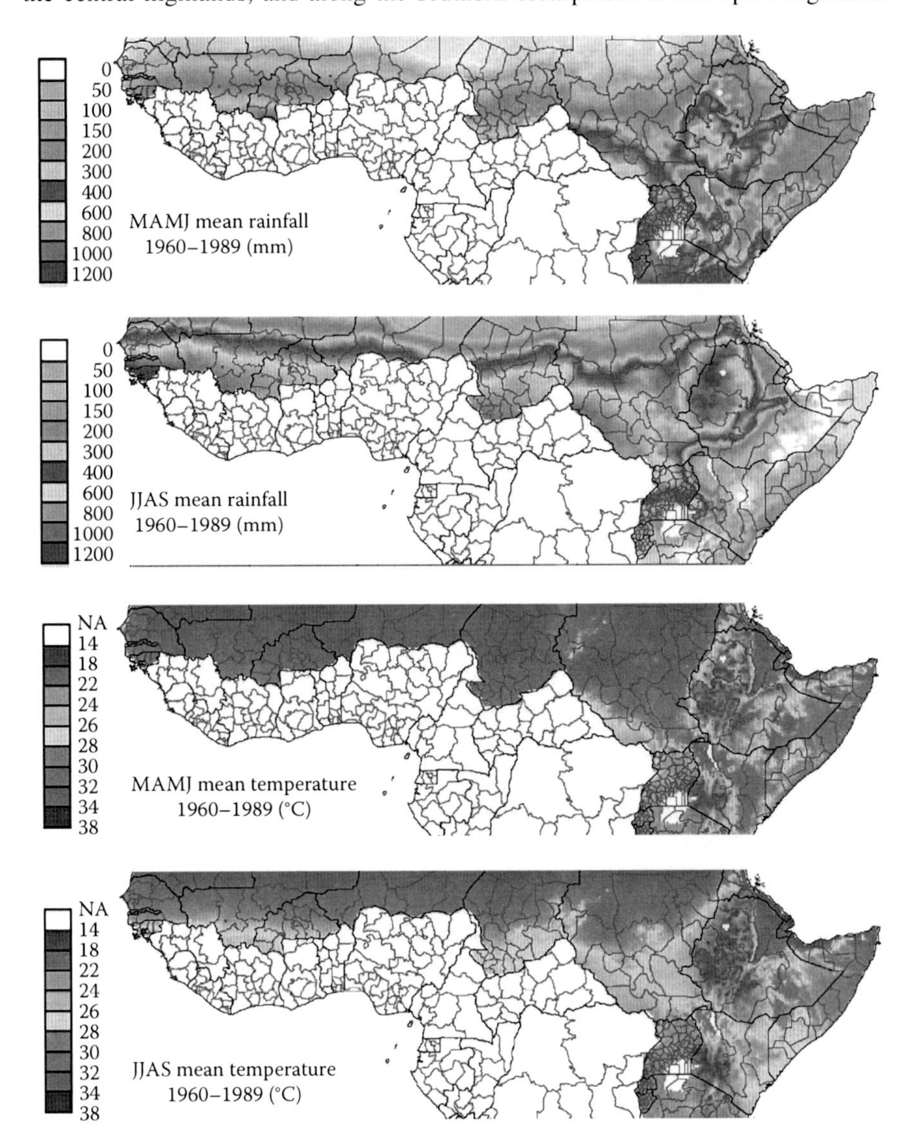

FIGURE 14.5 **(See color insert.)** FCLIM maps for average 1960–1989 seasonal rainfall and air temperature for the MAMJ and JJAS seasons.

These rains support the important Long and Belg cropping seasons in these countries. These regions also show up as cool islands in two mean temperature maps, with air temperatures of less than 24°C. These cool temperatures not only help reduce potential evaporation but also slow the maturation cycle of the crops, lengthening the required growing period. These long maturation periods can boost the accumulation of biomass but may also increase exposure to drought, since a long period of adequate crop water supply is required to produce optimal growing conditions. During JJAS, the intertropical front establishes itself north of the equator, and the Sahel receives most of its rains. Across the Sahel, temperatures decline between MAMJ and JJAS but still remain warm. A strong north–south temperature gradient appears during the JJAS period, with the southern edges of the Sahelian countries receiving the most rainfall and coolest air temperatures. A similar structure exists in the JJAS rainfall climatology, with the Sahel exhibiting very strong rainfall gradients. Exceptions to the latitudinal gradients in JJAS rainfall and temperature fields occur in Sudan and Ethiopia, where high mountains produce cooler, wetter conditions.

14.4.3 FEWS NET Trend Analysis Sigma Fields

When evaluating climate trends, two primary factors should be considered concerning (1) the magnitude of the observed trends and (2) the magnitude of the estimated trends relative to the underlying uncertainty of the interpolated fields. The latter factor is almost never considered and can often be obfuscated by the analysis of interpolated monthly or seasonal data. One simple way to evaluate these two components (trend magnitude and uncertainty) is to divide the interpolated trend fields by the standard error in the interpolation. The resulting unitless sigma images retain the sign of the underlying trend fields but are now expressed in units of standard errors (Figure 14.6). Values of 1, 2, and 3 correspond to the 85%, 98%, and 99.9% confidence levels, respectively, signifying the trend is statistically significant. Most regions covered in this analysis had sigma values with absolute values of greater than 2; thus, the signal-to-noise ratio for the trend analysis is high and our confidence in the spatial accuracy of the results high. This is surprising considering that the associated station densities were on the order of ~1 rainfall station every 5,000 km^2 and 1 temperature observing site every 40,000 km^2. The appropriate use of satellite fields helped achieve this result, reducing the geospatial random error and improving the signal-to-noise ratios. Note that other trend analysis products, such as those provided by the Climatic Research Unit (New et al., 1999, 2000), do not quantify the spatial uncertainty of trend estimates.

However, the sigma maps do not fully characterize all the sources of uncertainty. Undetected problems in the station observations and sampling uncertainties in the trend estimates are two unaccounted sources of uncertainty. Note also that the trend (and associated error) estimates are for areal averages and not point estimates. Some stations in "trend" areas will not exhibit trends, as shown by the station data in Figure 14.2. Based on these caveats, the sigma fields shown in Figure 14.6 suggest that the resulting trend analyses can be accepted with a high degree of confidence, either because of the high density of observations and reasonable levels of predictability in the rainfall trends, or because the trend signal is coherent (everywhere positive) and the spatial covariance pattern of the warming trends is relatively simple.

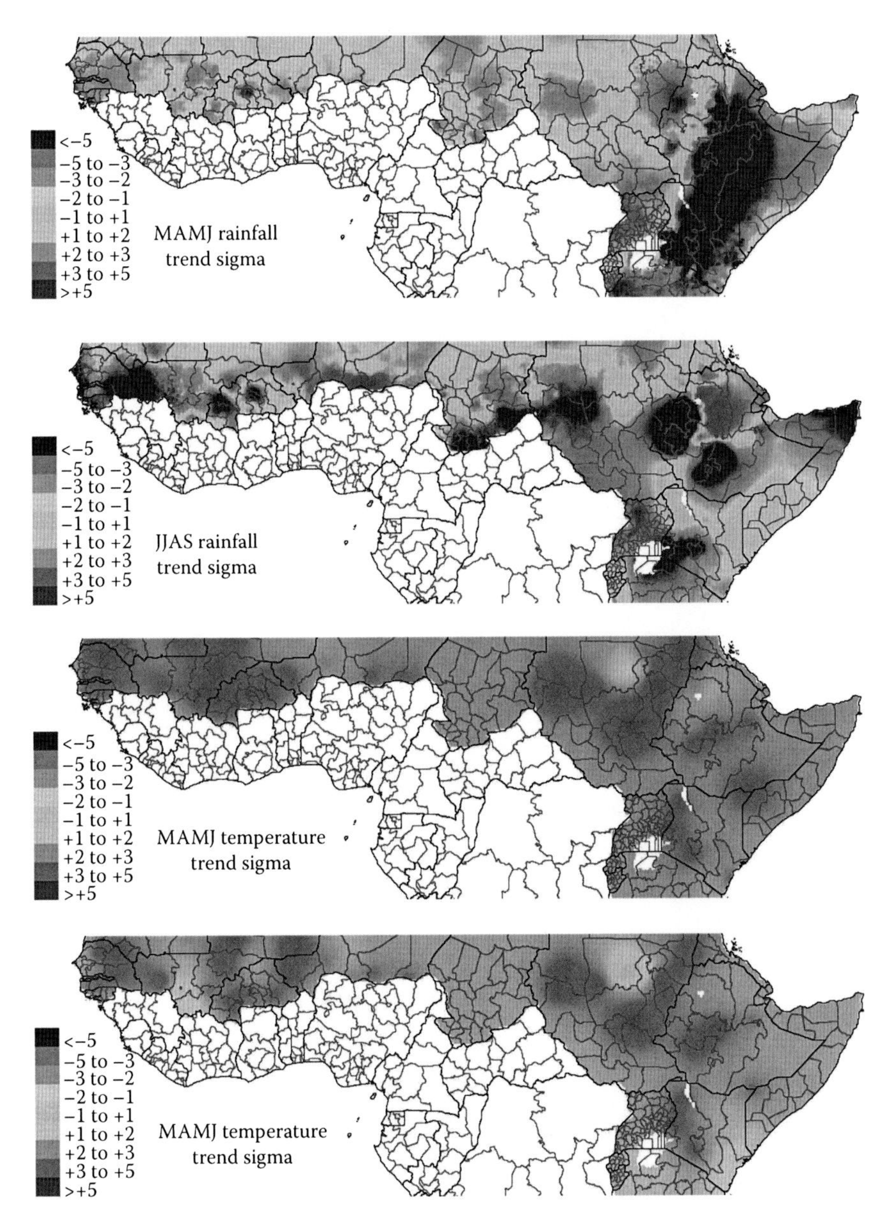

FIGURE 14.6 **(See color insert.)** Rainfall and temperature sigma fields. Sigma values are the estimated trend fields divided by the interpolation standard error.

14.4.4 FEWS NET Trend Analysis Results

Figure 14.7 shows the FTA maps for MAMJ and JJAS rainfall and air temperatures. For MAMJ rainfall, substantial rainfall declines (exceeding 20 mm per decade) are identified in central Kenya and south-central Ethiopia. A −20 mm per decade decline, between the 1960–1989 era and 2025, would be consistent with an overall

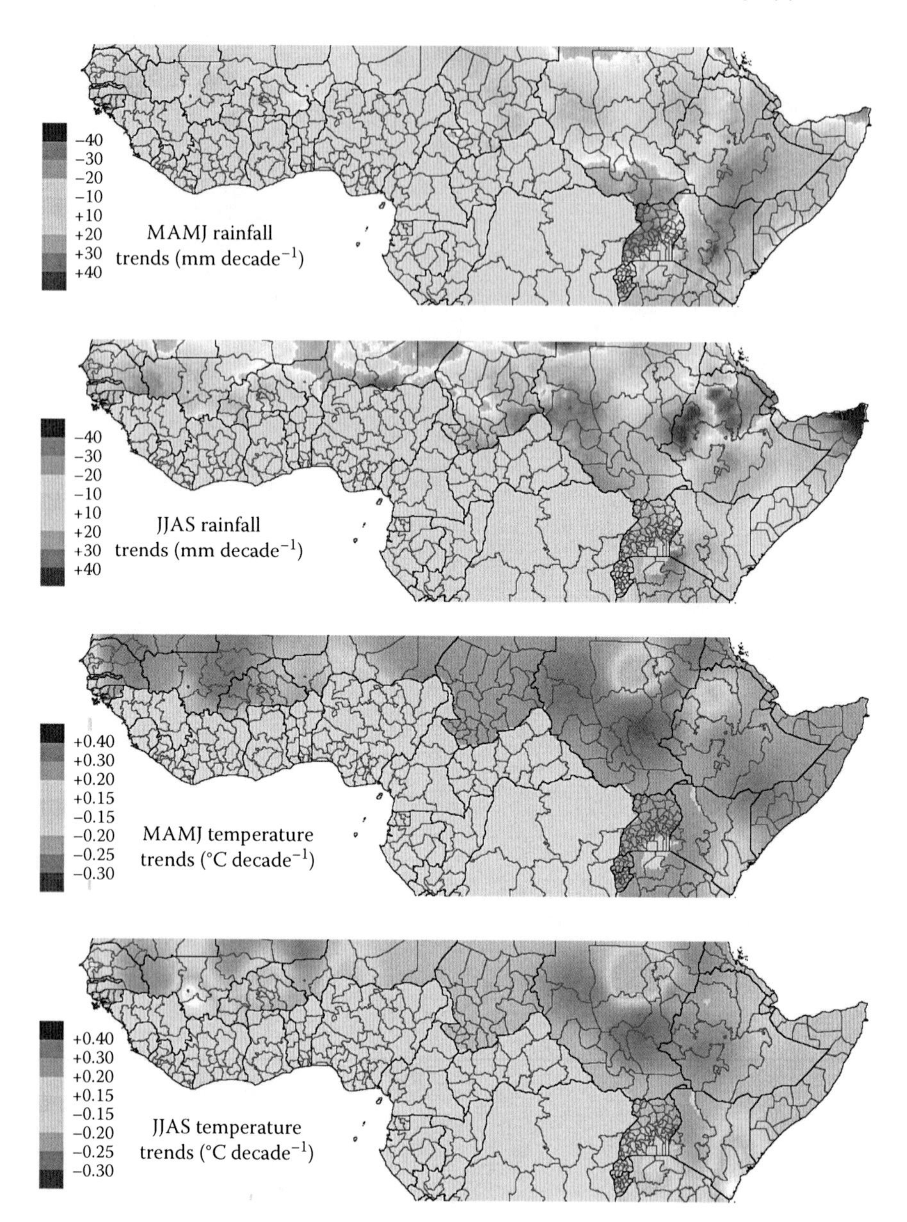

FIGURE 14.7 **(See color insert.)** The 1960–2009 rainfall and trend maps for MAMJ and JJAS.

rainfall reduction of −100 mm. In areas receiving an average of 400–600 mm of rainfall, this corresponds to a 17%–25% reduction in main growing seasonal rainfall. The station density in Ethiopia and Kenya is high, and the sigma fields for these regions are very low (often less than −7). As a result, the certainty surrounding these declines is high, and the potential impacts to the Ethiopian Belg and Kenyan

Long rain growing seasons are serious. The spatial pattern of JJAS rainfall trends is more complicated. Pockets of rainfall reduction appear near the border of Senegal and Mali, as well as southern Chad and Sudan. This drying is likely linked to drying in southwestern Ethiopia and appears to be accompanied by rainfall increases in northwestern Ethiopia. For southern Ethiopia and southern Sudan, the impact of these rainfall declines could be quite problematic, since they correspond with areas of fairly high population density. This will be discussed further in Section 14.4.5. Additional drying tendencies are observed over northern Somalia and parts of Uganda. Enhanced rainfall is observed over central Niger.

The MAMJ and JJAS temperature trends show values ranging from near zero to more than 0.4°C per decade. In general, the warming during the warmer MAMJ season is slightly greater than JJAS (Figure 14.7). Both the MAMJ and the JJAS warming trend exhibit similar east–west patterns. Warming is generally greater in Senegal–Mali and southern Sudan–Ethiopia than in Niger and northern Sudan and Ethiopia. The warming patterns tend to mirror the inverse of the rainfall trends. In some of the areas, the magnitude of the decadal temperature trends (up to 0.4°C per decade) approaches the interannual air temperature standard deviation and thus indicates large changes in climate.

14.4.5 FEWS NET Climate Impact Evaluations with Examples for Kenya, Ethiopia, and Sudan

Whether related to natural internal variations of the climate system or anthropogenic actions related to greenhouse gas emissions, low frequency climate trends can have serious impacts on food-insecure nations. To support the information needs of food security and development specialists, FEWS NET/USGS has created a new series of reports, *Informing Climate Change Adaptation*. These reports discuss the food security and climate adaptation implications of trend analyses, such as those presented in Figure 14.4, with the goal of better guiding development and disaster mitigation activities. These reports are disseminated through the USGS FEWS NET portal (http://earlywarning.usgs.gov/fews/reports.php). In previous research, we have suggested that the drying in parts of eastern Africa, Kenya, and Sudan is related to an ~1°C warming in the Indian Ocean. This warming is consistent and persistent over the 1950–present era, correlates strongly with global air temperature over the 1900–2009 era, and shows up as a pronounced "hockey stick" in related proxy data (coral and station air temperature data). It is also uniformly reproduced by all the IPCC climate models and several different precipitation time series (Williams and Funk, 2011) including the Global Precipitation Climatology Project (Adler et al., 2003), NCEP-DOE Reanalysis 2 (NCEP II, Kanamitsu et al., 2002), CPC Merged Analysis of Precipitation (Xie and Arkin, 1997), CPC Precipitation Reconstruction (Chen et al., 2003), and NOAA merged precipitation reconstruction (Smith et al., 2010). Both Indian Ocean sea surface temperatures and Kenyan, Ethiopian, and Sudanese rainfall and temperature trends exhibit considerable levels of persistence on decadal time scales. Thus, even in the absence of greenhouse gas forcings, momentum in the climate system seems likely to cause these trends to persist for at least the next 10–15 years.

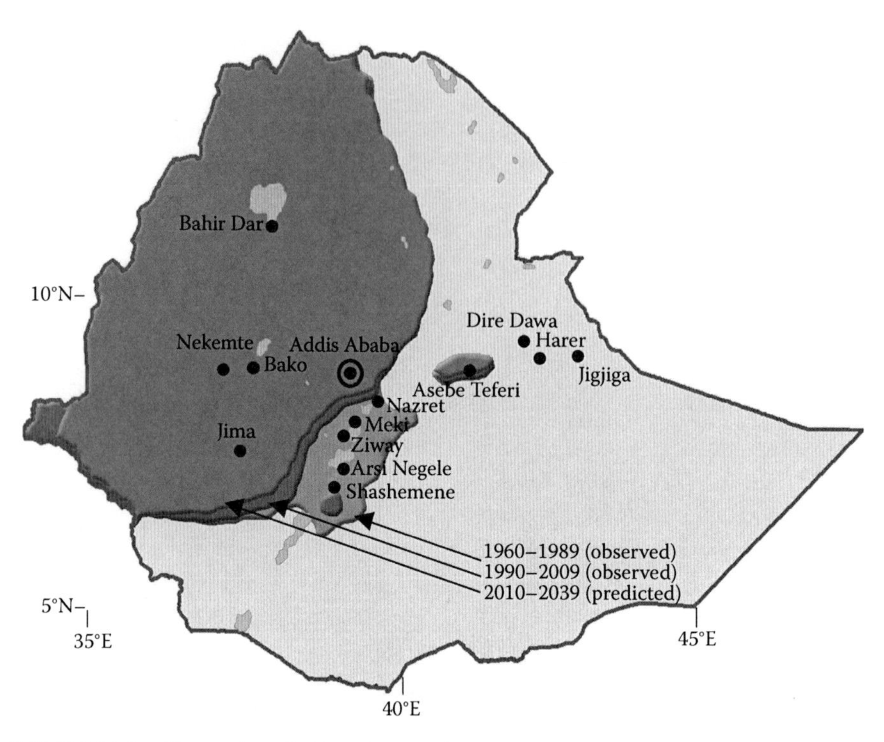

FIGURE 14.8 Change in suitable crop growing areas in south-central Ethiopia. The three overlain polygons show areas typically receiving enough rain to support a healthy crop (areas receiving, on average, more than 500 mm of JJAS rainfall), during 1960–1989, 1990–2009, and 2010–2039.

Figure 14.8 shows an example graphic for the main Ethiopian growing season (JJAS). The purpose of this graphic is to convey how increasingly frequent droughts may impact the area and location of prime crop growing areas. This transition over time is shown as the expansion or dilation of an isohyet, chosen to demarcate reasonably productive crop growing areas. In the example shown here, the 500 mm contour line was chosen, based on crop water demands and maps of productive crop growing areas (not shown). In Figure 14.8, the bottom (light brown) polygon depicts areas that, on average over 1960–1989, received enough rain to support a healthy crop (more than 500 mm of JJAS rainfall). The dark brown (middle) polygon shows the region that, on average, received more than 500 mm over 1990–2009. The combination of these two polygons, therefore, identifies where the rain gauge observations have exhibited a significant retreat in rainfall between 1960 and 2009, exposing the heavily farmed and populated area south of Addis Ababa to more frequent crop failures. This change in rainfall places increased climatic pressures on the precarious agricultural and pastoral livelihoods in this region. The orange (top) polygon shows the anticipated location of the 500 mm isohyets over the 2010–2039 time period. This contour line is produced by assuming a persistence of the local rate of change of rainfall (Figure 14.7). Interestingly, the retreat of the isohyets slows because of the steep topographic rainfall gradients (Figure 14.5). This map, thus, has two important adaptation messages. First, a large number (~16 million) of people in food-insecure southern Ethiopia have experienced lower rainfall and a greater number of drought events (Seleshi and Zanke, 2004). Second, the mountainous western half of the country is protected by the steep rainfall gradients, and the higher elevation rainfall is anticipated to be less negatively impacted. Efforts focused on increasing crop production in these moist regions are likely to be more successful, given the regions' much lower incidence of drought.

14.5 CONCLUSIONS

The study presented here demonstrates the contribution that satellite observations can make to traditional climate mapping applications. Although satellites provide indirect estimates of air temperature or rainfall, they have a tremendous ability to discriminate spatial gradients, distinguishing warm from cold and wet from dry. Rather than focus on satellites' representation of day-to-day variations, this work leverages correlations between satellite observations and in situ observations to produce high-quality maps of mean rainfall and temperature fields and trends. Because satellite fields relate directly to climate observations, they typically exhibit strong local correlations (as shown in Figure 14.4 and Table 14.2), with mean fields using these data as predictors being highly accurate (Table 14.3). This study found that satellite predictors also improved the accuracy of trend interpolations, and the inclusion of satellite data provided more accurate interpolation surfaces and the associated lower standard errors reduced spatial signal-to-noise ratios. Drought occurs at many temporal scales, and low frequency changes can be the hardest to map accurately; satellite data can help us map and mitigate these slowest of slow-onset disasters. This allows for confident assertions to be made over areas even where the density of observations is quite low. Accurate depictions of the mean climate and climate trends are important because they support the targeted identification of at-risk populations and a spatially aware approach to designing adaptation and development strategies. Our results suggest that the ever-increasing satellite record can contribute to these objectives by providing precise, spatially dense, and physically meaningful observations of climatic gradients.

Whether anthropogenic or caused by internal climate variations, monitoring and understanding recent low frequency climate changes will be a vital challenge for the twenty-first century. Demands for water and food will increase, while the earth's capacity to provide will remain relatively fixed. Effective climate observing systems will be one of the first lines of defense against climate shocks, and satellite systems can play a critical role. One bête noire, however, that has always plagued the satellite world has been the occurrence of temporal nonhomogeneities in the satellite record. Changes in platforms, sensors, and flight paths have all made piecing together a sterling record a difficult but worthy challenge. Although not suitable to all purposes, or intended as a replacement for a homogenous satellite record, the FCLIM and FTA methodology presented here leverages the tremendous wealth of the satellite record without being reliant on the satellites themselves for detecting trends. As population pressure and economic growth increase demands for water and food, the ability of satellites to accurately observe gradients of precipitation and temperature will help us to better map and manage our natural resources.

REFERENCES

Adler, R.F., J. Susskind, G.J. Huffman, A. Chang, R. Ferraro, P.-P. Xie, J. Janowiak, B. Rudolf, U. Schneider, S. Curtis, D. Bolvin, A. Gruber, J. Susskind, P. Arkin, and E. Nelkin. 2003. The version-2 global precipitation climatology project (GPCP) monthly precipitation analysis (1979–present). *Journal of Hydrometeorology* 4:1147–1167.

Brown, M.E. and C. Funk. 2008. Food security under climate change. *Science* 319:580–581.

Chen, M., P. Xie, J.E. Janowiak, P.A. Arkin, and T.M. Smith. 2003. Reconstruction of the oceanic precipitation from 1948 to the present. *AMS 14th Symposium on Global Change and Climate Variations*, Long Beach, CA. Boston, MA: American Meteorological Society.

Funk, C., A. Asfaw, P. Steffen, G. Senay, J. Rowland, and J. Verdin. 2003a. Estimating Meher crop production using rainfall in the 'Long Cycle' region of Ethiopia. FEWS NET Special Report. http://reliefweb.int/sites/reliefweb.int/files/resources/9EC256793FA1685C49256DB90003E3DC-fews-eth-06oct2.pdf (accessed on January 29, 2012).

Funk, C. and M. Brown. 2005. A maximum-to-minimum technique for making projections of NDVI in semi-arid Africa for food security early warning. *Remote Sensing of Environment* 101:249–256.

Funk, C. and M. Brown. 2009. Declining global per capita agricultural capacity production and warming oceans threaten food security. *Food Security* 1(3):271–289.

Funk, C., M. Dettinger, J.C. Michaelsen, J.P. Verdin, M.E. Brown, M. Barlow, and A. Hoell. 2008. Warming of the Indian Ocean threatens eastern and southern African food security but could be mitigated by agricultural development. *Proceedings of the National Academy* 105:11081–11086.

Funk, C., G. Husak, J. Michaelsen, T. Love, and D. Pedreros. 2007. Third generation rainfall climatologies: Satellite rainfall and topography provide a basis for smart interpolation. *Crop and Rangeland Monitoring Workshop*, Nairobi, Kenya, Extended Abstract. http://earlywarning.usgs.gov/fews/pubs/RecentDroughtTendenciesInEthiopia.pdf (accessed on January 29, 2012).

Funk, C. and J. Michaelsen. 2004. A simplified diagnostic model of orographic rainfall for enhancing satellite-based rainfall estimates in data poor regions. *Journal of Applied Meteorology* 43:1366–1378.

Funk, C., J. Michaelsen, J. Verdin, G. Artan, G. Husak, G. Senay, H. Gadain, and T. Magadzire. 2003b. The collaborative historical African rainfall model: Description and evaluation. *International Journal of Climatology* 23:47–66.

Funk, C., G. Senay, A. Asfaw, J. Verdin, J. Rowland, J. Michaelsen, G. Eilerts, D. Korecha, and R. Choularton. 2005. Recent drought tendencies in Ethiopia and equatorial-subtropical eastern Africa. FEWS NET Special Report. http://earlywarning.usgs.gov/fews/pubs/RecentDroughtTendenciesInEthiopia.pdf (accessed on January 29, 2012).

Funk, C. and J. Verdin. 2009. Real-time decision support systems: The Famine Early Warning System Network (Chapter 17). In *Satellite Rainfall Applications for Surface Hydrology*, G. MeKonnen and F. Hossain (eds.), New York: Springer-Verlag.

Janowiak, J.E., R.J. Joyce, and Y. Yarosh. 2001. A real-time global half-hourly pixel-resolution infrared dataset and its applications. *Bulletin of the American Meteorological Society* 82:205–217.

Kanamitsu, M., W. Ebisuzaki, J. Woollen, S.K. Yang, J.J. Hnilo, M. Fiorino, and G.L. Potter. 2002. NCEP-DOE AMIP-II reanalysis (R-2). *Bulletin of the American Meteorology Society* 83:1631–1643.

New, M., M. Hulme, and P.D. Jones. 1999. Representing twentieth century space-time climate variability. Part 1: Development of a 1961–90 mean monthly terrestrial climatology. *Journal of Climate* 12:829–856.

New, M., M. Hulme, and P.D. Jones. 2000. Representing twentieth century space-time climate variability. Part 2: Development of 1901–96 monthly grids of terrestrial surface climate. *Journal of Climate* 13:2217–2238.

Seleshi, Y. and U. Zanke. 2004. Recent changes in rainfall and rainy days in Ethiopia. *International Journal of Climatology* 24(8):973–983.

Smith, T.M., P.A. Arkin, M.R.P. Sapiano, and C.-Y. Chang. 2010. Merged statistical analyses of historical monthly precipitation anomalies beginning 1900. *Journal of Climate* 23:5755–5770.

Verdin, J., C. Funk, G. Senay, and R. Choularton. 2005. Climate science and famine early warning. *Philosophical Transactions of the Royal Society Biological Sciences* 360(1463):2155–2168.

Verdin, K.L. and S.K. Greenlee. 1996. Development of continental scale digital elevation models and extraction of hydrographic features. *Proceedings: Third International Conference on Integrating GIS and Environmental Modeling*, Santa Fe, NM.

Wan, Z. and J. Dozier. 1989. Land-surface temperature measurement from space: Physical principles and inverse modeling. *IEEE Transactions on Geoscience and Remote Sensing* 27(3):268–278.

Williams, P. and C. Funk. 2011. A westward extension of the warm pool leads to a westward extension of the Walker circulation, drying Eastern Africa. *Climate Dynamics* (DOI: 10.1007/s00382-010-0984-y).

Xie, P. and P.A. Arkin. 1997. A 17-year monthly analysis based on gauge observations, satellite estimates, and numerical model outputs. *Bulletin of the American Meteorological Society* 78(11):2539–2558.

15 Snow Cover Monitoring from Remote-Sensing Satellites

Possibilities for Drought Assessment

Cezar Kongoli, Peter Romanov, and Ralph Ferraro

CONTENTS

15.1 INTRODUCTION

Snow cover is an important earth surface characteristic because it influences partitioning of the surface radiation, energy, and hydrologic budgets. Snow is also an important source of moisture for agricultural crops and water supply in many higher latitude or mountainous areas. For instance, snowmelt provides approximately 50%–80% of the annual runoff in the western United States (Pagano and Garen, 2006) and Canadian Prairies (Gray et al., 1989; Fang and Pomeroy, 2007), which substantially impacts warm season hydrology. Limited soil moisture reserves from the winter period can result in agricultural drought (i.e., severe early growing season vegetation stress if rainfall deficits occur during that period), which can be prolonged or intensified well into the growing season if relatively dry conditions persist. Snow cover deficits can also result in hydrological drought (i.e., severe deficits in surface and subsurface water reserves including soil moisture, streamflow, reservoir and lake levels, and groundwater) since snowmelt runoff is the primary source of moisture to recharge these reserves for a wide range of agricultural, commercial, ecological, and municipal purposes. Semiarid regions that rely on snowmelt are especially vulnerable to winter moisture shortfalls since these areas are more likely to experience frequent droughts. In the Canadian Prairies, more than half the years of three decades (1910–1920, 1930–1939, and 1980–1989) were in drought. Wheaton et al. (2005) reported exceptionally low precipitation and low snow cover in the winter of 2000–2001, with the greatest anomalies of precipitation in Alberta and western Saskatchewan along with near-normal temperature in most of southern Canada. The reduced snowfall led to lower snow accumulation. A loss in agricultural production over Canada by an estimated $3.6 billion in 2001–2002 was attributed to this drought. Fang and Pomeroy (2008) analyzed the impacts of the most recent and severe drought of 1999/2004–2005 for part of the Canadian Prairies on the water supply of a wetland basin by using a physically based cold region hydrologic modeling system. Simulation results showed that much lower winter precipitation, less snow accumulation, and shorter snow cover duration were associated with much lower discharge from snowmelt runoff to the wetland area during much of the drought period of 1999/2004–2005 than during the nondrought period of 2005/2006.

Given the importance of snowmelt and its potential impact on water resources of snow-covered regions, the monitoring of snow is an integral part of water management in these regions. Hydrologically important snowpack measurements are the snow water equivalent (SWE), snow depth (SD), and snow-covered area (SCA). SWE refers to the amount of water (frozen and liquid) contained in the snowpack (calculated from SD multiplied by effective snow cover density). It is the most important snow parameter for snowmelt runoff forecasting before the onset of snowmelt season. For example, in the western United States and Alaska, the U.S. Department of Agriculture (USDA) Natural Resources Conservation Service (NRCS) maintains an automated SWE monitoring system called SNOwpack TELemetry (SNOTEL) and a network of manual snow courses that complement SNOTEL, which are used in combination with other climatological and streamflow data to create water supply forecasts (Pagano and Garren, 2006). SD is physically related to SWE and is routinely measured from ground-based meteorological stations. Daily observations of SD have been made dating from the late 1800s in a few of these countries (e.g., Switzerland, United States, former Soviet Union, and Finland). SCA indicates the spatial extent or fraction of land surface covered by snow. SCA monitoring can be used to estimate the snow duration, which has a stronger physical relationship with both SD and SWE. For example, in areas where the average winter temperature is below or close to the freezing point, moisture in the snowpack gradually accumulates throughout the winter season, and longer snow duration typically results in larger SWE by the beginning of spring snowmelt. Abnormally short duration of snow cover may be indicative of the lack of winter precipitation and therefore may be considered an early indicator of potential early spring drought.

The spatial distribution of ground-based stations measuring SD and SWE is generally skewed to low elevation regions of the Northern Hemisphere midlatitudes and to snow courses in mountainous regions. Also, most stations are located in relatively close proximity to urban areas for access purposes, and thus large geographic expanses may go unmonitored. In addition, the current tendency to reduce the number of manual stations and to replace manual stations with automated ones results in a continuous reduction in the amount of available ground-based information on snow, because some automated stations are not equipped with SD sensors. Satellite observations, on the other hand, can complement surface observations by providing information on the snow cover distribution and seasonal change with much-improved spatial coverage and temporal frequency that cannot be matched by in situ measurements.

In this chapter, the main satellite remote-sensing methods and applications for snow cover monitoring are reviewed, with an emphasis on the monitoring of SCA, SWE, and SD parameters. First, an overview of snow cover monitoring using optical imagery is provided, including a discussion of the main physical principles and sensor characteristics, operational products, and validation studies. Snow cover monitoring using passive microwave (MW) imagery is then summarized in a similar fashion, with additional discussion of data assimilation techniques used for improved monitoring of SWE and SD, which cannot be accomplished solely by optical or MW remote-sensing methods with sufficient accuracy. Multisensor blending techniques that utilize information from visible, infrared (IR), and MW observations are then described. Lastly, an example of applying remote sensing for SCA monitoring and early drought prediction is presented over the Ukraine.

15.2 SNOW MAPPING WITH OPTICAL SATELLITE OBSERVATIONS

15.2.1 Physical Principles and Sensor Characteristics

Snow is among the most "colorful" natural materials (Dozier et al., 2009) in that it possesses a strong spectral gradient in reflectance ranging from a high reflectance (albedo) in the visible wavelengths to low reflectance in the middle IR wavelengths (Wiscombe and Warren, 1980). Figure 15.1 shows plots of modeled spectral distribution of snow reflectance in the visible (0.4–0.7 μm) and near (0.7–1.3 μm) and middle (1.3–2 μm) IR wavelengths for a range of snow grain sizes. As shown, snow reflectance is high and relatively insensitive to grain size in the visible range, whereas reflectance decreases and its sensitivity to grain size increases dramatically in the near and middle IR ranges. This characteristic spectral response in the optical wavelengths distinguishes snow from most other natural surfaces (e.g., soil, water, and vegetation), as shown in Figure 15.2. In the far IR wavelengths (not shown), snow emits thermal radiation close to that of a blackbody, and, thus, its brightness temperature as observed by the satellite sensor depends mainly on the physical temperature of the top thin layer of the snowpack. In these wavelengths, the snow brightness temperature is relatively low, which is also useful information for snow identification.

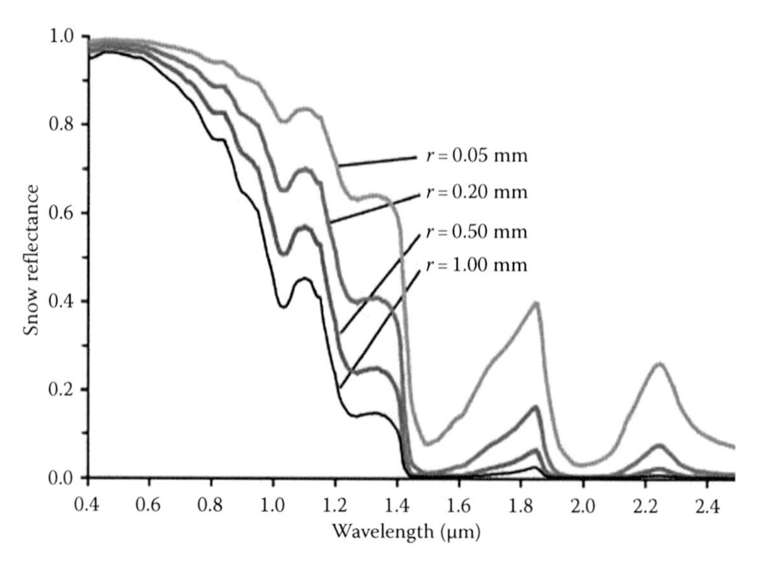

FIGURE 15.1 Snow reflectance spanning the visible, near-IR, and mid-IR wavelengths for a range of snow grain sizes (r). (From Dozier, J. and T.H. Painter, *Annu. Rev. Earth Planet. Sci.*, 32, 465, 2004.)

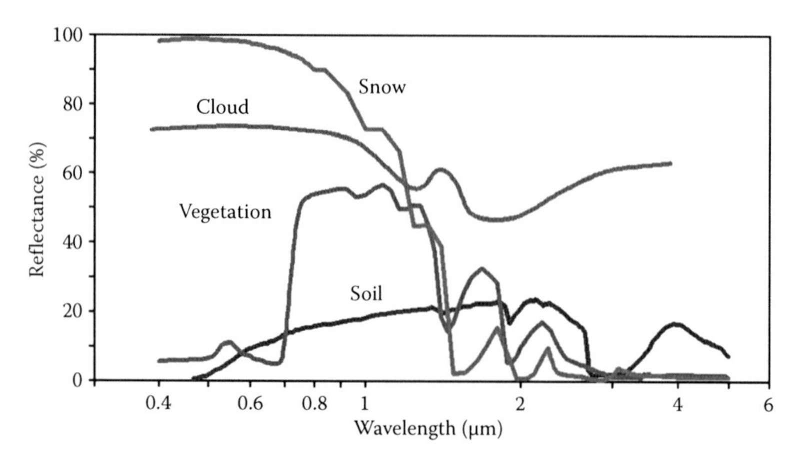

FIGURE 15.2 **(See color insert.)** Spectral reflectance for different land surface cover types including snow.

Most current meteorological satellite instruments collect observations in spectral regions centered in the visible at around 0.6 μm, the middle IR at 1.6 μm and 3.7–3.9 μm, and the thermal IR at 10–12 μm. Therefore, data collected from these instruments can generally be used to identify snow and map snow cover distribution. Snow retrievals from satellite optical measurements require clear sky conditions and sufficient daylight.

The purpose of satellite-based snow identification and mapping algorithms is to distinguish snow-covered land surface from snow-free land and clouds. Automated algorithms to identify snow usually incorporate a set of threshold tests that utilize satellite-observed reflectance and brightness temperature values in the spectral bands mentioned earlier, along with specific spectral indices. The Snow Index (SI), defined as a simple ratio of the reflectance in the visible (R_{vis}) and middle IR (R_{mir}), has been applied to identify snow in the algorithm described in Romanov et al. (2000). The snow detection algorithm of Hall et al. (2002) uses the normalized difference between reflectance in the visible and middle IR called the Normalized Difference Snow Index (NDSI):

$$NDSI = \frac{R_{vis} - R_{mir}}{R_{vis} + R_{mir}} \tag{15.1}$$

Clouds and snow-free land surfaces typically exhibit lower values of SI and NDSI than snow-covered land. In the snow-mapping algorithm of Hall et al. (2002), cloud-free pixels having an NDSI value >0.4, a visible reflectance >11%, and IR brightness temperature below 283 K are classified as snow covered.

The accuracy of snow cover maps can be affected by clouds that exhibit spectral features similar to snow, because such clouds may be misidentified as snow by the automated algorithm. Masking of snow cover by forest canopy changes the spectral reflectance of the scene by decreasing reflectance in the visible band and increasing it in the middle IR. Very dense boreal forests may mask the snow cover on the forest floor almost completely and therefore prevent proper snow detection and mapping with satellite data. The SD and SWE parameters are not typically retrieved from optical imagery since the snow reflectance is influenced by only a shallow layer of snow.

15.2.2 Primary Optical-Based Products

Optical measurements from earth-observing satellites present an important tool for monitoring SCA. The large number of satellite sensors providing measurements in the visible and IR spectral range, the relatively high spatial resolution of observations (1–4 km), and the relatively simple physical background of snow remote-sensing techniques are among the primary factors that explain widespread interest in using satellite optical data to develop snow products.

Identification of snow in satellite imagery by visual analysis and interpretation is the oldest snow-mapping technique. Since 1972, this approach has been routinely used by the National Oceanic and Atmospheric Administration (NOAA) to generate weekly maps of snow and ice distribution in the Northern Hemisphere. In 1999, the computer-based Interactive Multisensor Snow and Ice Mapping System (IMS) was implemented to facilitate image analysis by human analysts (Ramsay, 1998). This improved the nominal spatial resolution of the maps from 180 to 24 km and the temporal resolution from weekly to daily snow-mapping updates. In 2004, the spatial resolution of the IMS snow products was further increased to 4 km (Helfrich et al., 2007).

When mapping snow cover with IMS, analysts rely primarily on the visible imagery from polar-orbiting and geostationary satellites. The imagery from geostationary satellites is utilized in the form of animations, which help to distinguish moving clouds from snow. Quite often, analysts visually observe and map the distribution of snow cover through semitransparent clouds. This is an obvious advantage compared to automated techniques based on optical measurements (discussed later in this section) where practically any cloud prevents a reliable characterization of the state of the land surface. Since 2006, the upgraded IMS has had access to several automated snow and ice products generated at NOAA and the National Aeronautics and Space Administration (NASA), as well as surface in situ SD reports. Recently analysts also started using images from live-streaming web cameras throughout the world. Availability of these additional sources of information has substantially enhanced the potential of analysts to accurately reproduce the snow cover distribution, especially in the case of persistent cloud cover when application of satellite visible imagery is ineffective.

Owing to a simple and straightforward satellite image analysis technique and to the use of several additional sources of information on snow, interactive snow cover maps present a consistent, robust, and highly accurate product. IMS maps of snow and ice cover are considered the primary NOAA snow cover product and are incorporated in all global and mesoscale operational numerical weather prediction models run by NOAA's National Centers for Environmental Prediction (NCEP). High-spatial-resolution 4 km IMS maps are updated daily, making them potentially useful for various environmental and practical applications at regional and local scales, including drought monitoring. With more than 35 years of continuous snow cover monitoring, NOAA snow charts also present a unique source of information for global climate change studies (Frei and Robinson, 1999). It is important to note, however, that the changes in both frequency of map generation and spatial resolution have introduced inhomogeneity in the time series and thus substantially reduced the climatological value of the IMS-derived data set (Frei, 2009). Since 2007, NOAA interactive snow cover maps are produced at the National Ice Center (NIC). Daily IMS products are archived and are available from the National Snow and Ice Data Center (NSIDC).

In contrast to interactive snow-mapping techniques (similar to IMS), automated algorithms can better utilize the advantages of satellite observations, including high spatial resolution, multispectral sampling, and a frequent-repeat observation cycle. In the last two decades, data from polar-orbiting satellites have been most frequently used for monitoring global snow cover. Since 2000, NASA has produced snow cover maps from observations of the Moderate Resolution Imaging Spectroradiometer (MODIS) onboard the Terra and Aqua satellites (Hall et al., 2002). A suite of MODIS snow products includes global maps of snow cover distribution generated at daily, 16 day, and monthly time steps at a spatial resolution ranging from 500 m to 20 km (http://modis-250 m.nascom.nasa.gov/cgi-bin/browse/browse.cgi). Several algorithms have been developed and applied to identify and map snow cover using the Advanced Very High Resolution Radiometer (AVHRR) sensor onboard NOAA polar-orbiting satellites (e.g., Simpson et al., 1998; Baum and Trepte, 1999). In 2006, an automated algorithm to identify snow cover in NOAA AVHRR imagery was implemented at NOAA/NESDIS and used to produce daily global snow cover maps at a 4 km spatial resolution (http://www.star.nesdis.noaa.gov/smcd/emb/snow/HTML/snow.htm).

Continuous observations from AVHRR onboard different NOAA satellites have been available since the late 1970s. This extended time series of AVHRR data represents a valuable source of information for snow climatology and climate change studies.

The Canadian Center for Remote Sensing (CCRS) has consistently reprocessed historical AVHRR data for the time period from 1982 to 2005 to establish snow cover climatology over Canada at 1 km resolution (Khlopenkov and Trishchenko, 2007), but no attempts to expand these efforts globally have been reported to date.

Most current imaging instruments onboard geostationary satellites provide observations in the visible, middle IR, and thermal IR spectral bands and thus also allow for automated snow cover identification and mapping. The area coverage of geostationary satellites is generally limited to the area between 65° North and 65° South latitude, and thus they can only be used for snow monitoring in the midlatitudes. A particular advantage of geostationary satellites is their frequent views, available typically at 15 or 30 min intervals. Frequent observations provide more cloud-free views during the day and thus help to improve the effective area coverage of the daily snow product. A practical way to utilize multiple views from geostationary satellites and to reduce cloud contamination in the snow map is to apply a maximum temperature image compositing technique (e.g., Romanov et al., 2000). For every pixel of an image, the observation with the highest IR brightness temperature acquired during the day is identified and retained. Since cloudy areas are typically associated with lower IR brightness temperatures, the maximum temperature tends to be associated with the most cloud-free observations available during the day. Since 1999, this technique has been routinely used by NOAA/NESDIS for generating automated 4 km snow cover maps over North America. Observations from the Spinning Enhanced Visible and IR Imager (SEVIRI) onboard another geostationary satellite, Meteosat Second Generation (MSG), have also been applied to routinely generate snow maps over Europe and Northern Africa (Romanov and Tarpley, 2006; de Wildt et al., 2007).

15.2.3 Validation of Optical Products

A traditional technique for estimating the accuracy of satellite-based snow maps consists of direct comparison with in situ, synchronous surface SD measurements. Daily SD reports from first-order stations across the globe and additional stations within regional networks (e.g., U.S. Cooperative Network stations) provide the means for year-round evaluation of the accuracy of snow cover products in different physiogeographic regions.

Hall and Riggs (2007) analyzed the validation results of the MODIS snow cover products from studies over several locations around the world and concluded that the accuracy of the 0.5 km resolution snow maps generally exceeds 94%. This estimate closely corresponds to the results of Simic et al. (2004), where MODIS maps were compared to SD reports from Canadian first-order and climate monitoring network stations. However, the latter work demonstrated that the accuracy of MODIS maps, as well as the accuracy of other satellite snow remote-sensing products, substantially decreased to 80%–85% over densely forested areas. Masking and shadowing of snow cover by the tree canopy along with littering of the snow by tree debris substantially reduce the visible reflectance of the snow-covered land surface and thus complicate proper identification of snow in heavily forested areas. The accuracy of snow identification also degrades over shallow or patchy snow cover since portions of the snow-free land along with vegetation protrusions through the snowpack make the scene look "darker" in the visible spectral bands. Ault et al. (2006) reported only

41% correspondence of MODIS snow maps to surface observations when only a trace amount of snow cover (<10 mm) was reported on the ground. Snow maps generated from geostationary satellite data provide a similar level of accuracy as the polar satellite data-based products. Romanov et al. (2000) compared GOES-based snow maps with observations from U.S. Cooperative Network stations over the continental United States and found the two data sets to have an average agreement of 88%.

The estimated accuracy of interactive snow cover maps is similar or slightly less than the accuracy of automated products based on optical data (e.g., Romanov et al., 2000). It is important, however, for the comparison of reported accuracy of automated and interactive snow maps to be performed with care. The accuracy of automated optical products characterizes snow retrievals only over limited cloud-clear areas, whereas interactive snow maps provide continuous (gap-free) coverage of the area and thus are typically validated over the whole domain. Although a considerable number of snow map validation studies were conducted in the last decade, they do not provide comprehensive information on the accuracy of satellite-derived snow maps. The principal problem is the very sparse network of meteorological stations and in situ measurements at high latitudes where snow cover is most prevalent. Therefore, reported validation results characterize the accuracy of snow cover mapping primarily in the midlatitudes. In addition, practically all validation studies have focused on North America or Eurasia. Thus, there is considerable uncertainty with respect to the accuracy of snow cover mapping in the Southern Hemisphere. It is also important to note that results of comparisons with surface observations are not sufficient to make a justified conclusion on the comparative accuracy of different automated snow products derived from satellite optical measurements. The problem is that this approach does not provide any information on the accuracy of cloud identification. Algorithms utilizing a more conservative approach for the identification of clear-sky scenes tend to overestimate the cloud cover and are more likely to provide a higher accuracy in the mapped snow cover distribution. However, this approach also results in larger cloud gaps and thus provides information on the snow cover distribution over a smaller area than snow products derived with a less conservative clear-scene identification approach.

15.3 SNOW MAPPING WITH MICROWAVE SATELLITE OBSERVATIONS

15.3.1 Physical Principles and Instrument Characteristics

The fundamental reason for using radiation in the MW portion of the electromagnetic (EM) spectrum for mapping snow cover is that these wavebands provide unique information (Woodhouse, 2006) that complements visible remote-sensing methods. MW radiation can penetrate clouds, allowing time-continuous monitoring of snow cover distribution. MWs can also penetrate a deeper layer of snow cover and interact with snow grains in a characteristic fashion, allowing estimation of SD and SWE parameters (in addition to SCA), which is not possible with visible imagery. Since natural MW radiation can be observed by passive sensors and artificial MW radiation can be generated and sensed by active sensors, the measurements can be made during the day or night. However, there are some disadvantages to MW snow imaging. The longer MW wavelengths mean that large antennas (about a meter or longer) are required to achieve the spatial resolution appropriate for regional-scale studies (on the order of several kilometers). Active sensor systems such as the Synthetic Aperture Radar (SAR) instruments tend to be the heaviest, most power-consuming, and data-prolific space-borne instruments. MW sensors are also currently flown only aboard polar-orbiting satellites, which offer a much lower temporal resolution than optical sensors onboard geostationary satellites. In addition, as will be explained later, interpretation of MW imagery is often not as easy and straightforward as visible imagery.

MW radiation at wavelengths between 0.2 and 1.5 cm, or at frequencies between 160 and 20 GHz, has been shown to respond to snow cover ice grains in a characteristic fashion where the spectral emissivity and hence the brightness temperature as observed by a remote-sensing instrument decreases with increasing frequency. This characteristic spectral response is the result of the Rayleigh scattering behavior associated with snow grain sizes substantially smaller than the wavelength and its transition to Mie scattering behavior at larger grain sizes (for grain sizes comparable to the wavelength). Typical snow grains range between 0.2 and 0.5 mm in diameter for fine snow cover and 1 and 5 mm for coarse and very coarse grains. Figure 15.3 shows spectral MW measurements of snow and nonsnow materials made with a MW radiometer at 6, 10, 22, 37, and 94 GHz with a vertical polarization. All the listed snow types display a monotonic decrease in surface emissivity with increasing frequency, except for snow type 17 (bottom crust) at 94 and 37 GHz frequencies. The anomalous spectral response of snow type 17 (higher emissivity at 94 GHz than at 37 GHz) is explained by increased absorption (due to the presence of an ice layer)

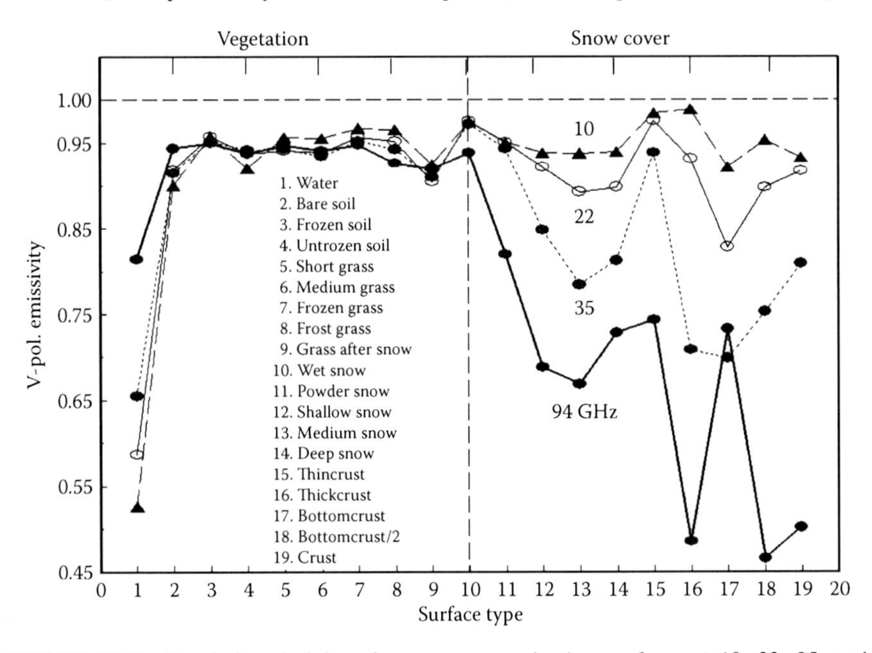

FIGURE 15.3 Spectral emissivity of snow cover and other surfaces at 10, 22, 35, and 94 GHz. (From Mätzler, C., *Meteor. Atmosph. Phys.*, 54(1–4), 241, 1994.)

TABLE 15.1
Snow Parameters That Affect Visible, Near-IR, IR, and MW Spectral Response

	Visible Solar Albedo	Near-IR Solar Albedo	Thermal IR Emissivity	MW Emissivity
Grain size	(+)	Yes		Yes
Zenith (or nadir) angle	(+)	Yes	Yes	Yes
Depth	Yes			Yes
Contaminants	Yes			
Liquid water content				Yes
Density				Yes
Temperature				Yes

+ Only if snowpack is thin or impurities are present.

or reduced scattering (due to the geometric optics limit for very large grain sizes) at higher frequencies. In contrast, water surfaces show an increase in emissivity with increasing frequency, which is typical of absorbing surfaces, whereas bare soils and vegetation show relatively small variations in emissivity with frequency.

Of particular importance for the retrieval of snow parameters is the observation that the MW spectral response shows dependence on a larger set of snow parameters than optical imagery (Table 15.1), which complicates the interpretation of MW imagery for snow identification and mapping. Among the snow parameters affecting the MW response, the most important are grain size, SD, SWE, and liquid water content. Dense vegetation can also attenuate MW radiation, particularly at higher frequencies (20 GHz and above), reducing the signal of the snow underneath. All other parameters being equal, an increase in SD or SWE is associated with a steeper emissivity gradient with frequency, because of increased scattering caused by a larger number of snow grains. Also, coarser-grained snow cover produces a steeper emissivity gradient. A small amount of liquid water in snow dramatically increases emission and reduces the scattering response and thus the ability to accurately map SCA, SD, and SWE over melting snow cover.

Table 15.2 lists the primary satellite passive MW sensors that are used for global mapping of snow cover properties, along with each instrument's average spatial resolution or the instantaneous field of view (IFOV). As shown, passive MW sensors have a coarse spatial resolution compared to optical sensors. This limits the utility of MW remote-sensing imagery for local-scale mapping of snow because smaller landscape features cannot be spatially resolved.

The earliest MW instrument used to generate snow maps is the Scanning Multichannel MW Radiometer (SMMR), flown on the Nimbus-7 Earth satellites and launched in 1978, followed by the Special Sensor MW/Imager (SSM/I) sensor on the U.S. Defense Meteorological Satellite Program (DMSP) satellites launched in 1987. The SMMR was an imaging dual polarized five-frequency radiometer (6, 10, 18, 21, and 37 GHz), while the SSM/I collects data at four frequencies (19, 22, 37,

TABLE 15.2
Main MW Sensors Used for Snow Mapping and Associated Data Characteristics

Overview of Passive MW Satellite Instruments	Mission	Availability	IFOV[a]	Comments
SMMR	Nimbus	25 Oct. 1978–20 Aug. 1987	18–27 km	
SSM/I	DMSP	July 1987–present	40–60 km	
SSMI/S	DMSP	18 Oct. 2003–present	31–4 km	Replaces SSM/I
AMSR	ADEOS-II	2003–present	20–30 km	
AMSR-E	EOS Aqua	4 May 2002–4 Oct. 2011	8–14 km	Modified from AMSR
AMSR2	GCOM-W	Planned		Future JAXA mission
AMSU	NOAA-N series METOP-A	NOAA-N series since 1998, METOP-A since 2004	48–72 km	

[a] Sizes at the instrument window channel frequency that occurs in the 30–40 GHz range, which is most frequently used for the retrieval of SD or SWE.

and 85 GHz). On the SSM/I, both vertical and horizontal polarizations are measured except at 22 GHz, for which only the vertical polarization is measured. More recent instruments include the Advanced MW Scanning Radiometer on the Earth Observing System (AMSR-E) flown on the NASA Aqua satellite, the Advanced MW Sounding Unit (AMSU) flown on the NOAA-N series and on the Meteorological Operational Satellite Programme (METOP-A), and the SSMI/S flown on DMSP. The AMSR-E is an imaging dual polarized six-frequency radiometer (6, 10, 18, 22, 37, and 89 GHz) with improved spectral and spatial resolution compared to SMMR. The AMSU contains both imaging and sounding channels in the 20–180 GHz frequency range. The sounding channels in the oxygen and water vapor absorption bands allow improved retrieval of important atmospheric parameters such as frozen precipitation (Kongoli et al., 2003). However, the AMSU lacks polarization information at the imaging channels that are useful for improved snow identification and especially snowmelt mapping. The SSMI/S instrument is a hybrid configuration between SSM/I and AMSU, containing both the SSM/I imaging and AMSU sounding channels.

15.3.2 Primary Microwave-Based Products

Most SCA retrieval algorithms from passive MW sensors are based on a decision-tree classification approach. A widely used algorithm is one developed by Grody (1991) and Grody and Basist (1996). Scattering surfaces (e.g., snow, deserts, rain, and frozen ground) and nonscattering surfaces (e.g., vegetation, bare soil, and water) are separated using brightness temperature-based scattering indices, followed by the application of additional brightness temperature-based thresholds to remove confounding factors (e.g., rain, frozen ground, and cold deserts). The algorithm was first applied to SMMR and SSM/I observations and later adopted for application to the AMSU (Grody et al., 2000; Kongoli et al., 2005, 2007).

A variety of SD and SWE algorithms exist, including static empirical (Künzi et al., 1982; Chang et al., 1987; Goodison, 1989; Kongoli et al., 2005, 2007), dynamic empirical (Josberger and Mognard, 2002; Foster et al., 2005), dynamic semiempirical (Kelly et al., 2003), and dynamic radiative transfer inversion (Pulliainen and Hallikainen, 2001) approaches. Empirical algorithms are simple linear regression functions that typically relate the difference in measured brightness temperature at two lower-frequency atmospheric window channels (i.e., 19 and 37 GHz for the SMMR, SSM/I, and AMSR-E, and 23 and 31 GHz for the AMSU instrument) to variations in measured SD or SWE. The static algorithms apply one set of regression coefficients, whereas the dynamic ones apply seasonally and spatially adjusted coefficients. Dynamic semiempirical algorithms use simple analytical expressions of SWE as a function of satellite brightness temperature derived from radiative transfer models and temporally and spatially varying snow parameters of grain size and snow density. The physically based algorithms use nonlinear iterative inversion techniques and radiative transfer snow emission models (Mätzler and Wiesman, 1999; Pulliainen et al., 1999; Wiesmann and Mätzler, 1999).

The longest time series of MW observations used for deriving snow cover data sets is a 30 year record of combined SMMR and SSM/I daily brightness temperature data starting from 1978, and available as a 25 km Equal Area Scalable Earth (EASE)-grid at NSIDC. Based on these data, Armstrong et al. (2007) have developed a 30 year global SWE and SCA climatology that is available at NSIDC. The algorithms to derive SCA are based on the Grody and Basist (1996) methodology, and the algorithms to derive SWE are based on the Chang et al. (1987) static empirical approach. Kelly et al. (2004b) also applied the Chang et al. (1987) method to AMSR-E data to develop daily, pentad, and monthly SWE products that are also available at NSIDC. The Canadian National Snow Information System for Water (NSISW), within the framework of "State of the Canadian Cryosphere" (SOCC, see www.socc.ca), applied a regionally adjusted static SWE algorithm (Goodison, 1989) to SMMR, SSM/I, and AMSR-E observations to develop SWE climatologies for the Canadian Prairies region from 1978 to present.

The utility of MW observations for the analysis of seasonal and interannual snow cover trends and variability has been demonstrated in several studies. Tedesco et al. (2009) applied a snowmelt detection algorithm to SMRR and SSM/I data over the 1978–2008 period to study pan-arctic terrestrial snowmelt trends and possible correlations with the Arctic Oscillation (AO). Melting was detected using a spatially and temporally dynamic algorithm based on the difference between daytime and nighttime brightness temperature values. Results over the 20 year period indicated statistically significant negative trends for melt onset and end dates (0.5 and 1 days/year earlier, respectively), as well as for the length of the melt season (0.6 day/year shorter). Results indicate that the AO index variability can explain up to 50% of the melt onset variability over Eurasia and only 10% over North America, which is consistent with spatial patterns of AO-related surface temperature changes.

Brodzik et al. (2006) investigated the extent and variability in SCA using a time series of visible and MW data starting from 1978 and found a decreasing trend in the Northern Hemisphere SCA from both methods. The strongest seasonal signal occurred from May to August, when both data sets indicated significant decreasing trends. The authors suggest that this pattern was physically related to increasing air temperatures during the period of maximum seasonal snowmelt over much of the Northern Hemisphere.

Che et al. (2008) conducted a study on the spatial and temporal distribution of seasonal SD derived from passive MW satellite remote-sensing data (SMMR from 1978 to 1987 and SSM/I from 1987 to 2006) in China using Grody and Basist's (1996) methodology to identify snow cover and a modified Chang et al. (1987) algorithm to compute SD. The algorithms were first validated with meteorological observations, considering the influences from vegetation, wet snow, precipitation, cold desert, and frozen ground. The modified SD algorithm was also dynamically adjusted based on the seasonal variation of grain size and snow density. The time series of SD estimates showed that the interannual SD variation was very significant, but the SCA of seasonal snow cover in the Northern Hemisphere exhibited a weak decrease over the same period with no clear trend in SCA change in China. However, SD over the Qinghai–Tibetan Plateau and northwestern China increased, while it weakly decreased in northeastern China. Overall, SD in China during the past three decades showed significant interannual variation with a weak increasing trend.

15.3.3 Validation of Microwave Products

Similar to optical imagery, the most widely used technique for validation of MW snow products is direct comparison with collocated SD or SWE surface observations. Accuracy assessment of SCA is also made with respect to optical imagery since the latter is considered more accurate and is available at higher resolution.

The mismatch of scale between point-based surface observations and the coarse satellite footprint of available passive MW satellite sensors is a recognized problem for SWE and SD algorithm calibration and validation. Validation of SWE is even more problematic due to the fact that the abundance and reliability of good quality in situ SWE data are less than those of SD (see, e.g., Kelly et al., 2004a). Studies suggest that the sampling density of "ground truth" in situ stations footprint-matched to the satellite sensor resolution has a substantial effect on algorithm error statistics. In a large-scale validation study, Dong et al. (2005) assessed this effect by calculating error statistics of a dynamic semiempirical algorithm as a function of the number of stations matched to the 0.5° by 0.5° SSMR footprint. The mean bias errors showed improvement from more than 40 mm SWE underestimation for pixels with only one station to <30 mm SWE underestimation for pixels with five or more stations, and the corresponding mean standard deviations decreased from 80 to 45 mm (Figure 15.4). It was suggested that the most likely reason for this improvement was that the increased number of stations yields an areal average estimate that is more compatible with the remote sensor. Chang et al. (2005) report that a density of 10 ground measurements in an SSM/I or AMSR-E footprint pixel is necessary to produce a sampling error of 5 cm SD or better.

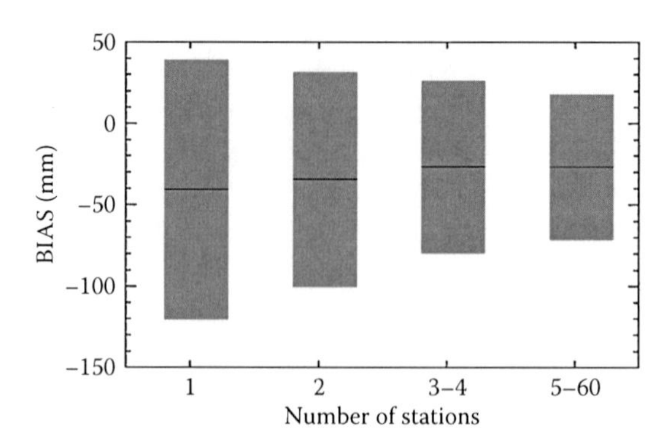

FIGURE 15.4 SWE error statistics (bias and standard deviation) as a function of the number of stations within the sensor estimate footprint. (From Dong, J. et al., *Remote Sens. Environ.*, 97(1), 68, 2005.)

Despite these issues, numerous validation studies have been conducted over the last 30 years to assess the accuracy of various MW-derived snow products, with the following results. SCA algorithms from passive MW sensors have demonstrated the ability to capture interannual and seasonal snow cover trends and variability (Armstrong and Brodzik, 2001), but regional and seasonal biases still exist. Negative biases occur for early-season shallow, melting, and forested snow covers. Positive biases occur for cold deserts, mainly attributed to desert soils scattering MW radiation similar to snow. Figure 15.5 shows maps of MW-derived SWE blended with the IMS-derived SCA taken as "ground truth" reference (maps a and c) and overlays of MW- and IMS-derived SCA (maps b and d) (Kongoli et al., 2007). Blue areas on maps b and d depict coincident SCAs, green areas depict missing MW-derived SCA, and brown areas depict false MW-derived SCA. Note the reduced extent of green areas in midwinter (map d) compared to early winter (map b), which is attributed to improved MW detection of aged, deeper snow covers. The brown areas over the Mongolian deserts are persistent and represent false MW-detected snow cover, attributed mainly to the scattering behavior of desert soils (Kongoli et al., 2007).

With respect to SWE, all passive MW products substantially underestimate wet snow, deeper snow (40 cm or greater), and snow under heavily forested terrain. As noted earlier, wet snow conditions reduce the MW scattering signal, which constitutes a fundamental physical limitation for mapping SWE over melting snow. Underestimation over deeper snow cover has been attributed to signal saturation at MW frequencies used for retrieval (Kongoli et al., 2007; Derksen, 2008), and over forest-covered snow because of the masking effect of forest canopy above the snow at frequencies of 20 GHz and above. Derksen et al. (2003) found that under these conditions, accuracy against in situ data and the ability to capture interannual variability weakened appreciably.

15.3.4 Data Assimilation Approaches for Estimating Snow Water Equivalent

Validation studies indicate that the MW-based SWE algorithms are not sufficiently accurate for operational applications. Data assimilation offers an opportunity to

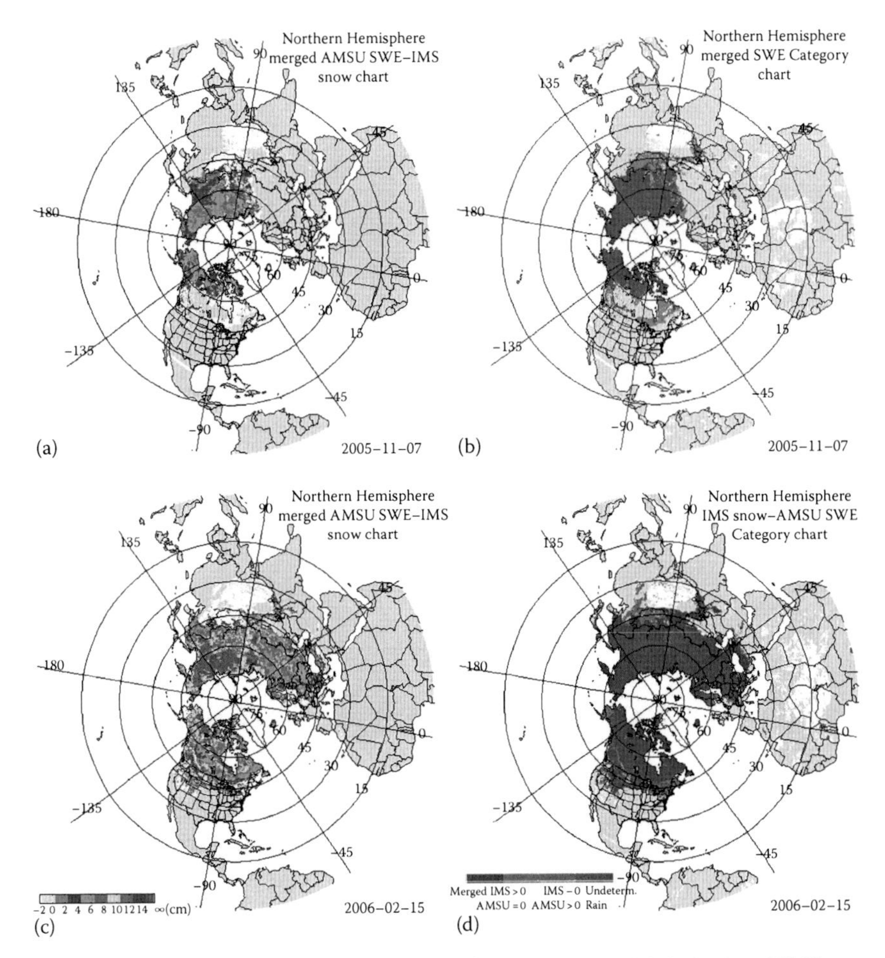

FIGURE 15.5 **(See color insert.)** Intercomparisons between MW-derived and IMS snow charts. (From Kongoli, C. et al., *Hydrol. Process.,* 21, 1597, 2007.) Maps a and c depict the MW-derived SWE blended with IMS-derived SCA for early and midwinter, respectively, and maps b and d depict SCA overlays of IMS and MW-derived products for early and midwinter, respectively. The SCA overlay maps (b and d) show coincident MW- and IMS-derived SCA in blue, MW-underestimated areas in green, and MW-overestimated SCA in brown (as compared to IMS taken as ground truth reference).

improve the performance of these algorithms by optimally merging information from remotely sensed and in situ observations or hydrologic model predictions.

A Bayesian assimilation technique (Pulliainen, 2006) was developed that weighs the space-borne MW data and SD interpolated from synoptic, near-real-time in situ observations with their estimated statistical accuracy. MW brightness temperature values were simulated using the Helsinki University of Technology (HUT) MW snow emission model (Pulliainen et al., 1999). The results obtained using SSM/I and AMSR-E data for northern Eurasia and Finland indicated that the employment of space-borne data using this assimilation technique improved SD and SWE retrieval accuracy (in 62% of the 3330 cases investigated) when compared with values interpolated from in situ observations. Another large-scale study of 26,063 samples by Loujus et al. (2010) reported preliminary results from this assimilation approach that were superior to those from static empirical algorithms of Chang et al. (1987). Root mean square error (RMSE) in SWE was 43 mm for Eurasia, compared to 63–73 mm from the empirical algorithms. The RMSE was further reduced to 33.5 mm when SWE values below 150 mm were analyzed. Despite improved error statistics, the assimilation algorithm still underestimates SWE in snow deeper than 150 mm, with the bias increasing as SWE values increase.

An assimilation scheme based on an ensemble Kalman filter (EnKF) to assimilate AMSR-E brightness temperatures into the Variable Infiltration Capacity (VIC) macroscale hydrology model has also been used (Andreadis and Lettenmaier, 2005). Brightness temperature values were simulated using a Dense Media Radiative Transfer (DMRT) model (Tsang et al., 2000). The magnitude of improvement when assimilating the AMSR-E data was small and appeared mostly when the peak seasonal SWE was relatively low. For cases of increasingly deeper snowpacks, the assimilation performance degraded when compared to hydrologically modeled SWE. As a result, a maximum SWE cutoff value was incorporated, and whenever the model-predicted SWE was over a snowpack saturation value of 240 mm, the AMSR-E observation was not assimilated. On average, the results improved using this cutoff threshold, compared to results from the original run when calculations were included over high snowpack.

Dong et al. (2007) assimilated SWE estimates derived from AMSR-E data directly into a three-layer snow hydrology model (Lynch-Stieglitz, 1994) using an extended Kalman filter (EKF) and compared multiyear model simulations with and without remotely sensed SWE assimilation with in situ SWE observations. The SWE estimates from assimilation were found to be superior to both the model simulation and remotely sensed estimates alone, except when model SWE estimates were >100 mm SWE early in the snow season.

15.4 SNOW COVER MONITORING USING SYNERGY OF OPTICAL AND MICROWAVE REMOTE-SENSING TECHNIQUES

Physical limitations inherent to automated optical and MW snow remote-sensing techniques result in their inability to provide continuous and accurate snow cover information at high spatial resolution under a variety of atmospheric, physiogeographic, and climatic conditions. This substantially reduces the value of snow cover products derived from individual satellite sensors and complicates their use in numerical model applications and in climate studies. In an attempt to improve satellite-based snow cover characterization, several techniques and algorithms have been proposed to combine snow cover observations in the optical and MW spectral bands. The principal objective of these techniques is to maximize advantages that optical and MW observations offer in an effort to provide continuous areal coverage of snow cover maps at the highest possible spatial and temporal resolution with the highest possible accuracy.

Armstrong et al. (2003) combined MODIS-based 8 day composited SCA maps with 8 day SWE maps derived from SSM/I data to produce an 8 day blended, 25 km resolution global SCA map. The technique integrated snow identified from the optical MODIS data with snow information from SSM/I to compensate for possible omissions of shallow or melting snow in the MW-based snow product. Foster et al. (2010) used the daily 5 km MODIS SCA product and the 25 km AMSR-E SWE retrievals to generate a daily blended SCA map. Because optical snow retrievals provide better accuracy and spatial resolution than the MW-based retrievals, the algorithm relies on MODIS data if available. Pixels that were cloud covered or did not have enough daylight to perform classification with MODIS data are filled in with AMSR-E retrievals. As a result, the output product presents a global binary (snow/no snow) 25 km map with no cloud-related coverage gaps. A similar approach to blending of MODIS and AMSR-E products has been used by Liang et al. (2008) to characterize daily changes of snow cover distribution over China.

A blending technique by Romanov et al. (2000) presents a more cautious approach to the use of MW-based snow retrievals from satellite observations. The algorithm combines observations from GOES Imager instruments onboard the GOES-East and GOES-West satellites with observations from SSM/I. Optical retrievals from the GOES imagers are used as the primary data source for the blended map, and MW data are added if the area was identified as "cloudy" by GOES. The principal difference of this approach from the one of Foster et al. (2010) is that MW observations classified as "snow-free land surface" are disregarded in the blending technique because of frequent omission of melting snow and shallow snow in the MW product. MW snow retrievals over mountains are also disregarded because of their tendency to confuse cold rocky surfaces with snow and thus to overestimate snow cover extent in high-altitude areas. At the final processing stage, pixels that remain "undetermined" in the current-day snow map are filled in with the data from the previous day's blended snow map. In 2000, this blending algorithm was implemented operationally at NOAA/NESDIS to generate 4 km daily snow maps over North America. In 2006, this snow-mapping system was upgraded by adding optical observations from NOAA AVHRR and MSG SEVIRI. The latter allowed the snow-mapping domain to be expanded from North America to the whole globe.

Since the blended snow cover map comprises both optical and MW snow retrievals, its overall accuracy typically ranges within the accuracy of the contributing products. In the cloud-clear portions of the imagery, the accuracy of the blended product and its effective spatial resolution is identical to the accuracy and spatial resolution of the optical SCA map. In cloudy areas, both the effective spatial resolution and the accuracy of the blended product degrade to the level corresponding to MW retrievals. The principal benefit of the blended product over the optical-based SCA maps is in its continuous area coverage, while providing a higher classification accuracy and better spatial resolution over clear-sky areas compared to MW-based products.

15.5 APPLICATION OF SATELLITE-BASED SNOW PRODUCTS FOR DROUGHT IDENTIFICATION AND MONITORING

The example provided in this section demonstrates a possible application of satellite-derived SCA maps to assess the spring snowmelt water availability and hence to facilitate the prediction of early-spring drought conditions. Note that a better assessment of spring snowmelt can be made by accurate knowledge of the SWE distribution before the onset of spring snowmelt, which would require application

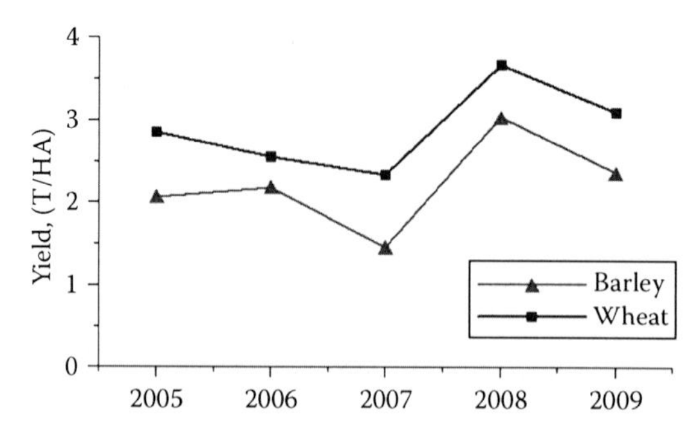

FIGURE 15.6 Yearly yield of barley and wheat in Ukraine, 2005–2009. Data were obtained from the USDA FAS Production, Supply and Distribution data set (online at http://www.fas.usda.gov/psdonline/psdQuery.aspx).

of satellite-derived SWE maps (not included in this section). However, as validation studies presented earlier have shown, the reliability of the available satellite-based operational SWE products for such assessments is questionable at this current time.

The case study is focused on Ukraine, which is one of the world's largest grain exporters of winter crops (primarily wheat and barley) that account for most of the country's total annual grain production. Snow cover plays an important role in winter crop production in Ukraine. The duration of the snow season ranges from several weeks in the southernmost part of the country to several months in the north. In addition to providing the source of snowmelt water for winter crop development in early spring, the snowpack also prevents these crops from frost and freeze damage during the winter. In spring 2007, a severe drought occurred in Ukraine, which seriously affected winter grain production. According to the data of USDA's Foreign Agricultural Service (FAS), in 2007, winter barley and wheat yields were approximately 10% and 30% less than in 2006 and 35% and 50% less than in 2008 (Figure 15.6). In an attempt to gain better insight into the origin of the drought and its possible relationship to the snow cover properties, we calculated and examined seasonal variations of the snow cover distribution in Ukraine. NOAA's IMS daily interactive snow cover maps were used to generate maps of snow cover duration for 10 consecutive winter seasons from 2000–2001 to 2008–2009. For every grid cell of the map, the duration of winter snow cover was estimated by calculating the number of days with snow cover within an annual snow season time period starting on August 1 and ending July 31 of the next year. The IMS product was applied in this study since NESDIS blended snow cover maps were not available before 2006.

Figure 15.7 presents IMS-derived estimates of the duration of snow cover in Ukraine during the 2006–2007 winter season as well as for two preceding and subsequent winters. As shown, the 2006–2007 winter season was characterized by substantially shorter duration of snow cover. Compared to the mean snow cover duration calculated for 10 winter seasons (2000–2009), in 2006–2007 the snow cover stayed on the ground 40–50 days less in the northern part of the country and about 20 days less in the south. In Ukraine, as well as in other midlatitude regions that are characterized by a persistent winter snow cover, SWE tends to gradually increase throughout the winter season. As a result, shorter (longer) snow cover duration results in a smaller (larger) amount

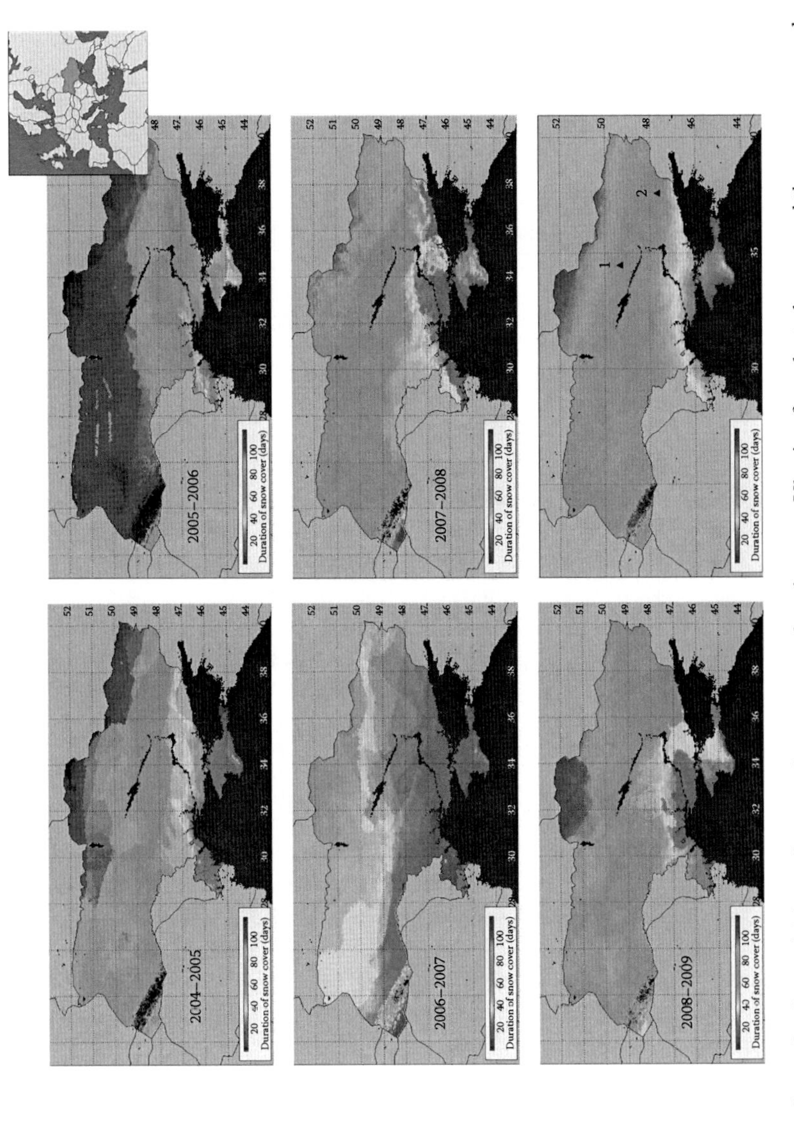

FIGURE 15.7 **(See color insert.)** Maps of seasonal snow cover duration over Ukraine for selected years and the mean seasonal snow cover duration for the years 2000–2009. The upper right image shows the location of Ukraine in Europe. Triangles in the map of the mean duration of snow cover (in the lower right) indicate the location of Poltava (1) and Donetsk (2).

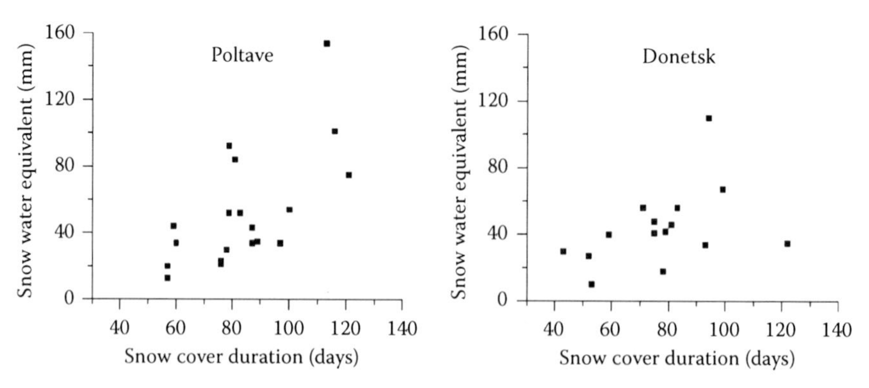

FIGURE 15.8 Scatter plots of snow cover duration and maximum SWE observed during a month-long period preceding the snowmelt for two locations in Ukraine: Poltava (49.36°N, 34.33°E) and Donetsk (48.04°N, 37.46°E). The location of the two stations is shown in Figure 15.7. The results are presented for the years 1967–1990. Snow cover duration and SWE data at the two stations were acquired from the Former Soviet Union Hydrological Snow Surveys data set available at NSIDC. (From Krenke, A., *Former Soviet Union Hydrological Snow Surveys, 1966–1996*, National Snow and Ice Data Center/World Data Center for Glaciology, Digital media, Boulder, CO, 1998.)

of water accumulated in the snowpack and released during the spring snowmelt, and, thus, snow duration can be used as a proxy for SWE available for spring snowmelt.

Figure 15.8 presents the statistics on the snow cover duration and the maximum seasonal SWE observed in Poltava (49.36°N, 34.33°E) and Donetsk (48.04°N, 37.46°E). The location of these stations is shown in Figure 15.7. These in situ observations were acquired from the Former Soviet Union Hydrological Snow Surveys data set available at NSIDC (Krenke, 1998) and cover the time period from 1967 to 1990. Both graphs in Figure 15.8 clearly demonstrate a decreasing trend in SWE by the end of the winter season with the decreasing duration of the seasonal snow cover. The analysis of IMS-derived maps of snow duration revealed that in the winter of 2006–2007 the duration of snow cover in Poltava and Donetsk was about three times less than normal and comprised 32 and 26 days, respectively. This anomalously short duration of the snow season gives a clear indication that the amount of moisture accumulated in the snowpack by the end of winter was also anomalously low. The shorter snow cover duration of the 2006–2007 winter season and the associated decline in snowmelt water may have contributed to the shortage of soil moisture for plant development in the spring and to drought conditions that occurred in early summer 2007. The early spring period spans the critical vegetative and reproductive growth stages of both winter wheat and barley that determine the final grain yield. Lower amounts of available soil moisture at the beginning of the 2007 growing season due to less spring snowmelt compared to the adjacent years are supported by soil moisture data derived from AMSR-E observations (Figure 15.9). Monthly soil moisture maps were generated from corresponding daily AMSR-E-based maps (AE_Land3 product) available at NSIDC (http://nsidc.org/data/ae_land3.html). As shown in Figure 15.9, the monthly average soil moisture values in the early spring months (March, April, and May) were noticeably lower in 2007 than in 2006. The largest soil moisture deficit in 2007 occurred in the central and eastern part of the country where most winter barley

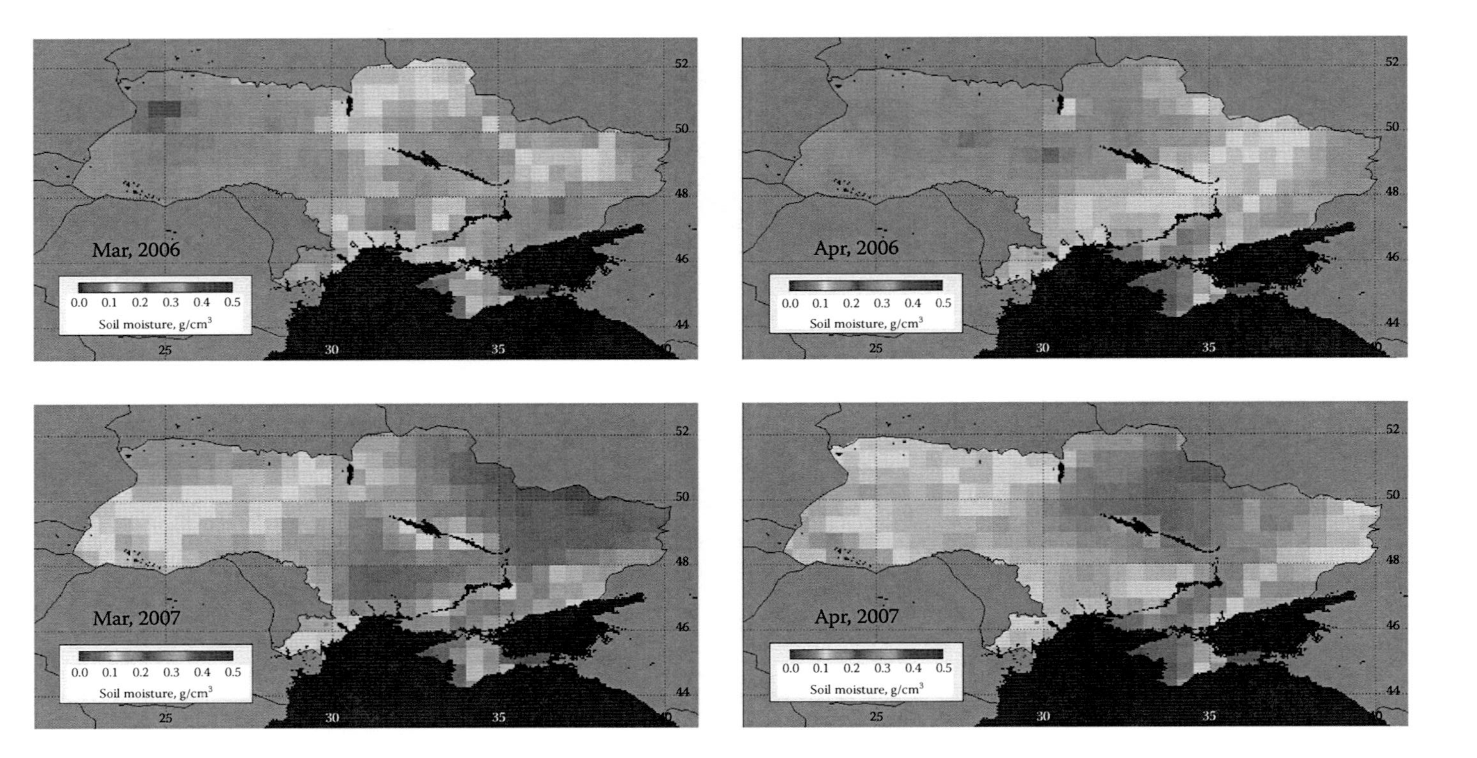

FIGURE 15.9 **(See color insert.)** Monthly average soil moisture over Ukraine for the months of March and April of 2006 and 2007. Soil moisture was retrieved from AMSR-E Aqua data.

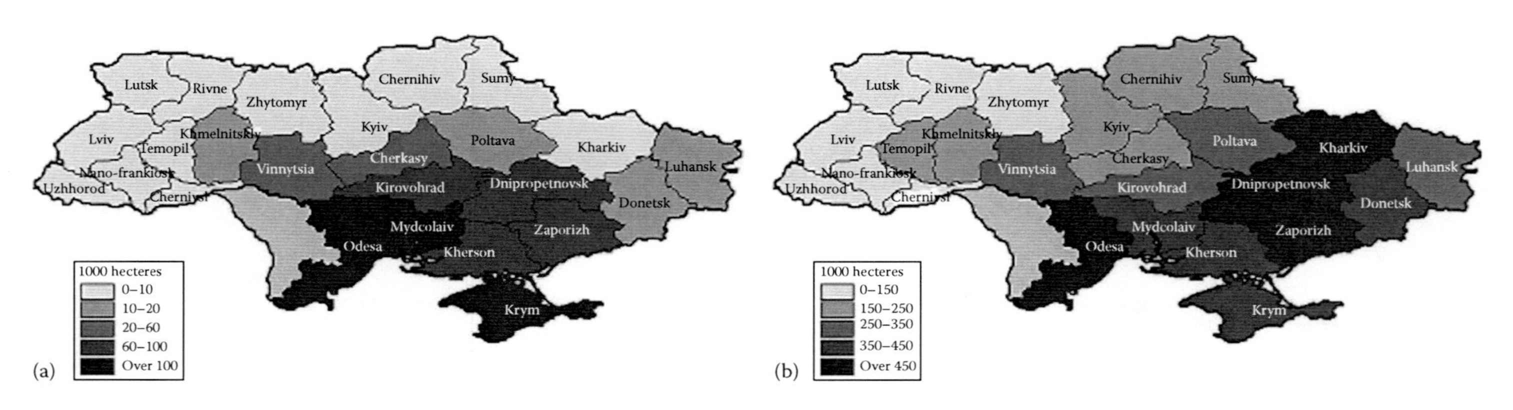

FIGURE 15.10 Area of winter barley (a) and winter wheat (b) in Ukraine (USDA FAS data).

and wheat are grown (see Figure 15.10). Lower levels of soil moisture in spring 2007 have also been confirmed by in situ measurements. In particular, according to reports from Poltava and Donetsk stations, the soil moisture in the top 20 cm of soil at the end of March 2007 was about 20% less than normal for these locations for this time of the year.

Although spring snowmelt presents an important source of moisture for crops at the beginning of the growing season, snow cover is by no means the only factor that has to be accounted for in winter crop condition monitoring and yield forecasting. Short duration of seasonal snow cover should be viewed only as indirect evidence of reduced winter precipitation that may contribute to the development of drought conditions later in spring. Other factors, such as soil moisture and precipitation in the preceding fall season and, most notably, the amount of liquid precipitation in early spring, also affect availability of soil moisture for crop development at the beginning of the growth season. As a result, snow cover products should be used in combination with other data sources reflecting other environmental conditions (e.g., precipitation, soil moisture, and vegetation health) to develop a more complete picture of early growing season drought conditions in the spring. A more detailed analysis of precipitation in Ukraine in winter 2006–2007 and spring 2007 has shown that the lack of water from snowmelt was not the only factor that caused drought conditions in 2007. Substantially lower amounts of precipitation were observed over most of the country later into the growing season, in March, April, and May. Time series of monthly precipitation amounts for two locations in Ukraine (see Figure 15.11)

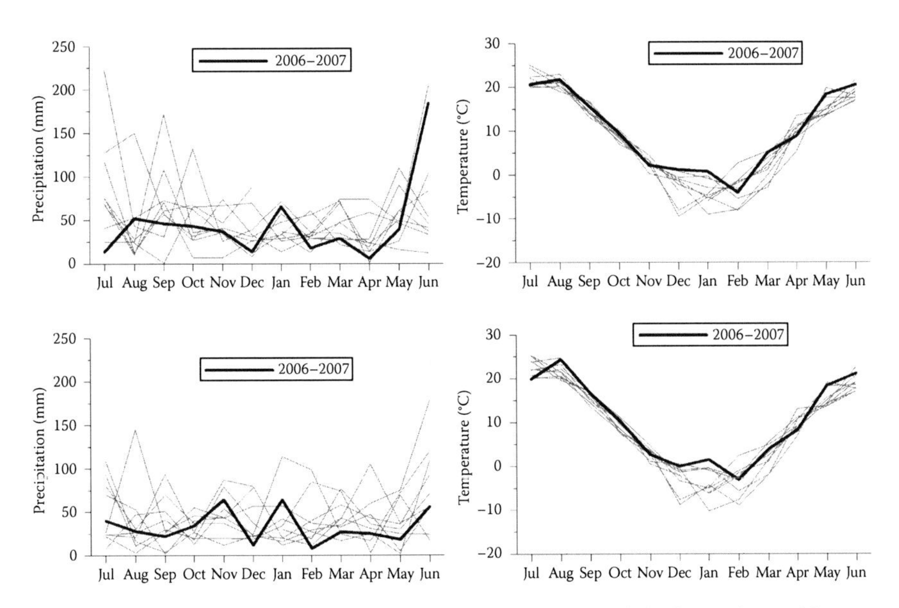

FIGURE 15.11 Seasonal time series of the monthly total precipitation and monthly mean air temperature at two stations in Ukraine, Poltava (upper row) and Donetsk (lower row), for the years 2000–2009. Precipitation and air temperature for the period from August 2006 to July 2007 are shown in black; data for all other years are shown in gray. Values of monthly total precipitation and monthly mean temperature were acquired from NCDC.

clearly demonstrate less than average precipitation amounts in December 2006 and in February–April 2007. Precipitation in January 2007 was somewhat higher than average, but apparently it was not sufficient to fully compensate for the precipitation shortage during the rest of the months. The time series of monthly mean temperature in Figure 15.11 show that the winter season of 2006–2007 was unusually warm. The mean temperature in December and January at both locations (Poltava and Donetsk) was above the freezing level, whereas in February it decreased to several degrees below 0°C. Relatively high wintertime temperatures give a reason to believe that no or very little freeze damage to winter crops occurred in the winter of 2006–2007.

15.6 CONCLUSIONS

In this chapter, a review of principal satellite remote-sensing methods for snow monitoring was provided, with an emphasis on the mapping of hydrologically important snow parameters of SWE, SD, and SCA. Satellite observations can complement in situ measurements by providing time- and space-continuous snow cover information, facilitating more accurate snowmelt runoff forecasting and drought predictions. Validation, accuracy assessment, and limitations of current remote sensing–based snow information received special attention. Optical remote sensing–based SCA products achieve a high accuracy (exceeding 94%) over areas not heavily forested and snowpack deeper than 1 cm. Given the relatively high spatial (1–4 km) and temporal (hourly for geostationary and daily for polar-orbiting sensors) resolution of optical satellite-based instruments, they can be used successfully to derive accurate snow cover distribution information for snowmelt runoff forecasting and drought assessment for many SCA of the world. This capability was demonstrated in a case study over Ukraine. An inherent limitation of optical remote-sensing snow cover information is that it is available only during the day and over cloud-free areas. MW-based snow cover information, on the other hand, is available during both day and night and under cloudy conditions to monitor snow cover distribution. In addition, MW imagery offers the capability of mapping SD and SWE, which are more physically related to snowmelt than SCA. Overall, passive MW-derived SCA using current operational algorithms shows agreement with observed seasonal and interannual trends, and thus has the ability to provide useful complementary winter moisture information to the optical-based sources. However, seasonal biases still exist. SCA is underestimated over melting snow, early season shallow snow, and forested snow covers and overestimated in cold desert soils. In addition, underestimations of SWE and SD occur for snow cover deeper than 40 cm. In order to maximize advantages of optical and passive MW imagery, several SCA blending techniques and products were also discussed. The principal benefit of the blended satellite products over the optical-based products is that they provide continuous SCA coverage, while achieving a higher classification accuracy and better spatial resolution over clear-sky areas compared to passive MW-based products. Recent blending (assimilation) approaches that combine MW-derived SWE and in situ or modeled SWE show encouraging results and performance superior to existing passive MW-based SWE approaches. However, none of the assimilation approaches improved performance over melting and deeper snow covers.

REFERENCES

Andreadis, K.M. and D.P. Lettenmaier. 2005. Assimilating passive microwave brightness temperature for snow water equivalent estimation. *American Meteorological Society Annual Meeting, 19th Conference on Hydrology*, San Diego, CA. Boston, MA: American Meteorological Society.

Armstrong, R.L. and M.J. Brodzik. 2001. Validation of passive microwave snow algorithms. *Remote Sensing and Hydrology 2000*, IAHS Publication No. 267:87–92.

Armstrong, R.L., M.J. Brodzik, K. Knowles, and M. Savoie. 2007. *Global Monthly EASE-Grid Snow Water Equivalent Climatology*. Boulder, CO: National Snow and Ice Data Center, Digital media.

Armstrong, R.L., M.J. Brodzik, M. Savoie, and K. Knowles. 2003. Enhanced hemispheric-scale snow mapping through the blending of optical and microwave satellite data. *Geophysical Research Abstracts* 5:12824.

Ault, T., K.P. Czajkowski, T. Benko, J. Coss, J. Struble, A. Spongberg, M. Templin, and C. Gross. 2006. Validation of the MODIS snow product and cloud mask using student and NWS cooperative station observations in the Lower Great Lakes Region. *Remote Sensing of Environment* 105:341–353.

Baum, B.A. and Q. Trepte. 1999. A grouped threshold approach for scene identification in AVHRR imagery. *Journal of Atmospheric and Oceanic Technology* 16:793–800.

Brodzik, M.J., R.L. Armstrong, E.C. Weatherhead, M.H. Savoie, K.W. Knowles, and D.A. Robinson. 2006. Regional trend analysis of satellite-derived snow extent and global temperature anomalies. American Geophysical Union Fall Meeting Supplement, *EOS Transactions* 87(52):Abstract U33A-0011.

Chang, A.T.C., J.L. Foster, and D.K. Hall. 1987. Nimbus - 7 SMMR derived global snow cover parameters. *Annals of Glaciology* 9:39–44.

Chang, A.T.C., R.E.J. Kelly, E.J. Josberger, R.L. Armstrong, and J.L. Foster. 2005. Analysis of gauge-measured and passive microwave derived snow depth variations of snowfields. *Journal of Hydrometeorology* 6:20–33.

Che, T., X. Li, R. Jin, R. Armstrong, and T. Zhang. 2008. Snow depth derived from passive microwave remote-sensing data. *Annals of Glaciology* 49:145–154.

Derksen, C. 2008. The contribution of AMSR-E 18.7 and 10.7 GHz measurements to improved boreal forest snow water equivalent retrievals. *Remote Sensing of Environment* 112:2700–2709.

Derksen, C., A. Walker, and B. Goodison. 2003. A comparison of 18 winter seasons of in situ and passive microwave-derived snow water equivalent estimates in Western Canada. *Remote Sensing of Environment* 88(3):271–282.

de Wildt, M., G. Seiz, and A. Gruen. 2007. Operational snow mapping using multitemporal Meteosat SEVIRI imagery. *Remote Sensing of Environment* 109:29–41.

Dong, J., J. Walker, P. Houser, and C. Sun. 2007. Scanning multichannel microwave radiometer snow water equivalent assimilation. *Journal of Geophysical Research* 112:D07108, doi:10.1029/2006JD007209.

Dong, J., J.P. Walker, and P.R. Houser. 2005. Factors affecting remotely sensed snow water equivalent uncertainty. *Remote Sensing of Environment* 97(1): 68–82.

Dozier, J., R.O. Green, A.W. Nolin, and T.H. Painter. 2009. Interpretation of snow properties from imaging spectrometry. *Remote Sensing of Environment* 113(1):S25–S37.

Dozier, J. and T.H. Painter. 2004. Multispectral and hyperspectral remote sensing of alpine snow properties. *Annual Review of Earth and Planetary Sciences* 32:465–494.

Fang, X. and J.W. Pomeroy. 2007. Snowmelt runoff sensitivity analysis to drought on the Canadian prairies. *Hydrological Processes* 21:2594–2609.

Fang, X. and J.W. Pomeroy. 2008. Drought impacts on Canadian prairie wetland snow hydrology. *Hydrological Processes* 22:2858–2873.

Foster, J.L., D.K. Hall, J.B. Eylander, G.A. Riggs, S.V. Nghiem, M. Tedesco, E.J. Kim, P.M. Montesano, R.E.J. Kelly, K.A. Casey and B. Choudhury. 2010. A blended global snow product using visible, passive microwave and scatterometer data. *International Journal of Remote Sensing* 32(5):1371–1395.

Foster, J.L., C.J. Sun, J.P. Walker, R. Kelly, A. Chang, J.R. Dong, and H. Powell. 2005. Quantifying the uncertainty in passive microwave snow water equivalent observations. *Remote Sensing of Environment* 94(2):187–203.

Frei, A. 2009. A new generation of satellite snow observations for large scale earth system studies. *Geography Compass* 3:879–902.

Frei, A. and D.A. Robinson. 1999. Northern Hemisphere snow extent: Regional variability 1972 to 1994. *International Journal of Climatology* 19:1535–1560.

Goodison, B.E. 1989. Determination of areal snow water equivalent on the Canadian prairies using passive microwave satellite data. *IEEE International Geoscience and Remote Sensing Symposium*, Vancouver, Canada.

Gray, D.M., J.W. Pomeroy, and R.J. Granger. 1989. Modelling snow transport, snowmelt and meltwater infiltration in open, northern regions. In *Northern Lakes and Rivers*, ed. W.C. Mackay, Occasional Publication No. 22, pp. 8–22. Edmonton, Alberta, Canada: Boreal Institute for Northern Studies, University of Alberta.

Grody, N.C. 1991. Classification of snow cover and precipitation using the special sensor microwave imager. *Journal of Geophysical Research* 96(D4):7423–7435.

Grody, N.C. and A.N. Basist. 1996. Global identification of snow cover using SSM/I measurements. *IEEE Transactions on Geoscience Remote Sensing* 34(1):237–249.

Grody, N.C., F. Weng, and R. Ferraro. 2000. Application of AMSU for obtaining hydrological parameters. In *Microwave Radiometry and Remote Sensing of the Earth's Surface and Atmosphere*, eds. P. Pampalono and S. Paloscia, pp. 339–351. Zeist, the Netherlands: Vision Sports Publishing.

Hall, D.K. and G.A. Riggs. 2007. Accuracy assessment of the MODIS snow-cover products. *Hydrological Processes* 21:1534–1547.

Hall, D.K., G.A. Riggs, V.V. Salomonson, N. DiGiromamo, and K.J. Bayr. 2002. MODIS snow-cover products. *Remote Sensing of Environment* 83:181–194.

Helfrich, S.R., D. McNamara, B.H. Ramsay, T. Baldwin, and T. Kasheta. 2007. Enhancements to, and forthcoming developments in the Interactive Multisensor Snow and Ice Mapping System (IMS). *Hydrological Processes* 21:1576–1586.

Josberger, E.G. and N.M. Mognard. 2002. A passive microwave snow depth algorithm with a proxy for snow metamorphism. *Hydrological Processes* 16(8):1557–1568.

Kelly, R.E.J., T.C. Alfred, J. Chang, J.L. Foster, and M. Tedesco. 2004b. Updated daily, 5-day and monthly. *AMSR-E/Aqua Monthly L3 Global Snow Water Equivalent EASE-Grids* V002. Boulder, CO: National Snow and Ice Data Center, Digital media.

Kelly, R.E., A.T. Chang, L. Tsang, and J.L. Foster. 2003. A prototype AMSR - E global snow area and snow depth algorithm. *IEEE Transactions on Geoscience and Remote Sensing* 41(2):230–242.

Kelly, R.E.J., N.A. Drake, and S.L. Barr. 2004a. *Spatial Modelling of the Terrestrial Environment.* Chichester, U.K.: John Wiley and Sons Ltd.

Khlopenkov, K.V. and A.P. Trishchenko. 2007. SPARC: New cloud, snow, and cloud shadow detection scheme for historical 1-km AVHRR data over Canada. *Journal of Atmospheric and Oceanic Technology* 24:322–343.

Kongoli, C., C. Dean, S. Helfrich, and R. Ferraro. 2007. Estimating the potential of a blended passive microwave-interactive multi-sensor snow water equivalent product for improved mapping of snow cover and estimations of snow water equivalent. *Hydrological Processes* 21:1597–1607.

Kongoli, C., R. Ferraro, P. Pellegrino, and H. Meng. 2005. Snow microwave products from the NOAA's Advanced Microwave Sounding Unit. *Proceedings of the 19th Conference on Hydrology,* San Diego, CA. Boston, MA: American Meteorological Society.

Kongoli, C., P. Pellegrino, R.R. Ferraro, N.C. Grody, and H. Meng. 2003. A new snowfall detection algorithm over land using measurements from the Advanced Microwave Sounding Unit (AMSU). *Geophysical Research Letters* 30(4):1756–1759.

Krenke, A. 1998. *Former Soviet Union Hydrological Snow Surveys, 1966-1996.* Edited by NSIDC. Boulder, CO: National Snow and Ice Data Center/World Data Center for Glaciology, Digital media.

Künzi, K.F., S. Patil, and H. Rott. 1982. Snow-cover parameters retrieved from Nimbus-7 Scanning Multichannel Microwave Radiometer (SMMR) data. *IEEE Transactions on Geoscience and Remote Sensing* 20(4):452–467.

Liang, T., X. Zhang, H. Xie, C. Wu, Q. Feng, X. Huang, and Q. Chen. 2008. Toward improved daily snow cover mapping with advanced combination of MODIS and AMSR-E measurements. *Remote Sensing of Environment* 112:3750–3761.

Loujus, K., J. Pulliainen, M. Takala, C. Derksen, H. Rott, T. Nagler, R. Solberg, A. Wiesmann, S. Metsamaki, E. Malnes, and B. Bojkov. 2010. Investigating the feasibility of the GLOBESNOW snow water equivalent data for climate research purposes. *Proceedings of IEEE International Geoscience and Remote Sensing Symposium*, Honolulu, HI.

Lynch-Stieglitz, M. 1994. The development and validation of a simple snow model for the GISS GCM. *Journal of Climate* 7:1842–1855.

Mätzler, C. 1994. Passive microwave signatures of landscapes in winter. *Meteorology and Atmospheric Physics* 54(1–4):241–260.

Mätzler, C. and A. Wiesmann. 1999. Extension of the microwave emission model of layered snowpacks to coarse-grained snow. *Remote Sensing of Environment* 70(3):317–325.

Pagano, T.C. and D.C. Garen. 2006. Integration of climate information and forecasts into western US water supply forecasts. In *Climate Variations, Climate Change, and Water Resources Engineering*, eds. J. D. Garbrecht and T. C. Piechota, pp. 86–103. Reston, VA: American Society of Civil Engineers.

Pulliainen, J. 2006. Mapping of snow water equivalent and snow depth in boreal and sub-arctic zones by assimilating space-borne microwave radiometer data and ground-based observations. *Remote Sensing of Environment* 101(2):257–269.

Pulliainen, J.T., J. Grandell, and M.T. Hallikainen. 1999. HUT snow emission model and its applicability to snow water equivalent retrieval. *IEEE Transactions on Geoscience and Remote Sensing* 37(3):1378–1390.

Pulliainen, J. and M. Hallikainen. 2001. Retrieval of regional snow water equivalent from space-borne passive microwave observations. *Remote Sensing of Environment* 75(1):76–85.

Ramsay, B. 1998. The interactive multisensor snow and ice mapping system. *Hydrological Processes* 12:1537–1546.

Romanov, P., G. Gutman, and I. Csiszar. 2000. Automated monitoring of snow over North America with multispectral satellite data. *Journal of Applied Meteorology* 39:1866–1880.

Romanov, P. and D. Tarpley. 2006. Monitoring snow cover over Europe with Meteosat SEVIRI. *Proceedings of the 2005 EUMETSAT Meteorological Satellite Conference*, pp. 282–287, Dubrovnik, Croatia.

Simic, A., R. Fernandes, R. Brown, P. Romanov, and W. Park. 2004. Validation of VEGETATION, MODIS, and GOES C SSM/I snow-cover products over Canada based on surface snow depth observations. *Hydrological Processes* 18(6):1089–1104.

Simpson, J.J., J.R. Stitt, and M. Sienko. 1998. Improved estimates of the areal extent of snow cover from AVHRR data. *Journal of Hydrology* 204:1–23.

Tedesco, M., M. Brodzik, R. Armstrong, M. Savoie, and J. Ramage. 2009. Pan arctic terrestrial snowmelt trends (1979-2008) from spaceborne passive microwave data and correlation with the Arctic Oscillation. *Geophysical Research Letters* 36:L21402, doi:10.1029/2009GL039672.

Tsang, L., C. Chen, A.T.C. Chang, J. Guo, and K. Ding. 2000. Dense media radiative transfer theory based on quasicrystalline approximation with applications to passive microwave remote sensing of snow. *Radio Science* 35(3):731–749.

Wheaton E., V. Wittrock, S. Kulshreshtha, G. Koshida, C. Grant, A. Chipanshi, and B. Bonsal. 2005. Lessons learned from the Canadian drought years 2001 and 2002: Synthesis report. Publication No. 11602-46E0. Prepared for Agriculture and Agri-Food Canada, SRC, Saskatoon, Saskatchewan.

Wiesmann, A. and C. Mätzler. 1999. Microwave emission model of layered snowpacks. *Remote Sensing of Environment* 70(3):307–316.

Wiscombe, W.J. and S.G. Warren. 1980. A model for the spectral albedo of snow. Part I: Pure snow. *Journal of the Atmospheric Sciences* 37:2712–2733.

Woodhouse, I. 2006. *Introduction to Microwave Remote Sensing*. Boca Raton, FL: CRC Press.

Part V

Summary

16 Future Opportunities and Challenges in Remote Sensing of Drought

Brian D. Wardlow, Martha C. Anderson, Justin Sheffield, Bradley D. Doorn, James P. Verdin, Xiwu Zhan, and Matthew Rodell

CONTENTS

16.1 INTRODUCTION

The value of satellite remote sensing for drought monitoring was first realized more than two decades ago with the application of Normalized Difference Vegetation Index (NDVI) data from the Advanced Very High Resolution Radiometer (AVHRR) for assessing the effect of drought on vegetation, as summarized by Anyamba and Tucker

(2012, Chapter 2). Other indices such as the Vegetation Health Index (VHI) (Kogan, 1995) were also developed during this time period and applied to AVHRR NDVI and brightness temperature data for routine global monitoring of drought conditions. These early efforts demonstrated the unique perspective that global imagers like AVHRR could provide for operational drought monitoring through near-daily, synoptic observations of earth's land surface. However, the advancement of satellite remote sensing for drought monitoring was limited by the relatively few spectral bands on operational global sensors such as AVHRR, along with a relatively short observational record.

Since the late 1990s, spectral coverage has been expanded with the launch of many new satellite-based remote sensing instruments, such as the Advanced Microwave Sounding Unit (AMSU), Tropical Rainfall Measuring Mission (TRMM), Moderate Resolution Imaging Spectroradiometer (MODIS), Medium Resolution Imaging Spectrometer (MERIS), Gravity Recovery and Climate Experiment (GRACE), and Advanced Microwave Scanning Radiometer-Earth Observation System (AMSR-E), which have provided an array of new earth observations and measurements to map, monitor, and estimate a wide range of environmental parameters relevant to drought monitoring. As illustrated by the various remote sensing methods presented in this book, many components of the hydrologic cycle can now be estimated and mapped from satellite imagery, providing the drought community with a more complete and comprehensive view of drought conditions. Precipitation in the form of both rainfall (Story, 2012, Chapter 12; AghaKouchak et al., 2012, Chapter 13; Funk et al., 2012, Chapter 14) and snow cover (Kongoli et al., 2012, Chapter 15) can now be monitored over large areas through a combination of ground-based radar and satellite-based optical and microwave observations. Variations in the flux of water from the land surface to the atmosphere can also be assessed by applying new innovative techniques to thermal and shortwave data to estimate evapotranspiration (ET) as demonstrated by Senay et al. (2012, Chapter 6), Anderson et al. (2012b, Chapter 7), and Marshall et al. (2012, Chapter 8). In addition, new perspectives on vegetation conditions are provided by hybrid vegetation indicators (Wardlow et al., 2012, Chapter 3; Tadesse et al., 2012, Chapter 4) and estimates of biophysical parameters such as the fraction of absorbed photosynthetic active radiation (fAPAR) (Rossi and Niemeyer, 2012, Chapter 5). Lastly, unprecedented insights into subsurface moisture conditions (i.e., soil moisture and groundwater) are possible through the analysis of microwave (Nghiem et al., 2012, Chapter 9) and gravity anomaly (Rodell, 2012, Chapter 11) data, as well as land data assimilation systems (LDASs) incorporating remotely sensed moisture signals (Sheffield et al., 2012, Chapter 10).

Remote sensing advancements like these will be essential to meeting the growing demand for tools that can provide accurate, timely, and integrated information on drought conditions to facilitate proactive decision making (NIDIS, 2007). Satellite-based approaches are key to addressing the significant gaps in spatial and temporal coverage of surface station instrument networks providing key observations (e.g., rainfall, snow, soil moisture, groundwater, and ET) over the United States and globally (NIDIS, 2007). Improved monitoring capabilities are particularly important and timely given recent increases in spatial extent, intensity, and duration of drought events observed in some regions of the world, as reported by the Intergovernmental

Panel on Climate Change (IPCC, 2007). Drought impacts in many food-insecure regions could potentially intensify over the next century as climatic changes alter patterns of hydrologic variables like evaporation, precipitation, air temperature, and snow cover (Burke et al., 2006; IPCC, 2007; USGCRP, 2009). Numerous national, regional, and global efforts such as the Famine Early Warning Systems Network (FEWS NET), National Integrated Drought Information System (NIDIS), and Group on Earth Observations (GEO), as well as regional drought centers (e.g., European Drought Observatory) and geospatial visualization and monitoring systems (e.g., NASA/USAID SERVIR), have been undertaken to improve drought monitoring and early warning throughout the world. Well-established and emerging remote sensing tools will be able to help fill important data and knowledge gaps (NIDIS, 2007; NRC, 2007) to address a wide range of drought-related issues including food security, water scarcity, and human health.

16.2 FUTURE OPPORTUNITIES FOR SPACE-BASED DROUGHT MONITORING

Over the next decade, further progress in the application of satellite remote sensing for drought monitoring is expected with the launch of several new instruments, many recommended by the 2007 Decadal Survey on Earth Science and Applications from Space (NRC, 2007). These instruments will enhance current capabilities for monitoring specific variables (e.g., soil moisture and terrestrial water storage) and will make possible unprecedented wide-area measurement of surface water elevation and extent. In addition, increased assimilation of satellite observations into LDASs is expected to significantly improve model estimates of hydrologic states, like soil moisture and stream flow, for drought monitoring. In this section, several of these planned satellite-based sensors will be reviewed, as well as the increased use of remotely sensed data in LDAS models, to highlight opportunities that exist to advance the contributions of remote sensing for operational drought monitoring and early warning.

16.2.1 Soil Moisture Active Passive

The Soil Moisture Active Passive (SMAP) mission, which is currently scheduled to be launched in the 2014–2015 time frame (NASA, 2010), is intended to provide global measurements of soil moisture and freeze/thaw state, with drought monitoring and prediction as targeted applications (Entekhabi et al., 2010). SMAP includes both an L-band radar and an L-band radiometer, which will allow global mapping of soil moisture at a 10 km spatial resolution with a 2–3 day revisit time under both clear and cloudy sky conditions. By combining coincident radiometer and radar measurements, SMAP will provide much higher spatial resolution soil moisture mapping capabilities than previous instruments such as AMSR-E (Nghiem et al., 2012, Chapter 9) and the European Soil Moisture and Ocean Salinity (SMOS; Kerr et al., 2010) mission, which generate products at 25 and 50 km resolutions, respectively. The SMAP 10 km soil moisture product will be achieved by combining higher accuracy but coarser spatial resolution (40 km) radiometer-based soil moisture retrievals with higher resolution radar data (1–3 km) that have lower retrieval accuracy. In addition, the integration of these two types of data enables soil moisture to be estimated under a wider range of vegetation conditions. Radar backscatter is highly influenced by land cover and only performs adequately in low vegetation conditions (Dubois et al., 1995), whereas the L-band radiometers have improved sensitivity to soil moisture under moderate vegetation cover. SMAP measurements will provide "direct" sensing of surface soil moisture in the top 5 cm of the soil profile (Entekhabi et al., 2010). Because many applications such as drought monitoring require information about soil moisture in the root zone, the SMAP mission will also provide estimates of soil moisture representative of a 1 m soil depth by merging SMAP observations with land surface model estimates in a soil moisture data assimilation system (Reichle, 2008).

16.2.2 Surface Water and Ocean Topography

The Surface Water and Ocean Topography (SWOT) mission is designed to provide water surface elevation (WSE) measurements for ocean and inland waters including lakes, reservoirs, rivers, and wetlands. The primary SWOT instrument will be an interferometric altimeter that utilizes a Ka-band synthetic aperture radar (SAR) interferometer with two antennas to measure WSE (Durand et al., 2010a). Over inland water, SWOT will directly measure the area and elevation of water inundation with a spatial resolution on the order of tens of meters. Specific observational goals include a maximum 10 day temporal resolution and the capability to spatially resolve rivers with widths >50–100 m and water bodies with areas >250 m^2. The vertical accuracy requirement of the WSE measurements is set at 10 cm over 1 km^2 area and river slope measurements within 1 cm per 1 km distance (Biancamaria et al., 2010). Several predecessor radar altimetry missions including Topex/Poseidon (Morris and Gill, 1994; Birkett, 1995, 1998; Maheu et al., 2003; Hwang et al., 2005; Birkett and Beckley, 2010; Cheng et al., 2010), Jason-1 and -2 (Shum et al., 2003), and the Environmental Satellite Radar Altimeter (ENVISAT RA; Frappart et al., 2006; Medina et al., 2008; Lee et al., 2011) have proven the science and utility of radar altimetry data for monitoring inland water dynamics (Alsdorf et al., 2007). Initial results in applying simulated SWOT data to characterize surface elevation changes of inland water bodies have been promising (Durand et al., 2008; Lee et al., 2010a). SWOT is anticipated to provide new information to better measure and understand the spatiotemporal dynamics of surface water (i.e., areal extent, storage, and discharge) globally, which would have tremendous benefit for many applications, including water resource management and hydrological drought monitoring and prediction. The SWOT mission is currently targeted to be launched in 2020 (NASA, 2010). In the meantime, several efforts have explored the expected accuracy of SWOT products for measuring change in storage of lakes (Lee et al., 2010a) and water surface profiles (Schumann et al., 2010), as well as the depth and discharge (Durand et al., 2010b) of rivers.

16.2.3 GRACE Follow-On and GRACE II

The GRACE mission has provided new insights into terrestrial water storage for drought monitoring as summarized by Rodell (2012, Chapter 11). GRACE is able to monitor groundwater changes, as well as variations in shallow and root-zone soil moisture content (Yeh et al., 2006; Rodell et al., 2007, 2009; Tiwari et al., 2009; Famiglietti et al., 2011). Outputs from GRACE related to surface and root-zone soil moisture and groundwater storage are now being applied in operational drought monitoring activities (Houborg et al., 2010; Rodell, 2012, Chapter 11). The National Research Council Decadal Survey (NRC, 2007) recommended that NASA launch an advanced technology gravimetry mission (GRACE II) toward the end of the decade. The proposed mission would replace the microwave interferometer with a laser and also fly at a lower altitude with a drag-free system, enabling perhaps an order of magnitude improvement in spatial resolution and making GRACE II directly applicable to a wider range of water resource characterization and management activities globally (NRC, 2007). However, taking into account the value of GRACE for many applications beyond drought monitoring, including measurement of ice sheet and glacier mass losses, groundwater depletion, and ocean bottom pressures, NASA has given preliminary approval to a GRACE Follow-On (GRACE FO) mission targeted for launch in 2016. GRACE FO would reduce an expected data gap after the terminus of GRACE and provide time for the technology developments required for GRACE II. The configuration of GRACE FO would be similar to GRACE, with incremental technological improvements that should afford some level of error reduction/increased spatial resolution.

16.2.4 Visible/Infrared Imager Radiometer Suite

The Visible/Infrared Imager Radiometer Suite (VIIRS) is the next-generation moderate resolution imaging radiometer launched onboard the National Polar-orbiting Operational Environmental Satellite System (NPOESS) Preparatory Project (NPP), platform in 2011, and scheduled for the subsequent series of Joint Polar Satellite System (JPSS) satellites (Lee et al., 2010b). The VIIRS instrument will be the operational successor to AVHRR and MODIS (Townshend and Justice, 2002), collecting data in 22 spectral bands spanning the visible as well as the near, middle, and thermal infrared (TIR) wavelength regions (Lee et al., 2006). The spectral bands of VIIRS were chosen primarily from two legacy instruments, AVHRR and MODIS, both of which have greatly contributed to drought monitoring tools, as discussed in earlier chapters. VIIRS is designed to provide daily global coverage at spatial resolutions of 370 m (5 bands) and 740 m (17 bands) (Lee et al., 2006). In addition, several data products, termed environmental data records (EDRs), will be produced from the VIIRS observations (Lee et al., 2006), including vegetation indices (NDVI and the Enhanced Vegetation Index, EVI) (Justice et al., 2010), land surface temperature (LST), and snow cover—each of which can be used for operational drought monitoring. The generation of VI and LST products from VIIRS is critical for extending the historical time series previously established with AVHRR and MODIS. The placement of a VIIRS instrument on the NPP platform was intended to provide continuity for observational data streams from instruments on NASA's Terra and Aqua platforms (Townshend and Justice, 2002). This will be followed by VIIRS instruments on a series of JPSS platforms that are planned to be operational into the 2023–2026 time period (Lee et al., 2010b). Future availability of VIIRS VI and LST data for drought monitoring will be needed for tools such as the Vegetation Drought Response Index (VegDRI, Brown et al., 2008; Wardlow et al., 2012, Chapter 3) and for the estimation of ET (Anderson et al., 2007; Senay et al., 2007) and ET-derived indices (Anderson et al., 2012a, 2012b, Chapter 7).

16.2.5 Landsat Data Continuity Mission

Along with recommendations for new missions, the NRC Decadal Survey also stressed the need for maintaining critical long-term earth surveillance programs such as Landsat. The Landsat satellite series has been routinely collecting global earth observations in the visible and near-infrared bands since 1972 (15–60 m spatial resolution) and the thermal band since 1982 (60–120 m spatial resolution). The resulting data set is the only long-term civilian archive of satellite imagery at scales of human influence, resolving individual farm fields, deforestation patterns, urban expansion, and other types of land use and land cover change. Landsat 7 (currently active) has functioned well past their expected lifetime. The Landsat Data Continuity Mission (LDCM) is scheduled for launch no earlier than January, 2013, and will carry the Operational Land Imager (OLI) that will continue six heritage shortwave bands as well as two new coastal and cirrus bands with a 30 m spatial resolution. LCDM will also carry the Thermal Infrared Sensor (TIRS) to collect TIR data in two channels to facilitate split-window atmospheric correction. TIRS will enable continued global mapping of ET and vegetation stress at field scale (Anderson et al., 2012a).

The revisit period of individual Landsat satellites (16 days or more, depending on cloud cover) is not ideal for operationally monitoring rapid changes in vegetation and moisture conditions associated with drought events. However, the Landsat series has the potential to provide 8 day (2 systems) or 5 day (3 systems) coverage if multiple systems are deployed concomitantly in staggered orbits. Increased temporal frequency could also be achieved by increasing the swath width of data collection. The potential also exists to fuse high-spatial/low-temporal information from Landsat with lower-spatial/daily maps from MODIS/VIIRS to produce daily maps of vegetation indices and ET at the Landsat scale (Gao et al., 2006; Anderson et al., 2012a). Such fused imagery could significantly address the growing need for high-resolution (subcounty) information about yield reduction and other drought impacts.

16.2.6 Next Generation Geostationary Satellites

Whereas polar orbiting satellites provide only periodic snapshots of land surface conditions (typically once per day or longer), diurnal variability in key environmental variables relating to surface water and energy balance (such as LST and solar radiation) can be readily observed with geostationary satellites, albeit at coarser spatial resolution. With advanced spacecraft and instrument technology, the Geostationary Operational Environmental Satellite "R" series (GOES-R) will replace the current GOES-N series to meet operational data needs of the National Oceanic and Atmospheric Administration (NOAA), providing higher spatial and temporal resolution over the full hemispherical disk covering North and South America. The launch of the first GOES-R series satellite is scheduled for 2015 (http://www.goes-r.gov/), carrying the Advanced Baseline Imager (ABI) instrument—a 16 band imager with 2 visible bands, 4 near-infrared bands, and 10 infrared (3.5–14 μm) bands (Schmit et al., 2005). Spatial resolutions of the ABI visible and IR channels will be 0.5 km and 1–2 km, respectively, compared with the 1 km and 4 km of the GOES-N series. In addition, the full disk coverage rate of ABI will be 5 min instead of the 30 min of the current GOES. Derived baseline data products with potential application for drought monitoring will include downward surface solar insolation, reflected solar insolation, and LST and will be generated with <1 h latency (http://www.goes-r.gov/products/baseline.html). The potential for developing ET and drought products from GOES-R ABI observations will be explored by the GOES-R Risk Reduction program (Guch and DeMaria, 2010).

Current geostationary coverage over Europe and Africa is provided by the Meteosat series operated by the European Organization for the Exploitation of Meteorological Satellites (EUMETSAT). Instruments on the Meteosat Second Generation (MSG) satellites, first launched in January 2004, include the Spinning Enhanced Visible and Infrared Imager (SEVIRI), with a total of 12 bands that generate images by scanning the earth every 15 min. The High Resolution Visible (HRV) band provides data at 1 km sampling, while the other bands sample at 3 km spatial resolution (http://www.eumetsat.int/Home/Main/Satellites/MeteosatSecondGeneration/). Data products generated from MSG SEVRI observations that are relevant to drought monitoring include downwelling surface longwave and shortwave fluxes, LST, leaf area index (LAI), and ET (http://www.eumetsat.int/Home/Main/DataProducts/Land/index.htm?l=en). For continuation of MSG data streams, two more MSG satellites are tentatively scheduled for launch in 2012 and 2014, and planning is underway for a Meteosat Third Generation (MTG) satellite series with improved spatiotemporal and spectral coverage (http://www.eumetsat.int/Home/Main/Satellites/MeteosatThirdGeneration/Instruments/).

To support global monitoring applications, several efforts are underway to assemble and often intercalibrate observations from an international system of geostationary satellites. These archives include the National Centers for Environmental Prediction (NCEP)/Climate Prediction Center (CPC) 4 km/3 h global infrared data set, the NOAA 10 km/3 h International Satellite Cloud Climatology Project (ISCCP) B1 data rescue project (Knapp, 2008), and the Global Monitoring for Environment and Security (GMES) 5 km/1 h Geoland2 project (Lacaze et al., 2010). Insolation and surface temperature products derived from these data can be used to drive global remote sensing models of ET and soil moisture status.

16.2.7 Global Precipitation Mission

The Global Precipitation Measurement (GPM) mission is the next-generation, dedicated precipitation measurement mission (Smith et al., 2004), which builds on the successful legacy of TRMM. TRMM is a joint venture by NASA-Japan Aerospace Exploration Agency (JAXA) and has provided unprecedented data on precipitation characteristics since its launch in 1997, surpassing its expected lifetime by many years. TRMM products have been used successfully as precipitation-related information inputs to regional drought monitoring, especially for less-developed regions with few sources of accurate rainfall measurements, supporting FEWS NET and the Princeton African Drought Monitor (Sheffield et al., 2008). Building on TRMM, the goal of GPM is to develop a next-generation remote sensing precipitation system that provides frequent, global, and accurate precipitation measurements for basic research, applications, and operational monitoring.

The GPM will consist of a "core" satellite that will carry a set of precipitation sensors, plus a constellation of satellites of "opportunity" that have a range of active and passive microwave (PMW) instruments, but, at the least, all will carry a PMW radiometer. The core satellite will be used as a calibration reference for the constellation of support satellites. The core satellite is being developed jointly by NASA/JAXA and is similar to the TRMM satellite. It will carry a dual frequency (Ku/Ka band) precipitation radar (DPR) and a high-resolution multichannel PMW rain radiometer (GPM Microwave Imager, GMI). The DPR will have higher sensitivity to light precipitation than the TRMM precipitation radar and will provide three-dimensional structure data of storm events. The GMI will for the first time use a set of frequencies to retrieve precipitation particle sizes that are crucial for making more accurate precipitation rate retrievals and will better distinguish between light, moderate, and heavy precipitation. The core satellite will be complemented by a European Space Agency (ESA) core satellite (European contribution to the Global Precipitation Mission, EGPM), which will have the capability to measure light to moderate drizzle and snowfall at mid- to high latitudes. The constellation will have a mixture of sun-synchronous and non-sun-synchronous orbits that will allow retrievals of precipitation rates globally, with approximately 3 h sampling for each point on earth for about 90% of the time. The expected launch date for the core satellite is 2013.

16.2.8 Other Proposed Sensors

Several other proposed satellite-based instruments hold potential for supporting operational drought monitoring in the future. The Earth Science Decadal Survey identified the Hyperspectral Infrared Imager (HyspIRI) and Snow and Cold Land Processes (SCLP) missions as two future efforts that are anticipated to provide relevant drought information. The HyspIRI mission (proposed launch in the 2020s) includes two instruments: a visible shortwave infrared (VSWIR) imaging spectrometer operating between 0.38 and 2.5 μm at a spatial scale of 60 m with a swath width of 145 km and a TIR multispectral scanner operating between 4 and 12 μm at a spatial scale of 60 m with a swath width of 600 km. In the current conceptual design, HyspIRI will collect data at 60 m spatial resolution and have a 5 day revisit interval for thermal acquisitions and 19 days for hyperspectral imaging. Early warning of drought is a specific societal benefit of HyspIRI listed in the Decadal Survey (NRC, 2007). Early signs of ecosystem changes due to drought stress, reflected in altered plant physiology (e.g., water and carbon flux changes), may be spectrally identifiable in the hyperspectral data provided by HyspIRI, while the high spatiotemporal resolution thermal imagery will be valuable for early detection of changes in plant water use due to soil moisture deficits, along with associated plant stress revealed in rising canopy temperatures.

The SCLP mission (proposed launch between 2016 and 2020) will consist of a combination of active (X- and Ku-band SARs) and passive (K- and Ku-band radiometers) microwave sensors to characterize snow cover and snow water equivalent, which should benefit drought monitoring by providing information about the expected soil moisture recharge and snowmelt runoff. The SCLP mission has been recommended for a low earth orbit to provide subkilometer spatial resolution with complementary higher resolution capabilities between 50 and 100 m and a temporal resolution of approximately 15 days to detect intraseasonal changes, with more frequent imaging capabilities (3–6 days) when needed (NRC, 2007).

16.2.9 Enhancement of Land Data Assimilation Systems with Remotely Sensed Data

The suite of existing and planned remotely sensed data products has the potential to provide a holistic view of drought and the hydrologic cycle in general. As these products give different and independent views of hydrologic states and fluxes, and the condition of vegetation, they provide complementary information that will help users robustly identify emerging droughts for different applications that rely on joint assessment of multiple variables (e.g., U.S. Drought Monitor, USDM). However, this potentially comes at a price because of inconsistencies in the assessments from individual products (Sheffield et al., 2009; Gao et al., 2010). This is in part because the products represent different quantities and scales and because of inherent inaccuracies in the instruments, algorithms, and assumptions. A desirable but challenging goal is to merge collections of products into a physically consistent representation of the land surface to enhance drought monitoring from a multivariate, multiuser perspective.

Sheffield et al. (2012, Chapter 10) described the potential to meet this challenge through direct insertion or assimilation of remote sensing products into the NLDAS land modeling and assimilation system. LDAS provide, through modeling, a physically consistent and continuous depiction of all major components of the land surface water cycle. However, model output can be sensitive to errors in the model input data (in particular, precipitation) and the model structure and parameterizations (e.g., rooting depth and soil moisture holding capacity). Assimilation enables integration of observational data, such as from remote sensing, into the modeling to correct for these errors. This improves the representation of important state variables, such as root-zone soil moisture, that are not directly available from remote sensing. Land data assimilation also provides a physically based framework for merging independent remote sensing products into spatially and temporally continuous fields of data.

Future improvements in land data assimilation for drought monitoring will leverage remote sensing products in a number of ways. A key enhancement goal is to improve the simulation of land surface conditions through the use of remotely sensed precipitation, especially in sparsely monitored or topographically complex regions. The planned GPM (Smith et al., 2004) will be a key factor to meet this goal with its global coverage at unprecedented spatial and temporal sampling (3 hourly and 0.25°). Assimilation of individual remote sensing products into land surface models has been demonstrated with GRACE data (Zaitchik et al., 2008), thermal (e.g., Crow et al., 2008) and microwave (e.g., Reichle et al., 2007) soil moisture retrievals, lake and river level altimetry (e.g., Andreadis et al., 2007), and microwave/visible snow retrievals (e.g., Dong et al., 2007). Assimilation of these products improves skill levels when compared to ground observations. The benefit to drought monitoring is beginning to be evaluated for large-scale applications (e.g., Bolten et al., 2010; Houborg et al., 2010). Furthermore, the complementary information (e.g., in terms of spatial and temporal coverage) contained in each of these assimilation products has not been exploited to its full potential through assimilation into a land model. Several promising avenues are emerging to do this, including joint assimilation of thermal-infrared and microwave soil moisture (Hain, 2010; Hain et al., 2011) and multiscale assimilation approaches (Pan et al., 2009). The latter approach combines products with complementary information at different scales—for example, coarse-resolution GRACE data and high-resolution thermal-infrared or microwave soil moisture.

16.3 CHALLENGES FOR THE REMOTE SENSING COMMUNITY

Over the past decade, the remote sensing community has made tremendous strides in advancing the application of satellite information for operational drought monitoring and early warning and is expected to continue to do so thanks to future missions including those highlighted in the previous section. The drought community is beginning to reap the benefits of these efforts through the array of new information generated from new types of satellite observations using advanced analytical tools and modeling approaches. However, several challenges need to be addressed to more fully integrate remote sensing data into routine drought monitoring and sustain the momentum established by the innovative tools that have recently emerged. Some of these key challenges in establishing remote sensing as a credible, and valuable, source of drought information are highlighted in the following.

16.3.1 Engagement of the User Community

Active engagement and communication between the remote sensing community and drought experts throughout the development of a new tool is important for the successful integration of satellite observation products into drought monitoring. The Earth Science Decadal Survey (NRC, 2007) emphasized this need for a stronger linkage between remote sensing scientists and end users to better define the data/information requirements for an application such as drought monitoring, as well as to improve the knowledge and capacity of users to apply these new types of satellite-derived products for their respective applications. Recommendations by drought experts regarding the type(s) of information, cartographic presentation, and update schedule should be acquired early in the process to guide product development. For drought applications, framing the current condition or state of a specific environmental variable (e.g., soil moisture) within a historical context (e.g., percent of historical average) is more important than delivery of the actual observation/estimate itself. Several examples of historic contextual products developed for drought monitoring were presented earlier in the book, including percentiles (Rodell, 2012, Chapter 11), standardized z-score anomalies (Anderson et al., 2012b, Chapter 7; Marshall et al., 2012, Chapter 8), percent change (Nghiem et al., 2012, Chapter 9), and percent of historical average (Anyamba and Tucker, 2012, Chapter 2). Appropriate cartographic color schemes for the maps are also important in order to present the information in a consistent format that is easily interpretable by drought experts. Examples were presented in this book by Wardlow et al. (2012, Chapter 3), Anderson et al. (2012b, Chapter 7), and Rodell (2012, Chapter 11) that apply a color palette used by the USDM in their maps; this palette has become commonly accepted in the drought community. Drought experts should also be actively engaged in the validation of remote sensing products. Their expert feedback and analysis is important for defining the performance and utility of a new tool or product for drought monitoring. These evaluation exercises also allow them to better understand the information being presented and how to best apply it, which will lead to more widespread integration into drought applications. Lastly, collaborative strategies need to be developed between the drought monitoring and remote sensing communities to optimally integrate satellite-based information into a coherent overall narrative that characterizes drought conditions in a meaningful way (NIDIS, 2007).

16.3.2 Accuracy Assessment

Accuracy assessment of remotely sensed information for drought monitoring is challenging, given the spatiotemporal complexity and differing sectoral definitions of drought (Wilhite and Glantz, 1985). Each drought event is unique, and its characteristics can vary in terms of onset, duration, intensity, and geographic extent. A number of qualitative and quantitative assessment techniques have been used, including comparisons with in situ measurements (e.g., soil moisture and precipitation), crop statistics (e.g., yields), and ground-level expert observations, as well as spatial and temporal pattern matching with other drought index maps (e.g., USDM). Although each technique has its own merits, there is no current consensus about the most appropriate set of methods to use for validation. The specific technique(s) selected is usually determined by the "ground truth" data that are available and the specific variable being validated. As a result, a "convergence of evidence" approach that collectively analyzes the findings from several assessment methods is needed to gain a more complete perspective of the accuracy and utility of a specific remote sensing tool or product for drought monitoring. Considerable work in the area of accuracy assessment is still needed in order to more fully realize the contribution of remote sensing for this application. In this section, some key points of emphasis to further advance the validation of remote sensing information for drought monitoring are discussed.

16.3.2.1 Extended and Near-Real-Time Assessment Campaigns

Given the complexity and spatiotemporal variability of drought, an extended accuracy assessment both temporally and geographically is needed to fully characterize the ability of a remote sensing tool to detect and monitor drought severity levels across diverse environments. Ideally, the assessments would be conducted in near real time to capitalize on the current drought information being reported (e.g., visual ground assessments, impacts, and media reports) and in situ data (e.g., rainfall, stream flow, and soil moisture) being collected. However, such an assessment campaign would require a considerable investment in time and resources. Retrospective analysis and case studies of previous drought events can be used to supplement a near-real-time assessment. However, many in situ observations and reports that would be useful for a thorough assessment are often difficult to locate or are not retained after a period of time. In addition, the relatively short operational period of a specific instrument can also limit the historical drought events that can be analyzed. Several sensors presented in this book have less than a decade of observations.

16.3.2.2 Consideration of Drought Impacts

Documentation of drought impacts on both natural and human systems is improving, allowing us to better quantify and understand the effects of drought that result from the complex interplay between a natural event (i.e., precipitation deficit) and the demand for water by humans and ecosystems (Wilhite et al., 2007). Several volunteer "citizen science" efforts such as the Drought Impact Reporter (DIR, http://droughtreporter.unl.edu/), the Community Collaborative Rain, Hail, and Snow Network (CoCoRAHS, http://www.cocorahs.org/), and National Phenology Network (NPN, http://www.usanpn.org/) have recently emerged to report both direct (e.g., reduced crop productivity) and indirect (e.g., reduced farm income) impacts of drought, resulting in a wealth of new ground-based drought information that could be utilized for validating remote sensing information. Although some impact data may not be directly comparable to remote sensing output in the way that traditional instrumental observations are (e.g., precipitation, soil moisture, or stream flow), they can provide a measure to verify relative changes and trends in drought conditions expressed in a satellite-derived product. In addition, linking these documented impacts with remote sensing–derived results is an important step in informing potential users of how these products relate to specific environmental condition(s) relevant to their application. Expansion of systematic impact archiving efforts outside the United States will be important for evaluating global drought monitoring tools.

16.3.3 Spatial Resolution and Scale

Higher spatial resolution drought information is being increasingly demanded in efforts to understand and address local-scale impacts on the ground, ranging from the county to field or parcel scale. A prime example is the USDM, which is a composite indicator of many data inputs (e.g., climate and hydrological indices and indicators and expert feedback) that depicts drought conditions as a series of severity contours in map form across the United States (Svoboda et al., 2002). The USDM was designated in the 2008 U.S. Farm Bill (Food, Conservation, and Energy Act of 2008 – H.R. 6124; http://www.usda.gov/documents/Bill_6124.pdf) as the primary tool to establish county-level agricultural producer eligibility for drought disaster assistance. Some data inputs used in USDM development lack the spatial resolution to discern local-scale drought patterns, making the depiction of county- to subcounty-scale drought a challenge. Despite this fact, USDM contours are used operationally by the U.S. Department of Agriculture (USDA) Farm Service Agency (FSA) to establish county-level eligibility for drought disaster assistance.

Eligibility is established for an entire county if the USDM assigns any area of the county with a specific drought designation over a specified time period; thus, the ability to more accurately characterize drought patterns and conditions at this critical subcounty spatial scale is needed to administer such a program.

Although in situ instrument networks providing essential climatic and hydrologic data to the USDM and other drought monitoring applications cannot themselves satisfy these spatial resolution requirements (NIDIS, 2007), satellite remote sensing can be employed to map effects of drought at these critical scales. Maps at 100 m spatial resolution from satellite imagery are capable of resolving drought impacts associated with individual cropped fields and other land parcels, whereas 1 km imagery is well suited for subcounty-level assessments. Such drought products will primarily use satellite imagery in the optical and thermal wavebands, where such spatial resolutions are achievable. They will focus on drought response variables, such as vegetation cover fraction/fraction of absorbed photosynthetically active radiation or ET, rather than climatic driver variables such as precipitation, which are typically more homogeneous on average at the 100–1000 m scale. For example, robust operational water stress mapping at the field scale would facilitate assessment of drought stress response as a function of crop type and phenological stage and would be invaluable for estimating end-of-season yields for different crops. In contrast, at the 10 km scale, it is difficult to establish the specific land-cover type most affected by an ongoing drought and what components of the landscape are predominantly contributing to the coarse-scale stress signal.

Innovative approaches are needed to generate spatially scalable drought indicators that could be applied globally at the 10–50 km scale to monitor regional food/water security, regionally at the 1 km scale to assist in national to continental decision-making activities (e.g., crop insurance payments, basin-level and transboundary water issues, and multinational monitoring tools such as the North American Drought Monitor), and locally at the 100 m parcel level for landscape-level applications (e.g., irrigation scheduling, rotational livestock grazing, and reservoir water management). In addition to supporting an array of drought applications, these spatially scalable drought indicators calculated over targeted drought-affected areas will improve our understanding of the linkages between moisture deficits and specific vegetation response, as well as the consistency in the drought information conveyed by the indicator calculated at various spatial resolutions. The challenge will be to develop robust drought indicators that are consistent in accuracy and over the extended period of record needed to define baseline normal conditions across this range of scales, particularly at the highest spatial resolutions where temporal sampling is much less frequent.

Such efforts would be benefited by a high spatiotemporal resolution, visible-NIR/thermal sensor such as NASA's HyspIRI, but as discussed in the following, data continuity requirements make such short-term NASA research missions of limited value for developing operational drought monitoring tools. Consequently, collaborations between remote sensing scientists and drought experts should be established in coordination with early-stage mission activities of new sensors. An example is the applications working group developed for the upcoming NASA SMAP mission (http://smap.jpl.nasa.gov/science/wgroups/applicWG/), tasked with testing and demonstrating the applicability of the new data sets for drought monitoring in order to justify potential operational follow-on missions to support these types of operational applications (as discussed in the next section). In addition, the development of disaggregation techniques similar to the Disaggregated Atmosphere Land Exchange Inverse (DisALEXI) model (Norman et al., 2003) and LDAS should be pursued (demonstrated by Rodell (2012, Chapter 11)) to enable data from coarser resolution, high repeat frequency satellite instruments (e.g., MODIS) to be applied at the higher spatial resolution of lower repeat frequency sensors (e.g., Landsat Enhanced Thematic Mapper) to meet the demands of global, continental, national, and subnational drought applications.

16.3.4 Long-Term Data Continuity

Long-term, sustained data records are essential for operational drought monitoring in order to provide a meaningful historical context to establish the relative severity of a current drought compared to previous events. From a climate perspective, 30 years is the accepted minimum length of an observational data record required to obtain a representative sample of the distributional characteristics (i.e., normal range of conditions or values) of key drought variables such as precipitation (Guttman, 1994; WMO, 2010). The observational records of most operational satellite-based instruments are much shorter. AVHRR is a primary exception, with a series of sensors onboard NOAA's family of polar orbiting platforms that have collected near-daily global image data since the early 1980s. However, many newer instruments such as MODIS and GRACE, which are providing data products that are increasingly being used for drought monitoring, have data records that approach or exceed a decade in length. It is critical that remote sensing observations and products essential for drought monitoring be identified and that long-term data continuity plans ensure their continued availability into the future. Long-term data continuity is vital for building the historical records necessary for anomaly detection, as well as maintaining a consistent and reliable data input for operational drought monitoring systems.

Some continuity efforts are underway, such as the development of VIIRS as an operational replacement for AVHRR and extension of the MODIS, and the LDCM to extend the 30+ year Landsat record (Wulder et al., 2008). Other planned missions such as GRACE-II and SMAP are also intended to continue observations of terrestrial water storage and soil moisture observations, respectively. A critical task for developing and extending a long-term, multisensor time series is intercalibration of data among the different instruments to develop a seamless long-term data stream that can be used for monitoring purposes. Prime examples of this are efforts to intercalibrate spectral data from the AVHRR instrument series to develop a consistent long-term NDVI time series (Tucker et al., 2005; Eidenshink, 2006) and the merging of the historical AVHRR NDVI data record with the more recent MODIS NDVI data (van Leeuwen et al., 2006) using a phenoregion-based translation approach (Gu et al., 2010).

Long-term data continuity is a challenge given the budgetary constraints of many space agencies and other organizations responsible for supporting the collection of satellite-based earth observations. International collaborations that leverage activities among these various groups may be necessary to collectively support the acquisition of critical earth observations (NRC, 2007) needed to support operational applications including drought monitoring worldwide. One such effort was highlighted by Nghiem et al. (2012, Chapter 9): the Oceansat-2 scatterometer from the Indian Space Agency is now being used to supplement the microwave data that had been provided by the NASA QSCAT instrument, which failed in November 2009, to estimate soil moisture conditions in support of drought monitoring. Similar collaborative efforts will be important in the future to support the global drought monitoring effort.

16.4 FINAL THOUGHTS

Drought is a common feature of climate throughout the world with a broad footprint of impacts influencing natural systems and many sectors of society. This natural hazard can further exacerbate many important challenges confronting society today, including food security, freshwater availability, and economic sustainability. As a result, there has been a paradigm shift in drought management from reactive, crisis-based approaches to more proactive, risk-based strategies to reduce societal vulnerability to drought (Wilhite and Pulwarty, 2005). Monitoring is a cornerstone of effective drought risk management, providing critical information to facilitate informed decision making to reduce risk and mitigate the effects of drought.

The satellite remote sensing community has been challenged and will continue to be tasked with providing unique data sets for assessing key components of the hydrologic cycle related to drought. Collectively, the potential of remote sensing to address this need is now beginning to be realized, as evidenced by the numerous new tools and techniques presented in this book. A full array of satellite-based information is now available to characterize precipitation inputs and surface and subsurface moisture conditions, providing a more complete picture of drought conditions than ever before available. The innovative techniques and new types of earth observation that are now being applied for drought monitoring have laid the groundwork for further innovations, as new tools mature and new data from the proposed missions highlighted in this chapter become available.

A strong emphasis has been placed on the effective application of information from remote sensing–based earth observations for "societal benefits" (NRC, 2007). The influence of drought cuts across many of the key societal benefit areas identified by both the Decadal Survey on Earth Science and Applications from Space (NRC, 2007) and the GEO Global Earth Observation System of Systems (GEOSS), including agriculture, climate, disasters, ecosystems, health, and water. Applications of satellite-based information support both scientific research and real-world decision making related to drought and its impacts (NRC, 2007). From a scientific standpoint, the suite of current and future remote sensing tools highlighted in this book will allow many components of the hydrologic cycle and environment (e.g., land use and land cover change) to be collectively analyzed using a systems approach to advance the science of drought monitoring and early warning. Remote sensing provides critical inputs for better understanding the spatiotemporal evolution and climatic drivers of droughts. Such research is necessary to build a strong scientific basis upon which drought risk management strategies and monitoring tools can be developed (Wilhite and Pulwarty, 2005). From a decision support perspective, the remote sensing scientist must be able to translate satellite-based earth observations and derivative products into useful, interpretable information for decision makers who often have nonscientific backgrounds. In order to improve capacity to use remote sensing–derived information in drought applications, drought experts and other decision makers should be involved in specifying their information requirements (accessibility, data types and formats, latency, and update frequency). Several chapter authors discussed tailoring data products from their tools based on feedback from the drought community, and similar efforts are encouraged to maximize the utilization of remote sensing observations in operational systems for drought monitoring and early warning.

REFERENCES

AghaKouchak, A., K. Hsu, S. Sorooshian, B. Inman, and X. Gao. 2012. Precipitation estimation from remotely sensing information using artificial neural networks: Application to drought monitoring and analysis. In *Remote Sensing of Drought: Innovative Monitoring Approaches*, eds. B.D. Wardlow, M.C. Anderson, and J.P. Verdin. Boca Raton, FL: CRC Press.

Alsdorf, D., E. Rodriguez, and D.P. Lettenmaier. 2007. Measuring surface water from space. *Reviews of Geophysics* 45(2):RG2002, doi: 10.1029/2006RG000197.

Anderson, M.C., R.G. Allen, A. Morse, and W.P. Kustas. 2012a. Use of Landsat thermal imagery in monitoring evapotranspiration and managing water resources. *Remote Sensing of Environment* (in press).

Anderson, M.C., C. Hain, B.D. Wardlow, A. Pimstein, J.R. Mecikalski, and W.P. Kustas. 2012b. A drought index based on thermal remote sensing of evapotranspiration. In *Remote Sensing of Drought: Innovative Monitoring Approaches*, eds. B.D. Wardlow, M.C. Anderson, and J.P. Verdin. Boca Raton, FL: CRC Press.

Anderson, M.C., J.M. Norman, J.R. Mecikalski, J.A. Otkin, and W.P. Kustas. 2007. A climatological study of evapotranspiration and moisture stress across the continental United States based on thermal remote sensing: 2. Surface moisture climatology. *Journal of Geophysical Research* 112(D11112):13, doi:10.1029/2006JD007507.

Andreadis, K.A., E.A. Clark, D.P. Lettenmaier, and D.E. Alsdorf. 2007. Prospects for river discharge and depth estimation through assimilation of swath-altimetry into a raster-based hydrodynamics model. *Geophysical Research Letters* 34:L10403, doi:10.1029/2007GL029721.

Anyamba, A. and C.J. Tucker. 2012. Historical perspective of AVHRR NDVI and vegetation drought monitoring. In *Remote Sensing of Drought: Innovative Monitoring Approaches*, eds. B.D. Wardlow, M.C. Anderson, and J.P. Verdin. Boca Raton, FL: CRC Press.

Biancamaria, S., K.M. Andreadis, M. Durand, E.A. Clark, E. Rodriguez, N.M. Mognard, D.E. Alsdorf, D.P. Lettenmaier, and Y. Oudin. 2010. Preliminary characterization of SWOT hydrology error budget and global capabilities. *IEEE Journal of Selected Topics in Applied Earth Observations and Remote Sensing* 3(1):6–19.

Birkett, C.M. 1995. The contribution of TOPEX/POSEIDON to the global monitoring of climatically sensitive lakes. *Journal of Geophysical Research – Oceans* 100(C12):179–204.

Birkett, C.M. 1998. Contribution of the TOPEX NASA radar altimeter to the global monitoring of large rivers and wetlands. *Water Resources Research* 34(5):1223–1239.

Birkett, C.M. and B. Beckley. 2010. Investigating the performance of the Jason-2/OSTM radar altimeter over lakes and reservoirs. *Marine Geodesy* 33(1):204–238.

Bolten, J.D., W.T. Crow, T.J. Jackson, X. Zhan, and C.A. Reynolds. 2010. Evaluating the utility of remotely-sensed soil moisture retrievals for operational agricultural drought monitoring. *IEEE Journal of Selected Topics in Applied Earth Observations and Remote Sensing* 3:57–66.

Brown, J.F., B.D Wardlow, T. Tadesse, M.J. Hayes, and B.C. Reed. 2008. The Vegetation Drought Response Index (VegDRI): A new integrated approach for monitoring drought stress in vegetation. *GIScience and Remote Sensing* 45(1):16–46.

Burke, E.J., S.J. Brown, and N. Christidis. 2006. Modeling the recent evolution of global drought and projections for the twenty first century with the Hadley Centre climate model. *Journal of Hydrometeorology* 7(5):1113–1125.

Cheng, K.-C., C.-Y. Kuo, H.-Z. Tseng, Y. Yi, and C.K. Shum. 2010. Lake surface height calibration of Jason-1 and Jason-2 over the Great Lakes. *Marine Geodesy* 33(1):186–203.

Crow, W.T., W.P. Kustas, and J.H. Prueger. 2008. Monitoring root-zone soil moisture through the assimilation of a thermal remote sensing-based soil moisture proxy into a water balance model. *Remote Sensing of Environment* 112:1268–1281.

Dong, J., J.P. Walker, P.R. Houser, and C. Sun. 2007. Scanning multichannel microwave radiometer snow water equivalent assimilation. *Journal of Geophysical Research* 112:D07108, doi:10.1029/2006JD007209.

Dubois, P., J. Van Zyl, and E. Engman. 1995. Measuring soil moisture with imaging radars. *IEEE Transactions on Geoscience and Remote Sensing* 33(4):915–926.

Durand, D., K.M. Andreadis, D.E. Alsdorf, D.P. Lettenmaier, D. Moller, and M. Wilson. 2008. Estimation of bathymetric depth and slope from data assimilation of swath altimetry into a hydrodynamic model. *Geophysical Research Letters* 35:L20401, doi:10.1029/2008GL034150.

Durand, M., L.L. Fu, D.P. Lettenmaier, D. Alsdorf, E. Rodríguez, and D. Esteban-Fernandez. 2010a. The Surface Water and Ocean Topography mission: Observing terrestrial surface water and oceanic submesoscale eddies. *Proceedings of the IEEE* 98(5):766–779.

Durand, M., E. Rodriguez, D.E. Alsdorf, and M. Trigg. 2010b. Estimating river depth from remote sensing swath interferometry measurements of river height, slope, and width. *IEEE Journal of Selected Topics in Applied Earth Observations and Remote Sensing* 3(1):20–31.

Eidenshink, J.C. 2006. A 16-year time-series of 1 lm AVHRR satellite data of the conterminous United States and Alaska. *Photogrammetric Engineering and Remote Sensing* 72:1027–1035.

Entekhabi, D., E.G. Njoku, P.E. O'Neill, K.H. Kellogg, W.T. Crow, W.N. Edelstein, J.K. Entin, S.D. Goodman, T.J. Jackson, J. Johnson, J. Kimball, J.R. Piepmeier, R.D. Koster, N. Martin, K.C. McDonald, M. Moghaddam, S. Moran, R. Reichle, J.C. Shi, M.W. Spencer, S.W. Thurman, L. Tsang, and J. Can Zyl. 2010. The Soil Moisture Active Passive (SMAP) mission. *Proceedings of the IEEE* 98(5):704–716.

Famiglietti, J.S., M. Lo, S.L. Ho, J. Bethune, K.J. Anderson, T.H. Syed, S.C. Swenson, C.R. de Linage, and M. Rodell. 2011. Satellites measure recent rates of groundwater depletion in California's Central Valley. *Geophysical Research Letters* 38:L03403, doi:10.1029/2010GL046442.

Frappart, F., S. Calmant, M. Cauhope, F. Seyler, and A. Cazenave. 2006. Preliminary results of ENVISAT RA-2-derived water levels validation over the Amazon basin. *Remote Sensing of Environment* 100:252–264.

Funk, C., J. Michaelsen, and M. Marshall. 2012. Mapping recent decadal climate variations in precipitation and temperature across eastern Africa and the Sahel. In *Remote Sensing of Drought: Innovative Monitoring Approaches*, eds. B.D. Wardlow, M.C. Anderson, and J.P. Verdin. Boca Raton, FL: CRC Press.

Gao, F., J. Masek, M. Schwaller, and F.G. Hall. 2006. On the blending of the Landsat and MODIS surface reflectance: Predicting daily Landsat surface reflectance. *IEEE Transactions on Geoscience and Remote Sensing* 44:2207–2218.

Gao, H., Q. Tang, C.R. Ferguson, E.F. Wood, and D. Lettenmaier. 2010. Estimating the water budget of major U.S. river basins via remote sensing. *International Journal of Remote Sensing* 31:3955–3978.

Gu, Y., J.F. Brown, T. Muira, W.J.D. van Leeuwen, and B.C. Reed. 2010. Phenological classification of the United States: A geographic framework for extending multi-sensor time-series data. *Remote Sensing* 2(2):526–544.

Guch, I. and M. DeMaria. 2010. GOES-R Risk Reduction program. In *6th Annual Symposium on Future National Operational Environmental Satellite Systems-NPOESS and GOES-R*, Atlanta, GA. January 16–21, 2010.

Guttman, N.D. 1994. On the sensitivity of sample L moments to sample size. *Journal of Climate* 7:1026–1029.

Hain, C.R. 2010. Developing a dual assimilation approach for thermal infrared and passive microwave soil moisture retrievals. PhD thesis, University of Alabama, Huntsville.

Hain, C.R., W.T. Crow, J.R. Mecikalski, M.C. Anderson, and T. Holmes. 2011. An intercomparison of available soil moisture estimates from thermal-infrared and passive microwave remote sensing and land-surface modeling. *Journal of Geophysical Research* (in press).

Houborg, R., M. Rodell, J. Lawrimore, B. Li, R. Reichle, R. Heim, M. Rosencrans, R. Tinker, J.S. Famiglietti, M. Svoboda, B. Wardlow, and B.F. Zaitchik. 2010. Using enhanced GRACE water storage data to improve drought detection by the U.S. and North American drought monitors. *IEEE International Geoscience and Remote Sensing Symposium Proceedings*, Honolulu, HI. July 25–30, pp. 710–713.

Hwang, C., M.-F. Peng, J. Ning, and C.-H. Sui. 2005. Lake level variations in China from TOPEX/Poseidon altimetry: Data quality assessment and links to precipitation and ENSO. *Geophysical Journal International* 161(1):1–11.

Intergovernmental Panel on Climate Change (IPCC). 2007. *Climate Change 2007: The Physical Basis. Contributions of Working Group I to the Fourth Assessment Report of the Intergovernmental Panel on Climate Change*, eds. S. Solomon, D. Qin, M. Manning, Z. Chen, M. Marquis, K.B. Averyt, M. Tignor, and H.L. Miller. Cambridge, U.K.: Cambridge University Press.

Justice, C.O., E. Vermote, J. Privette, and A. Sei. 2010. The evolution of U.S. moderate resolution optical land remote sensing from AVHRR to VIIRS. In *Land Remote Sensing and Global Environmental Change*, eds. B. Ramachandran, C.O. Justice, and M.J. Abrams, pp. 781–806. New York: Springer.

Kerr, Y.H., P. Waldteufel, J.P. Wigneron, S. Delwart, F. Cabot, J. Boutin, M.J. Escorihuela, J. Font, N. Reul, C. Gruthier, S. Enache Juglea, M.R. Drinkwater, A. Hahne, M. Martin-Neira, and S. Mecklenburg. 2010. The SMOS mission: New tool for monitoring key elements of the global water cycle. *Proceedings of the IEEE* 98(5):666–687.

Knapp, K.R. 2008. Calibration assessment of ISCCP geostationary infrared observations using HIRS. *Journal of Atmospheric and Oceanic Technology* 25(2):183–195.

Kogan, F.N. 1995. Application of vegetation index data and brightness temperature for drought detection. *Advances in Space Research* 11:91–100.

Kongoli, C., P. Romanov, and R. Ferraro. 2012. Snow cover monitoring from remote sensing satellites: Possibilities for drought assessment. In *Remote Sensing of Drought: Innovative Monitoring Approaches*, eds. B.D. Wardlow, M.C. Anderson, and J.P. Verdin. Boca Raton, FL: CRC Press.

Lacaze, R., G. Balsamo, F. Baret, B. Andrew, J. Calvet, F. Camacho, R.D'Andrimont, P. Philippe, B. Smets, H. Polive, K. Tansey, I. Trigo, W. Wagner, S. Freitas, H. Makhmara, V. Naeimi, and W. Marie. 2010. Geoland2—towards an operational GMES land monitoring core service; first results of the biogeophysical parameter core mapping service. *ISPRS Technical*

Commission VII Symposium – 100 Years ISPRS Advancing Remote Sensing Science, pp. 354–359, July 5–7. Vienna, Austria: International Society for Photogrammetry and Remote Sensing (ISPRS).

Lee, H., M. Durand, H.C. Jung, D. Alsdorf, C.K. Shum, and Y. Sheng. 2010a. Characterization of surface water storage changes in Arctic lakes using simulated SWOT measurements. *International Journal of Remote Sensing* 31(14):3931–3953.

Lee, T.E., S.D. Miller, F.J. Turk, C. Schueler, R. Julian, S. Deyo, P. Dills, and S. Wang. 2006. The NPOESS VIIRS day/night visible sensor. *Bulletin of the American Meteorological Society* 87(2):191–199.

Lee, T.F., C.S. Nelson, P. Dills, L. Peter, P. Riishogaard, A. Jones, L. Li, S. Miller, L.E. Flynn, G. Jedlovec, W. McCarty, C. Hoffman, and G. McWilliam. 2010b. NPOESS next-generation operational global Earth observations. *Bulletin of the American Meteorological Society* 91(6):727–740.

Lee, H., C.K. Shum, K.-H. Tseng, J.-Y. Guo, and C.-Y. Kuo. 2011. Present-day lake level variations from Envisat altimetry over the northeastern Qinghai-Tibetan Plateau: Links with precipitation and temperature. *Terrestrial, Atmospheric, and Oceanic Sciences* 22(2):169–175.

Maheu, C., A. Cazenave, and C.R. Mechoso. 2003. Water level fluctuations in the Plata Basin (South America) from Topex/Poseidon satellite altimetry. *Geophysical Research Letters* 30(3):1143, doi: 10.1029/2002GL016033.

Marshall, M.T., C. Funk, and J. Michaelsen. 2012. Agricultural drought monitoring in Kenya using evapotranspiration derived from remote sensing and reanalysis data. In *Remote Sensing of Drought: Innovative Monitoring Approaches*, eds. B.D. Wardlow, M.C. Anderson, and J.P. Verdin. Boca Raton, FL: CRC Press.

Medina, C.E., J. Gomez-Enri, J.J. Alonso, and P. Villares. 2008. Water level fluctuations derived from ENVISAT radar altimeter (RA-2) and in-situ measurements in a subtropical waterbody: Lake Izabal (Guatemala). *Remote Sensing of Environment* 112:3604–3617.

Morris, C.S. and S.K. Gill. 1994. Evaluation of the TOPEX/POSEIDON altimeter system over the Great Lakes. *Journal of Geophysical Research* 99(C12):527–539.

NASA (National Aeronautics and Space Administration). 2010. *Responding to the Challenge of Climate and Environmental Change: NASA's Plan for a Climate-Centric Architecture for Earth Observations and Applications from Space*. http://science.nasa.gov/media/medialibrary/2010/07/01/Climate_Architecture_Final.pdf (accessed on December 9, 2011).

Nghiem, S.V., B.D. Wardlow, D. Allured, M.D. Svoboda, D. LeComte, M. Rosencrans, S.K. Chan, and G. Neumann. 2012. Microwave remote sensing: Science and application. In *Remote Sensing of Drought: Innovative Monitoring Approaches*, eds. B.D. Wardlow, M.C. Anderson, and J.P. Verdin. Boca Raton, FL: CRC Press.

NIDIS (National Integrated Drought Information System). 2007. The national integrated drought information system implementation plan: A pathway for National resilience. Report of the NIDIS Implementation Team.

Norman, J.M., M.C. Anderson, W.P. Kustas, A.N. French, J.R. Mecikalski, R.D. Torn, G.R. Diak, T.J. Schmugge, and B.C.W. Tanner. 2003. Remote sensing of surface energy fluxes at 10^1-m pixel resolutions. *Water Resources Research* 39(1221):17, doi:10.1029/2002WR001775.

NRC (National Research Council). 2007. *Earth Science and Applications from Space: National Imperatives for the Next Decade and Beyond*. Washington, DC: The National Academies Press.

Pan, M., E. Wood, D. McLaughlin, D. Entekhabi, and L. Luo. 2009. A multiscale ensemble filtering system for hydrologic data assimilation: Part I, implementation and synthetic experiment. *Journal of Hydrometeorology* 10(3):794–806.

Reichle, R. 2008. Data assimilation methods in the Earth sciences. *Advances in Water Resources* 31:1411–1418.

Reichle, R.H., R.D. Koster, P. Liu, S.P.P. Mahanama, E.G. Njoku, and M. Owe. 2007. Comparison and assimilation of global soil moisture retrievals from the Advanced Microwave Scanning Radiometer for the Earth Observing System (AMSR-E) and the Scanning Multichannel Microwave Radiometer (SMMR). *Journal of Geophysical Research –Atmospheres* 112: D09108, doi:10.1029/2006JD008033.

Rodell, M. 2012. Satellite gravimetry applied to drought monitoring. In *Remote Sensing of Drought: Innovative Monitoring Approaches*, eds. B.D. Wardlow, M.C. Anderson, and J.P. Verdin. Boca Raton, FL: CRC Press.

Rodell, M., J. Chen, H. Kato, J.S. Famiglietti, J. Nigro, and C.R. Wilson. 2007. Estimating groundwater storage changes in the Mississippi River basin (USA) using GRACE. *Hydrogeology Journal* 15:159–166.

Rodell, M., I. Velicogna, and J.S. Famiglietti. 2009. Satellite-based estimates of groundwater depletion in India. *Nature* 460(20):999–1002.

Rossi, S. and S. Niemeyer. 2012. Drought monitoring with estimates of the fraction of absorbed photosynthetically-active radiation (fAPAR) derived from MERIS. In *Remote Sensing of Drought: Innovative Monitoring Approaches*, eds. B.D. Wardlow, M.C. Anderson, and J.P. Verdin. Boca Raton, FL: CRC Press.

Schmit, T.J., M.M. Gunshor, W.P. Menzel, J.J. Gurka, J. Li, and A.S. Bachmeier. 2005. Introducing the next-generation Advanced Baseline Imager on GOES-R. *Bulletin of the American Meteorological Society* 86:1079–1096.

Schumann, G., G. Di Baldassarre, D. Alsdorf, and P.D. Bates. 2010. Near real-time flood wave approximation on large rivers from space: Application to the River Po, Northern Italy. *Water Resources Research* 46:W05601.

Senay, G.B., S. Bohms, and J.P. Verdin. 2012. Remote sensing of evapotranspiration for operational drought monitoring using principles of water and energy balance. In *Remote Sensing of Drought: Innovative Monitoring Approaches*, eds. B.D. Wardlow, M.C. Anderson, and J.P. Verdin. Boca Raton, FL: CRC Press.

Senay, G.B., M. Budde, J.P. Verdin, and A.M. Melesse, 2007. A coupled remote sensing and simplified surface energy balance approach to estimate actual evapotranspiration from irrigated fields. *Sensors* 7:979–1000.

Sheffield, J., C.R. Ferguson, T.J. Troy, E.F. Wood, and M.F. McCabe. 2009. Closing the terrestrial water budget from satellite remote sensing. *Geophysical Research Letters* 36:L07403, doi:10.1029/2009GL037338.

Sheffield, J., E.F. Wood, D.P. Lettenmaier, and A. Lipponen. 2008. Experimental drought monitoring for Africa. *GEWEX News* 18(3):4–6.

Sheffield, J., Y. Xia, L. Luo, E.F. Wood, M. Ek, K.E. Mitchell, and NLDAS team. 2012. The North American land data assimilation system (NLDAS): A framework for merging model and satellite data for improved drought monitoring. In *Remote Sensing of Drought: Innovative Monitoring Approaches*, eds. B.D. Wardlow, M.C. Anderson, and J.P. Verdin. Boca Raton, FL: CRC Press.

Shum, C.K., Y. Yi, K. Cheng, C. Kino, A. Braun, S. Calmant, and D. Chambers. 2003. Calibration of Jason-1 altimeter over Lake Erie. *Marine Geodesy* 26:3–4.

Smith, E.A., G. Asrar, Y. Furuhama, A. Ginati, C. Kummerow, V. Levizzani, A. Mugnai, K. Nakamura, R. Adler, V. Casse, M. Cleave, M. Debois, J. Durning, J. Entin, P. Houser, T. Iguchi, R. Kakar, J. Kaye, M. Kojima, D. Lettenmaier, M. Luther, A. Mehta, P. Morel, T. Nakazawa, S. Neeck, K. Okamoto, R. Oki, G. Raju, M. Shepherd, E. Stocker, J. Testud, and E. Wood. 2004. International Global Precipitation Measurement (GPM) Program and Mission: An overview. In *Measuring Precipitation from Space: EURAINSAT and the Future*, eds. V. Levizzani and F.J. Turk, pp. 611–654. Dordrecht, the Netherlands: Kluwer Publishers.

Story, G.J. 2012. Estimating precipitation from WSR-88D observations and rain gauge data – potential for drought monitoring. In *Remote Sensing of Drought: Innovative Monitoring Approaches*, eds. B.D. Wardlow, M.C. Anderson, and J.P. Verdin. Boca Raton, FL: CRC Press.

Svoboda, M., D. LeComte, M. Hayes, R. Heim, K. Gleason, J. Angel, B. Rippey, R. Tinker, M. Palecki, D. Stooksbury, D. Miskus, and S. Stephens. 2002. The drought monitor. *Bulletin of the American Meteorological Society* 83(8):1181–1190.

Tadesse, T., B.D. Wardlow, M.D. Svoboda, and M.J. Hayes. 2012. The Vegetation Outlook (VegOut): Predicting remote sensing-based seasonal greenness. In *Remote Sensing of Drought: Innovative Monitoring Approaches*, eds. B.D. Wardlow, M.C. Anderson, and J.P. Verdin. Boca Raton, FL: CRC Press.

Tiwari, V. M., J. Wahr, and S. Swenson. 2009. Dwindling groundwater resources in northern India, from satellite gravity observations. *Geophysical Research Letters* 36:L18401, doi:10.1029/2009GL039401.

Townshend, J.R.G. and C.O. Justice. 2002. Towards operational monitoring of terrestrial systems by moderate-resolution remote sensing. *Remote Sensing of Environment* 83:351–359.

Tucker, C.J., J.E. Pinzon, M.E. Brown, D.A. Slayback, E.W. Pak, R. Mahoney, E.F. Vermote, and N. El Saleous. 2005. An extended AVHRR 8-km NDVI dataset compatible with MODIS and SPOT vegetation NDVI data. *International Journal of Remote Sensing* 26(20):4485–4498.

USGCRP. 2009. *Global Climate Change Impacts in the United States*, eds. T.R. Karl, J.M. Melillo, and T.C. Peterson. Cambridge, U.K.: Cambridge University Press.

van Leeuwen, W.J.D., B.J. Orr, S.E. Marsh, and S.M. Hermann. 2006. Multi-sensor NDVI data continuity: Uncertainties and implications for vegetation monitoring applications. *Remote Sensing of Environment* 100(1):67–81.

Wardlow, B.D., T. Tadesse, J.F. Brown, K. Callahan, S. Swain, and E. Hunt. 2012. The vegetation drought response index (VegDRI): An integration of satellite, climate, and biophysical data. In *Remote Sensing of Drought: Innovative Monitoring Approaches*, eds. B.D. Wardlow, M.C. Anderson, and J.P. Verdin. Boca Raton, FL: CRC Press.

Wilhite, D.A. and M.H. Glantz. 1985. Understanding the drought phenomenon: The role of definitions. *Water International* 10:111–120.

Wilhite, D.A. and R.S. Pulwarty. 2005. Drought and water crises: Lessons learned and the road ahead. In *Drought and Water Crises Science, Technology, and Management Issues*, ed. D.A. Wilhite, pp. 389–398. Boca Raton, FL: Taylor and Francis.

Wilhite, D.A., M.D. Svoboda, and M.J. Hayes. 2007. Understanding the complex impacts of drought: A key to enhancing drought mitigation and preparedness. *Water Resources Management* 21:763–774.

WMO (World Meteorological Organization). 2010. *Guide to Climatological Practices*. WMO Publication No. 100 (Third Edition). Geneva, Switzerland.

Wulder, M.A., J.C. White, S.N. Goward, J.G. Masek, J.R. Irons, M. Herold, W.B. Cohen, T.R. Loveland, and C.E. Woodcock. 2008. Landsat continuity: Issues and opportunities for land cover mapping. *Remote Sensing of Environment* 112(3):955–969.

Yeh, P.J.-F., S.C. Swenson, J.S. Famiglietti, and M. Rodell. 2006. Remote sensing of groundwater storage changes in Illinois using the Gravity Recovery and Climate Experiment (GRACE). *Water Resources Research* 42:W12203, doi:10.1029/2006WR005374.

Zaitchik, B.F., M. Rodell, and R.H. Reichle. 2008. Assimilation of GRACE terrestrial water storage data into a land surface model. *Journal of Hydrometeorology* 9:535–548.

Index

G

N

R

S

W